LINEAR SYSTEM
THEORY AND DESIGN

LINEAR SYSTEM THEORY AND DESIGN

Third Edition

Chi-Tsong Chen
State University of New York at Stony Brook

New York Oxford
OXFORD UNIVERSITY PRESS
1999

OXFORD UNIVERSITY PRESS

Oxford   New York
Athens   Auckland   Bangkok   Bogotá   Buenos Aires   Calcutta
Cape Town   Chennai   Dar es Salaam   Delhi   Florence   Hong Kong   Istanbul
Karachi   Kuala Lumpur   Madrid   Melbourne   Mexico City   Mumbai
Nairobi   Paris   São Paulo   Singapore   Taipei   Tokyo   Toronto   Warsaw

and associated companies in
Berlin   Ibadan

Copyright © 1999 by Oxford University Press, Inc.

Published by Oxford University Press, Inc.
198 Madison Avenue, New York, New York 10016

Oxford is a registered trademark of Oxford University Press

Library of Congress Cataloging-in-Publication Data
Chen, Chi-Tsong
   Linear system theory and design / by Chi-Tsong Chen. — 3rd ed.
      p.   cm. — (The Oxford series in electrical and computer engineering)
   Includes bibliographical references and index.
   ISBN 0-19-511777-8 (cloth).
      1. Linear systems.   2. System design.   I. Title.   II. Series.
   QA402.C44   1998
   629.8'32—dc21   97-35535
                                                                       CIP

Printing (last digit): 9 8 7 6 5 4 3 2 1

Printed in the United States of America
on acid-free paper

To
BIH-JAU

Contents

Chapter 8: State Feedback and State Estimators   *231*

Chapter 9: Pole Placement and Model Matching   *269*

Preface

This text is intended for use in senior/first-year graduate courses on linear systems and multivariable system design in electrical, mechanical, chemical, and aeronautical departments. It may also be useful to practicing engineers because it contains many design procedures. The mathematical background assumed is a working knowledge of linear algebra and the Laplace transform and an elementary knowledge of differential equations.

Linear system theory is a vast field. In this text, we limit our discussion to the conventional approaches of state-space equations and the polynomial fraction method of transfer matrices. The geometric approach, the abstract algebraic approach, rational fractions, and optimization are not discussed.

We aim to achieve two objectives with this text. The first objective is to use simple and efficient methods to develop results and design procedures. Thus the presentation is not exhaustive. For example, in introducing polynomial fractions, some polynomial theory such as the Smith–McMillan form and Bezout identities are not discussed. The second objective of this text is to enable the reader to employ the results to carry out design. Thus most results are discussed with an eye toward numerical computation. All design procedures in the text can be carried out using any software package that includes singular-value decomposition, QR decomposition, and the solution of linear algebraic equations and the Lyapunov equation. We give examples using MATLAB®, as the package[1] seems to be the most widely available.

This edition is a complete rewriting of the book *Linear System Theory and Design*, which was the expanded edition of *Introduction to Linear System Theory* published in 1970. Aside from, hopefully, a clearer presentation and a more logical development, this edition differs from the book in many ways:

- The order of Chapters 2 and 3 is reversed. In this edition, we develop mathematical descriptions of systems before reviewing linear algebra. The chapter on stability is moved earlier.
- This edition deals only with real numbers and foregoes the concept of *fields*. Thus it is mathematically less abstract than the original book. However, all results are still stated as theorems for easy reference.
- In Chapters 4 through 6, we discuss first the time-invariant case and then extend it to the time-varying case, instead of the other way around.

---

1. MATLAB is a registered trademark of the MathWorks, Inc., 24 Prime Park Way, Natick, MA 01760-1500. Phone: 508-647-7000, fax: 508-647-7001, E-mail: info@mathworks.com, http://www.mathworks.com.

- The discussion of discrete-time systems is expanded.

- In state-space design, Lyapunov equations are employed extensively and multivariable canonical forms are downplayed. This approach is not only easier for classroom presentation but also provides an attractive method for numerical computation.

- The presentation of the polynomial fraction method is streamlined. The method is equated with solving linear algebraic equations. We then discuss pole placement using a one-degree-of-freedom configuration, and model matching using a two-degree-of-freedom configuration.

- Examples using MATLAB are given throughout this new edition.

This edition is geared more for classroom use and engineering applications; therefore, many topics in the original book are deleted, including strict system equivalence, deterministic identification, computational issues, some multivariable canonical forms, and decoupling by state feedback. The polynomial fraction design in the input/output feedback (controller/estimator) configuration is deleted. Instead we discuss design in the two-parameter configuration. This configuration seems to be more suitable for practical application. The eight appendices in the original book are either incorporated into the text or deleted.

The logical sequence of all chapters is as follows:

$$Chap.\ 1\text{--}5\ \Rightarrow\ \begin{cases} Chap.\ 6\ \Rightarrow\ \begin{cases} Chap.\ 8 \\ Chap.\ 7 \end{cases} \\ Sec.\ 7.1\text{--}7.3\ \Rightarrow Sec.\ 9.1\text{--}9.3 \\ \qquad\qquad \Rightarrow Sec.\ 7.6\text{--}7.8\ \Rightarrow\ Sec.\ 9.4\text{--}9.5 \end{cases}$$

In addition, the material in Section 7.9 is needed to study Section 8.6.4. However, Section 8.6.4 may be skipped without loss of continuity. Furthermore, the concepts of controllability and observability in Chapter 6 are useful, but not essential for the material in Chapter 7. All minimal realizations in Chapter 7 can be checked using the concept of degrees, instead of checking controllability and observability. Therefore Chapters 6 and 7 are essentially independent.

This text provides more than enough material for a one-semester course. A one-semester course at Stony Brook covers Chapters 1 through 6, Sections 8.1–8.5, 7.1–7.2, and 9.1–9.3. Time-varying systems are not covered. Clearly, other arrangements are also possible for a one-semester course. A solutions manual is available from the publisher.

I am indebted to many people in revising this text. Professor Imin Kao and Mr. Juan Ochoa helped me with MATLAB. Professor Zongli Lin and Mr. T. Anantakrishnan read the whole manuscript and made many valuable suggestions. I am grateful to Dean Yacov Shamash, College of Engineering and Applied Sciences, SUNY at Stony Brook, for his encouragement. The revised manuscript was reviewed by Professor Harold Broberg, EET Department, Indiana Purdue University; Professor Peyman Givi, Department of Mechanical and Aerospace Engineering, State University of New York at Buffalo; Professor Mustafa Khammash, Department of Electrical and Computer Engineering, Iowa State University; and Professor B. Ross Barmish, Department of Electrical and Computer Engineering, University of Wisconsin. Their detailed and critical comments prompted me to restructure some sections and to include a number of mechanical vibration problems. I thank them all.

I am indebted to Mr. Bill Zobrist of Oxford University Press who persuaded me to undertake this revision. The people at Oxford University Press, including Krysia Bebick, Jasmine Urmeneta, Terri O'Prey, and Kristina Della Bartolomea were most helpful in this undertaking. Finally, I thank my wife, Bih-Jau, for her support during this revision.

Chi-Tsong Chen

LINEAR SYSTEM
THEORY AND DESIGN

Chapter

1

Introduction

1.1 Introduction

The study and design of physical systems can be carried out using empirical methods. We can apply various signals to a physical system and measure its responses. If the performance is not satisfactory, we can adjust some of its parameters or connect to it a compensator to improve its performance. This approach relies heavily on past experience and is carried out by trial and error and has succeeded in designing many physical systems.

Empirical methods may become unworkable if physical systems are complex or too expensive or too dangerous to be experimented on. In these cases, analytical methods become indispensable. The analytical study of physical systems consists of four parts: modeling, development of mathematical descriptions, analysis, and design. We briefly introduce each of these tasks.

The distinction between physical systems and models is basic in engineering. For example, circuits or control systems studied in any textbook are models of physical systems. A resistor with a constant resistance is a model; it will burn out if the applied voltage is over a limit. This power limitation is often disregarded in its analytical study. An inductor with a constant inductance is again a model; in reality, the inductance may vary with the amount of current flowing through it. Modeling is a very important problem, for the success of the design depends on whether the physical system is modeled properly.

A physical system may have different models depending on the questions asked. It may also be modeled differently in different operational ranges. For example, an electronic amplifier is modeled differently at high and low frequencies. A spaceship can be modeled as a particle in investigating its trajectory; however, it must be modeled as a rigid body in maneuvering. A spaceship may even be modeled as a flexible body when it is connected to a space station. In order to develop a suitable model for a physical system, a thorough understanding of the physical system and its operational range is essential. In this text, we will call a model of a physical system simply a *system*. Thus a physical system is a device or a collection of devices existing in the real world; a system is a model of a physical system.

Once a system (or model) is selected for a physical system, the next step is to apply various physical laws to develop mathematical equations to describe the system. For example, we apply Kirchhoff's voltage and current laws to electrical systems and Newton's law to mechanical systems. The equations that describe systems may assume many forms;

1

they may be linear equations, nonlinear equations, integral equations, difference equations, differential equations, or others. Depending on the problem under study, one form of equation may be preferable to another in describing the same system. In conclusion, a system may have different mathematical-equation descriptions just as a physical system may have many different models.

After a mathematical description is obtained, we then carry out analyses—quantitative and/or qualitative. In quantitative analysis, we are interested in the responses of systems excited by certain inputs. In qualitative analysis, we are interested in the general properties of systems, such as stability, controllability, and observability. Qualitative analysis is very important, because design techniques may often evolve from this study.

If the response of a system is unsatisfactory, the system must be modified. In some cases, this can be achieved by adjusting some parameters of the system; in other cases, compensators must be introduced. Note that the design is carried out on the model of the physical system. If the model is properly chosen, then the performance of the physical system should be improved by introducing the required adjustments or compensators. If the model is poor, then the performance of the physical system may not improve and the design is useless. Selecting a model that is close enough to a physical system and yet simple enough to be studied analytically is the most difficult and important problem in system design.

1.2 Overview

The study of systems consists of four parts: modeling, setting up mathematical equations, analysis, and design. Developing models for physical systems requires knowledge of the particular field and some measuring devices. For example, to develop models for transistors requires a knowledge of quantum physics and some laboratory setup. Developing models for automobile suspension systems requires actual testing and measurements; it cannot be achieved by use of pencil and paper. Computer simulation certainly helps but cannot replace actual measurements. Thus the modeling problem should be studied in connection with the specific field and cannot be properly covered in this text. In this text, we shall assume that models of physical systems are available to us.

The systems to be studied in this text are limited to linear systems. Using the concept of linearity, we develop in Chapter 2 that every linear system can be described by

$$\mathbf{y}(t) = \int_{t_0}^{t} \mathbf{G}(t, \tau)\mathbf{u}(\tau)\, d\tau \tag{1.1}$$

This equation describes the relationship between the input $\mathbf{u}$ and output $\mathbf{y}$ and is called the *input–output* or *external* description. If a linear system is lumped as well, then it can also be described by

$$\dot{\mathbf{x}}(t) = \mathbf{A}(t)\mathbf{x}(t) + \mathbf{B}(t)\mathbf{u}(t) \tag{1.2}$$

$$\mathbf{y}(t) = \mathbf{C}(t)\mathbf{x}(t) + \mathbf{D}(t)\mathbf{u}(t) \tag{1.3}$$

Equation (1.2) is a set of first-order differential equations and Equation (1.3) is a set of algebraic equations. They are called the *internal* description of linear systems. Because the vector $\mathbf{x}$ is called the *state*, the set of two equations is called the *state-space* or, simply, the *state* equation.

If a linear system has, in addition, the property of time invariance, then Equations (1.1) through (1.3) reduce to

$$\mathbf{y}(t) = \int_0^t \mathbf{G}(t - \tau)\mathbf{u}(\tau)\,d\tau \tag{1.4}$$

and

$$\dot{\mathbf{x}}(t) = \mathbf{A}\mathbf{x}(t) + \mathbf{B}\mathbf{u}(t) \tag{1.5}$$

$$\mathbf{y}(t) = \mathbf{C}\mathbf{x}(t) + \mathbf{D}\mathbf{u}(t) \tag{1.6}$$

For this class of linear time-invariant systems, the Laplace transform is an important tool in analysis and design. Applying the Laplace transform to (1.4) yields

$$\hat{\mathbf{y}}(s) = \hat{\mathbf{G}}(s)\hat{\mathbf{u}}(s) \tag{1.7}$$

where a variable with a circumflex denotes the Laplace transform of the variable. The function $\hat{\mathbf{G}}(s)$ is called the *transfer matrix*. Both (1.4) and (1.7) are input–output or external descriptions. The former is said to be in the time domain and the latter in the frequency domian.

Equations (1.1) through (1.6) are called continuous-time equations because their time variable t is a continuum defined at every time instant in $(-\infty, \infty)$. If the time is defined only at discrete instants, then the corresponding equations are called discrete-time equations. This text is devoted to the analysis and design centered around (1.1) through (1.7) and their discrete-time counterparts.

We briefly discuss the contents of each chapter. In Chapter 2, after introducing the aforementioned equations from the concepts of lumpedness, linearity, and time invariance, we show how these equations can be developed to describe systems. Chapter 3 reviews linear algebraic equations, the Lyapunov equation, and other pertinent topics that are essential for this text. We also introduce the Jordan form because it will be used to establish a number of results. We study in Chapter 4 solutions of the state-space equations in (1.2) and (1.5). Different analyses may lead to different state equations that describe the same system. Thus we introduce the concept of equivalent state equations. The basic relationship between state-space equations and transfer matrices is also established. Chapter 5 introduces the concepts of bounded-input bounded-output (BIBO) stability, marginal stability, and asymptotic stability. Every system must be designed to be stable; otherwise, it may burn out or disintegrate. Therefore stability is a basic system concept. We also introduce the Lyapunov theorem to check asymptotic stability.

Chapter 6 introduces the concepts of controllability and observability. They are essential in studying the internal structure of systems. A fundamental result is that the transfer matrix describes only the controllable and observable part of a state equation. Chapter 7 studies minimal realizations and introduces coprime polynomial fractions. We show how to obtain coprime fractions by solving sets of linear algebraic equations. The equivalence of controllable and observable state equations and coprime polynomial fractions is established.

The last two chapters discuss the design of time-invariant systems. We use controllable and observable state equations to carry out design in Chapter 8 and use coprime polynomial fractions in Chapter 9. We show that, under the controllability condition, all eigenvalues of a system can be arbitrarily assigned by introducing state feedback. If a state equation is observable, full-dimensional and reduced-dimensional state estimators, with any desired eigenvalues, can be constructed to generate estimates of the state. We also establish the separation property. In Chapter 9, we discuss pole placement, model matching, and their applications in tracking, disturbance rejection, and decoupling. We use the unity-feedback configuration in pole placement and the two-parameter configuration in model matching. In our design, no control performances such as rise time, settling time, and overshoot are considered; neither are constraints on control signals and on the degree of compensators. Therefore this is not a control text per se. However, all results are basic and useful in designing linear time-invariant control systems.

Chapter

Mathematical Descriptions of Systems

2.1 Introduction

The class of systems studied in this text is assumed to have some input terminals and output terminals as shown in Fig. 2.1. We assume that if an excitation or input is applied to the input terminals, a *unique* response or output signal can be measured at the output terminals. This unique relationship between the excitation and response, input and output, or cause and effect is essential in defining a system. A system with only one input terminal and only one output terminal is called a single-variable system or a single-input single-output (SISO) system. A system with two or more input terminals and/or two or more output terminals is called a multivariable system. More specifically, we can call a system a multi-input multi-output (MIMO) system if it has two or more input terminals and output terminals, a single-input multi-output (SIMO) system if it has one input terminal and two or more output terminals.

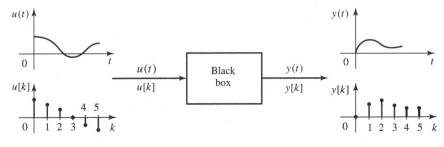

Figure 2.1 System.

A system is called a *continuous-time* system if it accepts continuous-time signals as its input and generates continuous-time signals as its output. The input will be denoted by lowercase italic $u(t)$ for single input or by boldface $\mathbf{u}(t)$ for multiple inputs. If the system has p input terminals, then $\mathbf{u}(t)$ is a $p \times 1$ vector or $\mathbf{u} = [u_1 \ u_2 \ \cdots \ u_p]'$, where the prime denotes the transpose. Similarly, the output will be denoted by $y(t)$ or $\mathbf{y}(t)$. The time t is assumed to range from $-\infty$ to ∞.

A system is called a *discrete-time* system if it accepts discrete-time signals as its input and generates discrete-time signals as its output. All discrete-time signals in a system will be assumed to have the same sampling period T. The input and output will be denoted by $u[k] := u(kT)$ and $y[k] := y(kT)$, where k denotes discrete-time instant and is an integer ranging from $-\infty$ to ∞. They become boldface for multiple inputs and multiple outputs.

2.1.1 Causality and Lumpedness

A system is called a *memoryless system* if its output $\mathbf{y}(t_0)$ depends only on the input applied at t_0; it is independent of the input applied before or after t_0. This will be stated succinctly as follows: current output of a memoryless system depends only on current input; it is independent of past and future inputs. A network that consists of only resistors is a memoryless system.

Most systems, however, have memory. By this we mean that the output at t_0 depends on $\mathbf{u}(t)$ for $t < t_0, t = t_0$, and $t > t_0$. That is, current output of a system with memory may depend on past, current, and future inputs.

A system is called a *causal* or *nonanticipatory* system if its current output depends on past and current inputs but not on future input. If a system is not causal, then its current output will depend on future input. In other words, a noncausal system can *predict* or *anticipate* what will be applied in the future. No physical system has such capability. Therefore every physical system is causal and causality is a necessary condition for a system to be built or implemented in the real world. This text studies only causal systems.

Current output of a causal system is affected by past input. How far back in time will the past input affect the current output? Generally, the time should go all the way back to minus infinity. In other words, the input from $-\infty$ to time t has an effect on $\mathbf{y}(t)$. Tracking $\mathbf{u}(t)$ from $t = -\infty$ is, if not impossible, very inconvenient. The concept of state can deal with this problem.

Definition 2.1 *The state* $\mathbf{x}(t_0)$ *of a system at time* t_0 *is the information at* t_0 *that, together with the input* $\mathbf{u}(t)$, *for* $t \geq t_0$, *determines uniquely the output* $\mathbf{y}(t)$ *for all* $t \geq t_0$.

By definition, if we know the state at t_0, there is no more need to know the input $\mathbf{u}(t)$ applied before t_0 in determining the output $\mathbf{y}(t)$ after t_0. Thus in some sense, the state summarizes the effect of past input on future output. For the network shown in Fig. 2.2, if we know the voltages $x_1(t_0)$ and $x_2(t_0)$ across the two capacitors and the current $x_3(t_0)$ passing through the inductor, then for any input applied on and after t_0, we can determine uniquely the output for $t \geq t_0$. Thus the state of the network at time t_0 is

Figure 2.2 Network with 3 state variables.

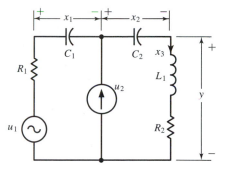

$$\mathbf{x}(t_0) = \begin{bmatrix} x_1(t_0) \\ x_2(t_0) \\ x_3(t_0) \end{bmatrix}$$

It is a 3×1 vector. The entries of $\mathbf{x}$ are called *state variables*. Thus, in general, we may consider the initial state simply as a set of initial conditions.

Using the state at t_0, we can express the input and output of a system as

$$\left. \begin{array}{c} \mathbf{x}(t_0) \\ \mathbf{u}(t), \ t \geq t_0 \end{array} \right\} \rightarrow \mathbf{y}(t), \ t \geq t_0 \qquad (2.1)$$

It means that the output is partly excited by the initial state at t_0 and partly by the input applied at and after t_0. In using (2.1), there is no more need to know the input applied before t_0 all the way back to $-\infty$. Thus (2.1) is easier to track and will be called a state-input–output pair.

A system is said to be *lumped* if its number of state variables is finite or its state is a finite vector. The network in Fig. 2.2 is clearly a lumped system; its state consists of three numbers. A system is called a *distributed* system if its state has infinitely many state variables. The transmission line is the most well known distributed system. We give one more example.

EXAMPLE 2.1 Consider the unit-time delay system defined by

$$y(t) = u(t - 1)$$

The output is simply the input delayed by one second. In order to determine $\{y(t), \ t \geq t_0\}$ from $\{u(t), \ t \geq t_0\}$, we need the information $\{u(t), \ t_0 - 1 \leq t < t_0\}$. Therefore the initial state of the system is $\{u(t), \ t_0 - 1 \leq t < t_0\}$. There are infinitely many points in $\{t_0 - 1 \leq t < t_0\}$. Thus the unit-time delay system is a distributed system.

2.2 Linear Systems

A system is called a *linear* system if for every t_0 and any two state-input–output pairs

$$\left. \begin{array}{c} \mathbf{x}_i(t_0) \\ \mathbf{u}_i(t), \ t \geq t_0 \end{array} \right\} \rightarrow \mathbf{y}_i(t), \ t \geq t_0$$

for $i = 1, \ 2,$ we have

$$\left.\begin{array}{l} \mathbf{x}_1(t_0) + \mathbf{x}_2(t_0) \\ \mathbf{u}_1(t) + \mathbf{u}_2(t), \quad t \geq t_0 \end{array}\right\} \rightarrow \mathbf{y}_1(t) + \mathbf{y}_2(t), \quad t \geq t_0 \quad \text{(additivity)}$$

and

$$\left.\begin{array}{l} \alpha\mathbf{x}_1(t_0) \\ \alpha\mathbf{u}_1(t), \quad t \geq t_0 \end{array}\right\} \rightarrow \alpha\mathbf{y}_1(t), \quad t \geq t_0 \quad \text{(homogeneity)}$$

for any real constant α. The first property is called the *additivity* property, the second, the *homogeneity* property. These two properties can be combined as

$$\left.\begin{array}{l} \alpha_1\mathbf{x}_1(t_0) + \alpha_2\mathbf{x}_2(t_0) \\ \alpha_1\mathbf{u}_1(t) + \alpha_2\mathbf{u}_2(t), \quad t \geq t_0 \end{array}\right\} \rightarrow \alpha_1\mathbf{y}_1(t) + \alpha_2\mathbf{y}_2(t), \quad t \geq t_0$$

for any real constants α_1 and α_2, and is called the *superposition property*. A system is called a nonlinear system if the superposition property does not hold.

If the input $\mathbf{u}(t)$ is identically zero for $t \geq t_0$, then the output will be excited exclusively by the initial state $\mathbf{x}(t_0)$. This output is called the *zero-input response* and will be denoted by $\mathbf{y}_{zi}$ or

$$\left.\begin{array}{l} \mathbf{x}(t_0) \\ \mathbf{u}(t) \equiv \mathbf{0}, \quad t \geq t_0 \end{array}\right\} \rightarrow \mathbf{y}_{zi}(t), \quad t \geq t_0$$

If the initial state $\mathbf{x}(t_0)$ is zero, then the output will be excited exclusively by the input. This output is called the *zero-state response* and will be denoted by $\mathbf{y}_{zs}$ or

$$\left.\begin{array}{l} \mathbf{x}(t_0) = \mathbf{0} \\ \mathbf{u}(t), \quad t \geq t_0 \end{array}\right\} \rightarrow \mathbf{y}_{zs}(t), \quad t \geq t_0$$

The additivity property implies

$$\text{Output due to } \begin{cases} \mathbf{x}(t_0) \\ \mathbf{u}(t), \quad t \geq t_0 \end{cases} = \text{output due to } \begin{cases} \mathbf{x}(t_0) \\ \mathbf{u}(t) \equiv \mathbf{0}, \quad t \geq t_0 \end{cases}$$

$$+ \text{ output due to } \begin{cases} \mathbf{x}(t_0) = \mathbf{0} \\ \mathbf{u}(t), \quad t \geq t_0 \end{cases}$$

or

Response = zero-input response + zero-state response

Thus the response of every linear system can be decomposed into the zero-state response and the zero-input response. Furthermore, the two responses can be studied separately and their sum yields the complete response. For nonlinear systems, the complete response can be very different from the sum of the zero-input response and zero-state response. Therefore we cannot separate the zero-input and zero-state responses in studying nonlinear systems.

If a system is linear, then the additivity and homogeneity properties apply to zero-state responses. To be more specific, if $\mathbf{x}(t_0) = \mathbf{0}$, then the output will be excited exclusively by the input and the state-input–output equation can be simplified as $\{\mathbf{u}_i \rightarrow \mathbf{y}_i\}$. If the system is linear, then we have $\{\mathbf{u}_1 + \mathbf{u}_2 \rightarrow \mathbf{y}_1 + \mathbf{y}_2\}$ and $\{\alpha\mathbf{u}_i \rightarrow \alpha\mathbf{y}_i\}$ for all α and all $\mathbf{u}_i$. A similar remark applies to zero-input responses of any linear system.

Input–output description   We develop a mathematical equation to describe the zero-state response of linear systems. In this study, the initial state is assumed implicitly to be zero and the output is excited exclusively by the input. We consider first SISO linear systems. Let $\delta_\Delta(t - t_1)$ be the pulse shown in Fig. 2.3. It has width Δ and height $1/\Delta$ and is located at time t_1. Then every input $u(t)$ can be approximated by a sequence of pulses as shown in Fig. 2.4. The pulse in Fig. 2.3 has height $1/\Delta$; thus $\delta_\Delta(t - t_i)\Delta$ has height 1 and the left-most pulse in Fig. 2.4 with height $u(t_i)$ can be expressed as $u(t_i)\delta_\Delta(t - t_1)\Delta$. Consequently, the input $u(t)$ can be expressed symbolically as

$$u(t) \approx \sum_i u(t_i)\delta_\Delta(t - t_i)\Delta$$

Let $g_\Delta(t, t_i)$ be the output at time t excited by the pulse $u(t) = \delta_\Delta(t - t_i)$ applied at time t_i. Then we have

$$\delta_\Delta(t - t_i) \rightarrow g_\Delta(t, t_i)$$

$$\delta_\Delta(t - t_i)u(t_i)\Delta \rightarrow g_\Delta(t, t_i)u(t_i)\Delta \quad \text{(homogeneity)}$$

$$\sum_i \delta_\Delta(t - t_i)u(t_i)\Delta \rightarrow \sum_i g_\Delta(t, t_i)u(t_i)\Delta \quad \text{(additivity)}$$

Thus the output $y(t)$ excited by the input $u(t)$ can be approximated by

$$y(t) \approx \sum_i g_\Delta(t, t_i)u(t_i)\Delta \tag{2.2}$$

Figure 2.3 Pulse at t_1.

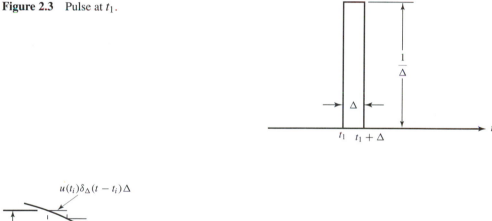

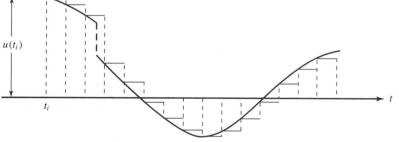

Figure 2.4 Approximation of input signal.

Now if Δ approaches zero, the pulse $\delta_\Delta(t-t_i)$ becomes an *impulse* at t_i, denoted by $\delta(t-t_i)$, and the corresponding output will be denoted by $g(t, t_i)$. As Δ approaches zero, the approximation in (2.2) becomes an equality, the summation becomes an integration, the discrete t_i becomes a continuum and can be replaced by τ, and Δ can be written as $d\tau$. Thus (2.2) becomes

$$y(t) = \int_{-\infty}^{\infty} g(t, \tau)u(\tau)\,d\tau \qquad (2.3)$$

Note that $g(t, \tau)$ is a function of two variables. The second variable denotes the time at which the impulse input is applied; the first variable denotes the time at which the output is observed. Because $g(t, \tau)$ is the response excited by an impulse, it is called the *impulse response*.

If a system is causal, the output will not appear before an input is applied. Thus we have

$$\text{Causal} \iff g(t, \tau) = 0 \quad \text{for } t < \tau$$

A system is said to be *relaxed* at t_0 if its initial state at t_0 is **0**. In this case, the output $y(t)$, for $t \geq t_0$, is excited exclusively by the input $u(t)$ for $t \geq t_0$. Thus the lower limit of the integration in (2.3) can be replaced by t_0. If the system is causal as well, then $g(t, \tau) = 0$ for $t < \tau$. Thus the upper limit of the integration in (2.3) can be replaced by t. In conclusion, every linear system that is causal and relaxed at t_0 can be described by

$$y(t) = \int_{t_0}^{t} g(t, \tau)u(\tau)\,d\tau \qquad (2.4)$$

In this derivation, the condition of lumpedness is not used. Therefore any lumped or distributed linear system has such an input–output description. This description is developed using only the additivity and homogeneity properties; therefore every linear system, be it an electrical system, a mechanical system, a chemical process, or any other system, has such a description.

If a linear system has p input terminals and q output terminals, then (2.4) can be extended to

$$\mathbf{y}(t) = \int_{t_0}^{t} \mathbf{G}(t, \tau)\mathbf{u}(\tau)\,d\tau \qquad (2.5)$$

where

$$\mathbf{G}(t, \tau) = \begin{bmatrix} g_{11}(t, \tau) & g_{12}(t, \tau) & \cdots & g_{1p}(t, \tau) \\ g_{21}(t, \tau) & g_{22}(t, \tau) & \cdots & g_{2p}(t, \tau) \\ \vdots & \vdots & & \vdots \\ g_{q1}(t, \tau) & g_{q2}(t, \tau) & \cdots & g_{qp}(t, \tau) \end{bmatrix}$$

and $g_{ij}(t, \tau)$ is the response at time t at the ith output terminal due to an impulse applied at time τ at the jth input terminal, the inputs at other terminals being identically zero. That is, $g_{ij}(t, \tau)$ is the impulse response between the jth input terminal and the ith output terminal. Thus **G** is called the *impulse response matrix* of the system. We stress once again that if a system is described by (2.5), the system is linear, relaxed at t_0, and causal.

State-space description   Every linear lumped system can be described by a set of equations of the form

$$\dot{\mathbf{x}}(t) = \mathbf{A}(t)\mathbf{x}(t) + \mathbf{B}(t)\mathbf{u}(t) \tag{2.6}$$

$$\mathbf{y}(t) = \mathbf{C}(t)\mathbf{x}(t) + \mathbf{D}(t)\mathbf{u}(t) \tag{2.7}$$

where $\dot{\mathbf{x}} := d\mathbf{x}/dt$.[1] For a p-input q-output system, $\mathbf{u}$ is a $p \times 1$ vector and $\mathbf{y}$ is a $q \times 1$ vector. If the system has n state variables, then $\mathbf{x}$ is an $n \times 1$ vector. In order for the matrices in (2.6) and (2.7) to be compatible, $\mathbf{A}$, $\mathbf{B}$, $\mathbf{C}$, and $\mathbf{D}$ must be $n \times n$, $n \times p$, $q \times n$, and $q \times p$ matrices. The four matrices are all functions of time or time-varying matrices. Equation (2.6) actually consists of a set of n first-order differential equations. Equation (2.7) consists of q algebraic equations. The set of two equations will be called an n-dimensional *state-space* equation or, simply, *state* equation. For distributed systems, the dimension is infinity and the two equations in (2.6) and (2.7) are not used.

The input–output description in (2.5) was developed from the linearity condition. The development of the state-space equation from the linearity condition, however, is not as simple and will not be attempted. We will simply accept it as a fact.

2.3 Linear Time-Invariant (LTI) Systems

A system is said to be *time invariant* if for every state-input–output pair

$$\left.\begin{array}{l} \mathbf{x}(t_0) \\ \mathbf{u}(t), \quad t \geq t_0 \end{array}\right\} \rightarrow \mathbf{y}(t), \quad t \geq t_0$$

and any T, we have

$$\left.\begin{array}{l} \mathbf{x}(t_0 + T) \\ \mathbf{u}(t - T), \quad t \geq t_0 + T \end{array}\right\} \rightarrow \mathbf{y}(t - T), \quad t \geq t_0 + T \quad \text{(time shifting)}$$

It means that if the initial state is shifted to time $t_0 + T$ and the same input waveform is applied from $t_0 + T$ instead of from t_0, then the output waveform will be the same except that it starts to appear from time $t_0 + T$. In other words, if the initial state and the input are the same, no matter at what time they are applied, the output waveform will always be the same. Therefore, for time-invariant systems, we can always assume, without loss of generality, that $t_0 = 0$. If a system is not time invariant, it is said to be *time varying*.

Time invariance is defined for systems, not for signals. Signals are mostly time varying. If a signal is time invariant such as $u(t) = 1$ for all t, then it is a very simple or a trivial signal. The characteristics of time-invariant systems must be independent of time. For example, the network in Fig. 2.2 is time invariant if R_i, C_i, and L_i are constants.

Some physical systems must be modeled as time-varying systems. For example, a burning rocket is a time-varying system, because its mass decreases rapidly with time. Although the performance of an automobile or a TV set may deteriorate over a long period of time, its characteristics do not change appreciable in the first couple of years. Thus a large number of physical systems can be modeled as time-invariant systems over a limited time period.

---

1. We use $A := B$ to denote that A, by definition, equals B. We use $A =: B$ to denote that B, by definition, equals A.

Input–output description   The zero-state response of a linear system can be described by (2.4). Now if the system is time invariant as well, then we have[2]

$$g(t, \tau) = g(t + T, \tau + T) = g(t - \tau, 0) = g(t - \tau)$$

for any T. Thus (2.4) reduces to

$$y(t) = \int_0^t g(t - \tau)u(\tau)\, d\tau = \int_0^t g(\tau)u(t - \tau)\, d\tau \tag{2.8}$$

where we have replaced t_0 by 0. The second equality can easily be verified by changing the variable. The integration in (2.8) is called a convolution integral. Unlike the time-varying case where g is a function of two variables, g is a function of a single variable in the time-invariant case. By definition $g(t) = g(t - 0)$ is the output at time t due to an impulse input applied at time 0. The condition for a linear time-invariant system to be causal is $g(t) = 0$ for $t < 0$.

EXAMPLE 2.2   The unit-time delay system studied in Example 2.1 is a device whose output equals the input delayed by 1 second. If we apply the impulse $\delta(t)$ at the input terminal, the output is $\delta(t - 1)$. Thus the impulse response of the system is $\delta(t - 1)$.

EXAMPLE 2.3   Consider the unity-feedback system shown in Fig. 2.5(a). It consists of a multiplier with gain a and a unit-time delay element. It is a SISO system. Let $r(t)$ be the input of the feedback system. If $r(t) = \delta(t)$, then the output is the impulse response of the feedback system and equals

$$g_f(t) = a\delta(t - 1) + a^2\delta(t - 2) + a^3\delta(t - 3) + \cdots = \sum_{i=1}^{\infty} a^i\delta(t - i) \tag{2.9}$$

Let $r(t)$ be any input with $r(t) \equiv 0$ for $t < 0$; then the output is given by

$$y(t) = \int_0^t g_f(t - \tau)r(\tau)\, d\tau = \sum_{i=1}^{\infty} a^i \int_0^t \delta(t - \tau - i)r(\tau)\, d\tau$$

$$= \sum_{i=1}^{\infty} a^i r(\tau)\Big|_{\tau = t - i} = \sum_{i=1}^{\infty} a^i r(t - i)$$

Because the unit-time delay system is distributed, so is the feedback system.

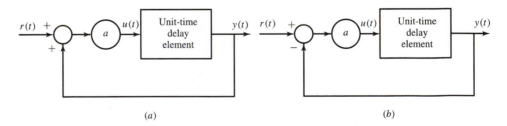

(a)   (b)

Figure 2.5   Positive and negative feedback systems.

2. Note that $g(t, \tau)$ and $g(t - \tau)$ are two different functions. However, for convenience, the same symbol g is used.

Transfer-function matrix   The Laplace transform is an important tool in the study of linear time-invariant (LTI) systems. Let $\hat{y}(s)$ be the Laplace transform of $y(t)$, that is,

$$\hat{y}(s) = \int_0^\infty y(t)e^{-st}dt$$

Throughout this text, we use a variable with a circumflex to denote the Laplace transform of the variable. For causal systems, we have $g(t) = 0$ for $t < 0$ or $g(t - \tau) = 0$ for $\tau > t$. Thus the upper integration limit in (2.8) can be replaced by ∞. Substituting (2.8) and interchanging the order of integrations, we obtain

$$\hat{y}(s) = \int_{t=0}^\infty \left(\int_{\tau=0}^\infty g(t - \tau)u(\tau)\,d\tau \right) e^{-s(t-\tau)}e^{-s\tau}dt$$

$$= \int_{\tau=0}^\infty \left(\int_{t=0}^\infty g(t - \tau)e^{-s(t-\tau)}\,dt \right) u(\tau)e^{-s\tau}d\tau$$

which becomes, after introducing the new variable $v = t - \tau$,

$$\hat{y}(s) = \int_{\tau=0}^\infty \left(\int_{v=-\tau}^\infty g(v)e^{-sv}\,dv \right) u(\tau)e^{-s\tau}\,d\tau$$

Again using the causality condition to replace the lower integration limit inside the parentheses from $v = -\tau$ to $v = 0$, the integration becomes independent of τ and the double integrations become

$$\hat{y}(s) = \int_{v=0}^\infty g(v)e^{-sv}\,dv \int_{\tau=0}^\infty u(\tau)e^{-s\tau}\,d\tau$$

or

$$\hat{y}(s) = \hat{g}(s)\hat{u}(s) \tag{2.10}$$

where

$$\hat{g}(s) = \int_0^\infty g(t)e^{-st}dt$$

is called the *transfer function* of the system. Thus the transfer function is the Laplace transform of the impulse response and, conversely, the impulse response is the inverse Laplace transform of the transfer function. We see that the Laplace transform transforms the convolution integral in (2.8) into the algebraic equation in (2.10). In analysis and design, it is simpler to use algebraic equations than to use convolutions. Thus the convolution in (2.8) will rarely be used in the remainder of this text.

For a p-input q-output system, (2.10) can be extended as

$$\begin{bmatrix} \hat{y}_1(s) \\ \hat{y}_2(s) \\ \vdots \\ \hat{y}_q(s) \end{bmatrix} = \begin{bmatrix} \hat{g}_{11}(s) & \hat{g}_{12}(s) & \cdots & \hat{g}_{1p}(s) \\ \hat{g}_{21}(s) & \hat{g}_{22}(s) & \cdots & \hat{g}_{2p}(s) \\ \vdots & \vdots & & \vdots \\ \hat{g}_{q1}(s) & \hat{g}_{q2}(s) & \cdots & \hat{g}_{qp}(s) \end{bmatrix} \begin{bmatrix} \hat{u}_1(s) \\ \hat{u}_2(s) \\ \vdots \\ \hat{u}_p(s) \end{bmatrix}$$

or

$$\hat{\mathbf{y}}(s) = \hat{\mathbf{G}}(s)\hat{\mathbf{u}}(s) \tag{2.11}$$

where $\hat{g}_{ij}(s)$ is the transfer function from the jth input to the ith output. The $q \times p$ matrix $\hat{\mathbf{G}}(s)$ is called the *transfer-function matrix* or, simply, *transfer matrix* of the system.

EXAMPLE 2.4 Consider the unit-time delay system studied in Example 2.2. Its impulse response is $\delta(t - 1)$. Therefore its transfer function is

$$\hat{g}(s) = \mathcal{L}[\delta(t - 1)] = \int_0^\infty \delta(t - 1)e^{-st}dt = e^{-st}\big|_{t=1} = e^{-s}$$

This transfer function is an irrational function of s.

EXAMPLE 2.5 Consider the feedback system shown in Fig. 2.5(a). The transfer function of the unit-time delay element is e^{-s}. The transfer function from r to y can be computed directly from the block diagram as

$$\hat{g}_f(s) = \frac{ae^{-s}}{1 - ae^{-s}} \tag{2.12}$$

This can also be obtained by taking the Laplace transform of the impulse response, which was computed in (2.9) as

$$g_f(t) = \sum_{i=1}^\infty a^i \delta(t - i)$$

Because $\mathcal{L}[\delta(t - i)] = e^{-is}$, the Laplace transform of $g_f(t)$ is

$$\hat{g}_f(s) = \mathcal{L}[g_f(t)] = \sum_{i=1}^\infty a^i e^{-is} = ae^{-s}\sum_{i=0}^\infty (ae^{-s})^i$$

Using

$$\sum_{i=0}^\infty r^i = \frac{1}{1 - r}$$

for $|r| < 1$, we can express the infinite series in closed form as

$$\hat{g}_f(s) = \frac{ae^{-s}}{1 - ae^{-s}}$$

which is the same as (2.12).

The transfer function in (2.12) is an irrational function of s. This is so because the feedback system is a distributed system. If a linear time-invariant system is lumped, then its transfer function will be a rational function of s. We study mostly lumped systems; thus the transfer functions we will encounter are mostly rational functions of s.

Every rational transfer function can be expressed as $\hat{g}(s) = N(s)/D(s)$, where $N(s)$ and $D(s)$ are polynomials of s. Let us use deg to denote the degree of a polynomial. Then $\hat{g}(s)$ can be classified as follows:

- $\hat{g}(s)$ proper $\Leftrightarrow$ deg $D(s) \geq$ deg $N(s) \Leftrightarrow \hat{g}(\infty) =$ zero or nonzero constant.

- $\hat{g}(s)$ strictly proper $\Leftrightarrow$ deg $D(s) >$ deg $N(s) \Leftrightarrow \hat{g}(\infty) = 0$.
- $\hat{g}(s)$ biproper $\Leftrightarrow$ deg $D(s) =$ deg $N(s) \Leftrightarrow \hat{g}(\infty) =$ nonzero constant.
- $\hat{g}(s)$ improper $\Leftrightarrow$ deg $D(s) <$ deg $N(s) \Leftrightarrow |\hat{g}(\infty)| = \infty$.

Improper rational transfer functions will amplify high-frequency noise, which often exists in the real world; therefore improper rational transfer functions rarely arise in practice.

A real or complex number λ is called a *pole* of the proper transfer function $\hat{g}(s) = N(s)/D(s)$ if $|\hat{g}(\lambda)| = \infty$; a *zero* if $\hat{g}(\lambda) = 0$. If $N(s)$ and $D(s)$ are *coprime*, that is, have no common factors of degree 1 or higher, then all roots of $N(s)$ are the zeros of $\hat{g}(s)$, and all roots of $D(s)$ are the poles of $\hat{g}(s)$. In terms of poles and zeros, the transfer function can be expressed as

$$\hat{g}(s) = k \frac{(s - z_1)(s - z_2) \cdots (s - z_m)}{(s - p_1)(s - p_2) \cdots (s - p_n)}$$

This is called the *zero-pole-gain* form. In MATLAB, this form can be obtained from the transfer function by calling `[z,p,k]= tf2zp(num,den)`.

A rational matrix $\hat{\mathbf{G}}(s)$ is said to be proper if its every entry is proper or if $\hat{\mathbf{G}}(\infty)$ is a zero or nonzero constant matrix; it is strictly proper if its every entry is strictly proper or if $\hat{\mathbf{G}}(\infty)$ is a zero matrix. If a rational matrix $\hat{\mathbf{G}}(s)$ is square and if both $\hat{\mathbf{G}}(s)$ and $\hat{\mathbf{G}}^{-1}(s)$ are proper, then $\hat{\mathbf{G}}(s)$ is said to be biproper. We call λ a pole of $\hat{\mathbf{G}}(s)$ if it is a pole of some entry of $\hat{\mathbf{G}}(s)$. Thus every pole of every entry of $\hat{\mathbf{G}}(s)$ is a pole of $\hat{\mathbf{G}}(s)$. There are a number of ways of defining zeros for $\hat{\mathbf{G}}(s)$. We call λ a *blocking zero* if it is a zero of every nonzero entry of $\hat{\mathbf{G}}(s)$. A more useful definition is the *transmission zero*, which will be introduced in Chapter 9.

State-space equation Every linear time-invariant lumped system can be described by a set of equations of the form

$$\dot{\mathbf{x}}(t) = \mathbf{A}\mathbf{x}(t) + \mathbf{B}\mathbf{u}(t)$$
$$\mathbf{y}(t) = \mathbf{C}\mathbf{x}(t) + \mathbf{D}\mathbf{u}(t)$$

(2.13)

For a system with p inputs, q outputs, and n state variables, $\mathbf{A}$, $\mathbf{B}$, $\mathbf{C}$, and $\mathbf{D}$ are, respectively, $n \times n$, $n \times p$, $q \times n$, and $q \times p$ constant matrices. Applying the Laplace transform to (2.13) yields

$$s\hat{\mathbf{x}}(s) - \mathbf{x}(0) = \mathbf{A}\hat{\mathbf{x}}(s) + \mathbf{B}\hat{\mathbf{u}}(s)$$
$$\hat{\mathbf{y}}(s) = \mathbf{C}\hat{\mathbf{x}}(s) + \mathbf{D}\hat{\mathbf{u}}(s)$$

which implies

$$\hat{\mathbf{x}}(s) = (s\mathbf{I} - \mathbf{A})^{-1}\mathbf{x}(0) + (s\mathbf{I} - \mathbf{A})^{-1}\mathbf{B}\hat{\mathbf{u}}(s) \tag{2.14}$$
$$\hat{\mathbf{y}}(s) = \mathbf{C}(s\mathbf{I} - \mathbf{A})^{-1}\mathbf{x}(0) + \mathbf{C}(s\mathbf{I} - \mathbf{A})^{-1}\mathbf{B}\hat{\mathbf{u}}(s) + \mathbf{D}\hat{\mathbf{u}}(s) \tag{2.15}$$

They are algebraic equations. Given $\mathbf{x}(0)$ and $\hat{\mathbf{u}}(s)$, $\hat{\mathbf{x}}(s)$ and $\hat{\mathbf{y}}(s)$ can be computed algebraically from (2.14) and (2.15). Their inverse Laplace transforms yield the time responses $\mathbf{x}(t)$ and $\mathbf{y}(t)$. The equations also reveal the fact that the response of a linear system can be decomposed as the zero-state response and the zero-input response. If the initial state $\mathbf{x}(0)$ is zero, then (2.15) reduces to

$$\hat{\mathbf{y}}(s) = [\mathbf{C}(s\mathbf{I} - \mathbf{A})^{-1}\mathbf{B} + \mathbf{D}]\hat{\mathbf{u}}(s)$$

Comparing this with (2.11) yields

$$\hat{\mathbf{G}}(s) = \mathbf{C}(s\mathbf{I} - \mathbf{A})^{-1}\mathbf{B} + \mathbf{D} \tag{2.16}$$

This relates the input–output (or transfer matrix) and state-space descriptions.

The functions `tf2ss` and `ss2tf` in MATLAB compute one description from the other. They compute only the SISO and SIMO cases. For example, `[num,den] = ss2tf(a,b,c, d,1)` computes the transfer matrix from the first input to all outputs or, equivalently, the first column of $\hat{\mathbf{G}}(s)$. If the last argument 1 in `ss2tf(a,b,c,d,1)` is replaced by 3, then the function generates the third column of $\hat{\mathbf{G}}(s)$.

To conclude this section, we mention that the Laplace transform is not used in studying linear time-varying systems. The Laplace transform of $g(t, \tau)$ is a function of two variables and $\mathcal{L}[\mathbf{A}(t)\mathbf{x}(t)] \neq \mathcal{L}[\mathbf{A}(t)]\mathcal{L}[\mathbf{x}(t)]$; thus the Laplace transform does not offer any advantage and is not used in studying time-varying systems.

2.3.1 Op-Amp Circuit Implementation

Every linear time-invariant (LTI) state-space equation can be implemented using an operational amplifier (op-amp) circuit. Figure 2.6 shows two standard op-amp circuit elements. All inputs are connected, through resistors, to the inverting terminal. Not shown are the grounded noninverting terminal and power supply. If the feedback branch is a resistor as shown in Fig. 2.6(a), then the output of the element is $-(ax_1 + bx_2 + cx_3)$. If the feedback branch is a capacitor with capacitance C and $RC = 1$ as shown in Fig. 2.6(b), and if the output is assigned as x, then $\dot{x} = -(av_1 + bv_2 + cv_3)$. We call the first element an *adder*; the second element, an *integrator*. Actually, the adder functions also as multipliers and the integrator functions also as multipliers and adder. If we use only one input, say, x_1, in Fig. 2.6(a), then the output equals $-ax_1$, and the element can be used as an *inverter* with gain a. Now we use an example to show that every LTI state-space equation can be implemented using the two types of elements in Fig. 2.6.

Consider the state equation

$$\begin{bmatrix} \dot{x}_1(t) \\ \dot{x}_2(t) \end{bmatrix} = \begin{bmatrix} 2 & -0.3 \\ 1 & -8 \end{bmatrix} \begin{bmatrix} x_1(t) \\ x_2(t) \end{bmatrix} + \begin{bmatrix} -2 \\ 0 \end{bmatrix} u(t) \tag{2.17}$$

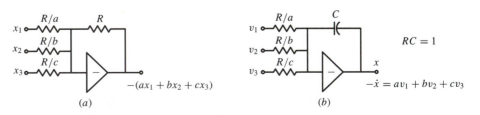

(a)   (b)

Figure 2.6   Two op-amp circuit elements.

$$y(t) = [-2 \ 3] \begin{bmatrix} x_1(t) \\ x_2(t) \end{bmatrix} + 5u(t) \tag{2.18}$$

It has dimension 2 and we need two integrators to implement it. We have the freedom in choosing the output of each integrator as $+x_i$ or $-x_i$. Suppose we assign the output of the left-hand-side (LHS) integrator as x_1 and the output of the right-hand-side (RHS) integrator as $-x_2$ as shown in Fig. 2.7. Then the input of the LHS integrator should be, from the first equation of (2.17), $-\dot{x}_1 = -2x_1 + 0.3x_2 + 2u$ and is connected as shown. The input of the RHS integrator should be $\dot{x}_2 = x_1 - 8x_2$ and is connected as shown. If the output of the adder is chosen as y, then its input should equal $-y = 2x_1 - 3x_2 - 5u$, and is connected as shown. Thus the state equation in (2.17) and (2.18) can be implemented as shown in Fig. 2.7. Note that there are many ways to implement the same equation. For example, if we assign the outputs of the two integrators in Fig. 2.7 as x_1 and x_2, instead of x_1 and $-x_2$, then we will obtain a different implementation.

In actual operational amplifier circuits, the range of signals is limited, usually 1 or 2 volts below the supplied voltage. If any signal grows outside the range, the circuit will saturate or burn out and the circuit will not behave as the equation dictates. There is, however, a way to deal with this problem, as we will discuss in Section 4.3.1.

2.4 Linearization

Most physical systems are nonlinear and time varying. Some of them can be described by the nonlinear differential equation of the form

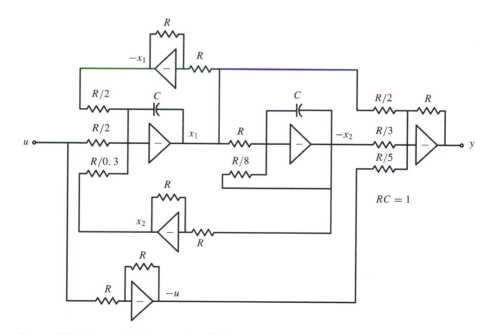

Figure 2.7   Op-amp implementation of (2.17) and (2.18).

$$\dot{\mathbf{x}}(t) = \mathbf{h}(\mathbf{x}(t), \mathbf{u}(t), t) \tag{2.19}$$

$$\mathbf{y}(t) = \mathbf{f}(\mathbf{x}(t), \mathbf{u}(t), t)$$

where $\mathbf{h}$ and $\mathbf{f}$ are nonlinear functions. The behavior of such equations can be very complicated and its study is beyond the scope of this text.

Some nonlinear equations, however, can be approximated by linear equations under certain conditions. Suppose for some input function $\mathbf{u}_o(t)$ and some initial state, $\mathbf{x}_o(t)$ is the solution of (2.19); that is,

$$\dot{\mathbf{x}}_o(t) = \mathbf{h}(\mathbf{x}_o(t), \mathbf{u}_o(t), t) \tag{2.20}$$

Now suppose the input is perturbed slightly to become $\mathbf{u}_o(t) + \bar{\mathbf{u}}(t)$ and the initial state is also perturbed only slightly. For some nonlinear equations, the corresponding solution may differ from $\mathbf{x}_o(t)$ only slightly. In this case, the solution can be expressed as $\mathbf{x}_o(t) + \bar{\mathbf{x}}(t)$ with $\bar{\mathbf{x}}(t)$ small for all t.[3] Under this assumption, we can expand (2.19) as

$$\dot{\mathbf{x}}_o(t) + \dot{\bar{\mathbf{x}}}(t) = \mathbf{h}(\mathbf{x}_o(t) + \bar{\mathbf{x}}(t), \mathbf{u}_o(t) + \bar{\mathbf{u}}(t), t)$$

$$= \mathbf{h}(\mathbf{x}_o(t), \mathbf{u}_o(t), t) + \frac{\partial \mathbf{h}}{\partial \mathbf{x}}\bar{\mathbf{x}} + \frac{\partial \mathbf{h}}{\partial \mathbf{u}}\bar{\mathbf{u}} + \cdots \tag{2.21}$$

where, for $\mathbf{h} = [h_1 \ h_2 \ h_3]'$, $\mathbf{x} = [x_1 \ x_2 \ x_3]'$, and $\mathbf{u} = [u_1 \ u_2]'$,

$$\mathbf{A}(t) := \frac{\partial \mathbf{h}}{\partial \mathbf{x}} := \begin{bmatrix} \partial h_1/\partial x_1 & \partial h_1/\partial x_2 & \partial h_1/\partial x_3 \\ \partial h_2/\partial x_1 & \partial h_2/\partial x_2 & \partial h_2/\partial x_3 \\ \partial h_3/\partial x_1 & \partial h_3/\partial x_2 & \partial h_3/\partial x_3 \end{bmatrix}$$

$$\mathbf{B}(t) := \frac{\partial \mathbf{h}}{\partial \mathbf{u}} := \begin{bmatrix} \partial h_1/\partial u_1 & \partial h_1/\partial u_2 \\ \partial h_2/\partial u_1 & \partial h_2/\partial u_2 \\ \partial h_3/\partial u_1 & \partial h_3/\partial u_2 \end{bmatrix}$$

They are called *Jacobians*. Because $\mathbf{A}$ and $\mathbf{B}$ are computed along the two time functions $\mathbf{x}_o(t)$ and $\mathbf{u}_o(t)$, they are, in general, functions of t. Using (2.20) and neglecting higher powers of $\bar{\mathbf{x}}$ and $\bar{\mathbf{u}}$, we can reduce (2.21) to

$$\dot{\bar{\mathbf{x}}}(t) = \mathbf{A}(t)\bar{\mathbf{x}}(t) + \mathbf{B}(t)\bar{\mathbf{u}}(t)$$

This is a linear state-space equation. The equation $\mathbf{y}(t) = \mathbf{f}(\mathbf{x}(t), \mathbf{u}(t), t)$ can be similarly linearized. This linearization technique is often used in practice to obtain linear equations.

2.5 Examples

In this section we use examples to illustrate how to develop transfer functions and state-space equations for physical systems.

EXAMPLE 2.6 Consider the mechanical system shown in Fig. 2.8. It consists of a block with mass m connected to a wall through a spring. We consider the applied force u to be the input

---

3. This is not true in general. For some nonlinear equations, a very small difference in initial states will generate completely different solutions, yielding the phenomenon of *chaos*.

and displacement y from the equilibrium to be the output. The friction between the floor and the block generally consists of three distinct parts: static friction, Coulomb friction, and viscous friction as shown in Fig. 2.9. Note that the horizontal coordinate is velocity $\dot{y} = dy/dt$. The friction is clearly not a linear function of the velocity. To simplify analysis, we disregard the static and Coulomb frictions and consider only the viscous friction. Then the friction becomes linear and can be expressed as $k_1 \dot{y}(t)$, where k_1 is the viscous friction coefficient. The characteristics of the spring are shown in Fig. 2.10; it is not linear. However, if the displacement is limited to $(y_1, \; y_2)$ as shown, then the spring can be considered to be linear and the spring force equals $k_2 y$, where k_2 is the spring constant. Thus the mechanical system can be modeled as a linear system under linearization and simplification.

Figure 2.8   Mechanical system.

Figure 2.9   Mechanical system.(a) Static and Coulomb frictions. (b) Viscous friction.

Figure 2.10   Characteristic of spring.

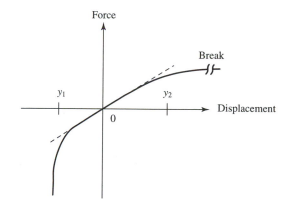

We apply Newton's law to develop an equation to describe the system. The applied force u must overcome the friction and the spring force. The remainder is used to accelerate the block. Thus we have

$$m\ddot{y} = u - k_1\dot{y} - k_2 y \qquad (2.22)$$

where $\ddot{y} = d^2 y(t)/dt^2$ and $\dot{y} = dy(t)/dt$. Applying the Laplace transform and assuming zero initial conditions, we obtain

$$ms^2\hat{y}(s) = \hat{u}(s) - k_1 s\hat{y}(s) - k_2\hat{y}(s)$$

which implies

$$\hat{y}(s) = \frac{1}{ms^2 + k_1 s + k_2}\hat{u}(s)$$

This is the input–output description of the system. Its transfer function is $1/(ms^2 + k_1 s + k_2)$. If $m = 1$, $k_1 = 3$, $k_2 = 2$, then the impulse response of the system is

$$g(t) = \mathcal{L}^{-1}\left[\frac{1}{s^2 + 3s + 2}\right] = \mathcal{L}^{-1}\left[\frac{1}{s+1} - \frac{1}{s+2}\right] = e^{-t} - e^{-2t}$$

and the convolution description of the system is

$$y(t) = \int_0^t g(t-\tau)u(\tau)\,d\tau = \int_0^t (e^{-(t-\tau)} - e^{-2(t-\tau)})u(\tau)\,d\tau$$

Next we develop a state-space equation to describe the system. Let us select the displacement and velocity of the block as state variables; that is, $x_1 = y$, $x_2 = \dot{y}$. Then we have, using (2.22),

$$\dot{x}_1 = x_2 \qquad m\dot{x}_2 = u - k_1 x_2 - k_2 x_1$$

They can be expressed in matrix form as

$$\begin{bmatrix} \dot{x}_1(t) \\ \dot{x}_2(t) \end{bmatrix} = \begin{bmatrix} 0 & 1 \\ -k_2/m & -k_1/m \end{bmatrix}\begin{bmatrix} x_1(t) \\ x_2(t) \end{bmatrix} + \begin{bmatrix} 0 \\ 1/m \end{bmatrix}u(t)$$

$$y(t) = [1 \ \ 0]\begin{bmatrix} x_1(t) \\ x_2(t) \end{bmatrix}$$

This state-space equation describes the system.

EXAMPLE 2.7   Consider the system shown in Fig. 2.11. It consists of two blocks, with masses m_1 and m_2, connected by three springs with spring constants k_i, $i = 1, 2, 3$. To simplify the discussion, we assume that there is no friction between the blocks and the floor. The applied

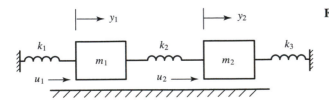

Figure 2.11   Spring-mass system.

force u_1 must overcome the spring forces and the remainder is used to accelerate the block, thus we have

$$u_1 - k_1 y_1 - k_2(y_1 - y_2) = m_1 \ddot{y}_1$$

or

$$m_1 \ddot{y}_1 + (k_1 + k_2)y_1 - k_2 y_2 = u_1 \tag{2.23}$$

For the second block, we have

$$m_2 \ddot{y}_2 - k_2 y_1 + (k_1 + k_2)y_2 = u_2 \tag{2.24}$$

They can be combined as

$$\begin{bmatrix} m_1 & 0 \\ 0 & m_2 \end{bmatrix} \begin{bmatrix} \ddot{y}_1 \\ \ddot{y}_2 \end{bmatrix} + \begin{bmatrix} k_1 + k_2 & -k_2 \\ -k_2 & k_1 + k_2 \end{bmatrix} \begin{bmatrix} y_1 \\ y_2 \end{bmatrix} = \begin{bmatrix} u_1 \\ u_2 \end{bmatrix}$$

This is a standard equation in studying vibration and is said to be in normal form. See Reference [18]. Let us define

$$x_1 := y_1 \qquad x_2 := \dot{y}_1 \qquad x_3 := y_2 \qquad x_4 := \dot{y}_2$$

Then we can readily obtain

$$\begin{bmatrix} \dot{x}_1 \\ \dot{x}_2 \\ \dot{x}_3 \\ \dot{x}_4 \end{bmatrix} = \begin{bmatrix} 0 & 1 & 0 & 0 \\ \dfrac{-(k_1 + k_2)}{m_1} & 0 & \dfrac{k_2}{m_1} & 0 \\ 0 & 0 & 0 & 1 \\ \dfrac{k_2}{m_2} & 0 & \dfrac{-(k_1 + k_2)}{m_1} & 0 \end{bmatrix} \begin{bmatrix} x_1 \\ x_2 \\ x_3 \\ x_4 \end{bmatrix} + \begin{bmatrix} 0 & 0 \\ \dfrac{1}{m_1} & 0 \\ 0 & 0 \\ 0 & \dfrac{1}{m_2} \end{bmatrix} \begin{bmatrix} u_1 \\ u_2 \end{bmatrix}$$

$$\mathbf{y} := \begin{bmatrix} y_1 \\ y_2 \end{bmatrix} = \begin{bmatrix} 1 & 0 & 0 & 0 \\ 0 & 0 & 1 & 0 \end{bmatrix} \mathbf{x}$$

This two-input two-output state equation describes the system in Fig. 2.11.

To develop its input–output description, we apply the Laplace transform to (2.23) and (2.24) and assume zero initial conditions to yield

$$m_1 s^2 \hat{y}_1(s) + (k_1 + k_2)\hat{y}_1(s) - k_2 \hat{y}_2(s) = \hat{u}_1(s)$$
$$m_2 s^2 \hat{y}_2(s) - k_2 \hat{y}_1(s) + (k_1 + k_2)\hat{y}_2(s) = \hat{u}_2(s)$$

From these two equations, we can readily obtain

$$\begin{bmatrix} \hat{y}_1(s) \\ \hat{y}_2(s) \end{bmatrix} = \begin{bmatrix} \dfrac{m_2 s^2 + k_1 + k_2}{d(s)} & \dfrac{k_2}{d(s)} \\ \dfrac{k_2}{d(s)} & \dfrac{m_1 s^2 + k_1 + k_2}{d(s)} \end{bmatrix} \begin{bmatrix} \hat{u}_1(s) \\ \hat{u}_2(s) \end{bmatrix}$$

where

$$d(s) := (m_1 s^2 + k_1 + k_2)(m_2 s^2 + k_1 + k_2) - k_2^2$$

This is the transfer-matrix description of the system. Thus what we will discuss in this text can be applied directly to study vibration.

EXAMPLE 2.8   Consider a cart with an inverted pendulum hinged on top of it as shown in Fig. 2.12. For simplicity, the cart and the pendulum are assumed to move in only one plane, and the friction, the mass of the stick, and the gust of wind are disregarded. The problem is to maintain the pendulum at the vertical position. For example, if the inverted pendulum is falling in the direction shown, the cart moves to the right and exerts a force, through the hinge, to push the pendulum back to the vertical position. This simple mechanism can be used as a model of a space vehicle on takeoff.

Let H and V be, respectively, the horizontal and vertical forces exerted by the cart on the pendulum as shown. The application of Newton's law to the linear movements yields

$$M\frac{d^2y}{dt^2} = u - H$$

$$H = m\frac{d^2}{dt^2}(y + l\sin\theta) = m\ddot{y} + ml\ddot{\theta}\cos\theta - ml(\dot{\theta})^2\sin\theta$$

$$mg - V = m\frac{d^2}{dt^2}(l\cos\theta) = ml[-\ddot{\theta}\sin\theta - (\dot{\theta})^2\cos\theta]$$

The application of Newton's law to the rotational movement of the pendulum around the hinge yields

$$mgl\sin\theta = ml\ddot{\theta}\cdot l + m\ddot{y}l\cos\theta$$

They are nonlinear equations. Because the design objective is to maintain the pendulum at the vertical position, it is reasonable to assume θ and $\dot{\theta}$ to be small. Under this assumption, we can use the approximation $\sin\theta = \theta$ and $\cos\theta = 1$. By retaining only the linear terms in θ and $\dot{\theta}$ or, equivalently, dropping the terms with θ^2, $(\dot{\theta})^2$, $\theta\dot{\theta}$, and $\theta\ddot{\theta}$, we obtain $V = mg$ and

$$M\ddot{y} = u - m\ddot{y} - ml\ddot{\theta}$$

$$g\theta = l\ddot{\theta} + \ddot{y}$$

which imply

$$M\ddot{y} = u - mg\theta \tag{2.25}$$

$$Ml\ddot{\theta} = (M + m)g\theta - u \tag{2.26}$$

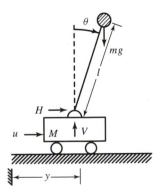

Figure 2.12   Cart with inverted pendulum.

Using these linearized equations, we now can develop the input–output and state-space descriptions. Applying the Laplace transform to (2.25) and (2.26) and assuming zero initial conditions, we obtain

$$Ms^2 \hat{y}(s) = \hat{u}(s) - mg\hat{\theta}(s)$$

$$Mls^2 \hat{\theta}(s) = (M + m)g\hat{\theta}(s) - \hat{u}(s)$$

From these equations, we can readily compute the transfer function $\hat{g}_{yu}(s)$ from u to y and the transfer function $\hat{g}_{\theta u}(s)$ from u to θ as

$$\hat{g}_{yu}(s) = \frac{s^2 - g}{s^2[Ms^2 - (M + m)g]}$$

$$\hat{g}_{\theta u}(s) = \frac{-1}{Ms^2 - (M + m)g}$$

To develop a state-space equation, we select state variables as $x_1 = y$, $x_2 = \dot{y}$, $x_3 = \theta$, and $x_4 = \dot{\theta}$. Then from this selection, (2.25), and (2.26) we can readily obtain

$$\begin{bmatrix} \dot{x}_1 \\ \dot{x}_2 \\ \dot{x}_3 \\ \dot{x}_4 \end{bmatrix} = \begin{bmatrix} 0 & 1 & 0 & 0 \\ 0 & 0 & -mg/M & 0 \\ 0 & 0 & 0 & 1 \\ 0 & 0 & (M+m)g/Ml & 0 \end{bmatrix} \begin{bmatrix} x_1 \\ x_2 \\ x_3 \\ x_4 \end{bmatrix} + \begin{bmatrix} 0 \\ 1/M \\ 0 \\ -1/Ml \end{bmatrix} u$$

$$y = [1\ 0\ 0\ 0]\mathbf{x} \tag{2.27}$$

This state equation has dimension 4 and describes the system when θ and $\dot{\theta}$ are very small.

EXAMPLE 2.9 A communication satellite of mass m orbiting around the earth is shown in Fig. 2.13. The altitude of the satellite is specified by $r(t)$, $\theta(t)$, and $\phi(t)$ as shown. The orbit can be controlled by three orthogonal thrusts $u_r(t)$, $u_\theta(t)$, and $u_\phi(t)$. The state, input, and output of the system are chosen as

$$\mathbf{x}(t) = \begin{bmatrix} r(t) \\ \dot{r}(t) \\ \theta(t) \\ \dot{\theta}(t) \\ \phi(t) \\ \dot{\phi}(t) \end{bmatrix} \qquad \mathbf{u}(t) = \begin{bmatrix} u_r(t) \\ u_\theta(t) \\ u_\phi(t) \end{bmatrix} \qquad \mathbf{y}(t) = \begin{bmatrix} r(t) \\ \theta(t) \\ \phi(t) \end{bmatrix}$$

Then the system can be shown to be described by

$$\dot{\mathbf{x}} = \mathbf{h}(\mathbf{x}, \mathbf{u}) = \begin{bmatrix} \dot{r} \\ r\dot{\theta}^2 \cos^2 \phi + r\dot{\phi}^2 - k/r^2 + u_r/m \\ \dot{\theta} \\ -2\dot{r}\dot{\theta}/r + 2\dot{\theta}\dot{\phi} \sin \phi / \cos \phi + u_\theta/mr \cos \phi \\ \dot{\phi} \\ -\dot{\theta}^2 \cos \phi \sin \phi - 2\dot{r}\dot{\phi}/r + u_\phi/mr \end{bmatrix} \tag{2.28}$$

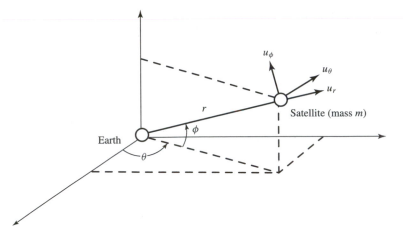

Figure 2.13   Satellite in orbit.

One solution, which corresponds to a circular equatorial orbit, is given by

$$\mathbf{x}_o(t) = [r_o \;\; 0 \;\; \omega_o t \;\; \omega_o \;\; 0 \;\; 0]' \qquad \mathbf{u}_o \equiv \mathbf{0}$$

with $r_o^3 \omega_o^2 = k$, a known physical constant. Once the satellite reaches the orbit, it will remain in the orbit as long as there are no disturbances. If the satellite deviates from the orbit, thrusts must be applied to push it back to the orbit. Define

$$\mathbf{x}(t) = \mathbf{x}_o(t) + \bar{\mathbf{x}}(t) \qquad \mathbf{u}(t) = \mathbf{u}_o(t) + \bar{\mathbf{u}}(t) \qquad \mathbf{y}(t) = \mathbf{y}_o + \bar{\mathbf{y}}(t)$$

If the perturbation is very small, then (2.28) can be linearized as

$$\dot{\bar{\mathbf{x}}}(t) = \left[\begin{array}{cccc:cc} 0 & 1 & 0 & 0 & 0 & 0 \\ 3\omega_o^2 & 0 & 0 & 2\omega_o r_o & 0 & 0 \\ 0 & 0 & 0 & 1 & 0 & 0 \\ 0 & \dfrac{-2\omega_o}{r_o} & 0 & 0 & 0 & 0 \\ \hdashline 0 & 0 & 0 & 0 & 0 & 1 \\ 0 & 0 & 0 & 0 & -\omega_o^2 & 0 \end{array} \right] \bar{\mathbf{x}}(t)$$

$$+ \left[\begin{array}{cc:c} 0 & 0 & 0 \\ \dfrac{1}{m} & 0 & 0 \\ 0 & 0 & 0 \\ 0 & \dfrac{1}{mr_o} & 0 \\ \hdashline 0 & 0 & 0 \\ 0 & 0 & \dfrac{1}{mr_o} \end{array} \right] \bar{\mathbf{u}}(t)$$

$$\bar{\mathbf{y}}(t) = \begin{bmatrix} 1 & 0 & 0 & 0 & \vdots & 0 & 0 \\ 0 & 0 & 1 & 0 & \vdots & 0 & 0 \\ \cdots & \cdots & \cdots & \cdots & \cdots & \cdots & \cdots \\ 0 & 0 & 0 & 0 & \vdots & 1 & 0 \end{bmatrix} \bar{\mathbf{x}}(t) \tag{2.29}$$

This six-dimensional state equation describes the system. In this equation, $\mathbf{A}$, $\mathbf{B}$, and $\mathbf{C}$ happen to be constant. If the orbit is an elliptic one, then they will be time varying. We note that the three matrices are all block diagonal. Thus the equation can be decomposed into two uncoupled parts, one involving r and θ, the other ϕ. Studying these two parts independently can simplify analysis and design.

EXAMPLE 2.10 In chemical plants, it is often necessary to maintain the levels of liquids. A simplified model of a connection of two tanks is shown in Fig. 2.14. It is assumed that under normal operation, the inflows and outflows of both tanks all equal Q and their liquid levels equal H_1 and H_2. Let u be inflow perturbation of the first tank, which will cause variations in liquid level x_1 and outflow y_1 as shown. These variations will cause level variation x_2 and outflow variation y in the second tank. It is assumed that

$$y_1 = \frac{x_1 - x_2}{R_1} \quad \text{and} \quad y = \frac{x_2}{R_2}$$

where R_i are the flow resistances and depend on the normal height H_1 and H_2. They can also be controlled by the valves. Changes of liquid levels are governed by

$$A_1 \, dx_1 = (u - y_1) \, dt \quad \text{and} \quad A_2 \, dx_2 = (y_1 - y) \, dt$$

where A_i are the cross sections of the tanks. From these equations, we can readily obtain

$$\dot{x}_1 = \frac{u}{A_1} - \frac{x_1 - x_2}{A_1 R_1}$$

$$\dot{x}_2 = \frac{x_1 - x_2}{A_2 R_1} - \frac{x_2}{A_2 R_2}$$

Thus the state-space description of the system is given by

$$\begin{bmatrix} \dot{x}_1 \\ \dot{x}_2 \end{bmatrix} = \begin{bmatrix} -1/A_2 R_2 & 1/A_1 R_1 \\ 1/A_2 R_1 & -(1/A_2 R_1 + 1/A_2 R_2) \end{bmatrix} \begin{bmatrix} x_1 \\ x_2 \end{bmatrix} + \begin{bmatrix} 1/A_1 \\ 0 \end{bmatrix} u$$

$$y = [0 \ \ 1/R_2] \mathbf{x}$$

Figure 2.14   Hydraulic tanks.

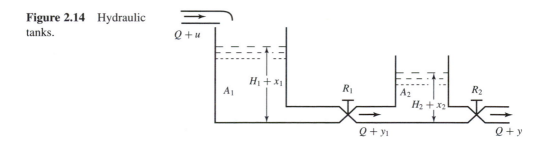

Its transfer function can be computed as

$$\hat{g}(s) = \frac{1}{A_1 A_2 R_1 R_2 s^2 + (A_1 R_1 + A_1 R_2 + A_2 R_2)s + 1}$$

2.5.1 RLC networks

In *RLC* networks, capacitors and inductors can store energy and are associated with state variables. If a capacitor voltage is assigned as a state variable x, then its current is $C\dot{x}$, where C is its capacitance. If an inductor current is assigned as a state variable x, then its voltage is $L\dot{x}$, where L is its inductance. Note that resistors are memoryless elements, and their currents or voltages should not be assigned as state variables. For most simple *RLC* networks, once state variables are assigned, their state equations can be developed by applying Kirchhoff's current and voltage laws, as the next example illustrates.

EXAMPLE 2.11 Consider the network shown in Fig. 2.15. We assign the C_i-capacitor voltages as x_i, $i = 1, 2$ and the inductor current as x_3. It is important to specify their polarities. Then their currents and voltage are, respectively, $C_1\dot{x}_1$, $C_2\dot{x}_2$, and $L\dot{x}_3$ with the polarities shown. From the figure, we see that the voltage across the resistor is $u - x_1$ with the polarity shown. Thus its current is $(u - x_1)/R$. Applying Kirchhoff's current law at node A yields $C_2\dot{x}_2 = x_3$; at node B it yields

$$\frac{u - x_1}{R} = C_1\dot{x}_1 + C_2\dot{x}_2 = C_1\dot{x}_1 + x_3$$

Thus we have

$$\dot{x}_1 = -\frac{x_1}{RC_1} - \frac{x_3}{C_1} + \frac{u}{RC_1}$$

$$\dot{x}_2 = \frac{1}{C_2}x_3$$

Appling Kirchhoff's voltage law to the right-hand-side loop yields $L\dot{x}_3 = x_1 - x_2$ or

$$\dot{x}_3 = \frac{x_1 - x_2}{L}$$

The output y is given by

$$y = L\dot{x}_3 = x_1 - x_2$$

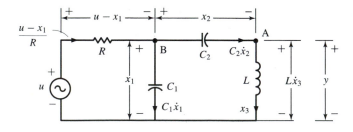

Figure 2.15 Network.

They can be combined in matrix form as

$$\dot{\mathbf{x}} = \begin{bmatrix} -1/RC_1 & 0 & -1/C_1 \\ 0 & 0 & 1/C_2 \\ 1/L & -1/L & 0 \end{bmatrix} \mathbf{x} + \begin{bmatrix} 1/RC_1 \\ 0 \\ 0 \end{bmatrix} u$$

$$y = [1 \quad -1 \quad 0]\mathbf{x} + 0 \cdot u$$

This three-dimensional state equation describes the network shown in Fig. 2.15.

The procedure used in the preceding example can be employed to develop state equations to describe simple *RLC* networks. The procedure is fairly simple: assign state variables and then use branch characteristics and Kirchhoff's laws to develop state equations. The procedure can be stated more systematically by using graph concepts, as we will introduce next. The procedure and subsequent Example 2.12, however, can be skipped without loss of continuity.

First we introduce briefly the concepts of tree, link, and cutset of a network. We consider only connected networks. Every capacitor, inductor, resistor, voltage source, and current source will be considered as a branch. Branches are connected at nodes. Thus a network can be considered to consist of only branches and nodes. A loop is a connection of branches starting from one point and coming back to the same point without passing any point twice. *The algebraic sum of all voltages along every loop is zero* (Kirchhoff's voltage law). The set of all branches connect to a node is called a *cutset*. More generally, a cutset of a connected network is any minimal set of branches so that the removal of the set causes the remaining network to be unconnected. For example, removing all branches connected to a node leaves the node unconnected to the remaining network. *The algebraic sum of all branch currents in every cutset is zero* (Kirchhoff's current law).

A *tree* of a network is defined as any connection of branches connecting all the nodes but containing no loops. A branch is called a *tree branch* if it is in the tree, a *link* if it is not. With respect to a chosen tree, every link has a unique loop, called the *fundamental loop*, in which the remaining loop branches are all tree branches. Every tree branch has a unique cutset, called the *fundamental cutset*, in which the remaining cutset branches are all links. In other words, a fundamental loop contains only one link and a fundamental cutset contains only one tree branch.

▶ **Procedure for developing state-space equations**[4]

1. Consider an *RLC* network. We first choose a *normal tree*. The branches of the normal tree are chosen in the order of voltage sources, capacitors, resistors, inductors, and current sources.

2. Assign the capacitor voltages in the normal tree and the inductor currents in the links as state variables. Capacitor voltages in the links and inductor currents in the normal tree are not assigned.

3. Express the voltage and current of every branch in terms of the state variables and, if necessary, the inputs by applying Kirchhoff's voltage law to fundamental loops and Kirchhoff's current law to fundamental cutsets.

---

4. The reader may skip this procedure and go directly to Example 2.13.

4. Apply Kirchhoff's voltage or current law to the fundamental loop or cutset of every branch that is assigned as a state variable.

EXAMPLE 2.12 Consider the network shown in Fig. 2.16. The normal tree is chosen as shown with heavy lines; it consists of the voltage source, two capacitors, and the 1-Ω resistor. The capacitor voltages in the normal tree and the inductor current in the link will be assigned as state variables. If the voltage across the 3-F capacitor is assigned as x_1, then its current is $3\dot{x}_1$. The voltage across the 1-F capacitor is assigned as x_2 and its current is $\dot{x}_2$. The current through the 2-H inductor is assigned as x_3 and its voltage is $2\dot{x}_3$. Because the 2-Ω resistor is a link, we use its fundamental loop to find its voltage as $u_1 - x_1$. Thus its current is $(u_1 - x_1)/2$. The 1-Ω resistor is a tree branch. We use its fundamental cutset to find its current as x_3. Thus its voltage is $1 \cdot x_3 = x_3$. This completes Step 3.

The 3-F capacitor is a tree branch and its fundamental cutset is as shown. The algebraic sum of the cutset currents is 0 or

$$\frac{u_1 - x_1}{2} - 3\dot{x}_1 + u_2 - x_3 = 0$$

which implies

$$\dot{x}_1 = -\tfrac{1}{6}x_1 - \tfrac{1}{3}x_3 + \tfrac{1}{6}u_1 + \tfrac{1}{3}u_2$$

The 1-F capacitor is a tree branch, and from its fundamental cutset we have $\dot{x}_2 - x_3 = 0$ or

$$\dot{x}_2 = x_3$$

The 2-H inductor is a link. The voltage along its fundamental loop is $2\dot{x}_3 + x_3 - x_1 + x_2 = 0$ or

$$\dot{x}_3 = \tfrac{1}{2}x_1 - \tfrac{1}{2}x_2 - \tfrac{1}{2}x_3$$

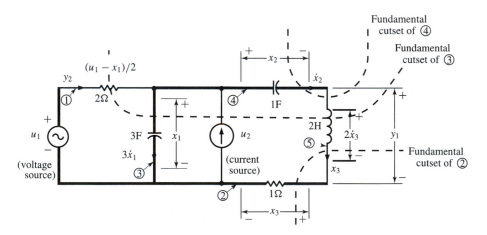

Figure 2.16 Network with two inputs.

They can be expressed in matrix form as

$$\dot{\mathbf{x}} = \begin{bmatrix} -\frac{1}{6} & 0 & -\frac{1}{3} \\ 0 & 0 & 1 \\ \frac{1}{2} & -\frac{1}{2} & -\frac{1}{2} \end{bmatrix} \mathbf{x} + \begin{bmatrix} \frac{1}{6} & \frac{1}{3} \\ 0 & 0 \\ 0 & 0 \end{bmatrix} \mathbf{u} \tag{2.30}$$

If we consider the voltage across the 2-H inductor and the current through the 2-Ω resistor as the outputs, then we have

$$y_1 = 2\dot{x}_3 = x_1 - x_2 - x_3 = [1 \quad -1 \quad -1]\mathbf{x}$$

and

$$y_2 = 0.5(u_1 - x_1) = [-0.5 \ 0 \ 0]\mathbf{x} + [0.5 \ 0]\mathbf{u}$$

They can be written in matrix form as

$$\mathbf{y} = \begin{bmatrix} 1 & -1 & -1 \\ -0.5 & 0 & 0 \end{bmatrix} \mathbf{x} + \begin{bmatrix} 0 & 0 \\ 0.5 & 0 \end{bmatrix} \mathbf{u} \tag{2.31}$$

Equations (2.30) and (2.31) are the state-space description of the network.

The transfer matrix of the network can be computed directly from the network or using the formula in (2.16):

$$\hat{\mathbf{G}}(s) = \mathbf{C}(s\mathbf{I} - \mathbf{A})^{-1}\mathbf{B} + \mathbf{D}$$

We will use MATLAB to compute this equation. We type

```
a=[-1/6 0 -1/3;0 0 1;0.5 -0.5 -0.5];b=[1/6 1/3;0 0;0 0];
c=[1 -1 -1;-0.5 0 0];d=[0 0;0.5 0];
[N1,d1]=ss2tf(a,b,c,d,1)
```

which yields

N1=

$$\begin{array}{cccc} 0.0000 & 0.1667 & -0.0000 & -0.0000 \\ 0.5000 & 0.2500 & 0.3333 & -0.0000 \end{array}$$

d1=

$$1.0000 \quad 0.6667 \quad 0.7500 \quad 0.0833$$

This is the first column of the transfer matrix. We repeat the computation for the second input. Thus the transfer matrix of the network is

$$\hat{\mathbf{G}}(s) = \begin{bmatrix} \dfrac{0.1667s^2}{s^3 + 0.6667s^2 + 0.75s + 0.083} & \dfrac{0.3333s^2}{s^3 + 0.6667s^2 + 0.75s + 0.0833} \\ \dfrac{0.5s^3 + 0.25s^2 + 0.3333s}{s^3 + 0.6667s^2 + 0.75s + 0.083} & \dfrac{-0.1667s^2 - 0.0833s - 0.0833}{s^3 + 0.6667s^2 + 0.75s + 0.0833} \end{bmatrix}$$

EXAMPLE 2.13   Consider the network shown in Fig. 2.17(a), where T is a tunnel diode with the characteristics shown in Fig. 2.17(b). Let x_1 be the voltage across the capacitor and x_2 be the current through the inductor. Then we have $v = x_1$ and

$$x_2(t) = C\dot{x}_1(t) + i(t) = C\dot{x}_1(t) + h(x_1(t))$$
$$L\dot{x}_2(t) = E - Rx_2(t) - x_1(t)$$

They can be arranged as

$$\dot{x}_1(t) = \frac{-h(x_1(t))}{C} + \frac{x_2(t)}{C}$$
$$\dot{x}_2(t) = \frac{-x_1(t) - Rx_2(t)}{L} + \frac{E}{L}$$

(2.32)

This set of nonlinear equations describes the network. Now if $x_1(t)$ is known to lie only inside the range (a, b) shown in Fig. 2.17(b), then $h(x_1(t))$ can be approximated by $h(x_1(t)) = x_1(t)/R_1$. In this case, the network can be reduced to the one in Fig. 2.17(c) and can be described by

$$\begin{bmatrix} \dot{x}_1 \\ \dot{x}_2 \end{bmatrix} = \begin{bmatrix} -1/CR_1 & 1/C \\ -1/L & -R/L \end{bmatrix} \begin{bmatrix} x_1 \\ x_2 \end{bmatrix} + \begin{bmatrix} 0 \\ 1/L \end{bmatrix} E$$

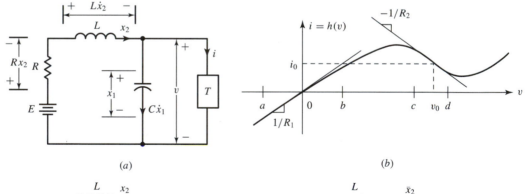

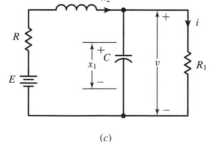

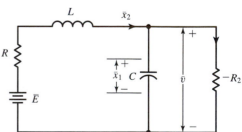

Figure 2.17   Network with a tunnel diode.

This is an LTI state-space equation. Now if $x_1(t)$ is known to lie only inside the range (c, d) shown in Fig. 2.17(b), we may introduce the variables $\bar{x}_1(t) = x_1(t) - v_o$, and $\bar{x}_2(t) = x_2(t) - i_o$ and approximate $h(x_1(t))$ as $i_o - \bar{x}_1(t)/R_2$. Substituting these into (2.32) yields

$$\begin{bmatrix} \dot{\bar{x}}_1 \\ \dot{\bar{x}}_2 \end{bmatrix} = \begin{bmatrix} 1/CR_2 & 1/C \\ -1/L & -R/L \end{bmatrix} \begin{bmatrix} \bar{x}_1 \\ \bar{x}_2 \end{bmatrix} + \begin{bmatrix} 0 \\ 1/L \end{bmatrix} \bar{E}$$

where $\bar{E} = E - v_o - Ri_o$. This equation is obtained by shifting the operating point from $(0, 0)$ to (v_o, i_o) and by linearization at (v_o, i_o). Because the two linearized equations are identical if $-R_2$ is replaced by R_1 and $\bar{E}$ by E, we can readily obtain its equivalent network shown in Fig. 2.17(d). Note that it is not obvious how to obtain the equivalent network from the original network without first developing the state equation.

2.6 Discrete-Time Systems

This section develops the discrete counterpart of continuous-time systems. Because most concepts in continuous-time systems can be applied directly to the discrete-time systems, the discussion will be brief.

The input and output of every discrete-time system will be assumed to have the same sampling period T and will be denoted by $\mathbf{u}[k] := \mathbf{u}(kT)$, $\mathbf{y}[k] := \mathbf{y}(kT)$, where k is an integer ranging from $-\infty$ to $+\infty$. A discrete-time system is causal if current output depends on current and past inputs. The state at time k_0, denoted by $\mathbf{x}[k_0]$, is the information at time instant k_0, which together with $\mathbf{u}[k]$, $k \geq k_0$, determines uniquely the output $\mathbf{y}[k]$, $k \geq k_0$. The entries of $\mathbf{x}$ are called state variables. If the number of state variables is finite, the discrete-time system is lumped; otherwise, it is distributed. Every continuous-time system involving time delay, as the ones in Examples 2.1 and 2.3, is a distributed system. In a discrete-time system, if the time delay is an integer multiple of the sampling period T, then the discrete-time system is a lumped system.

A discrete-time system is linear if the additivity and homogeneity properties hold. The response of every linear discrete-time system can be decomposed as

$$Response = zero\text{-}state\ response + zero\text{-}input\ response$$

and the zero-state responses satisfy the superposition property. So do the zero-input responses.

Input–output description   Let $\delta[k]$ be the *impulse sequence* defined as

$$\delta[k - m] = \begin{cases} 1 & \text{if } k = m \\ 0 & \text{if } k \neq m \end{cases}$$

where both k and m are integers, denoting sampling instants. It is the discrete counterpart of the impulse $\delta(t - t_1)$. The impulse $\delta(t - t_1)$ has zero width and infinite height and cannot be generated in practice; whereas the impulse sequence $\delta[k - m]$ can easily be generated. Let $u[k]$ be any input sequence. Then it can be expressed as

$$u[k] = \sum_{m=-\infty}^{\infty} u[m]\delta[k - m]$$

Let $g[k, m]$ be the output at time instant k excited by the impulse sequence applied at time instant m. Then we have

$$\delta[k - m] \rightarrow g[k, m]$$

$$\delta[k - m]u[m] \rightarrow g[k, m]u[m] \quad \text{(homogeneity)}$$

$$\sum_m \delta[k - m]u[m] \rightarrow \sum_m g[k, m]u[m] \quad \text{(additivity)}$$

Thus the output $y[k]$ excited by the input $u[k]$ equals

$$y[k] = \sum_{m=-\infty}^{\infty} g[k, m]u[m] \tag{2.33}$$

This is the discrete counterpart of (2.3) and its derivation is considerably simpler. The sequence $g[k, m]$ is called the *impulse response sequence*.

If a discrete-time system is causal, no output will appear before an input is applied. Thus we have

$$\text{Causal} \iff g[k, m] = 0, \quad \text{for } k < m$$

If a system is relaxed at k_0 and causal, then (2.33) can be reduced to

$$y[k] = \sum_{m=k_0}^{k} g[k, m]u[m] \tag{2.34}$$

as in (2.4).

If a linear discrete-time system is time invariant as well, then the time shifting property holds. In this case, the initial time instant can always be chosen as $k_0 = 0$ and (2.34) becomes

$$y[k] = \sum_{m=0}^{k} g[k - m]u[m] = \sum_{m=0}^{k} g[m]u[k - m] \tag{2.35}$$

This is the discrete counterpart of (2.8) and is called a *discrete convolution*.

The z-transform is an important tool in the study of LTI discrete-time systems. Let $\hat{y}(z)$ be the z-transform of $y[k]$ defined as

$$\hat{y}(z) := Z[y[k]] := \sum_{k=0}^{\infty} y[k]z^{-k} \tag{2.36}$$

We first replace the upper limit of the integration in (2.35) to ∞,[5] and then substitute it into (2.36) to yield

$$\hat{y}(z) = \sum_{k=0}^{\infty} \left(\sum_{m=0}^{\infty} g[k - m]u[m] \right) z^{-(k-m)} z^{-m}$$

$$= \sum_{m=0}^{\infty} \left(\sum_{k=0}^{\infty} g[k - m]z^{-(k-m)} \right) u[m]z^{-m}$$

$$= \left(\sum_{l=0}^{\infty} g[l]z^{-l} \right) \left(\sum_{m=0}^{\infty} u[m]z^{-m} \right) =: \hat{g}(z)\hat{u}(z)$$

---

5. This is permitted under the causality assumption.

where we have interchanged the order of summations, introduced the new variable $l = k - m$, and then used the fact that $g[l] = 0$ for $l < 0$ to make the inner summation independent of m. The equation

$$\hat{y}(z) = \hat{g}(z)\hat{u}(z) \tag{2.37}$$

is the discrete counterpart of (2.10). The function $\hat{g}(z)$ is the z-transform of the impulse response sequence $g[k]$ and is called the *discrete transfer function*. Both the discrete convolution and transfer function describe only zero-state responses.

EXAMPLE 2.14   Consider the unit-sampling-time delay system defined by

$$y[k] = u[k - 1]$$

The output equals the input delayed by one sampling period. Its impulse response sequence is $g[k] = \delta[k - 1]$ and its discrete transfer function is

$$\hat{g}(z) = Z[\delta[k - 1]] = z^{-1} = \frac{1}{z}$$

It is a rational function of z. Note that every continuous-time system involving time delay is a distributed system. This is not so in discrete-time systems.

EXAMPLE 2.15   Consider the discrete-time feedback system shown in Fig. 2.18(a). It is the discrete counterpart of Fig. 2.5(a). If the unit-sampling-time delay element is replaced by its transfer function z^{-1}, then the block diagram becomes the one in Fig. 2.18(b) and the transfer function from r to y can be computed as

$$\hat{g}(z) = \frac{az^{-1}}{1 - az^{-1}} = \frac{a}{z - a}$$

This is a rational function of z and is similar to (2.12). The transfer function can also be obtained by applying the z-transform to the impulse response sequence of the feedback system. As in (2.9), the impulse response sequence is

$$g_f[k] = a\delta[k - 1] + a^2\delta[k - 2] + \cdots = \sum_{m=1}^{\infty} a^m \delta[k - m]$$

The z-transform of $\delta[k - m]$ is z^{-m}. Thus the transfer function of the feedback system is

$$\hat{g}_f(z) = Z[g_f[k]] = az^{-1} + a^2 z^{-2} + a^3 z^{-3} + \cdots$$

$$= az^{-1} \sum_{m=0}^{\infty} (az^{-1})^m = \frac{az^{-1}}{1 - az^{-1}}$$

which yields the same result.

The discrete transfer functions in the preceding two examples are all rational functions of z. This may not be so in general. For example, if

$$g[k] = \begin{cases} 0 & \text{for } m \leq 0 \\ 1/k & \text{for } k = 1, 2, \ldots \end{cases}$$

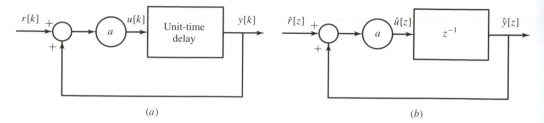

Figure 2.18 Discrete-time feedback system.

Then we have

$$\hat{g}(z) = z^{-1} + \tfrac{1}{2}z^{-2} + \tfrac{1}{3}z^{-3} + \cdots = -\ln(1 - z^{-1})$$

It is an irrational function of z. Such a system is a distributed system. We study in this text only lumped discrete-time systems and their discrete transfer functions are all rational functions of z.

Discrete rational transfer functions can be proper or improper. If a transfer function is improper such as $\hat{g}(z) = (z^2 + 2z - 1)/(z - 0.5)$, then

$$\frac{\hat{y}(z)}{\hat{u}(z)} = \frac{z^2 + 2z - 1}{z - 0.5}$$

which implies

$$y[k + 1] - 0.5y[k] = u[k + 2] + 2u[k + 1] - u[k]$$

or

$$y[k + 1] = 0.5y[k] + u[k + 2] + 2u[k + 1] - u[k]$$

It means that the output at time instant $k + 1$ depends on the input at time instant $k + 2$, a future input. Thus a discrete-time system described by an improper transfer function is not causal. We study only causal systems. Thus all discrete rational transfer functions will be proper. We mentioned earlier that we also study only proper rational transfer functions of s in the continuous-time case. The reason, however, is different. Consider $\hat{g}(s) = s$ or $y(t) = du(t)/dt$. It is a pure differentiator. If we define the differentiation as

$$y(t) = \frac{du(t)}{dt} = \lim_{\Delta \to 0} \frac{u(t + \Delta) - u(t)}{\Delta}$$

where $\Delta > 0$, then the output $y(t)$ depends on future input $u(t + \Delta)$ and the differentiator is not causal. However, if we define the differentiation as

$$y(t) = \frac{du(t)}{dt} = \lim_{\Delta \to 0} \frac{u(t) - u(t - \Delta)}{\Delta}$$

then the output $y(t)$ does not depend on future input and the differentiator is causal. Therefore in continuous-time systems, it is open to argument whether an improper transfer function represents a noncausal system. However, improper transfer functions of s will amplify high-frequency noise, which often exists in the real world. Therefore improper transfer functions are avoided in practice.

State-space equations   Every linear lumped discrete-time system can be described by

$$\mathbf{x}[k + 1] = \mathbf{A}[k]\mathbf{x}[k] + \mathbf{B}[k]\mathbf{u}[k]$$
$$\mathbf{y}[k] = \mathbf{C}[k]\mathbf{x}[k] + \mathbf{D}[k]\mathbf{u}[k]$$

(2.38)

where $\mathbf{A}, \mathbf{B}, \mathbf{C}$, and $\mathbf{D}$ are functions of k. If the system is time invariant as well, then (2.38) becomes

$$\mathbf{x}[k + 1] = \mathbf{A}\mathbf{x}[k] + \mathbf{B}\mathbf{u}[k]$$
$$\mathbf{y}[k] = \mathbf{C}\mathbf{x}[k] + \mathbf{D}\mathbf{u}[k]$$

(2.39)

where $\mathbf{A}, \mathbf{B}, \mathbf{C}$, and $\mathbf{D}$ are constant matrices. Let $\hat{\mathbf{x}}(z)$ be the z-transform of $\mathbf{x}[k]$ or

$$\hat{\mathbf{x}}(z) = Z[\mathbf{x}[k]] := \sum_{k=0}^{\infty} \mathbf{x}[k] z^{-k}$$

Then we have

$$Z[\mathbf{x}[k + 1]] = \sum_{k=0}^{\infty} \mathbf{x}[k + 1] z^{-k} = z \sum_{k=0}^{\infty} \mathbf{x}[k + 1] z^{-(k+1)}$$

$$= z \left[\sum_{l=1}^{\infty} \mathbf{x}[l] z^{-l} + \mathbf{x}[0] - \mathbf{x}[0] \right] = z(\hat{\mathbf{x}}(z) - \mathbf{x}[0])$$

Applying the z-transform to (2.39) yields

$$z\hat{\mathbf{x}}(z) - z\mathbf{x}[0] = \mathbf{A}\hat{\mathbf{x}}(z) + \mathbf{B}\hat{\mathbf{u}}(z)$$
$$\hat{\mathbf{y}}(z) = \mathbf{C}\hat{\mathbf{x}}(z) + \mathbf{D}\hat{\mathbf{u}}(z)$$

which implies

$$\hat{\mathbf{x}}(z) = (z\mathbf{I} - \mathbf{A})^{-1} z\mathbf{x}[0] + (z\mathbf{I} - \mathbf{A})^{-1} \mathbf{B}\hat{\mathbf{u}}(z)$$

(2.40)

$$\hat{\mathbf{y}}(z) = \mathbf{C}(z\mathbf{I} - \mathbf{A})^{-1} z\mathbf{x}[0] + \mathbf{C}(z\mathbf{I} - \mathbf{A})^{-1} \mathbf{B}\hat{\mathbf{u}}(z) + \mathbf{D}\hat{\mathbf{u}}(z)$$

(2.41)

They are the discrete counterparts of (2.14) and (2.15). Note that there is an extra z in front of $\mathbf{x}[0]$. If $\mathbf{x}[0] = \mathbf{0}$, then (2.41) reduces to

$$\hat{\mathbf{y}}(z) = [\mathbf{C}(z\mathbf{I} - \mathbf{A})^{-1}\mathbf{B} + \mathbf{D}]\hat{\mathbf{u}}(z)$$

(2.42)

Comparing this with the MIMO case of (2.37) yields

$$\hat{\mathbf{G}}(z) = \mathbf{C}(z\mathbf{I} - \mathbf{A})^{-1}\mathbf{B} + \mathbf{D}$$

(2.43)

This is the discrete counterpart of (2.16). If the Laplace transform variable s is replaced by the z-transform variable z, then the two equations are identical.

EXAMPLE 2.16 Consider a money market account in a brokerage firm. If the interest rate depends on the amount of money in the account, it is a nonlinear system. If the interest rate is the same no matter how much money is in the account, then it is a linear system. The account is a time-varying system if the interest rate changes with time; a time-invariant system if the interest rate is fixed. We consider here only the LTI case with interest rate $r = 0.015\%$ per day and compounded daily. The input $u[k]$ is the amount of money deposited into the account on the kth day and the output $y[k]$ is the total amount of money in the account at the end of the kth day. If we withdraw money, then $u[k]$ is negative.

If we deposit one dollar on the first day (that is, $u[0] = 1$) and nothing thereafter ($u[k] = 0, \ k = 1, \ 2, \ \ldots$), then $y[0] = u[0] = 1$ and $y[1] = 1 + 0.00015 = 1.00015$. Because the money is compounded daily, we have

$$y[2] = y[1] + y[1] \cdot 0.00015 = y[1] \cdot 1.00015 = (1.00015)^2$$

and, in general,

$$y[k] = (1.00015)^k$$

Because the input $\{1, \ 0, \ 0, \ \ldots\}$ is an impulse sequence, the output is, by definition, the impulse response sequence or

$$g[k] = (1.00015)^k$$

and the input–output description of the account is

$$y[k] = \sum_{m=0}^{k} g[k - m]u[m] = \sum_{m=0}^{k}(1.00015)^{k-m}u[m] \tag{2.44}$$

The discrete transfer function is the z-transform of the impulse response sequence or

$$\hat{g}(z) = Z[g[k]] = \sum_{k=0}^{\infty}(1.00015)^k z^{-k} = \sum_{k=0}^{\infty}(1.00015 z^{-1})^k$$

$$= \frac{1}{1 - 1.00015 z^{-1}} = \frac{z}{z - 1.00015} \tag{2.45}$$

Whenever we use (2.44) or (2.45), the initial state must be zero, or there is initially no money in the account.

Next we develop a state-space equation to describe the account. Suppose $y[k]$ is the total amount of money at the end of the kth day. Then we have

$$y[k + 1] = y[k] + 0.00015y[k] + u[k + 1] = 1.00015y[k] + u[k + 1] \tag{2.46}$$

If we define the state variable as $x[k] := y[k]$, then

$$x[k + 1] = 1.00015x[k] + u[k + 1]$$
$$y[k] = x[k] \tag{2.47}$$

Because of $u[k + 1]$, (2.47) is not in the standard form of (2.39). Thus we cannot select $x[k] := y[k]$ as a state variable. Next we select a different state variable as

$$x[k] := y[k] - u[k]$$

Substituting $y[k+1] = x[k+1] + u[k+1]$ and $y[k] = x[k] + u[k]$ into (2.46) yields

$$x[k+1] = 1.00015x[k] + 1.00015u[k]$$

$$y[k] = x[k] + u[k]$$

(2.48)

This is in the standard form and describes the money market account.

The linearization discussed for the continuous-time case can also be applied to the discrete-time case with only slight modification. Therefore its discussion will not be repeated.

2.7 Concluding Remarks

We introduced in this chapter the concepts of causality, lumpedness, linearity, and time invariance. Mathematical equations were then developed to describe causal systems, as summarized in the following.

System type	Internal description	External description
Distributed, linear		$\mathbf{y}(t) = \displaystyle\int_{t_0}^{t} \mathbf{G}(t, \tau)\mathbf{u}(\tau)\, d\tau$
Lumped, linear	$\dot{\mathbf{x}} = \mathbf{A}(t)\mathbf{x} + \mathbf{B}(t)\mathbf{u}$ $\mathbf{y} = \mathbf{C}(t)\mathbf{x} + \mathbf{D}(t)\mathbf{u}$	$\mathbf{y}(t) = \displaystyle\int_{t_0}^{t} \mathbf{G}(t, \tau)\mathbf{u}(\tau)\, d\tau$
Distributed, linear, time-invariant		$\mathbf{y}(t) = \displaystyle\int_{0}^{t} \mathbf{G}(t - \tau)\mathbf{u}(\tau)\, d\tau$ $\hat{\mathbf{y}}(s) = \hat{\mathbf{G}}(s)\hat{\mathbf{u}}(s),\ \hat{\mathbf{G}}(s)$ irrational
Lumped, linear, time-invariant	$\dot{\mathbf{x}} = \mathbf{A}\mathbf{x} + \mathbf{B}\mathbf{u}$ $\mathbf{y} = \mathbf{C}\mathbf{x} + \mathbf{D}\mathbf{u}$	$\mathbf{y}(t) = \displaystyle\int_{0}^{t} \mathbf{G}(t - \tau)\mathbf{u}(\tau)\, d\tau$ $\hat{\mathbf{y}}(s) = \hat{\mathbf{G}}(s)\hat{\mathbf{u}}(s),\ \hat{\mathbf{G}}(s)$ rational

Distributed systems cannot be described by finite-dimensional state-space equations. External description describes only zero-state responses; thus whenever we use the description, systems are implicitly assumed to be relaxed or their initial conditions are assumed to be zero.

We study in this text mainly lumped linear time-invariant systems. For this class of systems, we use mostly the time-domain description $(\mathbf{A}, \mathbf{B}, \mathbf{C}, \mathbf{D})$ in the internal description and the frequency-domain (Laplace-domain) description $\hat{\mathbf{G}}(s)$ in the external description. Furthermore, we will express every rational transfer matrix as a fraction of two polynomial matrices, as we will develop in the text. By so doing, all designs in the SISO case can be extended to the multivariable case.

The class of lumped linear time-invariant systems constitutes only a very small part of nonlinear and linear systems. For this small class of systems, we are able to give a complete treatment of analyses and syntheses. This study will form a foundation for studying more general systems.

PROBLEMS

2.1 Consider the memoryless systems with characteristics shown in Fig. 2.19, in which u denotes the input and y the output. Which of them is a linear system? Is it possible to introduce a new output so that the system in Fig. 2.19(b) is linear?

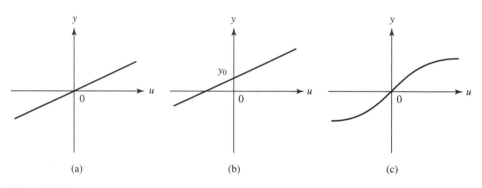

(a)   (b)   (c)

Figure 2.19

2.2 The impulse response of an ideal lowpass filter is given by

$$g(t) = 2\omega \frac{\sin 2\omega(t - t_0)}{2\omega(t - t_0)}$$

for all t, where ω and t_0 are constants. Is the ideal lowpass filter causal? Is it possible to build the filter in the real world?

2.3 Consider a system whose input u and output y are related by

$$y(t) = (P_\alpha u)(t) := \begin{cases} u(t) & \text{for } t \le \alpha \\ 0 & \text{for } t > \alpha \end{cases}$$

where α is a fixed constant. The system is called a *truncation operator*, which chops off the input after time α. Is the system linear? Is it time-invariant? Is it causal?

2.4 The input and output of an initially relaxed system can be denoted by $y = Hu$, where H is some mathematical operator. Show that if the system is causal, then

$$P_\alpha y = P_\alpha Hu = P_\alpha H P_\alpha u$$

where P_α is the truncation operator defined in Problem 2.3. Is it true $P_\alpha Hu = H P_\alpha u$?

2.5 Consider a system with input u and output y. Three experiments are performed on the system using the inputs $u_1(t)$, $u_2(t)$, and $u_3(t)$ for $t \ge 0$. In each case, the initial state $\mathbf{x}(0)$ at time $t = 0$ is the same. The corresponding outputs are denoted by y_1, y_2, and y_3. Which of the following statements are correct if $\mathbf{x}(0) \ne 0$?

1. If $u_3 = u_1 + u_2$, then $y_3 = y_1 + y_2$.

2. If $u_3 = 0.5(u_1 + u_2)$, then $y_3 = 0.5(y_1 + y_2)$.

3. If $u_3 = u_1 - u_2$, then $y_3 = y_1 - y_2$.

Which are correct if $\mathbf{x}(0) = \mathbf{0}$?

2.6 Consider a system whose input and output are related by

$$y(t) = \begin{cases} u^2(t)/u(t-1) & \text{if } u(t-1) \neq 0 \\ 0 & \text{if } u(t-1) = 0 \end{cases}$$

for all t. Show that the system satisfies the homogeneity property but not the additivity property.

2.7 Show that if the additivity property holds, then the homogeneity property holds for all rational numbers α. Thus if a system has some "continuity" property, then additivity implies homogeneity.

2.8 Let $g(t, \tau) = g(t+\alpha, \tau+\alpha)$ for all t, τ, and α. Show that $g(t, \tau)$ depends only on $t - \tau$. [*Hint:* Define $x = t + \tau$ and $y = t - \tau$ and show that $\partial g(t, \tau)/\partial x = 0$.]

2.9 Consider a system with impulse response as shown in Fig. 2.20(a). What is the zero-state response excited by the input $u(t)$ shown in Fig. 2.20(b).

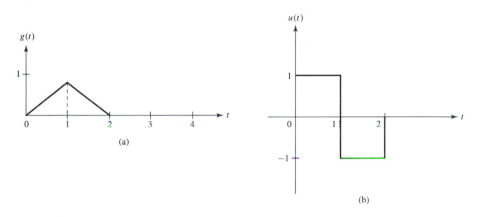

(a)

(b)

Figure 2.20

2.10 Consider a system described by

$$\ddot{y} + 2\dot{y} - 3y = \dot{u} - u$$

What are the transfer function and the impulse response of the system?

2.11 Let $\bar{y}(t)$ be the unit-step response of a linear time-invariant system. Show that the impulse response of the system equals $d\bar{y}(t)/dt$.

2.12 Consider a two-input and two-output system described by

$$D_{11}(p)y_1(t) + D_{12}(p)y_2(t) = N_{11}(p)u_1(t) + N_{12}(p)u_2(t)$$

$$D_{21}(p)y_1(t) + D_{22}(p)y_2(t) = N_{21}(p)u_1(t) + N_{22}(p)u_2(t)$$

where N_{ij} and D_{ij} are polynomials of $p := d/dt$. What is the transfer matrix of the system?

2.13 Consider the feedback systems shown in Fig. 2.5. Show that the unit-step responses of the positive-feedback system are as shown in Fig. 2.21(a) for $a = 1$ and in Fig. 2.21(b) for $a = 0.5$. Show also that the unit-step responses of the negative-feedback system are as shown in Figs. 2.21(c) and 2.21(d), respectively, for $a = 1$ and $a = 0.5$.

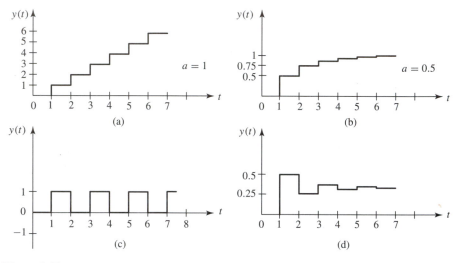

Figure 2.21

2.14 Draw an op-amp circuit diagram for

$$\dot{\mathbf{x}} = \begin{bmatrix} -2 & 4 \\ 0 & 5 \end{bmatrix} \mathbf{x} + \begin{bmatrix} 2 \\ -4 \end{bmatrix} u$$
$$y = [3 \quad 10]\mathbf{x} - 2u$$

2.15 Find state equations to describe the pendulum systems in Fig. 2.22. The systems are useful to model one- or two-link robotic manipulators. If θ, θ_1, and θ_2 are very small, can you consider the two systems as linear?

2.16 Consider the simplified model of an aircraft shown in Fig. 2.23. It is assumed that the aircraft is in an equilibrium state at the pitched angle θ_0, elevator angle u_0, altitude h_0, and cruising speed v_0. It is assumed that small deviations of θ and u from θ_0 and u_0 generate forces $f_1 = k_1\theta$ and $f_2 = k_2u$ as shown in the figure. Let m be the mass of the aircraft, I the moment of inertia about the center of gravity P, $b\dot{\theta}$ the aerodynamic damping, and h the deviation of the altitude from h_0. Find a state equation to describe the system. Show also that the transfer function from u to h, by neglecting the effect of I, is

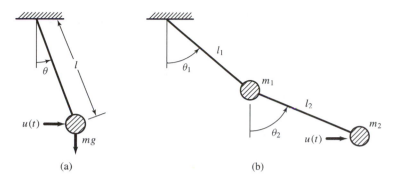

Figure 2.22

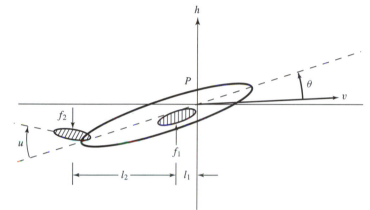

Figure 2.23

$$\hat{g}(s) = \frac{\hat{h}(s)}{\hat{u}(s)} = \frac{k_1 k_2 l_2 - k_2 bs}{ms^2(bs + k_1 l_1)}$$

2.17 The soft landing phase of a lunar module descending on the moon can be modeled as shown in Fig. 2.24. The thrust generated is assumed to be proportional to $\dot{m}$, where m is the mass of the module. Then the system can be described by $m\ddot{y} = -k\dot{m} - mg$, where g is the gravity constant on the lunar surface. Define state variables of the system as $x_1 = y, x_2 = \dot{y}, x_3 = m$, and $u = \dot{m}$. Find a state-space equation to describe the system.

2.18 Find the transfer functions from u to y_1 and from y_1 to y of the hydraulic tank system shown in Fig. 2.25. Does the transfer function from u to y equal the product of the two transfer functions? Is this also true for the system shown in Fig. 2.14? [*Answer:* No, because of the *loading* problem in the two tanks in Fig. 2.14. The loading problem is an important issue in developing mathematical equations to describe composite systems. See Reference [7].]

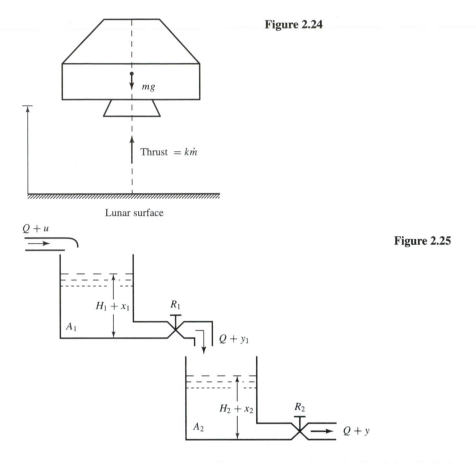

Figure 2.24

mg

Thrust $= k\dot{m}$

Lunar surface

Figure 2.25

$Q + u$

$H_1 + x_1$

A_1

R_1

$Q + y_1$

$H_2 + x_2$

A_2

R_2

$Q + y$

2.19 Find a state equation to describe the network shown in Fig. 2.26. Find also its transfer function.

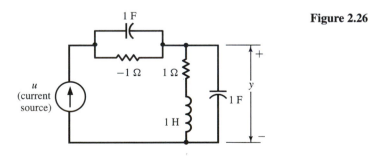

Figure 2.26

1 F

$-1\,\Omega$ $1\,\Omega$

u
(current
source)

1 F

y

1 H

+

−

2.20 Find a state equation to describe the network shown in Fig. 2.2. Compute also its transfer matrix.

2.21 Consider the mechanical system shown in Fig. 2.27. Let I denote the moment of inertia of the bar and block about the hinge. It is assumed that the angular displacement θ is very small. An external force u is applied to the bar as shown. Let y be the displacement of the block, with mass m_2, from equilibrium. Find a state-space equation to describe the system. Find also the transfer function from u to y.

Figure 2.27

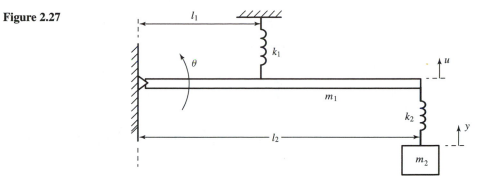

Chapter

3

Linear Algebra

3.1 Introduction

This chapter reviews a number of concepts and results in linear algebra that are essential in the study of this text. The topics are carefully selected, and only those that will be used subsequently are introduced. Most results are developed intuitively in order for the reader to better grasp the ideas. They are stated as theorems for easy reference in later chapters. However, no formal proofs are given.

As we saw in the preceding chapter, all parameters that arise in the real world are real numbers. Therefore we deal only with real numbers, unless stated otherwise, throughout this text. Let $\mathbf{A}$, $\mathbf{B}$, $\mathbf{C}$, and $\mathbf{D}$ be, respectively, $n \times m$, $m \times r$, $l \times n$, and $r \times p$ real matrices. Let $\mathbf{a}_i$ be the ith column of $\mathbf{A}$, and $\mathbf{b}_j$ the jth row of $\mathbf{B}$. Then we have

$$\mathbf{AB} = [\mathbf{a}_1 \ \mathbf{a}_2 \ \cdots \ \mathbf{a}_m] \begin{bmatrix} \mathbf{b}_1 \\ \mathbf{b}_2 \\ \vdots \\ \mathbf{b}_m \end{bmatrix} = \mathbf{a}_1 \mathbf{b}_1 + \mathbf{a}_2 \mathbf{b}_2 + \cdots + \mathbf{a}_m \mathbf{b}_m \tag{3.1}$$

$$\mathbf{CA} = \mathbf{C}[\mathbf{a}_1 \ \mathbf{a}_2 \ \cdots \ \mathbf{a}_m] = [\mathbf{Ca}_1 \ \mathbf{Ca}_2 \ \cdots \ \mathbf{Ca}_m] \tag{3.2}$$

and

$$\mathbf{BD} = \begin{bmatrix} \mathbf{b}_1 \\ \mathbf{b}_2 \\ \vdots \\ \mathbf{b}_m \end{bmatrix} \mathbf{D} = \begin{bmatrix} \mathbf{b}_1 \mathbf{D} \\ \mathbf{b}_2 \mathbf{D} \\ \vdots \\ \mathbf{b}_m \mathbf{D} \end{bmatrix} \tag{3.3}$$

These identities can easily be verified. Note that $\mathbf{a}_i \mathbf{b}_i$ is an $n \times r$ matrix; it is the product of an $n \times 1$ column vector and a $1 \times r$ row vector. The product $\mathbf{b}_i \mathbf{a}_i$ is not defined unless $n = r$; it becomes a scalar if $n = r$.

3.2 Basis, Representation, and Orthonormalization

Consider an n-dimensional real linear space, denoted by $\mathcal{R}^n$. Every vector in $\mathcal{R}^n$ is an n-tuple of real numbers such as

$$\mathbf{x} = \begin{bmatrix} x_1 \\ x_2 \\ \vdots \\ x_n \end{bmatrix}$$

To save space, we write it as $\mathbf{x} = [x_1 \ x_2 \ \cdots \ x_n]'$, where the prime denotes the transpose.

The set of vectors $\{\mathbf{x}_1, \mathbf{x}_2, \ldots, \mathbf{x}_m\}$ in $\mathcal{R}^n$ is said to be *linearly dependent* if there exist real numbers $\alpha_1, \alpha_2, \ldots, \alpha_m$, not all zero, such that

$$\alpha_1 \mathbf{x}_1 + \alpha_2 \mathbf{x}_2 + \cdots + \alpha_m \mathbf{x}_m = \mathbf{0} \tag{3.4}$$

If the only set of α_i for which (3.4) holds is $\alpha_1 = 0, \alpha_2 = 0, \ldots, \alpha_m = 0$, then the set of vectors $\{\mathbf{x}_1, \mathbf{x}_2, \ldots, \mathbf{x}_m\}$ is said to be *linearly independent*.

If the set of vectors in (3.4) is linearly dependent, then there exists at least one α_i, say, α_1, that is different from zero. Then (3.4) implies

$$\mathbf{x}_1 = -\frac{1}{\alpha_1}[\alpha_2 \mathbf{x}_2 + \alpha_3 \mathbf{x}_3 + \cdots + \alpha_m \mathbf{x}_m]$$
$$=: \beta_2 \mathbf{x}_2 + \beta_3 \mathbf{x}_3 + \cdots + \beta_m \mathbf{x}_m$$

where $\beta_i = -\alpha_i/\alpha_1$. Such an expression is called a linear combination.

The *dimension* of a linear space can be defined as the maximum number of linearly independent vectors in the space. Thus in $\mathcal{R}^n$, we can find at most n linearly independent vectors.

Basis and representation   A set of linearly independent vectors in $\mathcal{R}^n$ is called a *basis* if every vector in $\mathcal{R}^n$ can be expressed as a unique linear combination of the set. In $\mathcal{R}^n$, any set of n linearly independent vectors can be used as a basis. Let $\{\mathbf{q}_1, \mathbf{q}_2, \ldots, \mathbf{q}_n\}$ be such a set. Then every vector $\mathbf{x}$ can be expressed uniquely as

$$\mathbf{x} = \alpha_1 \mathbf{q}_1 + \alpha_2 \mathbf{q}_2 + \cdots + \alpha_n \mathbf{q}_n \tag{3.5}$$

Define the $n \times n$ square matrix

$$\mathbf{Q} := [\mathbf{q}_1 \ \mathbf{q}_2 \ \cdots \ \mathbf{q}_n] \tag{3.6}$$

Then (3.5) can be written as

$$\mathbf{x} = \mathbf{Q} \begin{bmatrix} \alpha_1 \\ \alpha_2 \\ \vdots \\ \alpha_n \end{bmatrix} =: \mathbf{Q}\bar{\mathbf{x}} \tag{3.7}$$

We call $\bar{\mathbf{x}} = [\alpha_1 \ \alpha_2 \ \cdots \ \alpha_n]'$ the *representation* of the vector $\mathbf{x}$ with respect to the basis $\{\mathbf{q}_1, \mathbf{q}_2, \ldots, \mathbf{q}_n\}$.

We will associate with every $\mathcal{R}^n$ the following *orthonormal basis*:

$$\mathbf{i}_1 = \begin{bmatrix} 1 \\ 0 \\ 0 \\ \vdots \\ 0 \\ 0 \end{bmatrix}, \ \mathbf{i}_2 = \begin{bmatrix} 0 \\ 1 \\ 0 \\ \vdots \\ 0 \\ 0 \end{bmatrix}, \ \cdots, \ \mathbf{i}_{n-1} = \begin{bmatrix} 0 \\ 0 \\ 0 \\ \vdots \\ 1 \\ 0 \end{bmatrix}, \ \mathbf{i}_n = \begin{bmatrix} 0 \\ 0 \\ 0 \\ \vdots \\ 0 \\ 1 \end{bmatrix} \tag{3.8}$$

With respect to this basis, we have

$$\mathbf{x} := \begin{bmatrix} x_1 \\ x_2 \\ \vdots \\ x_n \end{bmatrix} = x_1\mathbf{i}_1 + x_2\mathbf{i}_2 + \cdots + x_n\mathbf{i}_n = \mathbf{I}_n \begin{bmatrix} x_1 \\ x_2 \\ \vdots \\ x_n \end{bmatrix}$$

where $\mathbf{I}_n$ is the $n \times n$ unit matrix. In other words, the representation of any vector $\mathbf{x}$ with respect to the orthonormal basis in (3.8) equals itself.

EXAMPLE 3.1 Consider the vector $\mathbf{x} = [1 \ 3]'$ in $\mathcal{R}^2$ as shown in Fig. 3.1, The two vectors $\mathbf{q}_1 = [3 \ 1]'$ and $\mathbf{q}_2 = [2 \ 2]'$ are clearly linearly independent and can be used as a basis. If we draw from $\mathbf{x}$ two lines in parallel with $\mathbf{q}_2$ and $\mathbf{q}_1$, they intersect at $-\mathbf{q}_1$ and $2\mathbf{q}_2$ as shown. Thus the representation of $\mathbf{x}$ with respect to $\{\mathbf{q}_1, \mathbf{q}_2\}$ is $[-1 \ 2]'$. This can also be verified from

$$\mathbf{x} = \begin{bmatrix} 1 \\ 3 \end{bmatrix} = [\mathbf{q}_1 \ \mathbf{q}_2] \begin{bmatrix} -1 \\ 2 \end{bmatrix} = \begin{bmatrix} 3 & 2 \\ 1 & 2 \end{bmatrix} \begin{bmatrix} -1 \\ 2 \end{bmatrix}$$

To find the representation of $\mathbf{x}$ with respect to the basis $\{\mathbf{q}_2, \mathbf{i}_2\}$, we draw from $\mathbf{x}$ two lines in parallel with $\mathbf{i}_2$ and $\mathbf{q}_2$. They intersect at $0.5\mathbf{q}_2$ and $2\mathbf{i}_2$. Thus the representation of $\mathbf{x}$ with respect to $\{\mathbf{q}_2, \mathbf{i}_2\}$ is $[0.5 \ 2]'$. (Verify.)

Norms of vectors The concept of *norm* is a generalization of length or magnitude. Any real-valued function of $\mathbf{x}$, denoted by $||\mathbf{x}||$, can be defined as a norm if it has the following properties:

1. $||\mathbf{x}|| \geq 0$ for every $\mathbf{x}$ and $||\mathbf{x}|| = 0$ if and only if $\mathbf{x} = \mathbf{0}$.

2. $||\alpha\mathbf{x}|| = |\alpha|||\mathbf{x}||$, for any real α.

3. $||\mathbf{x}_1 + \mathbf{x}_2|| \leq ||\mathbf{x}_1|| + ||\mathbf{x}_2||$ for every $\mathbf{x}_1$ and $\mathbf{x}_2$.

Figure 3.1 Different representations of vector **x**.

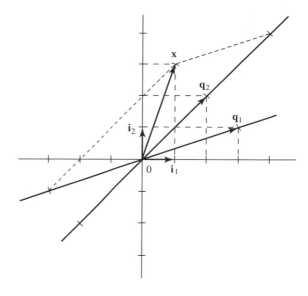

The last inequality is called the *triangular inequality*.

Let $\mathbf{x} = [x_1 \ x_2 \ \cdots \ x_n]'$. Then the norm of $\mathbf{x}$ can be chosen as any one of the following:

$$||\mathbf{x}||_1 := \sum_{i=1}^{n} |x_i|$$

$$||\mathbf{x}||_2 := \sqrt{\mathbf{x}'\mathbf{x}} = \left(\sum_{i=1}^{n} |x_i|^2 \right)^{1/2}$$

$$||\mathbf{x}||_\infty := \max_i |x_i|$$

They are called, respectively, 1-norm, 2- or Euclidean norm, and infinite-norm. The 2-norm is the length of the vector from the origin. We use exclusively, unless stated otherwise, the Euclidean norm and the subscript 2 will be dropped.

In MATLAB, the norms just introduced can be obtained by using the functions `norm(x,1)`, `norm(x,2)` = `norm(x)`, and `norm(x,inf)`.

Orthonormalization   A vector $\mathbf{x}$ is said to be normalized if its Euclidean norm is 1 or $\mathbf{x}'\mathbf{x} = 1$. Note that $\mathbf{x}'\mathbf{x}$ is scalar and $\mathbf{xx}'$ is $n \times n$. Two vectors $\mathbf{x}_1$ and $\mathbf{x}_2$ are said to be *orthogonal* if $\mathbf{x}_1'\mathbf{x}_2 = \mathbf{x}_2'\mathbf{x}_1 = 0$. A set of vectors $\mathbf{x}_i$, $i = 1, 2, \ldots, m$, is said to be *orthonormal* if

$$\mathbf{x}_i'\mathbf{x}_j = \begin{cases} 0 & \text{if } i \neq j \\ 1 & \text{if } i = j \end{cases}$$

Given a set of linearly independent vectors $\mathbf{e}_1$, $\mathbf{e}_2$, $\ldots$, $\mathbf{e}_m$, we can obtain an orthonormal set using the procedure that follows:

$$\mathbf{u}_1 := \mathbf{e}_1 \qquad\qquad \mathbf{q}_1 := \mathbf{u}_1/\|\mathbf{u}_1\|$$
$$\mathbf{u}_2 := \mathbf{e}_2 - (\mathbf{q}_1'\mathbf{e}_2)\mathbf{q}_1 \qquad \mathbf{q}_2 := \mathbf{u}_2/\|\mathbf{u}_2\|$$
$$\vdots$$
$$\mathbf{u}_m := \mathbf{e}_m - \sum_{k=1}^{m-1}(\mathbf{q}_k'\mathbf{e}_m)\mathbf{q}_k \qquad \mathbf{q}_m := \mathbf{u}_m/\|\mathbf{u}_m\|$$

The first equation normalizes the vector $\mathbf{e}_1$ to have norm 1. The vector $(\mathbf{q}_1'\mathbf{e}_2)\mathbf{q}_1$ is the projection of the vector $\mathbf{e}_2$ along $\mathbf{q}_1$. Its subtraction from $\mathbf{e}_2$ yields the vertical part $\mathbf{u}_2$. It is then normalized to 1 as shown in Fig. 3.2. Using this procedure, we can obtain an orthonormal set. This is called the *Schmidt orthonormalization procedure*.

Let $\mathbf{A} = [\mathbf{a}_1 \ \mathbf{a}_2 \ \cdots \ \mathbf{a}_m]$ be an $n \times m$ matrix with $m \le n$. If all columns of $\mathbf{A}$ or $\{\mathbf{a}_i, \ i = 1, 2, \ldots, m\}$ are orthonormal, then

$$\mathbf{A}'\mathbf{A} = \begin{bmatrix} \mathbf{a}_1' \\ \mathbf{a}_2' \\ \vdots \\ \mathbf{a}_m' \end{bmatrix} [\mathbf{a}_1 \ \mathbf{a}_2 \ \cdots \ \mathbf{a}_m] = \begin{bmatrix} 1 & 0 & \cdots & 0 \\ 0 & 1 & \cdots & 0 \\ \vdots & \vdots & \ddots & \vdots \\ 0 & 0 & \cdots & 1 \end{bmatrix} = \mathbf{I}_m$$

where $\mathbf{I}_m$ is the unit matrix of order m. Note that, in general, $\mathbf{A}\mathbf{A}' \ne \mathbf{I}_n$. See Problem 3.4.

3.3 Linear Algebraic Equations

Consider the set of linear algebraic equations

$$\mathbf{A}\mathbf{x} = \mathbf{y} \qquad\qquad (3.9)$$

where $\mathbf{A}$ and $\mathbf{y}$ are, respectively, $m \times n$ and $m \times 1$ real matrices and $\mathbf{x}$ is an $n \times 1$ vector. The matrices $\mathbf{A}$ and $\mathbf{y}$ are given and $\mathbf{x}$ is the unknown to be solved. Thus the set actually consists of m equations and n unknowns. The number of equations can be larger than, equal to, or smaller than the number of unknowns.

We discuss the existence condition and general form of solutions of (3.9). The *range space* of $\mathbf{A}$ is defined as all possible linear combinations of all columns of $\mathbf{A}$. The *rank* of $\mathbf{A}$ is

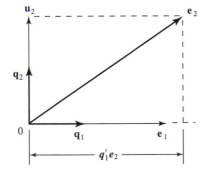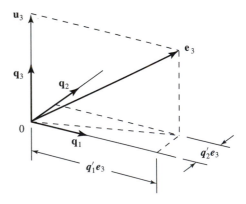

Figure 3.2 Schmidt orthonormization procedure.

defined as the dimension of the range space or, equivalently, the number of linearly independent columns in $\mathbf{A}$. A vector $\mathbf{x}$ is called a *null vector* of $\mathbf{A}$ if $\mathbf{A}\mathbf{x} = \mathbf{0}$. The *null space* of $\mathbf{A}$ consists of all its null vectors. The *nullity* is defined as the maximum number of linearly independent null vectors of $\mathbf{A}$ and is related to the rank by

$$\text{Nullity } (\mathbf{A}) = \text{ number of columns of } \mathbf{A} - \text{ rank } (\mathbf{A}) \tag{3.10}$$

EXAMPLE 3.2 Consider the matrix

$$\mathbf{A} = \begin{bmatrix} 0 & 1 & 1 & 2 \\ 1 & 2 & 3 & 4 \\ 2 & 0 & 2 & 0 \end{bmatrix} =: [\mathbf{a}_1 \ \mathbf{a}_2 \ \mathbf{a}_3 \ \mathbf{a}_4] \tag{3.11}$$

where $\mathbf{a}_i$ denotes the ith column of $\mathbf{A}$. Clearly $\mathbf{a}_1$ and $\mathbf{a}_2$ are linearly independent. The third column is the sum of the first two columns or $\mathbf{a}_1 + \mathbf{a}_2 - \mathbf{a}_3 = \mathbf{0}$. The last column is twice the second column, or $2\mathbf{a}_2 - \mathbf{a}_4 = \mathbf{0}$. Thus $\mathbf{A}$ has two linearly independent columns and has rank 2. The set $\{\mathbf{a}_1, \ \mathbf{a}_2\}$ can be used as a basis of the range space of $\mathbf{A}$.

Equation (3.10) implies that the nullity of $\mathbf{A}$ is 2. It can readily be verified that the two vectors

$$\mathbf{n}_1 = \begin{bmatrix} 1 \\ 1 \\ -1 \\ 0 \end{bmatrix} \qquad \mathbf{n}_2 = \begin{bmatrix} 0 \\ 2 \\ 0 \\ -1 \end{bmatrix} \tag{3.12}$$

meet the condition $\mathbf{A}\mathbf{n}_i = \mathbf{0}$. Because the two vectors are linearly independent, they form a basis of the null space.

The rank of $\mathbf{A}$ is defined as the number of linearly independent columns. It also equals the number of linearly independent rows. Because of this fact, if $\mathbf{A}$ is $m \times n$, then

$$\text{rank}(\mathbf{A}) \leq \min(m, \ n)$$

In MATLAB, the range space, null space, and rank can be obtained by calling the functions `orth`, `null`, and `rank`. For example, for the matrix in (3.11), we type

```
a-[0  1 1 2;1 2 3 4;2 0 2 0];
rank(a)
```

which yields 2. Note that MATLAB computes ranks by using singular-value decomposition (svd), which will be introduced later. The svd algorithm also yields the range and null spaces of the matrix. The MATLAB function `R=orth(a)` yields[1]

```
Ans      R=
```

$$\begin{matrix} 0.3782 & -0.3084 \\ 0.8877 & -0.1468 \\ 0.2627 & 0.9399 \end{matrix} \tag{3.13}$$

---

1. This is obtained using MATLAB Version 5. Earlier versions may yield different results.

The two columns of R form an orthonormal basis of the range space. To check the orthonormality, we type R′ *R, which yields the unity matrix of order 2. The two columns in R are not obtained from the basis $\{a_1, a_2\}$ in (3.11) by using the Schmidt orthonormalization procedure; they are a by-product of svd. However, the two bases should span the same range space. This can be verified by typing

```
rank([a1 a2 R])
```

which yields 2. This confirms that $\{a_1, a_2\}$ span the same space as the two vectors of R. We mention that the rank of a matrix can be very sensitive to roundoff errors and imprecise data. For example, if we use the five-digit display of R in (3.13), the rank of [a1 a2 R] is 3. The rank is 2 if we use the R stored in MATLAB, which uses 16 digits plus exponent.

The null space of (3.11) can be obtained by typing null(a), which yields

$$
\begin{array}{ll}
\text{Ans} \qquad \text{N=} \\[6pt]
0.3434 & -0.5802 \\
0.8384 & 0.3395 \\
-0.3434 & 0.5802 \\
-0.2475 & -0.4598
\end{array}
\tag{3.14}
$$

The two columns are an orthonormal basis of the null space spanned by the two vectors $\{n_1, n_2\}$ in (3.12). All discussion for the range space applies here. That is, rank([n1 n2 N]) yields 3 if we use the five-digit display in (3.14). The rank is 2 if we use the N stored in MATLAB.

With this background, we are ready to discuss solutions of (3.9). We use ρ to denote the rank of a matrix.

▶ **Theorem 3.1**

1. Given an $m \times n$ matrix A and an $m \times 1$ vector y, an $n \times 1$ solution x exists in $Ax = y$ if and only if y lies in the range space of A or, equivalently,

$$
\rho(A) = \rho([A \ \ y])
$$

where $[A \ \ y]$ is an $m \times (n + 1)$ matrix with y appended to A as an additional column.

2. Given A, a solution x exists in $Ax = y$ for every y, if and only if A has rank m (full row rank).

The first statement follows directly from the definition of the range space. If A has full row rank, then the rank condition in (1) is always satisfied for every y. This establishes the second statement.

▶ **Theorem 3.2 (Parameterization of all solutions)**

Given an $m \times n$ matrix A and an $m \times 1$ vector y, let x_p be a solution of $Ax = y$ and let $k := n - \rho(A)$ be the nullity of A. If A has rank n (full column rank) or $k = 0$, then the solution x_p is unique. If $k > 0$, then for every real α_i, $i = 1, 2, \ldots, k$, the vector

$$\mathbf{x} = \mathbf{x}_p + \alpha_1 \mathbf{n}_1 + \cdots + \alpha_k \mathbf{n}_k \tag{3.15}$$

is a solution of $\mathbf{Ax} = \mathbf{y}$, where $\{\mathbf{n}_1, \ldots, \mathbf{n}_k\}$ is a basis of the null space of $\mathbf{A}$.

Substituting (3.15) into $\mathbf{Ax} = \mathbf{y}$ yields

$$\mathbf{Ax}_p + \sum_{i=1}^{k} \alpha_i \mathbf{An}_i = \mathbf{Ax}_p + \mathbf{0} = \mathbf{y}$$

Thus, for every α_i, (3.15) is a solution. Let $\bar{\mathbf{x}}$ be a solution or $\mathbf{A\bar{x}} = \mathbf{y}$. Subtracting this from $\mathbf{Ax}_p = \mathbf{y}$ yields

$$\mathbf{A}(\bar{\mathbf{x}} - \mathbf{x}_p) = \mathbf{0}$$

which implies that $\bar{\mathbf{x}} - \mathbf{x}_p$ is in the null space. Thus $\bar{\mathbf{x}}$ can be expressed as in (3.15). This establishes Theorem 3.2.

EXAMPLE 3.3 Consider the equation

$$\mathbf{Ax} = \begin{bmatrix} 0 & 1 & 1 & 2 \\ 1 & 2 & 3 & 4 \\ 2 & 0 & 2 & 0 \end{bmatrix} \mathbf{x} =: [\mathbf{a}_1 \ \mathbf{a}_2 \ \mathbf{a}_3 \ \mathbf{a}_4]\mathbf{x} = \begin{bmatrix} -4 \\ -8 \\ 0 \end{bmatrix} = \mathbf{y} \tag{3.16}$$

This $\mathbf{y}$ clearly lies in the range space of $\mathbf{A}$ and $\mathbf{x}_p = [0 \ -4 \ 0 \ 0]'$ is a solution. A basis of the null space of $\mathbf{A}$ was shown in (3.12). Thus the general solution of (3.16) can be expressed as

$$\mathbf{x} = \mathbf{x}_p + \alpha_1 \mathbf{n}_1 + \alpha_2 \mathbf{n}_2 = \begin{bmatrix} 0 \\ -4 \\ 0 \\ 0 \end{bmatrix} + \alpha_1 \begin{bmatrix} 1 \\ 1 \\ -1 \\ 0 \end{bmatrix} + \alpha_2 \begin{bmatrix} 0 \\ 2 \\ 0 \\ -1 \end{bmatrix} \tag{3.17}$$

for any real α_1 and α_2.

In application, we will also encounter $\mathbf{xA} = \mathbf{y}$, where the $m \times n$ matrix $\mathbf{A}$ and the $1 \times n$ vector $\mathbf{y}$ are given, and the $1 \times m$ vector $\mathbf{x}$ is to be solved. Applying Theorems 3.1 and 3.2 to the transpose of the equation, we can readily obtain the following result.

▶ **Corollary 3.2**

1. Given an $m \times n$ matrix $\mathbf{A}$, a solution $\mathbf{x}$ exists in $\mathbf{xA} = \mathbf{y}$, for any $\mathbf{y}$, if and only if $\mathbf{A}$ has full column rank.

2. Given an $m \times n$ matrix $\mathbf{A}$ and an $1 \times n$ vector $\mathbf{y}$, let $\mathbf{x}_p$ be a solution of $\mathbf{xA} = \mathbf{y}$ and let $k = m - \rho(\mathbf{A})$. If $k = 0$, the solution $\mathbf{x}_p$ is unique. If $k > 0$, then for any α_i, $i = 1, 2, \ldots, k$, the vector

$$\mathbf{x} = \mathbf{x}_p + \alpha_1 \mathbf{n}_1 + \cdots + \alpha_k \mathbf{n}_k$$

is a solution of $\mathbf{xA} = \mathbf{y}$, where $\mathbf{n}_i \mathbf{A} = \mathbf{0}$ and the set $\{\mathbf{n}_1, \ldots, \mathbf{n}_k\}$ is linearly independent.

In MATLAB, the solution of $\mathbf{Ax} = \mathbf{y}$ can be obtained by typing $\mathbf{A}\backslash\mathbf{y}$. Note the use of backslash, which denotes matrix left division. For example, for the equation in (3.16), typing

```
a=[0 1 1 2;1 2 3 4;2 0 2 0];y=[-4;-8;0];
a\y
```

yields $[0 \ -4 \ 0 \ 0]'$. The solution of $\mathbf{xA} = \mathbf{y}$ can be obtained by typing $\mathbf{y}/\mathbf{A}$. Here we use slash, which denotes matrix right division.

Determinant and inverse of square matrices   The rank of a matrix is defined as the number of linearly independent columns or rows. It can also be defined using the determinant. The determinant of a 1×1 matrix is defined as itself. For $n = 2, 3, \ldots$, the determinant of $n \times n$ square matrix $\mathbf{A} = [a_{ij}]$ is defined recursively as, for any chosen j,

$$\det \mathbf{A} = \sum_{i}^{n} a_{ij} c_{ij} \tag{3.18}$$

where a_{ij} denotes the entry at the ith row and jth column of $\mathbf{A}$. Equation (3.18) is called the *Laplace expansion*. The number c_{ij} is the *cofactor* corresponding to a_{ij} and equals $(-1)^{i+j} \det M_{ij}$, where M_{ij} is the $(n-1) \times (n-1)$ submatrix of $\mathbf{A}$ by deleting its ith row and jth column. If $\mathbf{A}$ is diagonal or triangular, then $\det \mathbf{A}$ equals the product of all diagonal entries.

The determinant of any $r \times r$ submatrix of $\mathbf{A}$ is called a *minor* of order r. Then the rank can be defined as the largest order of all nonzero minors of $\mathbf{A}$. In other words, if $\mathbf{A}$ has rank r, then there is at least one nonzero minor of order r, and every minor of order larger than r is zero. A square matrix is said to be *nonsingular* if its determinant is nonzero. Thus a nonsingular square matrix has full rank and all its columns (rows) are linearly independent.

The *inverse* of a nonsingular square matrix $\mathbf{A} = [a_{ij}]$ is denoted by $\mathbf{A}^{-1}$. The inverse has the property $\mathbf{AA}^{-1} = \mathbf{A}^{-1}\mathbf{A} = \mathbf{I}$ and can be computed as

$$\mathbf{A}^{-1} = \frac{\text{Adj } \mathbf{A}}{\det \mathbf{A}} = \frac{1}{\det \mathbf{A}} [c_{ij}]' \tag{3.19}$$

where c_{ij} is the cofactor. If a matrix is singular, its inverse does not exist. If $\mathbf{A}$ is 2×2, then we have

$$\mathbf{A}^{-1} := \begin{bmatrix} a_{11} & a_{12} \\ a_{21} & a_{22} \end{bmatrix}^{-1} = \frac{1}{a_{11}a_{22} - a_{12}a_{21}} \begin{bmatrix} a_{22} & -a_{12} \\ -a_{21} & a_{11} \end{bmatrix} \tag{3.20}$$

Thus the inverse of a 2×2 matrix is very simple: interchanging diagonal entries and changing the sign of off-diagonal entries (without changing position) and dividing the resulting matrix by the determinant of $\mathbf{A}$. In general, using (3.19) to compute the inverse is complicated. If $\mathbf{A}$ is triangular, it is simpler to compute its inverse by solving $\mathbf{AA}^{-1} = \mathbf{I}$. Note that the inverse of a triangular matrix is again triangular. The MATLAB function `inv` computes the inverse of $\mathbf{A}$.

▶ **Theorem 3.3**

Consider $\mathbf{Ax} = \mathbf{y}$ with $\mathbf{A}$ square.

1. If $\mathbf{A}$ is nonsingular, then the equation has a unique solution for every $\mathbf{y}$ and the solution equals $\mathbf{A}^{-1}\mathbf{y}$. In particular, the only solution of $\mathbf{Ax} = \mathbf{0}$ is $\mathbf{x} = \mathbf{0}$.

2. The homogeneous equation $\mathbf{Ax} = \mathbf{0}$ has nonzero solutions if and only if $\mathbf{A}$ is singular. The number of linearly independent solutions equals the nullity of $\mathbf{A}$.

3.4 Similarity Transformation

Consider an $n \times n$ matrix $\mathbf{A}$. It maps $\mathcal{R}^n$ into itself. If we associate with $\mathcal{R}^n$ the orthonormal basis $\{\mathbf{i}_1, \mathbf{i}_2, \ldots, \mathbf{i}_n\}$ in (3.8), then the ith column of $\mathbf{A}$ is the representation of $\mathbf{A}\mathbf{i}_i$ with respect to the orthonormal basis. Now if we select a different set of basis $\{\mathbf{q}_1, \mathbf{q}_2, \ldots, \mathbf{q}_n\}$, then the matrix $\mathbf{A}$ has a different representation $\bar{\mathbf{A}}$. It turns out that *the ith column of $\bar{\mathbf{A}}$ is the representation of $\mathbf{A}\mathbf{q}_i$ with respect to the basis* $\{\mathbf{q}_1, \mathbf{q}_2, \ldots, \mathbf{q}_n\}$. This is illustrated by the example that follows.

EXAMPLE 3.4 Consider the matrix

$$\mathbf{A} = \begin{bmatrix} 3 & 2 & -1 \\ -2 & 1 & 0 \\ 4 & 3 & 1 \end{bmatrix} \tag{3.21}$$

Let $\mathbf{b} = [0 \ \ 0 \ \ 1]'$. Then we have

$$\mathbf{Ab} = \begin{bmatrix} -1 \\ 0 \\ 1 \end{bmatrix}, \quad \mathbf{A}^2\mathbf{b} = \mathbf{A}(\mathbf{Ab}) = \begin{bmatrix} -4 \\ 2 \\ -3 \end{bmatrix}, \quad \mathbf{A}^3\mathbf{b} = \mathbf{A}(\mathbf{A}^2\mathbf{b}) = \begin{bmatrix} -5 \\ 10 \\ -13 \end{bmatrix}$$

It can be verified that the following relation holds:

$$\mathbf{A}^3\mathbf{b} = 17\mathbf{b} - 15\mathbf{Ab} + 5\mathbf{A}^2\mathbf{b} \tag{3.22}$$

Because the three vectors $\mathbf{b}$, $\mathbf{Ab}$, and $\mathbf{A}^2\mathbf{b}$ are linearly independent, they can be used as a basis. We now compute the representation of $\mathbf{A}$ with respect to the basis. It is clear that

$$\mathbf{A}(\mathbf{b}) = [\mathbf{b} \ \ \mathbf{Ab} \ \ \mathbf{A}^2\mathbf{b}] \begin{bmatrix} 0 \\ 1 \\ 0 \end{bmatrix}$$

$$\mathbf{A}(\mathbf{Ab}) = [\mathbf{b} \ \ \mathbf{Ab} \ \ \mathbf{A}^2\mathbf{b}] \begin{bmatrix} 0 \\ 0 \\ 1 \end{bmatrix}$$

$$\mathbf{A}(\mathbf{A}^2\mathbf{b}) = [\mathbf{b} \ \ \mathbf{Ab} \ \ \mathbf{A}^2\mathbf{b}] \begin{bmatrix} 17 \\ -15 \\ 5 \end{bmatrix}$$

where the last equation is obtained from (3.22). Thus the representation of $\mathbf{A}$ with respect to the basis $\{\mathbf{b}, \mathbf{Ab}, \mathbf{A}^2\mathbf{b}\}$ is

$$\bar{\mathbf{A}} = \begin{bmatrix} 0 & 0 & 17 \\ 1 & 0 & -15 \\ 0 & 1 & 5 \end{bmatrix} \tag{3.23}$$

The preceding discussion can be extended to the general case. Let $\mathbf{A}$ be an $n \times n$ matrix. If there exists an $n \times 1$ vector $\mathbf{b}$ such that the n vectors $\mathbf{b}, \mathbf{Ab}, \ldots, \mathbf{A}^{n-1}\mathbf{b}$ are linearly independent and if

$$\mathbf{A}^n = \beta_1 \mathbf{b} + \beta_2 \mathbf{Ab} + \cdots + \beta_n \mathbf{A}^{n-1}\mathbf{b}$$

then the representation of $\mathbf{A}$ with respect to the basis $\{\mathbf{b}, \ \mathbf{Ab}, \ \ldots, \ \mathbf{A}^{n-1}\mathbf{b}\}$ is

$$\bar{\mathbf{A}} = \begin{bmatrix} 0 & 0 & \cdots & 0 & \beta_1 \\ 1 & 0 & \cdots & 0 & \beta_2 \\ 0 & 1 & \cdots & 0 & \beta_3 \\ \vdots & \vdots & \ddots & \vdots & \vdots \\ 0 & 0 & \cdots & 0 & \beta_{n-1} \\ 0 & 0 & \cdots & 1 & \beta_n \end{bmatrix} \tag{3.24}$$

This matrix is said to be in a *companion* form.

Consider the equation

$$\mathbf{Ax} = \mathbf{y} \tag{3.25}$$

The square matrix $\mathbf{A}$ maps $\mathbf{x}$ in $\mathcal{R}^n$ into $\mathbf{y}$ in $\mathcal{R}^n$. With respect to the basis $\{\mathbf{q}_1, \ \mathbf{q}_2, \ \ldots, \ \mathbf{q}_n\}$, the equation becomes

$$\bar{\mathbf{A}}\bar{\mathbf{x}} = \bar{\mathbf{y}} \tag{3.26}$$

where $\bar{\mathbf{x}}$ and $\bar{\mathbf{y}}$ are the representations of $\mathbf{x}$ and $\mathbf{y}$ with respect to the basis $\{\mathbf{q}_1, \ \mathbf{q}_2, \ \ldots, \ \mathbf{q}_n\}$. As discussed in (3.7), they are related by

$$\mathbf{x} = \mathbf{Q}\bar{\mathbf{x}} \qquad \mathbf{y} = \mathbf{Q}\bar{\mathbf{y}}$$

with

$$\mathbf{Q} = [\mathbf{q}_1 \ \mathbf{q}_2 \ \cdots \ \mathbf{q}_n] \tag{3.27}$$

an $n \times n$ nonsingular matrix. Substituting these into (3.25) yields

$$\mathbf{AQ}\bar{\mathbf{x}} = \mathbf{Q}\bar{\mathbf{y}} \quad \text{or} \quad \mathbf{Q}^{-1}\mathbf{AQ}\bar{\mathbf{x}} = \bar{\mathbf{y}} \tag{3.28}$$

Comparing this with (3.26) yields

$$\bar{\mathbf{A}} = \mathbf{Q}^{-1}\mathbf{AQ} \quad \text{or} \quad \mathbf{A} = \mathbf{Q}\bar{\mathbf{A}}\mathbf{Q}^{-1} \tag{3.29}$$

This is called the *similarity transformation* and $\mathbf{A}$ and $\bar{\mathbf{A}}$ are said to be *similar*. We write (3.29) as

$$\mathbf{AQ} = \mathbf{Q}\bar{\mathbf{A}}$$

or

$$\mathbf{A}[\mathbf{q}_1 \ \mathbf{q}_2 \ \cdots \ \mathbf{q}_n] = [\mathbf{Aq}_1 \ \mathbf{Aq}_2 \ \cdots \ \mathbf{Aq}_n] = [\mathbf{q}_1 \ \mathbf{q}_2 \ \cdots \ \mathbf{q}_n]\bar{\mathbf{A}}$$

This shows that the ith column of $\bar{\mathbf{A}}$ is indeed the representation of $\mathbf{Aq}_i$ with respect to the basis $\{\mathbf{q}_1, \mathbf{q}_2, \ldots, \mathbf{q}_n\}$.

3.5 Diagonal Form and Jordan Form

A square matrix $\mathbf{A}$ has different representations with respect to different sets of basis. In this section, we introduce a set of basis so that the representation will be diagonal or block diagonal.

A real or complex number λ is called an *eigenvalue* of the $n \times n$ real matrix $\mathbf{A}$ if there exists a nonzero vector $\mathbf{x}$ such that $\mathbf{Ax} = \lambda\mathbf{x}$. Any nonzero vector $\mathbf{x}$ satisfying $\mathbf{Ax} = \lambda\mathbf{x}$ is called a (right) *eigenvector* of $\mathbf{A}$ associated with eigenvalue λ. In order to find the eigenvalue of $\mathbf{A}$, we write $\mathbf{Ax} = \lambda\mathbf{x} = \lambda\mathbf{Ix}$ as

$$(\mathbf{A} - \lambda\mathbf{I})\mathbf{x} = \mathbf{0} \tag{3.30}$$

where $\mathbf{I}$ is the unit matrix of order n. This is a homogeneous equation. If the matrix $(\mathbf{A} - \lambda\mathbf{I})$ is nonsingular, then the only solution of (3.30) is $\mathbf{x} = \mathbf{0}$ (Theorem 3.3). Thus in order for (3.30) to have a nonzero solution $\mathbf{x}$, the matrix $(\mathbf{A} - \lambda\mathbf{I})$ must be singular or have determinant 0. We define

$$\Delta(\lambda) = \det(\lambda\mathbf{I} - \mathbf{A})$$

It is a monic polynomial of degree n with real coefficients and is called the *characteristic polynomial* of $\mathbf{A}$. A polynomial is called monic if its leading coefficient is 1. If λ is a root of the characteristic polynomial, then the determinant of $(\mathbf{A} - \lambda\mathbf{I})$ is 0 and (3.30) has at least one nonzero solution. Thus every root of $\Delta(\lambda)$ is an eigenvalue of $\mathbf{A}$. Because $\Delta(\lambda)$ has degree n, the $n \times n$ matrix $\mathbf{A}$ has n eigenvalues (not necessarily all distinct).

We mention that the matrices

$$\begin{bmatrix} 0 & 0 & 0 & -\alpha_4 \\ 1 & 0 & 0 & -\alpha_3 \\ 0 & 1 & 0 & -\alpha_2 \\ 0 & 0 & 1 & -\alpha_1 \end{bmatrix} \qquad \begin{bmatrix} -\alpha_1 & -\alpha_2 & -\alpha_3 & -\alpha_4 \\ 1 & 0 & 0 & 0 \\ 0 & 1 & 0 & 0 \\ 0 & 0 & 1 & 0 \end{bmatrix}$$

and their transposes

$$\begin{bmatrix} 0 & 1 & 0 & 0 \\ 0 & 0 & 1 & 0 \\ 0 & 0 & 0 & 1 \\ -\alpha_4 & -\alpha_3 & -\alpha_2 & -\alpha_1 \end{bmatrix} \qquad \begin{bmatrix} -\alpha_1 & 1 & 0 & 0 \\ -\alpha_2 & 0 & 1 & 0 \\ -\alpha_3 & 0 & 0 & 1 \\ -\alpha_4 & 0 & 0 & 0 \end{bmatrix}$$

all have the following characteristic polynomial:

$$\Delta(\lambda) = \lambda^4 + \alpha_1\lambda^3 + \alpha_2\lambda^2 + \alpha_3\lambda + \alpha_4$$

These matrices can easily be formed from the coefficients of $\Delta(\lambda)$ and are called *companion-form* matrices. The companion-form matrices will arise repeatedly later. The matrix in (3.24) is in such a form.

Eigenvalues of A are all distinct   Let λ_i, $i = 1, 2, \ldots, n$, be the eigenvalues of $\mathbf{A}$ and be all distinct. Let $\mathbf{q}_i$ be an eigenvector of $\mathbf{A}$ associated with λ_i; that is, $\mathbf{Aq}_i = \lambda_i \mathbf{q}_i$. Then the set of eigenvectors $\{\mathbf{q}_1, \mathbf{q}_2, \ldots, \mathbf{q}_n\}$ is linearly independent and can be used as a basis. Let $\hat{\mathbf{A}}$ be the representation of $\mathbf{A}$ with respect to this basis. Then the first column of $\hat{\mathbf{A}}$ is the representation of $\mathbf{Aq}_1 = \lambda_1 \mathbf{q}_1$ with respect to $\{\mathbf{q}_1, \mathbf{q}_2, \ldots, \mathbf{q}_n\}$. From

$$\mathbf{Aq}_1 = \lambda_1 \mathbf{q}_1 = [\mathbf{q}_1 \ \mathbf{q}_2 \ \cdots \ \mathbf{q}_n] \begin{bmatrix} \lambda_1 \\ 0 \\ 0 \\ \vdots \\ 0 \end{bmatrix}$$

we conclude that the first column of $\hat{\mathbf{A}}$ is $[\lambda_1 \ 0 \ \cdots \ 0]'$. The second column of $\hat{\mathbf{A}}$ is the representation of $\mathbf{Aq}_2 = \lambda_2 \mathbf{q}_2$ with respect to $\{\mathbf{q}_1, \mathbf{q}_2, \ldots, \mathbf{q}_n\}$, that is, $[0 \ \lambda_1 \ 0 \ \cdots \ 0]'$. Proceeding forward, we can establish

$$\hat{\mathbf{A}} = \begin{bmatrix} \lambda_1 & 0 & 0 & \cdots & 0 \\ 0 & \lambda_2 & 0 & \cdots & 0 \\ 0 & 0 & \lambda_3 & \cdots & 0 \\ \vdots & \vdots & \vdots & & \vdots \\ 0 & 0 & 0 & \cdots & \lambda_n \end{bmatrix} \tag{3.31}$$

This is a diagonal matrix. Thus we conclude that every matrix with distinct eigenvalues has a diagonal matrix representation by using its eigenvectors as a basis. Different orderings of eigenvectors will yield different diagonal matrices for the same $\mathbf{A}$.

If we define

$$\mathbf{Q} = [\mathbf{q}_1 \ \mathbf{q}_2 \ \cdots \ \mathbf{q}_n] \tag{3.32}$$

then the matrix $\hat{\mathbf{A}}$ equals

$$\hat{\mathbf{A}} = \mathbf{Q}^{-1}\mathbf{AQ} \tag{3.33}$$

as derived in (3.29). Computing (3.33) by hand is not simple because of the need to compute the inverse of $\mathbf{Q}$. However, if we know $\hat{\mathbf{A}}$, then we can verify (3.33) by checking $\mathbf{Q}\hat{\mathbf{A}} = \mathbf{AQ}$.

EXAMPLE 3.5   Consider the matrix

$$\mathbf{A} = \begin{bmatrix} 0 & 0 & 0 \\ 1 & 0 & 2 \\ 0 & 1 & 1 \end{bmatrix}$$

Its characteristic polynomial is

$$\Delta(\lambda) = \det(\lambda\mathbf{I} - \mathbf{A}) = \det \begin{bmatrix} \lambda & 0 & 0 \\ -1 & \lambda & -2 \\ 0 & -1 & \lambda - 1 \end{bmatrix}$$

$$= \lambda[\lambda(\lambda - 1) - 2] = (\lambda - 2)(\lambda + 1)\lambda$$

Thus $\mathbf{A}$ has eigenvalues 2, -1, and 0. The eignevector associated with $\lambda = 2$ is any nonzero solution of

$$(\mathbf{A} - 2\mathbf{I})\mathbf{q}_1 = \begin{bmatrix} -2 & 0 & 0 \\ 1 & -2 & 2 \\ 0 & 1 & -1 \end{bmatrix} \mathbf{q}_1 = \mathbf{0}$$

Thus $\mathbf{q}_1 = [0 \ 1 \ 1]'$ is an eigenvector associated with $\lambda = 2$. Note that the eigenvector is not unique, $[0 \ \alpha \ \alpha]'$ for any nonzero real α can also be chosen as an eigenvector. The eigenvector associated with $\lambda = -1$ is any nonzero solution of

$$(\mathbf{A} - (-1)\mathbf{I})\mathbf{q}_2 = \begin{bmatrix} 1 & 0 & 0 \\ 1 & 1 & 2 \\ 0 & 1 & 2 \end{bmatrix} \mathbf{q}_2 = \mathbf{0}$$

which yields $\mathbf{q}_2 = [0 \ -2 \ 1]'$. Similarly, the eigenvector associated with $\lambda = 0$ can be computed as $\mathbf{q}_3 = [2 \ 1 \ -1]'$. Thus the representation of $\mathbf{A}$ with respect to $\{\mathbf{q}_1, \mathbf{q}_2, \mathbf{q}_3\}$ is

$$\hat{\mathbf{A}} = \begin{bmatrix} 2 & 0 & 0 \\ 0 & -1 & 0 \\ 0 & 0 & 0 \end{bmatrix} \tag{3.34}$$

It is a diagonal matrix with eigenvalues on the diagonal. This matrix can also be obtained by computing

$$\hat{\mathbf{A}} = \mathbf{Q}^{-1}\mathbf{A}\mathbf{Q}$$

with

$$\mathbf{Q} = [\mathbf{q}_1 \ \mathbf{q}_2 \ \mathbf{q}_3] = \begin{bmatrix} 0 & 0 & 2 \\ 1 & -2 & 1 \\ 1 & 1 & -1 \end{bmatrix} \tag{3.35}$$

However, it is simpler to verify $\mathbf{Q}\hat{\mathbf{A}} = \mathbf{A}\mathbf{Q}$ or

$$\begin{bmatrix} 0 & 0 & 2 \\ 1 & -2 & 1 \\ 1 & 1 & -1 \end{bmatrix} \begin{bmatrix} 2 & 0 & 0 \\ 0 & -1 & 0 \\ 0 & 0 & 0 \end{bmatrix} = \begin{bmatrix} 0 & 0 & 0 \\ 1 & 0 & 2 \\ 0 & 1 & 1 \end{bmatrix} \begin{bmatrix} 0 & 0 & 2 \\ 1 & -2 & 1 \\ 1 & 1 & -1 \end{bmatrix}$$

The result in this example can easily be obtained using MATLAB. Typing

$$\texttt{a=[0 0 0;1 0 2;0 1 1]; [q,d]=eig(a)}$$

yields

$$q = \begin{bmatrix} 0 & 0 & 0.8186 \\ 0.7071 & 0.8944 & 0.4082 \\ 0.7071 & -0.4472 & -0.4082 \end{bmatrix} \qquad d = \begin{bmatrix} 2 & 0 & 0 \\ 0 & -1 & 0 \\ 0 & 0 & 0 \end{bmatrix}$$

where d is the diagonal matrix in (3.34). The matrix q is different from the $\mathbf{Q}$ in (3.35); but their corresponding columns differ only by a constant. This is due to nonuniqueness of eigenvectors and every column of q is normalized to have norm 1 in MATLAB. If we type $\texttt{eig(a)}$ without the left-hand-side argument, then MATLAB generates only the three eigenvalues $\texttt{2, -1, 0}$.

We mention that eigenvalues in MATLAB are *not* computed from the characteristic polynomial. Computing the characteristic polynomial using the Laplace expansion and then computing its roots are not numerically reliable, especially when there are repeated roots. Eigenvalues are computed in MATLAB directly from the matrix by using similarity transformations. Once all eigenvalues are computed, the characteristic polynomial equals $\prod(\lambda - \lambda_i)$. In MATLAB, typing `r=eig(a);poly(r)` yields the characteristic polynomial.

EXAMPLE 3.6 Consider the matrix

$$\mathbf{A} = \begin{bmatrix} -1 & 1 & 1 \\ 0 & 4 & -13 \\ 0 & 1 & 0 \end{bmatrix}$$

Its characteristic polynomial is $(\lambda+1)(\lambda^2-4\lambda+13)$. Thus $\mathbf{A}$ has eigenvalues -1, $2\pm3j$. Note that complex conjugate eigenvalues must appear in pairs because $\mathbf{A}$ has only real coefficients. The eigenvectors associated with -1 and $2+3j$ are, respectively, $[1 \ 0 \ 0]'$ and $[j \ -3+2j \ j]'$. The eigenvector associated with $\lambda = 2 - 3j$ is $[-j \ -3-2j \ -j]'$, the complex conjugate of the eigenvector associated with $\lambda = 2 + 3j$. Thus we have

$$\mathbf{Q} = \begin{bmatrix} 1 & j & -j \\ 0 & -3+2j & -3-2j \\ 0 & j & j \end{bmatrix} \quad \text{and} \quad \hat{\mathbf{A}} = \begin{bmatrix} -1 & 0 & 0 \\ 0 & 2+3j & 0 \\ 0 & 0 & 2-3j \end{bmatrix} \quad (3.36)$$

The MATLAB function `[q,d]=eig(a)` yields

$$q = \begin{bmatrix} 1 & 0.2582j & -0.2582j \\ 0 & -0.7746+0.5164j & -0.7746-0.5164j \\ 0 & 0.2582j & -0.2582j \end{bmatrix}$$

and

$$d = \begin{bmatrix} -1 & 0 & 0 \\ 0 & 2+3j & 0 \\ 0 & 0 & 2-3j \end{bmatrix}$$

All discussion in the preceding example applies here.

Complex eigenvalues Even though the data we encounter in practice are all real numbers, complex numbers may arise when we compute eigenvalues and eigenvectors. To deal with this problem, we must extend real linear spaces into complex linear spaces and permit all scalars such as α_i in (3.4) to assume complex numbers. To see the reason, we consider

$$\mathbf{Av} = \begin{bmatrix} 1 & 1+j \\ 1-j & 2 \end{bmatrix} \mathbf{v} = \mathbf{0} \quad (3.37)$$

If we restrict $\mathbf{v}$ to real vectors, then (3.37) has no nonzero solution and the two columns of $\mathbf{A}$ are linearly independent. However, if $\mathbf{v}$ is permitted to assume complex numbers, then $\mathbf{v} = [-2 \ 1 - j]'$ is a nonzero solution of (3.37). Thus the two columns of $\mathbf{A}$ are linearly dependent and $\mathbf{A}$ has rank 1. This is the rank obtained in MATLAB. Therefore, whenever complex eigenvalues arise, we consider complex linear spaces and complex scalars and transpose is replaced by complex-conjugate transpose. By so doing, all concepts and results developed for real vectors and matrices can be applied to complex vectors and matrices. Incidentally, the diagonal matrix with complex eigenvalues in (3.36) can be transformed into a very useful real matrix as we will discuss in Section 4.3.1.

Eigenvalues of A are not all distinct   An eigenvalue with multiplicity 2 or higher is called a *repeated* eigenvalue. In contrast, an eigenvalue with multiplicity 1 is called a *simple* eigenvalue. If A has only simple eigenvalues, it always has a diagonal-form representation. If A has repeated eigenvalues, then it may not have a diagonal form representation. However, it has a block-diagonal and triangular-form representation as we will discuss next.

Consider an $n \times n$ matrix A with eigenvalue λ and multiplicity n. In other words, A has only one distinct eigenvalue. To simplify the discussion, we assume $n = 4$. Suppose the matrix $(A - \lambda I)$ has rank $n - 1 = 3$ or, equivalently, nullity 1; then the equation

$$(A - \lambda I)q = 0$$

has only one independent solution. Thus A has only one eigenvector associated with λ. We need $n - 1 = 3$ more linearly independent vectors to form a basis for $\mathcal{R}^4$. The three vectors q_2, q_3, q_4 will be chosen to have the properties $(A - \lambda I)^2 q_2 = 0$, $(A - \lambda I)^3 q_3 = 0$, and $(A - \lambda I)^4 q_4 = 0$.

A vector v is called a *generalized eigenvector* of grade n if

$$(A - \lambda I)^n v = 0$$

and

$$(A - \lambda I)^{n-1} v \neq 0$$

If $n = 1$, they reduce to $(A - \lambda I)v = 0$ and $v \neq 0$ and v is an ordinary eigenvector. For $n = 4$, we define

$$v_4 := v$$
$$v_3 := (A - \lambda I)v_4 = (A - \lambda I)v$$
$$v_2 := (A - \lambda I)v_3 = (A - \lambda I)^2 v$$
$$v_1 := (A - \lambda I)v_2 = (A - \lambda I)^3 v$$

They are called a chain of generalized eigenvectors of length $n = 4$ and have the properties $(A - \lambda I)v_1 = 0$, $(A - \lambda I)^2 v_2 = 0$, $(A - \lambda I)^3 v_3 = 0$, and $(A - \lambda I)^4 v_4 = 0$. These vectors, as generated, are automatically linearly independent and can be used as a basis. From these equations, we can readily obtain

$$Av_1 = \lambda v_1$$
$$Av_2 = v_1 + \lambda v_2$$
$$Av_3 = v_2 + \lambda v_3$$
$$Av_4 = v_3 + \lambda v_4$$

Then the representation of $\mathbf{A}$ with respect to the basis $\{\mathbf{v}_1, \ \mathbf{v}_2, \mathbf{v}_3, \ \mathbf{v}_4\}$ is

$$
\mathbf{J} := \begin{bmatrix} \lambda & 1 & 0 & 0 \\ 0 & \lambda & 1 & 0 \\ 0 & 0 & \lambda & 1 \\ 0 & 0 & 0 & \lambda \end{bmatrix} \tag{3.38}
$$

We verify this for the first and last columns. The first column of $\mathbf{J}$ is the representation of $\mathbf{A}\mathbf{v}_1 = \lambda \mathbf{v}_1$ with respect to $\{\mathbf{v}_1, \mathbf{v}_2, \mathbf{v}_3, \mathbf{v}_4\}$, which is $[\lambda \ 0 \ 0 \ 0]'$. The last column of $\mathbf{J}$ is the representation of $\mathbf{A}\mathbf{v}_4 = \mathbf{v}_3 + \lambda \mathbf{v}_4$ with respect to $\{\mathbf{v}_1, \mathbf{v}_2, \mathbf{v}_3, \mathbf{v}_4\}$, which is $[0 \ 0 \ 1 \ \lambda]'$. This verifies the representation in (3.38). The matrix $\mathbf{J}$ has eigenvalues on the diagonal and 1 on the superdiagonal. If we reverse the order of the basis, then the 1's will appear on the subdiagonal. The matrix is called a *Jordan* block of order $n = 4$.

If $(\mathbf{A} - \lambda \mathbf{I})$ has rank $n - 2$ or, equivalently, nullity 2, then the equation

$$(\mathbf{A} - \lambda \mathbf{I})\mathbf{q} = \mathbf{0}$$

has two linearly independent solutions. Thus $\mathbf{A}$ has two linearly independent eigenvectors and we need $(n - 2)$ generalized eigenvectors. In this case, there exist two chains of generalized eigenvectors $\{\mathbf{v}_1, \mathbf{v}_2, \dots, \mathbf{v}_k\}$ and $\{\mathbf{u}_1, \mathbf{u}_2, \dots, \mathbf{u}_l\}$ with $k + l = n$. If $\mathbf{v}_1$ and $\mathbf{u}_1$ are linearly independent, then the set of n vectors $\{\mathbf{v}_1, \dots, \mathbf{v}_k, \mathbf{u}_1, \dots, \mathbf{u}_l\}$ is linearly independent and can be used as a basis. With respect to this basis, the representation of $\mathbf{A}$ is a block diagonal matrix of form

$$\hat{\mathbf{A}} = \mathrm{diag}\{\mathbf{J}_1, \ \mathbf{J}_2\}$$

where $\mathbf{J}_1$ and $\mathbf{J}_2$ are, respectively, Jordan blocks of order k and l.

Now we discuss a specific example. Consider a 5×5 matrix $\mathbf{A}$ with repeated eigenvalue λ_1 with multiplicity 4 and simple eigenvalue λ_2. Then there exists a nonsingular matrix $\mathbf{Q}$ such that

$$\hat{\mathbf{A}} = \mathbf{Q}^{-1}\mathbf{A}\mathbf{Q}$$

assumes one of the following forms

$$
\hat{\mathbf{A}}_1 = \begin{bmatrix} \lambda_1 & 1 & 0 & 0 & 0 \\ 0 & \lambda_1 & 1 & 0 & 0 \\ 0 & 0 & \lambda_1 & 1 & 0 \\ 0 & 0 & 0 & \lambda_1 & 0 \\ 0 & 0 & 0 & 0 & \lambda_2 \end{bmatrix} \qquad
\hat{\mathbf{A}}_2 = \begin{bmatrix} \lambda_1 & 1 & 0 & 0 & 0 \\ 0 & \lambda_1 & 1 & 0 & 0 \\ 0 & 0 & \lambda_1 & 0 & 0 \\ 0 & 0 & 0 & \lambda_1 & 0 \\ 0 & 0 & 0 & 0 & \lambda_2 \end{bmatrix}
$$

$$
\hat{\mathbf{A}}_3 = \begin{bmatrix} \lambda_1 & 1 & 0 & 0 & 0 \\ 0 & \lambda_1 & 0 & 0 & 0 \\ 0 & 0 & \lambda_1 & 1 & 0 \\ 0 & 0 & 0 & \lambda_1 & 0 \\ 0 & 0 & 0 & 0 & \lambda_2 \end{bmatrix} \qquad
\hat{\mathbf{A}}_4 = \begin{bmatrix} \lambda_1 & 1 & 0 & 0 & 0 \\ 0 & \lambda_1 & 0 & 0 & 0 \\ 0 & 0 & \lambda_1 & 0 & 0 \\ 0 & 0 & 0 & \lambda_1 & 0 \\ 0 & 0 & 0 & 0 & \lambda_2 \end{bmatrix}
$$

$$
\hat{\mathbf{A}}_5 = \begin{bmatrix} \lambda_1 & 0 & 0 & 0 & 0 \\ 0 & \lambda_1 & 0 & 0 & 0 \\ 0 & 0 & \lambda_1 & 0 & 0 \\ 0 & 0 & 0 & \lambda_1 & 0 \\ 0 & 0 & 0 & 0 & \lambda_2 \end{bmatrix} \tag{3.39}
$$

The first matrix occurs when the nullity of $(\mathbf{A} - \lambda_1\mathbf{I})$ is 1. If the nullity is 2, then $\hat{\mathbf{A}}$ has two Jordan blocks associated with λ_1; it may assume the form in $\hat{\mathbf{A}}_2$ or in $\hat{\mathbf{A}}_3$. If $(\mathbf{A} - \lambda_1\mathbf{I})$ has nullity 3, then $\hat{\mathbf{A}}$ has three Jordan blocks associated with λ_1 as shown in $\hat{\mathbf{A}}_4$. Certainly, the positions of the Jordan blocks can be changed by changing the order of the basis. If the nullity is 4, then $\hat{\mathbf{A}}$ is a diagonal matrix as shown in $\hat{\mathbf{A}}_5$. All these matrices are triangular and block diagonal with Jordan blocks on the diagonal; they are said to be in Jordan form. A diagonal matrix is a degenerated Jordan form; its Jordan blocks all have order 1. If $\mathbf{A}$ can be diagonalized, we can use [q,d]=eig(a) to generate $\mathbf{Q}$ and $\hat{\mathbf{A}}$ as shown in Examples 3.5 and 3.6. If $\mathbf{A}$ cannot be diagonized, $\mathbf{A}$ is said to be *defective* and [q,d]=eig(a) will yield an incorrect solution. In this case, we may use the MATLAB function [q,d]=jordan(a). However, jordan will yield a correct result only if $\mathbf{A}$ has integers or ratios of small integers as its entries.

Jordan-form matrices are triangular and block diagonal and can be used to establish many general properties of matrices. For example, because $\det(\mathbf{CD}) = \det\mathbf{C}\det\mathbf{D}$ and $\det\mathbf{Q}\det\mathbf{Q}^{-1} = \det\mathbf{I} = 1$, from $\mathbf{A} = \mathbf{Q}\hat{\mathbf{A}}\mathbf{Q}^{-1}$, we have

$$\det\mathbf{A} = \det\mathbf{Q}\det\hat{\mathbf{A}}\det\mathbf{Q}^{-1} = \det\hat{\mathbf{A}}$$

The determinant of $\hat{\mathbf{A}}$ is the product of all diagonal entries or, equivalently, all eigenvalues of $\mathbf{A}$. Thus we have

$$\det\mathbf{A} = \text{ product of all eigenvalues of } \mathbf{A}$$

which implies that $\mathbf{A}$ *is nonsingular if and only if it has no zero eigenvalue.*

We discuss a useful property of Jordan blocks to conclude this section. Consider the Jordan block in (3.38) with order 4. Then we have

$$(\mathbf{J} - \lambda\mathbf{I}) = \begin{bmatrix} 0 & 1 & 0 & 0 \\ 0 & 0 & 1 & 0 \\ 0 & 0 & 0 & 1 \\ 0 & 0 & 0 & 0 \end{bmatrix} \quad (\mathbf{J} - \lambda\mathbf{I})^2 = \begin{bmatrix} 0 & 0 & 1 & 0 \\ 0 & 0 & 0 & 1 \\ 0 & 0 & 0 & 0 \\ 0 & 0 & 0 & 0 \end{bmatrix}$$

$$(\mathbf{J} - \lambda\mathbf{I})^3 = \begin{bmatrix} 0 & 0 & 0 & 1 \\ 0 & 0 & 0 & 0 \\ 0 & 0 & 0 & 0 \\ 0 & 0 & 0 & 0 \end{bmatrix} \tag{3.40}$$

and $(\mathbf{J} - \lambda\mathbf{I})^k = \mathbf{0}$ for $k \geq 4$. This is called *nilpotent*.

3.6 Functions of a Square Matrix

This section studies functions of a square matrix. We use Jordan form extensively because many properties of functions can almost be visualized in terms of Jordan form. We study first polynomials and then general functions of a square matrix.

Polynomials of a square matrix   Let $\mathbf{A}$ be a square matrix. If k is a positive integer, we define

$$\mathbf{A}^k := \mathbf{A}\mathbf{A}\cdots\mathbf{A} \quad (k \text{ terms})$$

and $\mathbf{A}^0 = \mathbf{I}$. Let $f(\lambda)$ be a polynomial such as $f(\lambda) = \lambda^3 + 2\lambda^2 - 6$ or $(\lambda + 2)(4\lambda - 3)$. Then $f(\mathbf{A})$ is defined as

$$f(\mathbf{A}) = \mathbf{A}^3 + 2\mathbf{A}^2 - 6\mathbf{I} \quad \text{or} \quad f(\mathbf{A}) = (\mathbf{A} + 2\mathbf{I})(4\mathbf{A} - 3\mathbf{I})$$

If $\mathbf{A}$ is block diagonal, such as

$$\mathbf{A} = \begin{bmatrix} \mathbf{A}_1 & \mathbf{0} \\ \mathbf{0} & \mathbf{A}_2 \end{bmatrix}$$

where $\mathbf{A}_1$ and $\mathbf{A}_2$ are square matrices of any order, then it is straightforward to verify

$$\mathbf{A}^k = \begin{bmatrix} \mathbf{A}_1^k & \mathbf{0} \\ \mathbf{0} & \mathbf{A}_2^k \end{bmatrix} \quad \text{and} \quad f(\mathbf{A}) = \begin{bmatrix} f(\mathbf{A}_1) & \mathbf{0} \\ \mathbf{0} & f(\mathbf{A}_2) \end{bmatrix} \tag{3.41}$$

Consider the similarity transformation $\hat{\mathbf{A}} = \mathbf{Q}^{-1}\mathbf{A}\mathbf{Q}$ or $\mathbf{A} = \mathbf{Q}\hat{\mathbf{A}}\mathbf{Q}^{-1}$. Because

$$\mathbf{A}^k = (\mathbf{Q}\hat{\mathbf{A}}\mathbf{Q}^{-1})(\mathbf{Q}\hat{\mathbf{A}}\mathbf{Q}^{-1}) \cdots (\mathbf{Q}\hat{\mathbf{A}}\mathbf{Q}^{-1}) = \mathbf{Q}\hat{\mathbf{A}}^k\mathbf{Q}^{-1}$$

we have

$$f(\mathbf{A}) = \mathbf{Q}f(\hat{\mathbf{A}})\mathbf{Q}^{-1} \quad \text{or} \quad f(\hat{\mathbf{A}}) = \mathbf{Q}^{-1}f(\mathbf{A})\mathbf{Q} \tag{3.42}$$

A *monic* polynomial is a polynomial with 1 as its leading coefficient. The *minimal polynomial* of $\mathbf{A}$ is defined as the monic polynomial $\psi(\lambda)$ of least degree such that $\psi(\mathbf{A}) = \mathbf{0}$. Note that the $\mathbf{0}$ is a zero matrix of the same order as $\mathbf{A}$. A direct consequence of (3.42) is that $f(\mathbf{A}) = \mathbf{0}$ if and only if $f(\hat{\mathbf{A}}) = \mathbf{0}$. Thus $\mathbf{A}$ and $\hat{\mathbf{A}}$ or, more general, all similar matrices have the same minimal polynomial. Computing the minimal polynomial directly from $\mathbf{A}$ is not simple (see Problem 3.25); however, if the Jordan-form representation of $\mathbf{A}$ is available, the minimal polynomial can be read out by inspection.

Let λ_i be an eigenvalue of $\mathbf{A}$ with multiplicity n_i. That is, the characteristic polynomial of $\mathbf{A}$ is

$$\Delta(\lambda) = \det(\lambda\mathbf{I} - \mathbf{A}) = \prod_i (\lambda - \lambda_i)^{n_i}$$

Suppose the Jordan form of $\mathbf{A}$ is known. Associated with each eigenvalue, there may be one or more Jordan blocks. The *index* of λ_i, denoted by $\bar{n}_i$, is defined as the largest order of all Jordan blocks associated with λ_i. Clearly we have $\bar{n}_i \leq n_i$. For example, the multiplicities of λ_1 in all five matrices in (3.39) are 4; their indices are, respectively, 4, 3, 2, 2, and 1. The multiplicities and indices of λ_2 in all five matrices in (3.39) are all 1. Using the indices of all eigenvalues, the minimal polynomial can be expressed as

$$\psi(\lambda) = \prod_i (\lambda - \lambda_i)^{\bar{n}_i}$$

with degree $\bar{n} = \sum \bar{n}_i \leq \sum n_i = n = $ dimension of $\mathbf{A}$. For example, the minimal polynomials of the five matrices in (3.39) are

$$\psi_1 = (\lambda - \lambda_1)^4(\lambda - \lambda_2) \qquad \psi_2 = (\lambda - \lambda_1)^3(\lambda - \lambda_2)$$

$$\psi_3 = (\lambda - \lambda_1)^2(\lambda - \lambda_2) \qquad \psi_4 = (\lambda - \lambda_1)^2(\lambda - \lambda_2)$$

$$\psi_5 = (\lambda - \lambda_1)(\lambda - \lambda_2)$$

Their characteristic polynomials, however, all equal

$$\Delta(\lambda) = (\lambda - \lambda_1)^4(\lambda - \lambda_2)$$

We see that the minimal polynomial is a factor of the characteristic polynomial and has a degree less than or equal to the degree of the characteristic polynomial. Clearly, if all eigenvalues of $\mathbf{A}$ are distinct, then the minimal polynomial equals the characteristic polynomial.

Using the nilpotent property in (3.40), we can show that

$$\psi(\mathbf{A}) = \mathbf{0}$$

and that no polynomial of lesser degree meets the condition. Thus $\psi(\lambda)$ as defined is the minimal polynomial.

Theorem 3.4 (Cayley-Hamilton theorem)

Let

$$\Delta(\lambda) = \det(\lambda\mathbf{I} - \mathbf{A}) = \lambda^n + \alpha_1\lambda^{n-1} + \cdots + \alpha_{n-1}\lambda + \alpha_n$$

be the characteristic polynomial of $\mathbf{A}$. Then

$$\Delta(\mathbf{A}) = \mathbf{A}^n + \alpha_1\mathbf{A}^{n-1} + \cdots + \alpha_{n-1}\mathbf{A} + \alpha_n\mathbf{I} = \mathbf{0} \qquad (3.43)$$

In words, a matrix satisfies its own characteristic polynomial. Because $n_i \geq \bar{n}_i$, the characteristic polynomial contains the minimal polynomial as a factor or $\Delta(\lambda) = \psi(\lambda)h(\lambda)$ for some polynomial $h(\lambda)$. Because $\psi(\mathbf{A}) = \mathbf{0}$, we have $\Delta(\mathbf{A}) = \psi(\mathbf{A})h(\mathbf{A}) = \mathbf{0} \cdot h(\mathbf{A}) = \mathbf{0}$. This establishes the theorem. The Cayley–Hamilton theorem implies that $\mathbf{A}^n$ can be written as a linear combination of $\{\mathbf{I}, \mathbf{A}, \ldots, \mathbf{A}^{n-1}\}$. Multiplying (3.43) by $\mathbf{A}$ yields

$$\mathbf{A}^{n+1} + \alpha_1\mathbf{A}^n + \cdots + \alpha_{n-1}\mathbf{A}2 + \alpha_n\mathbf{A} = \mathbf{0} \cdot \mathbf{A} = \mathbf{0}$$

which implies that $\mathbf{A}^{n+1}$ can be written as a linear combination of $\{\mathbf{A}, \mathbf{A}^2, \ldots, \mathbf{A}^n\}$, which, in turn, can be written as a linear combination of $\{\mathbf{I}, \mathbf{A}, \ldots, \mathbf{A}^{n-1}\}$. Proceeding forward, we conclude that, for any polynomial $f(\lambda)$, no matter how large its degree is, $f(\mathbf{A})$ can always be expressed as

$$f(\mathbf{A}) = \beta_0\mathbf{I} + \beta_1\mathbf{A} + \cdots + \beta_{n-1}\mathbf{A}^{n-1} \qquad (3.44)$$

for some β_i. In other words, every polynomial of an $n \times n$ matrix $\mathbf{A}$ can be expressed as a linear combination of $\{\mathbf{I}, \mathbf{A}, \ldots, \mathbf{A}^{n-1}\}$. If the minimal polynomial of $\mathbf{A}$ with degree $\bar{n}$ is available, then every polynomial of $\mathbf{A}$ can be expressed as a linear combination of $\{\mathbf{I}, \mathbf{A}, \ldots, \mathbf{A}^{\bar{n}-1}\}$. This is a better result. However, because $\bar{n}$ may not be available, we discuss in the following only (3.44) with the understanding that all discussion applies to $\bar{n}$.

One way to compute (3.44) is to use long division to express $f(\lambda)$ as

$$f(\lambda) = q(\lambda)\Delta(\lambda) + h(\lambda) \qquad (3.45)$$

where $q(\lambda)$ is the quotient and $h(\lambda)$ is the remainder with degree less than n. Then we have

$$f(\mathbf{A}) = q(\mathbf{A})\Delta(\mathbf{A}) + h(\mathbf{A}) = q(\mathbf{A})\mathbf{0} + h(\mathbf{A}) = h(\mathbf{A})$$

Long division is not convenient to carry out if the degree of $f(\lambda)$ is much larger than the degree of $\Delta(\lambda)$. In this case, we may solve $h(\lambda)$ directly from (3.45). Let

$$h(\lambda) := \beta_0 + \beta_1\lambda + \cdots + \beta_{n-1}\lambda^{n-1}$$

where the n unknowns β_i are to be solved. If all n eigenvalues of $\mathbf{A}$ are distinct, these β_i can be solved from the n equations

$$f(\lambda_i) = q(\lambda_i)\Delta(\lambda_i) + h(\lambda_i) = h(\lambda_i)$$

for $i = 1, 2, \ldots, n$. If $\mathbf{A}$ has repeated eigenvalues, then (3.45) must be differentiated to yield additional equations. This is stated as a theorem.

▶ **Theorem 3.5**

We are given $f(\lambda)$ and an $n \times n$ matrix $\mathbf{A}$ with characteristic polynomial

$$\Delta(\lambda) = \prod_{i=1}^{m}(\lambda - \lambda_i)^{n_i}$$

where $n = \sum_{i=1}^{m} n_i$. Define

$$h(\lambda) := \beta_0 + \beta_1\lambda + \cdots + \beta_{n-1}\lambda^{n-1}$$

It is a polynomial of degree $n - 1$ with n unknown coefficients. These n unknowns are to be solved from the following set of n equations:

$$f^{(l)}(\lambda_i) = h^{(l)}(\lambda_i) \quad \text{for } l = 0, 1, \ldots, n_i - 1 \quad \text{and} \quad i = 1, 2, \ldots, m$$

where

$$f^{(l)}(\lambda_i) := \frac{d^l f(\lambda)}{d\lambda^l}\bigg|_{\lambda=\lambda_i}$$

and $h^{(l)}(\lambda_i)$ is similarly defined. Then we have

$$f(\mathbf{A}) = h(\mathbf{A})$$

and $h(\lambda)$ is said to equal $f(\lambda)$ on the spectrum of $\mathbf{A}$.

EXAMPLE 3.7 Compute $\mathbf{A}^{100}$ with

$$\mathbf{A} = \begin{bmatrix} 0 & 1 \\ -1 & -2 \end{bmatrix}$$

In other words, given $f(\lambda) = \lambda^{100}$, compute $f(\mathbf{A})$. The characteristic polynomial of $\mathbf{A}$ is $\Delta(\lambda) = \lambda^2 + 2\lambda + 1 = (\lambda + 1)^2$. Let $h(\lambda) = \beta_0 + \beta_1\lambda$. On the spectrum of $\mathbf{A}$, we have

$$f(-1) = h(-1): \qquad (-1)^{100} = \beta_0 - \beta_1$$
$$f'(-1) = h'(-1): \qquad 100 \cdot (-1)^{99} = \beta_1$$

Thus we have $\beta_1 = -100$, $\beta_0 = 1 + \beta_1 = -99$, $h(\lambda) = -99 - 100\lambda$, and

$$\mathbf{A}^{100} = \beta_0 \mathbf{I} + \beta_1 \mathbf{A} = -99\mathbf{I} - 100\mathbf{A}$$

$$= -99\begin{bmatrix} 1 & 0 \\ 0 & 1 \end{bmatrix} - 100\begin{bmatrix} 0 & 1 \\ -1 & -2 \end{bmatrix} = \begin{bmatrix} -199 & -100 \\ 100 & 101 \end{bmatrix}$$

Clearly $\mathbf{A}^{100}$ can also be obtained by multiplying $\mathbf{A}$ 100 times. However, it is simpler to use Theorem 3.5.

Functions of a square matrix   Let $f(\lambda)$ be any function, not necessarily a polynomial. One way to define $f(\mathbf{A})$ is to use Theorem 3.5. Let $h(\lambda)$ be a polynomial of degree $n - 1$, where n is the order of $\mathbf{A}$. We solve the coefficients of $h(\lambda)$ by equating $f(\lambda) = h(\lambda)$ on the spectrum of $\mathbf{A}$. Then $f(\mathbf{A})$ is defined as $h(\mathbf{A})$.

EXAMPLE 3.8 Let

$$\mathbf{A}_1 = \begin{bmatrix} 0 & 0 & -2 \\ 0 & 1 & 0 \\ 1 & 0 & 3 \end{bmatrix}$$

Compute $e^{\mathbf{A}_1 t}$. Or, equivalently, if $f(\lambda) = e^{\lambda t}$, what is $f(\mathbf{A}_1)$?

The characteristic polynomial of $\mathbf{A}_1$ is $(\lambda - 1)^2(\lambda - 2)$. Let $h(\lambda) = \beta_0 + \beta_1 \lambda + \beta_2 \lambda^2$. Then

$$\begin{aligned} f(1) = h(1): & \qquad e^t = \beta_0 + \beta_1 + \beta_2 \\ f'(1) = h'(1): & \qquad te^t = \beta_1 + 2\beta_2 \\ f(2) = h(2): & \qquad e^{2t} = \beta_0 + 2\beta_1 + 4\beta_2 \end{aligned}$$

Note that, in the second equation, the differentiation is with respect to λ, not t. Solving these equations yields $\beta_0 = -2te^t + e^{2t}$, $\beta_1 = 3te^t + 2e^t - 2e^{2t}$, and $\beta_2 = e^{2t} - e^t - te^t$. Thus we have

$$e^{\mathbf{A}_1 t} = h(\mathbf{A}_1) = (-2te^t + e^{2t})\mathbf{I} + (3te^t + 2e^t - 2e^{2t})\mathbf{A}_1$$

$$+ (e^{2t} - e^t - te^t)\mathbf{A}_1^2 = \begin{bmatrix} 2e^t - e^{2t} & 0 & 2e^t - 2e^{2t} \\ 0 & e^t & 0 \\ e^{2t} - e^t & 0 & 2e^{2t} - e^t \end{bmatrix}$$

EXAMPLE 3.9 Let

$$\mathbf{A}_2 = \begin{bmatrix} 0 & 2 & -2 \\ 0 & 1 & 0 \\ 1 & -1 & 3 \end{bmatrix}$$

Compute $e^{\mathbf{A}_2 t}$. The characteristic polynomial of $\mathbf{A}_2$ is $(\lambda - 1)^2(\lambda - 2)$, which is the same as for $\mathbf{A}_1$. Hence we have the same $h(\lambda)$ as in Example 3.8. Consequently, we have

$$e^{\mathbf{A}_2 t} = h(\mathbf{A}_2) = \begin{bmatrix} 2e^t - e^{2t} & 2te^t & 2e^t - 2e^{2t} \\ 0 & e^t & 0 \\ e^{2t} - e^t & -te^t & 2e^{2t} - e^t \end{bmatrix}$$

EXAMPLE 3.10 Consider the Jordan block of order 4:

$$
\hat{\mathbf{A}} =
\begin{bmatrix}
\lambda_1 & 1 & 0 & 0 \\
0 & \lambda_1 & 1 & 0 \\
0 & 0 & \lambda_1 & 1 \\
0 & 0 & 0 & \lambda_1
\end{bmatrix}
\tag{3.46}
$$

Its characteristic polynomial is $(\lambda - \lambda_1)^4$. Although we can select $h(\lambda)$ as $\beta_0 + \beta_1\lambda + \beta_2\lambda^2 + \beta_3\lambda^3$, it is computationally simpler to select $h(\lambda)$ as

$$
h(\lambda) = \beta_0 + \beta_1(\lambda - \lambda_1) + \beta_2(\lambda - \lambda_1) + \beta_3(\lambda - \lambda_1)^3
$$

This selection is permitted because $h(\lambda)$ has degree $(n - 1) = 3$ and $n = 4$ independent unknowns. The condition $f(\lambda) = h(\lambda)$ on the spectrum of $\hat{\mathbf{A}}$ yields immediately

$$
\beta_0 = f(\lambda_1), \quad \beta_1 = f'(\lambda_1), \quad \beta_2 = \frac{f''(\lambda_1)}{2!}, \quad \beta_3 = \frac{f^{(3)}(\lambda_1)}{3!}
$$

Thus we have

$$
f(\hat{\mathbf{A}}) = f(\lambda_1)\mathbf{I} + \frac{f'(\lambda_1)}{1!}(\hat{\mathbf{A}} - \lambda_1\mathbf{I}) + \frac{f''(\lambda_1)}{2!}(\hat{\mathbf{A}} - \lambda_1\mathbf{I})^2 + \frac{f^{(3)}(\lambda_1)}{3!}(\hat{\mathbf{A}} - \lambda_1\mathbf{I})^3
$$

Using the special forms of $(\hat{\mathbf{A}} - \lambda_1\mathbf{I})^k$ as discussed in (3.40), we can readily obtain

$$
f(\hat{\mathbf{A}}) =
\begin{bmatrix}
f(\lambda_1) & f'(\lambda_1)/1! & f''(\lambda_1)/2! & f^{(3)}(\lambda_1)/3! \\
0 & f(\lambda_1) & f'(\lambda_1)/1! & f''(\lambda_1)/2! \\
0 & 0 & f(\lambda_1) & f'(\lambda_1)/1! \\
0 & 0 & 0 & f(\lambda_1)
\end{bmatrix}
\tag{3.47}
$$

If $f(\lambda) = e^{\lambda t}$, then

$$
e^{\hat{\mathbf{A}}t} =
\begin{bmatrix}
e^{\lambda_1 t} & te^{\lambda_1 t} & t^2 e^{\lambda_1 t}/2! & t^3 e^{\lambda_1 t}/3! \\
0 & e^{\lambda_1 t} & te^{\lambda_1 t} & t^2 e^{\lambda_1 t}/2! \\
0 & 0 & e^{\lambda_1 t} & te^{\lambda_1 t} \\
0 & 0 & 0 & e^{\lambda_1 t}
\end{bmatrix}
\tag{3.48}
$$

Because functions of $\mathbf{A}$ are defined through polynomials of $\mathbf{A}$, Equations (3.41) and (3.42) are applicable to functions.

EXAMPLE 3.11 Consider

$$
\mathbf{A} =
\begin{bmatrix}
\lambda_1 & 1 & 0 & 0 & 0 \\
0 & \lambda_1 & 1 & 0 & 0 \\
0 & 0 & \lambda_1 & 0 & 0 \\
0 & 0 & 0 & \lambda_2 & 1 \\
0 & 0 & 0 & 0 & \lambda_2
\end{bmatrix}
$$

It is block diagonal and contains two Jordan blocks. If $f(\lambda) = e^{\mathbf{A}t}$, then (3.41) and (3.48) imply

$$e^{\mathbf{A}t} = \begin{bmatrix} e^{\lambda_1 t} & te^{\lambda_1 t} & t^2 e^{\lambda_1 t}/2! & 0 & 0 \\ 0 & e^{\lambda_1 t} & te^{\lambda_1 t} & 0 & 0 \\ 0 & 0 & e^{\lambda_1 t} & 0 & 0 \\ 0 & 0 & 0 & e^{\lambda_2 t} & te^{\lambda_2 t} \\ 0 & 0 & 0 & 0 & e^{\lambda_2 t} \end{bmatrix}$$

If $f(\lambda) = (s - \lambda)^{-1}$, then (3.41) and (3.47) imply

$$(s\mathbf{I} - \mathbf{A})^{-1} = \begin{bmatrix} \dfrac{1}{(s - \lambda_1)} & \dfrac{1}{(s - \lambda_1)^2} & \dfrac{1}{(s - \lambda_1)^3} & 0 & 0 \\ 0 & \dfrac{1}{(s - \lambda_1)} & \dfrac{1}{(s - \lambda_1)^2} & 0 & 0 \\ 0 & 0 & \dfrac{1}{(s - \lambda_1)} & 0 & 0 \\ 0 & 0 & 0 & \dfrac{1}{(s - \lambda_2)} & \dfrac{1}{(s - \lambda_2)^2} \\ 0 & 0 & 0 & 0 & \dfrac{1}{(s - \lambda_2)} \end{bmatrix} \quad (3.49)$$

Using power series The function of $\mathbf{A}$ was defined using a polynomial of finite degree. We now give an alternative definition by using an infinite power series. Suppose $f(\lambda)$ can be expressed as the power series

$$f(\lambda) = \sum_{i=0}^{\infty} \beta_i \lambda^i$$

with the radius of convergence ρ. If all eigenvalues of $\mathbf{A}$ have magnitudes less than ρ, then $f(\mathbf{A})$ can be defined as

$$f(\mathbf{A}) = \sum_{i=0}^{\infty} \beta_i \mathbf{A}^i \quad (3.50)$$

Instead of proving the equivalence of this definition and the definition based on Theorem 3.5, we use (3.50) to derive (3.47).

EXAMPLE 3.12 Consider the Jordan-form matrix $\hat{\mathbf{A}}$ in (3.46). Let

$$f(\lambda) = f(\lambda_1) + f'(\lambda_1)(\lambda - \lambda_1) + \frac{f''(\lambda_1)}{2!}(\lambda - \lambda_1)^2 + \cdots$$

then

$$f(\hat{\mathbf{A}}) = f(\lambda_1)\mathbf{I} + f'(\lambda_1)(\hat{\mathbf{A}} - \lambda_1\mathbf{I}) + \cdots + \frac{f^{(n-1)}(\lambda_1)}{(n-1)!}(\hat{\mathbf{A}} - \lambda_1\mathbf{I})^{n-1} + \cdots$$

Because $(\hat{\mathbf{A}} - \lambda_1\mathbf{I})^k = \mathbf{0}$ for $k \geq n = 4$ as discussed in (3.40), the infinite series reduces immediately to (3.47). Thus the two definitions lead to the same function of a matrix.

The most important function of $\mathbf{A}$ is the exponential function $e^{\mathbf{A}t}$. Because the Taylor series

$$e^{\lambda t} = 1 + \lambda t + \frac{\lambda^2 t^2}{2!} + \cdots + \frac{\lambda^n t^n}{n!} + \cdots$$

converges for all finite λ and t, we have

$$e^{\mathbf{A}t} = \mathbf{I} + t\mathbf{A} + \frac{t^2}{2!}\mathbf{A}^2 + \cdots = \sum_{k=0}^{\infty} \frac{1}{k!}t^k\mathbf{A}^k \tag{3.51}$$

This series involves only multiplications and additions and may converge rapidly; therefore it is suitable for computer computation. We list in the following the program in MATLAB that computes (3.51) for $t = 1$:

```
Function E=expm2(A)
E=zeros(size(A));
F=eye(size(A));
k=1;
while norm(E+F-E,1)>0
      E=E+F;
      F=A*F/k;
      k=k+1;
end
```

In the program, E denotes the partial sum and F is the next term to be added to E. The first line defines the function. The next two lines initialize E and F. Let c_k denote the kth term of (3.51) with $t = 1$. Then we have $c_{k+1} = (\mathbf{A}/k)c_k$ for $k = 1, 2, \ldots$. Thus we have F = A * F/k. The computation stops if the 1-norm of E + F − E, denoted by norm(E + F − E, 1), is rounded to 0 in computers. Because the algorithm compares F and E, not F and 0, the algorithm uses norm(E + F − E, 1) instead of norm(F,1). Note that norm(a,1) is the 1-norm discussed in Section 3.2 and will be discussed again in Section 3.9. We see that the series can indeed be programmed easily. To improve the computed result, the techniques of scaling and squaring can be used. In MATLAB, the function expm2 uses (3.51). The function expm or expm1, however, uses the so-called Padé approximation. It yields comparable results as expm2 but requires only about half the computing time. Thus expm is preferred to expm2. The function expm3 uses Jordan form, but it will yield an incorrect solution if a matrix is not diagonalizable. If a closed-form solution of $e^{\mathbf{A}t}$ is needed, we must use Theorem 3.5 or Jordan form to compute $e^{\mathbf{A}t}$.

We derive some important properties of $e^{\mathbf{A}t}$ to conclude this section. Using (3.51), we can readily verify the next two equalities

$$e^{\mathbf{0}} = \mathbf{I} \tag{3.52}$$

$$e^{\mathbf{A}(t_1+t_2)} = e^{\mathbf{A}t_1}e^{\mathbf{A}t_2} \tag{3.53}$$

$$[e^{\mathbf{A}t}]^{-1} = e^{-\mathbf{A}t} \tag{3.54}$$

To show (3.54), we set $t_2 = -t_1$. Then (3.53) and (3.52) imply

$$e^{\mathbf{A}t_1}e^{-\mathbf{A}t_1} = e^{\mathbf{A}\cdot 0} = e^{\mathbf{0}} = \mathbf{I}$$

which implies (3.54). Thus the inverse of $e^{\mathbf{A}t}$ can be obtained by simply changing the sign of t. Differentiating term by term of (3.51) yields

$$\frac{d}{dt}e^{\mathbf{A}t} = \sum_{k=1}^{\infty} \frac{1}{(k-1)!}t^{k-1}\mathbf{A}^k$$

$$= \mathbf{A}\left(\sum_{k=0}^{\infty} \frac{1}{k!}t^k\mathbf{A}^k\right) = \left(\sum_{k=0}^{\infty} \frac{1}{k!}t^k\mathbf{A}^k\right)\mathbf{A}$$

Thus we have

$$\frac{d}{dt}e^{\mathbf{A}t} = \mathbf{A}e^{\mathbf{A}t} = e^{\mathbf{A}t}\mathbf{A} \tag{3.55}$$

This is an important equation. We mention that

$$e^{(\mathbf{A}+\mathbf{B})t} \neq e^{\mathbf{A}t}e^{\mathbf{B}t} \tag{3.56}$$

The equality holds only if $\mathbf{A}$ and $\mathbf{B}$ commute or $\mathbf{AB} = \mathbf{BA}$. This can be verified by direct substitution of (3.51).

The Laplace transform of a function $f(t)$ is defined as

$$\hat{f}(s) := \mathcal{L}[f(t)] = \int_0^{\infty} f(t)e^{-st}dt$$

It can be shown that

$$\mathcal{L}\left[\frac{t^k}{k!}\right] = s^{-(k+1)}$$

Taking the Laplace transform of (3.51) yields

$$\mathcal{L}[e^{\mathbf{A}t}] = \sum_{k=0}^{\infty} s^{-(k+1)}\mathbf{A}^k = s^{-1}\sum_{k=0}^{\infty}(s^{-1}\mathbf{A})^k$$

Because the infinite series

$$\sum_{k=0}^{\infty}(s^{-1}\lambda)^k = 1 + s^{-1}\lambda + s^{-2}\lambda^2 + \cdots = (1 - s^{-1}\lambda)^{-1}$$

converges for $|s^{-1}\lambda| < 1$, we have

$$s^{-1}\sum_{k=0}^{\infty}(s^{-1}\mathbf{A})^k = s^{-1}\mathbf{I} + s^{-2}\mathbf{A} + s^{-3}\mathbf{A}^2 + \cdots$$

$$= s^{-1}(\mathbf{I} - s^{-1}\mathbf{A})^{-1} = \left[s(\mathbf{I} - s^{-1}\mathbf{A})\right]^{-1} = (s\mathbf{I} - \mathbf{A})^{-1} \tag{3.57}$$

and

$$\mathcal{L}[e^{At}] = (s\mathbf{I} - \mathbf{A})^{-1} \tag{3.58}$$

Although in the derivation of (3.57) we require s to be sufficiently large so that all eigenvalues of $s^{-1}\mathbf{A}$ have magnitudes less than 1, Equation (3.58) actually holds for all s except at the eigenvalues of $\mathbf{A}$. Equation (3.58) can also be established from (3.55). Because $\mathcal{L}[df(t)/dt] = s\mathcal{L}[f(t)] - f(0)$, applying the Laplace transform to (3.55) yields

$$s\mathcal{L}[e^{At}] - e^0 = \mathbf{A}\mathcal{L}[e^{At}]$$

or

$$(s\mathbf{I} - \mathbf{A})\mathcal{L}[e^{At}] = e^0 = \mathbf{I}$$

which implies (3.58).

3.7 Lyapunov Equation

Consider the equation

$$\mathbf{A}\mathbf{M} + \mathbf{M}\mathbf{B} = \mathbf{C} \tag{3.59}$$

where $\mathbf{A}$ and $\mathbf{B}$ are, respectively, $n \times n$ and $m \times m$ constant matrices. In order for the equation to be meaningful, the matrices $\mathbf{M}$ and $\mathbf{C}$ must be of order $n \times m$. The equation is called the *Lyapunov* equation.

The equation can be written as a set of standard linear algebraic equations. To see this, we assume $n = 3$ and $m = 2$ and write (3.59) explicitly as

$$\begin{bmatrix} a_{11} & a_{12} & a_{13} \\ a_{21} & a_{22} & a_{23} \\ a_{31} & a_{32} & a_{33} \end{bmatrix} \begin{bmatrix} m_{11} & m_{12} \\ m_{21} & m_{22} \\ m_{31} & m_{32} \end{bmatrix} + \begin{bmatrix} m_{11} & m_{12} \\ m_{21} & m_{22} \\ m_{31} & m_{32} \end{bmatrix} \begin{bmatrix} b_{11} & b_{12} \\ b_{21} & b_{22} \end{bmatrix}$$
$$= \begin{bmatrix} c_{11} & c_{12} \\ c_{21} & c_{22} \\ c_{31} & c_{32} \end{bmatrix}$$

Multiplying them out and then equating the corresponding entries on both sides of the equality, we obtain

$$\begin{bmatrix} a_{11}+b_{11} & a_{12} & a_{13} & b_{21} & 0 & 0 \\ a_{21} & a_{22}+b_{11} & a_{23} & 0 & b_{21} & 0 \\ a_{31} & a_{32} & a_{33}+b_{11} & 0 & 0 & b_{21} \\ b_{12} & 0 & 0 & a_{11}+b_{22} & a_{12} & a_{13} \\ 0 & b_{12} & 0 & a_{21} & a_{22}+b_{22} & a_{23} \\ 0 & 0 & b_{12} & a_{31} & a_{32} & a_{33}+b_{22} \end{bmatrix}$$

$$\times \begin{bmatrix} m_{11} \\ m_{21} \\ m_{31} \\ m_{12} \\ m_{22} \\ m_{32} \end{bmatrix} = \begin{bmatrix} c_{11} \\ c_{21} \\ c_{31} \\ c_{12} \\ c_{22} \\ c_{32} \end{bmatrix} \tag{3.60}$$

This is indeed a standard linear algebraic equation. The matrix on the preceding page is a square matrix of order $n \times m = 3 \times 2 = 6$.

Let us define $\mathcal{A}(\mathbf{M}) := \mathbf{AM} + \mathbf{MB}$. Then the Lyapunov equation can be written as $\mathcal{A}(\mathbf{M}) = \mathbf{C}$. It maps an nm-dimensional linear space into itself. A scalar η is called an eigenvalue of $\mathcal{A}$ if there exists a nonzero $\mathbf{M}$ such that

$$\mathcal{A}(\mathbf{M}) = \eta\mathbf{M}$$

Because $\mathcal{A}$ can be considered as a square matrix of order nm, it has nm eigenvalues η_k, for $k = 1, 2, \ldots, nm$. It turns out

$$\eta_k = \lambda_i + \mu_j \qquad \text{for } i = 1, 2, \ldots, n; \quad j = 1, 2, \ldots, m$$

where λ_i, $i = 1, 2, \ldots, n$, and μ_j, $j = 1, 2, \ldots, m$, are, respectively, the eigenvalues of $\mathbf{A}$ and $\mathbf{B}$. In other words, the eigenvalues of $\mathcal{A}$ are all possible sums of the eigenvalues of $\mathbf{A}$ and $\mathbf{B}$.

We show intuitively why this is the case. Let $\mathbf{u}$ be an $n \times 1$ right eigenvector of $\mathbf{A}$ associated with λ_i; that is, $\mathbf{Au} = \lambda_i\mathbf{u}$. Let $\mathbf{v}$ be a $1 \times m$ left eigenvector of $\mathbf{B}$ associated with μ_j; that is, $\mathbf{vB} = \mathbf{v}\mu_j$. Applying $\mathcal{A}$ to the $n \times m$ matrix $\mathbf{uv}$ yields

$$\mathcal{A}(\mathbf{uv}) = \mathbf{Auv} + \mathbf{uvB} = \lambda_i\mathbf{uv} + \mathbf{uv}\mu_j = (\lambda_i + \mu_j)\mathbf{uv}$$

Because both $\mathbf{u}$ and $\mathbf{v}$ are nonzero, so is the matrix $\mathbf{uv}$. Thus $(\lambda_i + \mu_j)$ is an eigenvalue of $\mathcal{A}$.

The determinant of a square matrix is the product of all its eigenvalues. Thus a matrix is nonsingular if and only if it has no zero eigenvalue. If there are no i and j such that $\lambda_i + \mu_j = 0$, then the square matrix in (3.60) is nonsingular and, for every $\mathbf{C}$, there exists a unique $\mathbf{M}$ satisfying the equation. In this case, the Lyapunov equation is said to be nonsingular. If $\lambda_i + \mu_j = 0$ for some i and j, then for a given $\mathbf{C}$, solutions may or may not exist. If $\mathbf{C}$ lies in the range space of $\mathcal{A}$, then solutions exist and are not unique. See Problem 3.32.

The MATLAB function `m=lyap(a,b,-c)` computes the solution of the Lyapunov equation in (3.59).

3.8 Some Useful Formulas

This section discusses some formulas that will be needed later. Let $\mathbf{A}$ and $\mathbf{B}$ be $m \times n$ and $n \times p$ constant matrices. Then we have

$$\rho(\mathbf{AB}) \leq \min(\rho(\mathbf{A}), \rho(\mathbf{B})) \tag{3.61}$$

where ρ denotes the rank. This can be argued as follows. Let $\rho(\mathbf{B}) = \alpha$. Then $\mathbf{B}$ has α linearly independent rows. In $\mathbf{AB}$, $\mathbf{A}$ operates on the rows of $\mathbf{B}$. Thus the rows of $\mathbf{AB}$ are linear combinations of the rows of **B**. Thus **AB** has at most α linearly independent rows. In **AB**, **B** operates on the columns of **A**. Thus if **A** has β linearly independent columns, then **AB** has at most β linearly independent columns. This establishes (3.61). Consequently, if $\mathbf{A} = \mathbf{B}_1\mathbf{B}_2\mathbf{B}_3 \cdots$, then the rank of **A** is equal to or smaller than the smallest rank of $\mathbf{B}_i$.

Let **A** be $m \times n$ and let **C** and **D** be any $n \times n$ and $m \times m$ nonsingular matrices. Then we have

$$\rho(\mathbf{AC}) = \rho(\mathbf{A}) = \rho(\mathbf{DA}) \tag{3.62}$$

In words, the rank of a matrix will not change after pre- or postmultiplying by a nonsingular matrix. To show (3.62), we define

$$\mathbf{P} := \mathbf{AC} \tag{3.63}$$

Because $\rho(\mathbf{A}) \leq \min(m, n)$ and $\rho(\mathbf{C}) = n$, we have $\rho(\mathbf{A}) \leq \rho(\mathbf{C})$. Thus (3.61) implies

$$\rho(\mathbf{P}) \leq \min(\rho(\mathbf{A}), \rho(\mathbf{C})) \leq \rho(\mathbf{A})$$

Next we write (3.63) as $\mathbf{A} = \mathbf{PC}^{-1}$. Using the same argument, we have $\rho(\mathbf{A}) \leq \rho(\mathbf{P})$. Thus we conclude $\rho(\mathbf{P}) = \rho(\mathbf{A})$. A consequence of (3.62) is that the rank of a matrix will not change by elementary operations. Elementary operations are (1) multiplying a row or a column by a nonzero number, (2) interchanging two rows or two columns, and (3) adding the product of one row (column) and a number to another row (column). These operations are the same as multiplying nonsingular matrices. See Reference [6, p. 542].

Let **A** be $m \times n$ and **B** be $n \times m$. Then we have

$$\det(\mathbf{I}_m + \mathbf{AB}) = \det(\mathbf{I}_n + \mathbf{BA}) \tag{3.64}$$

where $\mathbf{I}_m$ is the unit matrix of order m. To show (3.64), let us define

$$\mathbf{N} = \begin{bmatrix} \mathbf{I}_m & \mathbf{A} \\ \mathbf{0} & \mathbf{I}_n \end{bmatrix} \qquad \mathbf{Q} = \begin{bmatrix} \mathbf{I}_m & \mathbf{0} \\ -\mathbf{B} & \mathbf{I}_n \end{bmatrix} \qquad \mathbf{P} = \begin{bmatrix} \mathbf{I}_m & -\mathbf{A} \\ \mathbf{B} & \mathbf{I}_n \end{bmatrix}$$

We compute

$$\mathbf{NP} = \begin{bmatrix} \mathbf{I}_m + \mathbf{AB} & \mathbf{0} \\ \mathbf{B} & \mathbf{I}_n \end{bmatrix}$$

and

$$\mathbf{QP} = \begin{bmatrix} \mathbf{I}_m & -\mathbf{A} \\ \mathbf{0} & \mathbf{I}_n + \mathbf{BA} \end{bmatrix}$$

Because **N** and **Q** are block triangular, their determinants equal the products of the determinant of their block-diagonal matrices or

$$\det \mathbf{N} = \det \mathbf{I}_m \cdot \det \mathbf{I}_n = 1 = \det \mathbf{Q}$$

Likewise, we have

$$\det(\mathbf{NP}) = \det(\mathbf{I}_m + \mathbf{AB}) \qquad \det(\mathbf{QP}) = \det(\mathbf{I}_n + \mathbf{BA})$$

Because

$$\det(\mathbf{NP}) = \det \mathbf{N} \det \mathbf{P} = \det \mathbf{P}$$

and

$$\det(\mathbf{QP}) = \det \mathbf{Q} \det \mathbf{P} = \det \mathbf{P}$$

we conclude $\det(\mathbf{I}_m + \mathbf{AB}) = \det(\mathbf{I}_n + \mathbf{BA})$.

In $\mathbf{N}$, $\mathbf{Q}$, and $\mathbf{P}$, if $\mathbf{I}_n$, $\mathbf{I}_m$, and $\mathbf{B}$ are replaced, respectively, by $\sqrt{s}\mathbf{I}_n$, $\sqrt{s}\mathbf{I}_m$, and $-\mathbf{B}$, then we can readily obtain

$$s^n \det(s\mathbf{I}_m - \mathbf{AB}) = s^m \det(s\mathbf{I}_n - \mathbf{BA}) \tag{3.65}$$

which implies, for $n = m$ or for $n \times n$ square matrices $\mathbf{A}$ and $\mathbf{B}$,

$$\det(s\mathbf{I}_n - \mathbf{AB}) = \det(s\mathbf{I}_n - \mathbf{BA}) \tag{3.66}$$

They are useful formulas.

3.9 Quadratic Form and Positive Definiteness

An $n \times n$ real matrix $\mathbf{M}$ is said to be *symmetric* if its transpose equals itself. The scalar function $\mathbf{x}'\mathbf{Mx}$, where $\mathbf{x}$ is an $n \times 1$ real vector and $\mathbf{M}' = \mathbf{M}$, is called a *quadratic form*. We show that all eigenvalues of symmetric $\mathbf{M}$ are real.

The eigenvalues and eigenvectors of real matrices can be complex as shown in Example 3.6. Therefore we must allow $\mathbf{x}$ to assume complex numbers for the time being and consider the scalar function $\mathbf{x}^*\mathbf{Mx}$, where $\mathbf{x}^*$ is the complex conjugate transpose of $\mathbf{x}$. Taking the complex conjugate transpose of $\mathbf{x}^*\mathbf{Mx}$ yields

$$(\mathbf{x}^*\mathbf{Mx})^* = \mathbf{x}^*\mathbf{M}^*\mathbf{x} = \mathbf{x}^*\mathbf{M}'\mathbf{x} = \mathbf{x}^*\mathbf{Mx}$$

where we have used the fact that the complex conjugate transpose of a real $\mathbf{M}$ reduces to simply the transpose. Thus $\mathbf{x}^*\mathbf{Mx}$ is real for any complex $\mathbf{x}$. This assertion is not true if $\mathbf{M}$ is not symmetric. Let λ be an eigenvalue of $\mathbf{M}$ and $\mathbf{v}$ be its eigenvector; that is, $\mathbf{Mv} = \lambda\mathbf{v}$. Because

$$\mathbf{v}^*\mathbf{Mv} = \mathbf{v}^*\lambda\mathbf{v} = \lambda(\mathbf{v}^*\mathbf{v})$$

and because both $\mathbf{v}^*\mathbf{Mv}$ and $\mathbf{v}^*\mathbf{v}$ are real, the eigenvalue λ must be real. This shows that all eigenvalues of symmetric $\mathbf{M}$ are real. After establishing this fact, we can return our study to exclusively real vector $\mathbf{x}$.

We claim that every symmetric matrix can be diagonalized using a similarity transformation even it has repeated eigenvalue λ. To show this, we show that there is no generalized eigenvector of grade 2 or higher. Suppose $\mathbf{x}$ is a generalized eigenvector of grade 2 or

$$(\mathbf{M} - \lambda\mathbf{I})^2\mathbf{x} = \mathbf{0} \tag{3.67}$$

$$(\mathbf{M} - \lambda\mathbf{I})\mathbf{x} \neq \mathbf{0} \tag{3.68}$$

Consider

$$[(\mathbf{M} - \lambda\mathbf{I})\mathbf{x}]'(\mathbf{M} - \lambda\mathbf{I})\mathbf{x} = \mathbf{x}'(\mathbf{M}' - \lambda\mathbf{I}')(\mathbf{M} - \lambda\mathbf{I})\mathbf{x} = \mathbf{x}'(\mathbf{M} - \lambda\mathbf{I})^2\mathbf{x}$$

which is nonzero according to (3.68) but is zero according to (3.67). This is a contradiction. Therefore the Jordan form of $\mathbf{M}$ has no Jordan block of order 2. Similarly, we can show that the Jordan form of $\mathbf{M}$ has no Jordan block of order 3 or higher. Thus we conclude that there exists a nonsingular $\mathbf{Q}$ such that

$$\mathbf{M} = \mathbf{QDQ}^{-1} \tag{3.69}$$

where $\mathbf{D}$ is a diagonal matrix with real eigenvalues of $\mathbf{M}$ on the diagonal.

A square matrix $\mathbf{A}$ is called an *orthogonal matrix* if all columns of $\mathbf{A}$ are orthonormal. Clearly $\mathbf{A}$ is nonsingular and we have

$$\mathbf{A}'\mathbf{A} = \mathbf{I} \quad \text{and} \quad \mathbf{A}^{-1} = \mathbf{A}'$$

which imply $\mathbf{AA}' = \mathbf{AA}^{-1} = \mathbf{I} = \mathbf{A}'\mathbf{A}$. Thus the inverse of an orthogonal matrix equals its transpose. Consider (3.69). Because $\mathbf{D}' = \mathbf{D}$ and $\mathbf{M}' = \mathbf{M}$, (3.69) equals its own transpose or

$$\mathbf{QDQ}^{-1} = [\mathbf{QDQ}^{-1}]' = [\mathbf{Q}^{-1}]'\mathbf{DQ}'$$

which implies $\mathbf{Q}^{-1} = \mathbf{Q}'$ and $\mathbf{Q}'\mathbf{Q} = \mathbf{QQ}' = \mathbf{I}$. Thus $\mathbf{Q}$ is an orthogonal matrix; its columns are orthonormalized eigenvectors of $\mathbf{M}$. This is summarized as a theorem.

▶ **Theorem 3.6**

For every real symmetric matrix $\mathbf{M}$, there exists an orthogonal matrix $\mathbf{Q}$ such that

$$\mathbf{M} = \mathbf{QDQ}' \quad \text{or} \quad \mathbf{D} = \mathbf{Q}'\mathbf{MQ}$$

where $\mathbf{D}$ is a diagonal matrix with the eigenvalues of $\mathbf{M}$, which are all real, on the diagonal.

A symmetric matrix $\mathbf{M}$ is said to be *positive definite*, denoted by $\mathbf{M} > 0$, if $\mathbf{x}'\mathbf{Mx} > 0$ for every nonzero $\mathbf{x}$. It is *positive semidefinite*, denoted by $\mathbf{M} \geq 0$, if $\mathbf{x}'\mathbf{Mx} \geq 0$ for every nonzero $\mathbf{x}$. If $\mathbf{M} > 0$, then $\mathbf{x}'\mathbf{Mx} = 0$ if and only if $\mathbf{x} = \mathbf{0}$. If $\mathbf{M}$ is positive semidefinite, then there exists a nonzero $\mathbf{x}$ such that $\mathbf{x}'\mathbf{Mx} = 0$. This property will be used repeatedly later.

▶ **Theorem 3.7**

A symmetric $n \times n$ matrix $\mathbf{M}$ is positive definite (positive semidefinite) if and only if any one of the following conditions holds.

1. Every eigenvalue of $\mathbf{M}$ is positive (zero or positive).
2. All the *leading* principal minors of $\mathbf{M}$ are positive (all the principal minors of $\mathbf{M}$ are zero or positive).
3. There exists an $n \times n$ nonsingular matrix $\mathbf{N}$ (an $n \times n$ singular matrix $\mathbf{N}$ or an $m \times n$ matrix $\mathbf{N}$ with $m < n$) such that $\mathbf{M} = \mathbf{N}'\mathbf{N}$.

Condition (1) can readily be proved by using Theorem 3.6. Next we consider Conditon (3). If $\mathbf{M} = \mathbf{N}'\mathbf{N}$, then

$$\mathbf{x}'\mathbf{Mx} = \mathbf{x}'\mathbf{N}'\mathbf{Nx} = (\mathbf{Nx})'(\mathbf{Nx}) = ||\mathbf{Nx}||_2^2 \geq 0$$

for any $\mathbf{x}$. If $\mathbf{N}$ is nonsingular, the only $\mathbf{x}$ to make $\mathbf{Nx} = \mathbf{0}$ is $\mathbf{x} = \mathbf{0}$. Thus $\mathbf{M}$ is positive definite. If $\mathbf{N}$ is singular, there exists a nonzero $\mathbf{x}$ to make $\mathbf{Nx} = \mathbf{0}$. Thus $\mathbf{M}$ is positive semidefinite. For a proof of Condition (2), see Reference [10].

We use an example to illustrate the principal minors and leading principal minors. Consider

$$\mathbf{M} = \begin{bmatrix} m_{11} & m_{12} & m_{13} \\ m_{21} & m_{22} & m_{23} \\ m_{31} & m_{32} & m_{33} \end{bmatrix}$$

Its principal minors are m_{11}, m_{22}, m_{33},

$$\det \begin{bmatrix} m_{11} & m_{12} \\ m_{21} & m_{22} \end{bmatrix}, \quad \det \begin{bmatrix} m_{11} & m_{13} \\ m_{31} & m_{33} \end{bmatrix}, \quad \det \begin{bmatrix} m_{22} & m_{23} \\ m_{32} & m_{33} \end{bmatrix}$$

and $\det \mathbf{M}$. Thus the principal minors are the determinants of all submatrices of $\mathbf{M}$ whose diagonals coincide with the diagonal of $\mathbf{M}$. The leading principal minors of $\mathbf{M}$ are

$$m_{11}, \quad \det \begin{bmatrix} m_{11} & m_{12} \\ m_{21} & m_{22} \end{bmatrix}, \quad \text{and} \quad \det \mathbf{M}$$

Thus the leading principal minors of $\mathbf{M}$ are the determinants of the submatrices of $\mathbf{M}$ obtained by deleting the last k columns and last k rows for $k = 2, 1, 0$.

▶ **Theorem 3.8**

1. An $m \times n$ matrix $\mathbf{H}$, with $m \geq n$, has rank n, if and only if the $n \times n$ matrix $\mathbf{H}'\mathbf{H}$ has rank n or $\det(\mathbf{H}'\mathbf{H}) \neq 0$.

2. An $m \times n$ matrix $\mathbf{H}$, with $m \leq n$, has rank m, if and only if the $m \times m$ matrix $\mathbf{HH}'$ has rank m or $\det(\mathbf{HH}') \neq 0$.

The symmetric matrix $\mathbf{H}'\mathbf{H}$ is always positive semidefinite. It becomes positive definite if $\mathbf{H}'\mathbf{H}$ is nonsingular. We give a proof of this theorem. The argument in the proof will be used to establish the main results in Chapter 6; therefore the proof is spelled out in detail.

 Proof: *Necessity*: The condition $\rho(\mathbf{H}'\mathbf{H}) = n$ implies $\rho(\mathbf{H}) = n$. We show this by contradiction. Suppose $\rho(\mathbf{H}'\mathbf{H}) = n$ but $\rho(\mathbf{H}) < n$. Then there exists a nonzero vector $\mathbf{v}$ such that $\mathbf{Hv} = \mathbf{0}$, which implies $\mathbf{H}'\mathbf{Hv} = \mathbf{0}$. This contradicts $\rho(\mathbf{H}'\mathbf{H}) = n$. Thus $\rho(\mathbf{H}'\mathbf{H}) = n$ implies $\rho(\mathbf{H}) = n$.

Sufficiency: The condition $\rho(\mathbf{H}) = n$ implies $\rho(\mathbf{H}'\mathbf{H}) = n$. Suppose not, or $\rho(\mathbf{H}'\mathbf{H}) < n$; then there exists a nonzero vector $\mathbf{v}$ such that $\mathbf{H}'\mathbf{Hv} = \mathbf{0}$, which implies $\mathbf{v}'\mathbf{H}'\mathbf{Hv} = 0$ or

$$0 = \mathbf{v}'\mathbf{H}'\mathbf{Hv} = (\mathbf{Hv})'(\mathbf{Hv}) = ||\mathbf{Hv}||_2^2$$

Thus we have $\mathbf{Hv} = \mathbf{0}$. This contradicts the hypotheses that $\mathbf{v} \neq 0$ and $\rho(\mathbf{H}) = n$. Thus $\rho(\mathbf{H}) = $ implies $\rho(\mathbf{H}'\mathbf{H}) = n$. This establishes the first part of Theorem 3.8. The second part can be established similarly.   Q.E.D.

We discuss the relationship between the eigenvalues of $\mathbf{H}'\mathbf{H}$ and those of $\mathbf{HH}'$. Because both $\mathbf{H}'\mathbf{H}$ and $\mathbf{HH}'$ are symmetric and positive semidefinite, their eigenvalues are real and nonnegative (zero or positive). If $\mathbf{H}$ is $m \times n$, then $\mathbf{H}'\mathbf{H}$ has n eigenvalues and $\mathbf{H}\mathbf{H}'$ has m eigenvalues. Let $\mathbf{A} = \mathbf{H}$ and $\mathbf{B} = \mathbf{H}'$. Then (3.65) becomes

$$\det(s\mathbf{I}_m - \mathbf{H}\mathbf{H}') = s^{m-n}\det(s\mathbf{I}_n - \mathbf{H}'\mathbf{H}) \tag{3.70}$$

This implies that the characteristic polynomials of $\mathbf{H}\mathbf{H}'$ and $\mathbf{H}'\mathbf{H}$ differ only by s^{m-n}. Thus we conclude that $\mathbf{H}\mathbf{H}'$ and $\mathbf{H}'\mathbf{H}$ have the same nonzero eigenvalues but may have different numbers of zero eigenvalues. Furthermore, they have at most $\bar{n} := \min(m, n)$ number of nonzero eigenvalues.

3.10 Singular-Value Decomposition

Let $\mathbf{H}$ be an $m \times n$ real matrix. Define $\mathbf{M} := \mathbf{H}'\mathbf{H}$. Clearly $\mathbf{M}$ is $n \times n$, symmetric, and semidefinite. Thus all eigenvalues of $\mathbf{M}$ are real and nonnegative (zero or positive). Let r be the number of its positive eigenvalues. Then the eigenvalues of $\mathbf{M} = \mathbf{H}'\mathbf{H}$ can be arranged as

$$\lambda_1^2 \geq \lambda_2^2 \geq \cdots \lambda_r^2 > 0 = \lambda_{r+1} = \cdots = \lambda_n$$

Let $\bar{n} := \min(m, n)$. Then the set

$$\lambda_1 \geq \lambda_2 \geq \cdots \lambda_r > 0 = \lambda_{r+1} = \cdots = \lambda_{\bar{n}}$$

is called the *singular values* of $\mathbf{H}$. The singular values are usually arranged in descending order in magnitude.

EXAMPLE 3.13   Consider the 2×3 matrix

$$\mathbf{H} = \begin{bmatrix} -4 & -1 & 2 \\ 2 & 0.5 & -1 \end{bmatrix}$$

We compute

$$\mathbf{M} = \mathbf{H}'\mathbf{H} = \begin{bmatrix} 20 & 5 & -10 \\ 5 & 1.25 & -2.5 \\ -10 & -2.5 & 5 \end{bmatrix}$$

and compute its characteristic polynomial as

$$\det(\lambda\mathbf{I} - \mathbf{M}) = \lambda^3 - 26.25\lambda^2 = \lambda^2(\lambda - 26.25)$$

Thus the eigenvalues of $\mathbf{H}'\mathbf{H}$ are 26.25, 0, and 0, and the singular values of $\mathbf{H}$ are $\sqrt{26.25} = 5.1235$ and 0. Note that the number of singular values equals $\min(n, m)$.

In view of (3.70), we can also compute the singular values of $\mathbf{H}$ from the eigenvalues of $\mathbf{H}\mathbf{H}'$. Indeed, we have

$$\bar{\mathbf{M}} := \mathbf{H}\mathbf{H}' = \begin{bmatrix} 21 & -10.5 \\ -10.5 & 5.25 \end{bmatrix}$$

and

$$\det(\lambda\mathbf{I} - \bar{\mathbf{M}}) = \lambda^2 - 26.25\lambda = \lambda(\lambda - 26.25)$$

Thus the eigenvalues of $\mathbf{H}\mathbf{H}'$ are 26.25 and 0 and the singular values of $\mathbf{H}'$ are 5.1235 and 0. We see that the eigenvalues of $\mathbf{H}'\mathbf{H}$ differ from those of $\mathbf{H}\mathbf{H}'$ only in the number of zero eigenvalues and the singular values of $\mathbf{H}$ equal the singular values of $\mathbf{H}'$.

For $\mathbf{M} = \mathbf{H'H}$, there exists, following Theorem 3.6, an orthogonal matrix $\mathbf{Q}$ such that

$$\mathbf{Q'H'HQ} = \mathbf{D} =: \mathbf{S'S} \tag{3.71}$$

where $\mathbf{D}$ is an $n \times n$ diagonal matrix with λ_i^2 on the diagonal. The matrix $\mathbf{S}$ is $m \times n$ with the singular values λ_i on the diagonal. Manipulation on (3.71) will lead eventually to the theorem that follows.

▶ **Theorem 3.9 (Singular-value decomposition)**

Every $m \times n$ matrix $\mathbf{H}$ can be transformed into the form

$$\mathbf{H} = \mathbf{RSQ'}$$

with $\mathbf{R'R} = \mathbf{RR'} = \mathbf{I}_m$, $\mathbf{Q'Q} = \mathbf{QQ'} = \mathbf{I}_n$, and $\mathbf{S}$ being $m \times n$ with the singular values of $\mathbf{H}$ on the diagonal.

The columns of $\mathbf{Q}$ are orthonormalized eigenvectors of $\mathbf{H'H}$ and the columns of $\mathbf{R}$ are orthonormalized eigenvectors of $\mathbf{HH'}$. Once $\mathbf{R}$, $\mathbf{S}$, and $\mathbf{Q}$ are computed, the rank of $\mathbf{H}$ equals the number of nonzero singular values. If the rank of $\mathbf{H}$ is r, the first r columns of $\mathbf{R}$ are an orthonormal basis of the range space of $\mathbf{H}$. The last $(n - r)$ columns of $\mathbf{Q}$ are an orthonormal basis of the null space of $\mathbf{H}$. Although computing singular-value decomposition is time consuming, it is very reliable and gives a quantitative measure of the rank. Thus it is used in MATLAB to compute the rank, range space, and null space. In MATLAB, the singular values of $\mathbf{H}$ can be obtained by typing s=svd(H). Typing [R,S,Q]=svd(H) yields the three matrices in the theorem. Typing orth(H) and null(H) yields, respectively, orthonormal bases of the range space and null space of $\mathbf{H}$. The function null will be used repeatedly in Chapter 7.

EXAMPLE 3.14 Consider the matrix in (3.11). We type

```
a=[0 1 1 2;1 2 3 4;2 0 2 0];
[r,s,q]=svd(a)
```

which yield

$$r = \begin{bmatrix} 0.3782 & -0.3084 & 0.8729 \\ 0.8877 & -0.1468 & -0.4364 \\ 0.2627 & 0.9399 & 0.2182 \end{bmatrix} \quad s = \begin{bmatrix} 6.1568 & 0 & 0 & 0 \\ 0 & 2.4686 & 0 & 0 \\ 0 & 0 & 0 & 0 \end{bmatrix}$$

$$q = \begin{bmatrix} 0.2295 & 0.7020 & 0.3434 & -0.5802 \\ 0.3498 & -0.2439 & 0.8384 & 0.3395 \\ 0.5793 & 0.4581 & -0.3434 & 0.5802 \\ 0.6996 & -0.4877 & -0.2475 & -0.4598 \end{bmatrix}$$

Thus the singular values of the matrix $\mathbf{A}$ in (3.11) are 6.1568, 2.4686, and 0. The matrix has two nonzero singular values, thus its rank is 2 and, consequently, its nullity is $4 - \rho(\mathbf{A}) = 2$. The first two columns of r are the orthonormal basis in (3.13) and the last two columns of q are the orthonormal basis in (3.14).

3.11 Norms of Matrices

The concept of norms for vectors can be extended to matrices. This concept is needed in Chapter 5. Let $\mathbf{A}$ be an $m \times n$ matrix. The norm of $\mathbf{A}$ can be defined as

$$||\mathbf{A}|| = \sup_{\mathbf{x} \neq 0} \frac{||\mathbf{A}\mathbf{x}||}{||\mathbf{x}||} = \sup_{||\mathbf{x}||=1} ||\mathbf{A}\mathbf{x}|| \tag{3.72}$$

where sup stands for supremum or the least upper bound. This norm is defined through the norm of $\mathbf{x}$ and is therefore called an *induced norm*. For different $||\mathbf{x}||$, we have different $||\mathbf{A}||$. For example, if the 1-norm $||\mathbf{x}||_1$ is used, then

$$||\mathbf{A}||_1 = \max_j \left(\sum_{i=1}^{m} |a_{ij}| \right) = \text{largest column absolute sum}$$

where a_{ij} is the ijth element of $\mathbf{A}$. If the Euclidean norm $||\mathbf{x}||_2$ is used, then

$$||\mathbf{A}||_2 = \text{ largest singular value of } \mathbf{A}$$
$$= (\text{largest eigenvalue of } \mathbf{A}'\mathbf{A})^{1/2}$$

If the infinite-norm $||\mathbf{x}||_\infty$ is used, then

$$||\mathbf{A}||_\infty = \max_i \left(\sum_{j=1}^{n} |a_{ij}| \right) = \text{largest row absolute sum}$$

These norms are all different for the same $\mathbf{A}$. For example, if

$$\mathbf{A} = \begin{bmatrix} 3 & 2 \\ -1 & 0 \end{bmatrix}$$

then $||\mathbf{A}||_1 = 3 + |-1| = 4$, $||\mathbf{A}||_2 = 3.7$, and $||\mathbf{A}||_\infty = 3 + 2 = 5$, as shown in Fig. 3.3. The MATLAB functions `norm(a,1)`, `norm(a,2)=norm(a)`, and `norm(a,inf)` compute the three norms.

The norm of matrices has the following properties:

$$||\mathbf{A}\mathbf{x}|| \leq ||\mathbf{A}||||\mathbf{x}||$$
$$||\mathbf{A} + \mathbf{B}|| \leq ||\mathbf{A}|| + ||\mathbf{B}||$$
$$||\mathbf{A}\mathbf{B}|| \leq ||\mathbf{A}||||\mathbf{B}||$$

PROBLEMS

The reader should try first to solve all problems involving numerical numbers by hand and then verify the results using MATLAB or any software.

3.1 Consider Fig. 3.1. What is the representation of the vector $\mathbf{x}$ with respect to the basis $\{\mathbf{q}_1, \mathbf{i}_2\}$? What is the representation of $\mathbf{q}_1$ with respect to $\{\mathbf{i}_2, \mathbf{q}_2\}$?

3.2 What are the 1-norm, 2-norm, and infinite-norm of the vectors

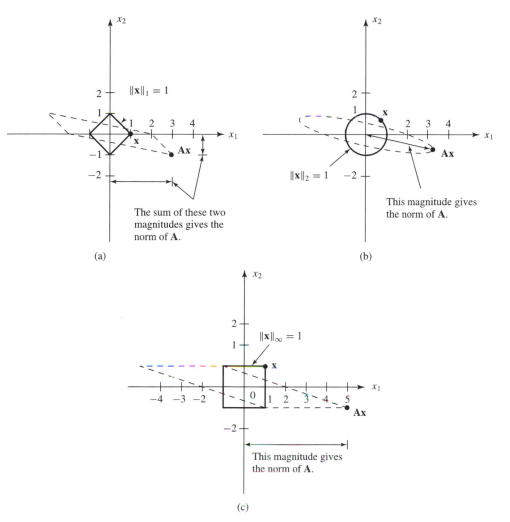

Figure 3.3   Different norms of **A**.

$$\mathbf{x}_1 = \begin{bmatrix} 2 \\ -3 \\ 1 \end{bmatrix} \qquad \mathbf{x}_2 = \begin{bmatrix} 1 \\ 1 \\ 1 \end{bmatrix}$$

3.3   Find two orthonormal vectors that span the same space as the two vectors in Problem 3.2.

3.4   Consider an $n \times m$ matrix **A** with $n \geq m$. If all columns of **A** are orthonormal, then $\mathbf{A}'\mathbf{A} = \mathbf{I}_m$. What can you say about $\mathbf{A}\mathbf{A}'$?

3.5   Find the ranks and nullities of the following matrices:

$$A_1 = \begin{bmatrix} 0 & 1 & 0 \\ 0 & 0 & 0 \\ 0 & 0 & 1 \end{bmatrix} \qquad A_2 = \begin{bmatrix} 4 & 1 & -1 \\ 3 & 2 & 0 \\ 1 & 1 & 0 \end{bmatrix} \qquad A_3 = \begin{bmatrix} 1 & 2 & 3 & 4 \\ 0 & -1 & -2 & 2 \\ 0 & 0 & 0 & 1 \end{bmatrix}$$

3.6 Find bases of the range spaces and null spaces of the matrices in Problem 3.5.

3.7 Consider the linear algebraic equation

$$\begin{bmatrix} 2 & -1 \\ -3 & 3 \\ -1 & 2 \end{bmatrix} \mathbf{x} = \begin{bmatrix} 1 \\ 0 \\ 1 \end{bmatrix} = \mathbf{y}$$

It has three equations and two unknowns. Does a solution $\mathbf{x}$ exist in the equation? Is the solution unique? Does a solution exist if $\mathbf{y} = [1 \ 1 \ 1]'$?

3.8 Find the general solution of

$$\begin{bmatrix} 1 & 2 & 3 & 4 \\ 0 & -1 & -2 & 2 \\ 0 & 0 & 0 & 1 \end{bmatrix} \mathbf{x} = \begin{bmatrix} 3 \\ 2 \\ 1 \end{bmatrix}$$

How many parameters do you have?

3.9 Find the solution in Example 3.3 that has the smallest Euclidean norm.

3.10 Find the solution in Problem 3.8 that has the smallest Euclidean norm.

3.11 Consider the equation

$$\mathbf{x}[n] = \mathbf{A}^n x[0] + \mathbf{A}^{n-1} \mathbf{b} u[0] + \mathbf{A}^{n-2} \mathbf{b} u[1] + \cdots + \mathbf{A} \mathbf{b} u[n-2] + \mathbf{b} u[n-1]$$

where $\mathbf{A}$ is an $n \times n$ matrix and $\mathbf{b}$ is an $n \times 1$ column vector. Under what conditions on $\mathbf{A}$ and $\mathbf{b}$ will there exist $u[0], u[1], \ldots, u[n-1]$ to meet the equation for any $\mathbf{x}[n]$ and $\mathbf{x}[0]$? [*Hint:* Write the equation in the form

$$\mathbf{x}[n] - \mathbf{A}^n \mathbf{x}[0] = [\mathbf{b} \ \mathbf{Ab} \ \cdots \ \mathbf{A}^{n-1}\mathbf{b}] \begin{bmatrix} u[n-1] \\ u[n-2] \\ \vdots \\ u[0] \end{bmatrix}]$$

3.12 Given

$$A = \begin{bmatrix} 2 & 1 & 0 & 0 \\ 0 & 2 & 1 & 0 \\ 0 & 0 & 2 & 0 \\ 0 & 0 & 0 & 1 \end{bmatrix} \qquad \mathbf{b} = \begin{bmatrix} 0 \\ 0 \\ 1 \\ 1 \end{bmatrix} \qquad \bar{\mathbf{b}} = \begin{bmatrix} 1 \\ 2 \\ 3 \\ 1 \end{bmatrix}$$

what are the representations of $\mathbf{A}$ with respect to the basis $\{\mathbf{b}, \mathbf{Ab}, \mathbf{A}^2\mathbf{b}, \mathbf{A}^3\mathbf{b}\}$ and the basis $\{\bar{\mathbf{b}}, \mathbf{A}\bar{\mathbf{b}}, \mathbf{A}^2\bar{\mathbf{b}}, \mathbf{A}^3\bar{\mathbf{b}}\}$, respectively? (Note that the representations are the same!)

3.13 Find Jordan-form representations of the following matrices:

$$\mathbf{A}_1 = \begin{bmatrix} 1 & 4 & 10 \\ 0 & 2 & 0 \\ 0 & 0 & 3 \end{bmatrix} \qquad \mathbf{A}_2 = \begin{bmatrix} 0 & 1 & 0 \\ 0 & 0 & 1 \\ -2 & -4 & -3 \end{bmatrix}$$

$$\mathbf{A}_3 = \begin{bmatrix} 1 & 0 & -1 \\ 0 & 1 & 0 \\ 0 & 0 & 2 \end{bmatrix} \qquad \mathbf{A}_4 = \begin{bmatrix} 0 & 4 & 3 \\ 0 & 20 & 16 \\ 0 & -25 & -20 \end{bmatrix}$$

Note that all except $\mathbf{A}_4$ can be diagonalized.

3.14 Consider the companion-form matrix

$$\mathbf{A} = \begin{bmatrix} -\alpha_1 & -\alpha_2 & -\alpha_3 & -\alpha_4 \\ 1 & 0 & 0 & 0 \\ 0 & 1 & 0 & 0 \\ 0 & 0 & 1 & 0 \end{bmatrix}$$

Show that its characteristic polynomial is given by

$$\Delta(\lambda) = \lambda^4 + \alpha_1 \lambda^3 + \alpha_2 \lambda^2 + \alpha_3 \lambda + \alpha_4$$

Show also that if λ_i is an eigenvalue of $\mathbf{A}$ or a solution of $\Delta(\lambda) = 0$, then $[\lambda_i^3 \ \lambda_i^2 \ \lambda_i \ 1]'$ is an eigenvector of $\mathbf{A}$ associated with λ_i.

3.15 Show that the *Vandermonde* determinant

$$\begin{bmatrix} \lambda_1^3 & \lambda_2^3 & \lambda_3^3 & \lambda_4^3 \\ \lambda_1^2 & \lambda_2^2 & \lambda_3^2 & \lambda_4^2 \\ \lambda_1 & \lambda_2 & \lambda_3 & \lambda_4 \\ 1 & 1 & 1 & 1 \end{bmatrix}$$

equals $\prod_{1 \le i < j \le 4}(\lambda_j - \lambda_i)$. Thus we conclude that the matrix is nonsingular or, equivalently, the eigenvectors are linearly independent if all eigenvalues are distinct.

3.16 Show that the companion-form matrix in Problem 3.14 is nonsingular if and only if $\alpha_4 \ne 0$. Under this assumption, show that its inverse equals

$$\mathbf{A}^{-1} = \begin{bmatrix} 0 & 1 & 0 & 0 \\ 0 & 0 & 1 & 0 \\ 0 & 0 & 0 & 1 \\ -1/\alpha_4 & -\alpha_1/\alpha_4 & -\alpha_2/\alpha_4 & -\alpha_3/\alpha_4 \end{bmatrix}$$

3.17 Consider

$$\mathbf{A} = \begin{bmatrix} \lambda & \lambda T & \lambda T^2/2 \\ 0 & \lambda & \lambda T \\ 0 & 0 & \lambda \end{bmatrix}$$

with $\lambda \ne 0$ and $T > 0$. Show that $[0 \ 0 \ 1]'$ is a generalized eigenvector of grade 3 and the three columns of

$$\mathbf{Q} = \begin{bmatrix} \lambda^2 T^2 & \lambda T^2 & 0 \\ 0 & \lambda T & 0 \\ 0 & 0 & 1 \end{bmatrix}$$

constitute a chain of generalized eigenvectors of length 3. Verify

$$\mathbf{Q}^{-1}\mathbf{A}\mathbf{Q} = \begin{bmatrix} \lambda & 1 & 0 \\ 0 & \lambda & 1 \\ 0 & 0 & \lambda \end{bmatrix}$$

3.18 Find the characteristic polynomials and the minimal polynomials of the following matrices:

$$\begin{bmatrix} \lambda_1 & 1 & 0 & 0 \\ 0 & \lambda_1 & 1 & 0 \\ 0 & 0 & \lambda_1 & 0 \\ 0 & 0 & 0 & \lambda_2 \end{bmatrix} \qquad \begin{bmatrix} \lambda_1 & 1 & 0 & 0 \\ 0 & \lambda_1 & 1 & 0 \\ 0 & 0 & \lambda_1 & 0 \\ 0 & 0 & 0 & \lambda_1 \end{bmatrix}$$

$$\begin{bmatrix} \lambda_1 & 1 & 0 & 0 \\ 0 & \lambda_1 & 0 & 0 \\ 0 & 0 & \lambda_1 & 0 \\ 0 & 0 & 0 & \lambda_1 \end{bmatrix} \qquad \begin{bmatrix} \lambda_1 & 0 & 0 & 0 \\ 0 & \lambda_1 & 0 & 0 \\ 0 & 0 & \lambda_1 & 0 \\ 0 & 0 & 0 & \lambda_1 \end{bmatrix}$$

3.19 Show that if λ is an eigenvalue of $\mathbf{A}$ with eigenvector $\mathbf{x}$, then $f(\lambda)$ is an eigenvalue of $f(\mathbf{A})$ with the same eigenvector $\mathbf{x}$.

3.20 Show that an $n \times n$ matrix has the property $\mathbf{A}^k = \mathbf{0}$ for $k \geq m$ if and only if $\mathbf{A}$ has eigenvalues 0 with multiplicity n and index m or less. Such a matrix is called a *nilpotent* matrix.

3.21 Given

$$\mathbf{A} = \begin{bmatrix} 1 & 1 & 0 \\ 0 & 0 & 1 \\ 0 & 0 & 1 \end{bmatrix}$$

find $\mathbf{A}^{10}$, $\mathbf{A}^{103}$, and $e^{\mathbf{A}t}$.

3.22 Use two different methods to compute $e^{\mathbf{A}t}$ for $\mathbf{A}_1$ and $\mathbf{A}_4$ in Problem 3.13.

3.23 Show that functions of the same matrix commute; that is,

$$f(\mathbf{A})g(\mathbf{A}) = g(\mathbf{A})f(\mathbf{A})$$

Consequently we have $\mathbf{A}e^{\mathbf{A}t} = e^{\mathbf{A}t}\mathbf{A}$.

3.24 Let

$$\mathbf{C} = \begin{bmatrix} \lambda_1 & 0 & 0 \\ 0 & \lambda_2 & 0 \\ 0 & 0 & \lambda_3 \end{bmatrix}$$

Find a matrix $\mathbf{B}$ such that $e^{\mathbf{B}} = \mathbf{C}$. Show that if $\lambda_i = 0$ for some i, then $\mathbf{B}$ does not exist. Let

$$\mathbf{C} = \begin{bmatrix} \lambda & 1 & 0 \\ 0 & \lambda & 0 \\ 0 & 0 & \lambda \end{bmatrix}$$

Find a $\mathbf{B}$ such that $e^{\mathbf{B}} = \mathbf{C}$. Is it true that, for any nonsingular $\mathbf{C}$, there exists a matrix $\mathbf{B}$ such that $e^{\mathbf{B}} = \mathbf{C}$?

3.25 Let

$$(s\mathbf{I} - \mathbf{A})^{-1} = \frac{1}{\Delta(s)} \text{Adj}\,(s\mathbf{I} - \mathbf{A})$$

and let $m(s)$ be the monic greatest common divisor of all entries of Adj $(s\mathbf{I} - \mathbf{A})$. Verify for the matrix $\mathbf{A}_3$ in Problem 3.13 that the minimal polynominal of $\mathbf{A}$ equals $\Delta(s)/m(s)$.

3.26 Define

$$(s\mathbf{I} - \mathbf{A})^{-1} := \frac{1}{\Delta(s)} \left[\mathbf{R}_0 s^{n-1} + \mathbf{R}_1 s^{n-2} + \cdots + \mathbf{R}_{n-2}s + \mathbf{R}_{n-1} \right]$$

where

$$\Delta(s) := \det(s\mathbf{I} - \mathbf{A}) := s^n + \alpha_1 s^{n-1} + \alpha_2 s^{n-2} + \cdots + \alpha_n$$

and $\mathbf{R}_i$ are constant matrices. This definition is valid because the degree in s of the adjoint of $(s\mathbf{I} - \mathbf{A})$ is at most $n - 1$. Verify

$$\alpha_1 = -\frac{\text{tr}(\mathbf{AR}_0)}{1} \qquad \mathbf{R}_0 = \mathbf{I}$$

$$\alpha_2 = -\frac{\text{tr}(\mathbf{AR}_1)}{2} \qquad \mathbf{R}_1 = \mathbf{AR}_0 + \alpha_1\mathbf{I} = \mathbf{A} + \alpha_1\mathbf{I}$$

$$\alpha_3 = -\frac{\text{tr}(\mathbf{AR}_2)}{3} \qquad \mathbf{R}_2 = \mathbf{AR}_1 + \alpha_2\mathbf{I} = \mathbf{A}^2 + \alpha_1\mathbf{A} + \alpha_2\mathbf{I}$$

$$\vdots$$

$$\alpha_{n-1} = -\frac{\text{tr}(\mathbf{AR}_{n-2})}{n-1} \qquad \mathbf{R}_{n-1} = \mathbf{AR}_{n-2} + \alpha_{n-1}\mathbf{I} = \mathbf{A}^{n-1} + \alpha_1\mathbf{A}^{n-2}$$

$$+ \cdots + \alpha_{n-2}\mathbf{A} + \alpha_{n-1}\mathbf{I}$$

$$\alpha_n = -\frac{\text{tr}(\mathbf{AR}_{n-1})}{n} \qquad \mathbf{0} = \mathbf{AR}_{n-1} + \alpha_n\mathbf{I}$$

where tr stands for the *trace* of a matrix and is defined as the sum of all its diagonal entries. This process of computing α_i and $\mathbf{R}_i$ is called the *Leverrier algorithm*.

3.27 Use Problem 3.26 to prove the Cayley–Hamilton theorem.

3.28 Use Problem 3.26 to show

$$(s\mathbf{I} - \mathbf{A})^{-1} = \frac{1}{\Delta(s)} \left[\mathbf{A}^{n-1} + (s + \alpha_1)\mathbf{A}^{n-2} + (s^2 + \alpha_1 s + \alpha_2)\mathbf{A}^{n-3} \right.$$

$$+ \cdots + (s^{n-1} + \alpha_1 s^{n-2} + \cdots + \alpha_{n-1})\mathbf{I}\Big]$$

3.29 Let all eigenvalues of $\mathbf{A}$ be distinct and let $\mathbf{q}_i$ be a right eigenvector of $\mathbf{A}$ associated with λ_i; that is, $\mathbf{A}\mathbf{q}_i = \lambda_i \mathbf{q}_i$. Define $\mathbf{Q} = [\mathbf{q}_1 \ \mathbf{q}_2 \ \cdots \ \mathbf{q}_n]$ and define

$$\mathbf{P} := \mathbf{Q}^{-1} =: \begin{bmatrix} \mathbf{p}_1 \\ \mathbf{p}_2 \\ \vdots \\ \mathbf{p}_n \end{bmatrix}$$

where $\mathbf{p}_i$ is the ith row of $\mathbf{P}$. Show that $\mathbf{p}_i$ is a left eigenvector of $\mathbf{A}$ associated with λ_i; that is, $\mathbf{p}_i \mathbf{A} = \lambda_i \mathbf{p}_i$.

3.30 Show that if all eigenvalues of $\mathbf{A}$ are distinct, then $(s\mathbf{I} - \mathbf{A})^{-1}$ can be expressed as

$$(s\mathbf{I} - \mathbf{A})^{-1} = \sum \frac{1}{s - \lambda_i} \mathbf{q}_i \mathbf{p}_i$$

where $\mathbf{q}_i$ and $\mathbf{p}_i$ are right and left eigenvectors of $\mathbf{A}$ associated with λ_i.

3.31 Find the $\mathbf{M}$ to meet the Lyapunov equation in (3.59) with

$$\mathbf{A} = \begin{bmatrix} 0 & 1 \\ -2 & -2 \end{bmatrix} \qquad \mathbf{B} = 3 \qquad \mathbf{C} = \begin{bmatrix} 3 \\ 3 \end{bmatrix}$$

What are the eigenvalues of the Lyapunov equation? Is the Lyapunov equation singular? Is the solution unique?

3.32 Repeat Problem 3.31 for

$$\mathbf{A} = \begin{bmatrix} 0 & 1 \\ -1 & -2 \end{bmatrix} \qquad \mathbf{B} = 1 \qquad \mathbf{C}_1 = \begin{bmatrix} 3 \\ 3 \end{bmatrix} \qquad \mathbf{C}_2 = \begin{bmatrix} 3 \\ -3 \end{bmatrix}$$

with two different $\mathbf{C}$.

3.33 Check to see if the following matrices are positive definite or semidefinite:

$$\begin{bmatrix} 2 & 3 & 2 \\ 3 & 1 & 0 \\ 2 & 0 & 2 \end{bmatrix} \qquad \begin{bmatrix} 0 & 0 & -1 \\ 0 & 0 & 0 \\ -1 & 0 & 2 \end{bmatrix} \qquad \begin{bmatrix} a_1 a_1 & a_1 a_2 & a_1 a_3 \\ a_2 a_1 & a_2 a_2 & a_2 a_3 \\ a_3 a_1 & a_3 a_2 & a_3 a_3 \end{bmatrix}$$

3.34 Compute the singular values of the following matrices:

$$\begin{bmatrix} -1 & 0 & 1 \\ 2 & -1 & 0 \end{bmatrix} \qquad \begin{bmatrix} -1 & 2 \\ 2 & 4 \end{bmatrix}$$

3.35 If $\mathbf{A}$ is symmetric, what is the relationship between its eigenvalues and singular values?

3.36 Show

$$
\det \left(\mathbf{I}_n + \begin{bmatrix} a_1 \\ a_2 \\ \vdots \\ a_n \end{bmatrix} [b_1\ b_2\ \cdots\ b_n] \right) = 1 + \sum_{m=1}^{n} a_m b_m
$$

3.37 Show (3.65).

3.38 Consider $\mathbf{Ax} = \mathbf{y}$, where $\mathbf{A}$ is $m \times n$ and has rank m. Is $(\mathbf{A'A})^{-1}\mathbf{A'y}$ a solution? If not, under what condition will it be a solution? Is $\mathbf{A'}(\mathbf{AA'})^{-1}\mathbf{y}$ a solution?

Chapter

4

State-Space Solutions and Realizations

4.1 Introduction

We showed in Chapter 2 that linear systems can be described by convolutions and, if lumped, by state-space equations. This chapter discusses how to find their solutions. First we discuss briefly how to compute solutions of the input-output description. There is no simple analytical way of computing the convolution

$$y(t) = \int_{\tau=t_0}^{t} g(t, \tau)u(\tau)\, d\tau$$

The easiest way is to compute it numerically on a digital computer. Before doing so, the equation must be discretized. One way is to discretize it as

$$y(k\Delta) = \sum_{m=k_0}^{k} g(k\Delta, m\Delta)u(m\Delta)\Delta \tag{4.1}$$

where Δ is called the integration step size. This is basically the discrete convolution discussed in (2.34). This discretization is the easiest but yields the least accurate result for the same integration step size. For other integration methods, see, for example, Reference [17].

For the linear time-invariant (LTI) case, we can also use $\hat{y}(s) = \hat{g}(s)\hat{u}(s)$ to compute the solution. If a system is distributed, $\hat{g}(s)$ will not be a rational function of s. Except for some special cases, it is simpler to compute the solution directly in the time domain as in (4.1). If the system is lumped, $\hat{g}(s)$ will be a rational function of s. In this case, if the Laplace transform of $u(t)$ is also a rational function of s, then the solution can be obtained by taking the inverse Laplace transform of $\hat{g}(s)\hat{u}(s)$. This method requires computing poles, carrying out partial fraction expansion, and then using a Laplace transform table. These can be carried out using the MATLAB functions `roots` and `residue`. However, when there are repeated poles, the computation may become very sensitive to small changes in the data, including roundoff errors; therefore computing solutions using the Laplace transform is not a viable method on digital computers. A better method is to transform transfer functions into state-space equations and then compute the solutions. This chapter discusses solutions of state equations, how to transform transfer functions into state equations, and other related topics. We discuss first the time-invariant case and then the time-varying case.

4.2 Solution of LTI State Equations

Consider the linear time-invariant (LTI) state-space equation

$$\dot{\mathbf{x}}(t) = \mathbf{A}\mathbf{x}(t) + \mathbf{B}\mathbf{u}(t) \tag{4.2}$$

$$\mathbf{y}(t) = \mathbf{C}\mathbf{x}(t) + \mathbf{D}\mathbf{u}(t) \tag{4.3}$$

where $\mathbf{A}$, $\mathbf{B}$, $\mathbf{C}$, and $\mathbf{D}$ are, respectively, $n \times n$, $n \times p$, $q \times n$, and $q \times p$ constant matrices. The problem is to find the solution excited by the initial state $\mathbf{x}(0)$ and the input $\mathbf{u}(t)$. The solution hinges on the exponential function of $\mathbf{A}$ studied in Section 3.6. In particular, we need the property in (3.55) or

$$\frac{d}{dt}e^{\mathbf{A}t} = \mathbf{A}e^{\mathbf{A}t} = e^{\mathbf{A}t}\mathbf{A}$$

to develop the solution. Premultiplying $e^{-\mathbf{A}t}$ on both sides of (4.2) yields

$$e^{-\mathbf{A}t}\dot{\mathbf{x}}(t) - e^{-\mathbf{A}t}\mathbf{A}\mathbf{x}(t) = e^{-\mathbf{A}t}\mathbf{B}\mathbf{u}(t)$$

which implies

$$\frac{d}{dt}\left(e^{-\mathbf{A}t}\mathbf{x}(t)\right) = e^{-\mathbf{A}t}\mathbf{B}\mathbf{u}(t)$$

Its integration from 0 to t yields

$$e^{-\mathbf{A}\tau}\mathbf{x}(\tau)\Big|_{\tau=0}^{t} = \int_0^t e^{-\mathbf{A}\tau}\mathbf{B}\mathbf{u}(\tau)\,d\tau$$

Thus we have

$$e^{-\mathbf{A}t}\mathbf{x}(t) - e^{0}\mathbf{x}(0) = \int_0^t e^{-\mathbf{A}\tau}\mathbf{B}\mathbf{u}(\tau)\,d\tau \tag{4.4}$$

Because the inverse of $e^{-\mathbf{A}t}$ is $e^{\mathbf{A}t}$ and $e^{0} = \mathbf{I}$ as discussed in (3.54) and (3.52), (4.4) implies

$$\mathbf{x}(t) = e^{\mathbf{A}t}\mathbf{x}(0) + \int_0^t e^{\mathbf{A}(t-\tau)}\mathbf{B}\mathbf{u}(\tau)\,d\tau \tag{4.5}$$

This is the solution of (4.2).

It is instructive to verify that (4.5) is the solution of (4.2). To verify this, we must show that (4.5) satisfies (4.2) and the initial condition $\mathbf{x}(t) = \mathbf{x}(0)$ at $t = 0$. Indeed, at $t = 0$, (4.5) reduces to

$$\mathbf{x}(0) = e^{\mathbf{A} \cdot 0} \mathbf{x}(0) = e^0 \mathbf{x}(0) = \mathbf{I}\mathbf{x}(0) = \mathbf{x}(0)$$

Thus (4.5) satisfies the initial condition. We need the equation

$$\frac{\partial}{\partial t} \int_{t_0}^t f(t, \tau) \, d\tau = \int_{t_0}^t \left(\frac{\partial}{\partial t} f(t, \tau) \right) d\tau + f(t, \tau)|_{\tau=t} \tag{4.6}$$

to show that (4.5) satisfies (4.2). Differentiating (4.5) and using (4.6), we obtain

$$\dot{\mathbf{x}}(t) = \frac{d}{dt} \left[e^{\mathbf{A}t} \mathbf{x}(0) + \int_0^t e^{\mathbf{A}(t-\tau)} \mathbf{B}\mathbf{u}(\tau) \, d\tau \right]$$

$$= \mathbf{A}e^{\mathbf{A}t} \mathbf{x}(0) + \int_0^t \mathbf{A}e^{\mathbf{A}(t-\tau)} \mathbf{B}\mathbf{u}(\tau) \, d\tau + e^{\mathbf{A}(t-\tau)} \mathbf{B}\mathbf{u}(\tau)\big|_{\tau=t}$$

$$= \mathbf{A} \left(e^{\mathbf{A}t} \mathbf{x}(0) + \int_0^t e^{\mathbf{A}(t-\tau)} \mathbf{B}\mathbf{u}(\tau) \, d\tau \right) + e^{\mathbf{A} \cdot 0} \mathbf{B}\mathbf{u}(t)$$

which becomes, after substituting (4.5),

$$\dot{\mathbf{x}}(t) = \mathbf{A}\mathbf{x}(t) + \mathbf{B}\mathbf{u}(t)$$

Thus (4.5) meets (4.2) and the initial condition $\mathbf{x}(0)$ and is the solution of (4.2).

Substituting (4.5) into (4.3) yields the solution of (4.3) as

$$\mathbf{y}(t) = \mathbf{C}e^{\mathbf{A}t} \mathbf{x}(0) + \mathbf{C} \int_0^t e^{\mathbf{A}(t-\tau)} \mathbf{B}\mathbf{u}(\tau) \, d\tau + \mathbf{D}\mathbf{u}(t) \tag{4.7}$$

This solution and (4.5) are computed directly in the time domain. We can also compute the solutions by using the Laplace transform. Applying the Laplace transform to (4.2) and (4.3) yields, as derived in (2.14) and (2.15),

$$\hat{\mathbf{x}}(s) = (s\mathbf{I} - \mathbf{A})^{-1}[\mathbf{x}(0) + \mathbf{B}\hat{\mathbf{u}}(s)]$$

$$\hat{\mathbf{y}}(s) = \mathbf{C}(s\mathbf{I} - \mathbf{A})^{-1}[\mathbf{x}(0) + \mathbf{B}\hat{\mathbf{u}}(s)] + \mathbf{D}\hat{\mathbf{u}}(s)$$

Once $\hat{\mathbf{x}}(s)$ and $\hat{\mathbf{y}}(s)$ are computed algebraically, their inverse Laplace transforms yield the time-domain solutions.

We now give some remarks concerning the computation of $e^{\mathbf{A}t}$. We discussed in Section 3.6 three methods of computing functions of a matrix. They can all be used to compute $e^{\mathbf{A}t}$:

1. Using Theorem 3.5: First, compute the eigenvalues of $\mathbf{A}$; next, find a polynomial $h(\lambda)$ of degree $n - 1$ that equals $e^{\lambda t}$ on the spectrum of $\mathbf{A}$; then $e^{\mathbf{A}t} = h(\mathbf{A})$.

2. Using Jordan form of $\mathbf{A}$: Let $\mathbf{A} = \mathbf{Q}\hat{\mathbf{A}}\mathbf{Q}^{-1}$; then $e^{\mathbf{A}t} = \mathbf{Q}e^{\hat{\mathbf{A}}t}\mathbf{Q}^{-1}$, where $\hat{\mathbf{A}}$ is in Jordan form and $e^{\hat{\mathbf{A}}t}$ can readily be obtained by using (3.48).

3. Using the infinite power series in (3.51): Although the series will not, in general, yield a closed-form solution, it is suitable for computer computation, as discussed following (3.51).

In addition, we can use (3.58) to compute $e^{\mathbf{A}t}$, that is,

$$e^{\mathbf{A}t} = \mathcal{L}^{-1}(s\mathbf{I} - \mathbf{A})^{-1} \tag{4.8}$$

The inverse of $(s\mathbf{I} - \mathbf{A})$ is a function of $\mathbf{A}$; therefore, again, we have many methods to compute it:

1. Taking the inverse of $(s\mathbf{I} - \mathbf{A})$.
2. Using Theorem 3.5.
3. Using $(s\mathbf{I} - \mathbf{A})^{-1} = \mathbf{Q}(s\mathbf{I} - \hat{\mathbf{A}})^{-1}\mathbf{Q}^{-1}$ and (3.49).
4. Using the infinite power series in (3.57).
5. Using the Leverrier algorithm discussed in Problem 3.26.

EXAMPLE 4.1 We use Methods 1 and 2 to compute $(s\mathbf{I} - \mathbf{A})^{-1}$, where

$$\mathbf{A} = \begin{bmatrix} 0 & -1 \\ 1 & -2 \end{bmatrix}$$

Method 1: We use (3.20) to compute

$$(s\mathbf{I} - \mathbf{A})^{-1} = \begin{bmatrix} s & 1 \\ -1 & s+2 \end{bmatrix}^{-1} = \frac{1}{s^2 + 2s + 1}\begin{bmatrix} s+2 & -1 \\ 1 & s \end{bmatrix}$$

$$= \begin{bmatrix} (s+2)/(s+1)^2 & -1/(s+1)^2 \\ 1/(s+1)^2 & s/(s+)^2 \end{bmatrix}$$

Method 2: The eigenvalues of $\mathbf{A}$ are $-1, -1$. Let $h(\lambda) = \beta_0 + \beta_1\lambda$. If $h(\lambda)$ equals $f(\lambda) := (s - \lambda)^{-1}$ on the spectrum of $\mathbf{A}$, then

$$f(-1) = h(-1): \quad (s+1)^{-1} = \beta_0 - \beta_1$$
$$f'(-1) = h'(-1): \quad (s+1)^{-2} = \beta_1$$

Thus we have

$$h(\lambda) = [(s+1)^{-1} + (s+1)^{-2}] + (s+1)^{-2}\lambda$$

and

$$(s\mathbf{I} - \mathbf{A})^{-1} = h(\mathbf{A}) = [(s+1)^{-1} + (s+1)^{-2}]\mathbf{I} + (s+1)^{-2}\mathbf{A}$$

$$= \begin{bmatrix} (s+2)/(s+1)^2 & -1/(s+1)^2 \\ 1/(s+1)^2 & s/(s+)^2 \end{bmatrix}$$

EXAMPLE 4.2 Consider the equation

$$\dot{\mathbf{x}}(t) = \begin{bmatrix} 0 & -1 \\ 1 & -2 \end{bmatrix}\mathbf{x}(t) + \begin{bmatrix} 0 \\ 1 \end{bmatrix}u(t)$$

Its solution is

$$\mathbf{x}(t) = e^{\mathbf{A}t}\mathbf{x}(0) + \int_0^t e^{\mathbf{A}(t-\tau)}\mathbf{B}u(\tau)\,d\tau$$

The matrix function $e^{\mathbf{A}t}$ is the inverse Laplace transform of $(s\mathbf{I} - \mathbf{A})^{-1}$, which was computed in the preceding example. Thus we have

$$e^{\mathbf{A}t} = \mathcal{L}^{-1} \begin{bmatrix} \dfrac{s+2}{(s+1)^2} & \dfrac{-1}{(s+1)^2} \\ \dfrac{1}{(s+1)^2} & \dfrac{s}{(s+2)^2} \end{bmatrix} = \begin{bmatrix} (1+t)e^{-t} & -te^{-t} \\ te^{-t} & (1-t)e^{-t} \end{bmatrix}$$

and

$$\mathbf{x}(t) = \begin{bmatrix} (1+t)e^{-t} & -te^{-t} \\ te^{-t} & (1-t)e^{-t} \end{bmatrix} \mathbf{x}(0) + \begin{bmatrix} -\int_0^t (t-\tau)e^{-(t-\tau)}u(\tau)\,d\tau \\ \int_0^t [1-(t-\tau)]e^{-(t-\tau)}u(\tau)\,d\tau \end{bmatrix}$$

We discuss a general property of the zero-input response $e^{\mathbf{A}t}\mathbf{x}(0)$. Consider the second matrix in (3.39). Then we have

$$e^{\mathbf{A}t} = \mathbf{Q} \begin{bmatrix} e^{\lambda_1 t} & te^{\lambda_1 t} & t^2 e^{\lambda_1 t}/2 & 0 & 0 \\ 0 & e^{\lambda_1 t} & te^{\lambda_1 t} & 0 & 0 \\ 0 & 0 & e^{\lambda_1 t} & 0 & 0 \\ 0 & 0 & 0 & e^{\lambda_1 t} & 0 \\ 0 & 0 & 0 & 0 & e^{\lambda_2 t} \end{bmatrix} \mathbf{Q}^{-1}$$

Every entry of $e^{\mathbf{A}t}$ and, consequently, of the zero-input response is a linear combination of terms $\{e^{\lambda_1 t}, \ te^{\lambda_1 t}, \ t^2 e^{\lambda_1 t}, \ e^{\lambda_2 t}\}$. These terms are dictated by the eigenvalues and their indices. In general, if $\mathbf{A}$ has eigenvalue λ_1 with index $\bar{n}_1$, then every entry of $e^{\mathbf{A}t}$ is a linear combination of

$$e^{\lambda_1 t} \quad te^{\lambda_1 t} \quad \cdots \quad t^{\bar{n}_1 - 1} e^{\lambda_1 t}$$

Every such term is *analytic* in the sense that it is infinitely differentiable and can be expanded in a Taylor series at every t. This is a nice property and will be used in Chapter 6.

If every eigenvalue, simple or repeated, of $\mathbf{A}$ has a negative real part, then every zero-input response will approach zero as $t \to \infty$. If $\mathbf{A}$ has an eigenvalue, simple or repeated, with a positive real part, then most zero-input responses will grow unbounded as $t \to \infty$. If $\mathbf{A}$ has some eigenvalues with zero real part and all with index 1 and the remaining eigenvalues all have negative real parts, then no zero-input response will grow unbounded. However, if the index is 2 or higher, then some zero-input response may become unbounded. For example, if $\mathbf{A}$ has eigenvalue 0 with index 2, then $e^{\mathbf{A}t}$ contains the terms $\{1, \ t\}$. If a zero-input response contains the term t, then it will grow unbounded.

4.2.1 Discretization

Consider the continuous-time state equation

$$\dot{\mathbf{x}}(t) = \mathbf{A}\mathbf{x}(t) + \mathbf{B}u(t) \tag{4.9}$$

$$\mathbf{y}(t) = \mathbf{C}\mathbf{x}(t) + \mathbf{D}u(t) \tag{4.10}$$

If the set of equations is to be computed on a digital computer, it must be discretized. Because

$$\dot{\mathbf{x}}(t) = \lim_{T \to 0} \frac{\mathbf{x}(t+T) - \mathbf{x}(t)}{T}$$

we can approximate (4.9) as

$$\mathbf{x}(t+T) = \mathbf{x}(t) + \mathbf{A}\mathbf{x}(t)T + \mathbf{B}\mathbf{u}(t)T \tag{4.11}$$

If we compute $\mathbf{x}(t)$ and $\mathbf{y}(t)$ only at $t = kT$ for $k = 0, 1, \ldots$, then (4.11) and (4.10) become

$$\mathbf{x}((k+1)T) = (\mathbf{I} + T\mathbf{A})\mathbf{x}(kT) + T\mathbf{B}\mathbf{u}(kT)$$

$$\mathbf{y}(kT) = \mathbf{C}\mathbf{x}(kT) + \mathbf{D}\mathbf{u}(kT)$$

This is a discrete-time state-space equation and can easily be computed on a digital computer. This discretization is the easiest to carry out but yields the least accurate results for the same T. We discuss next a different discretization.

If an input $\mathbf{u}(t)$ is generated by a digital computer followed by a digital-to-analog converter, then $\mathbf{u}(t)$ will be piecewise constant. This situation often arises in computer control of control systems. Let

$$\mathbf{u}(t) = \mathbf{u}(kT) =: \mathbf{u}[k] \qquad \text{for } kT \le t < (k+1)T \tag{4.12}$$

for $k = 0, 1, 2, \ldots$. This input changes values only at discrete-time instants. For this input, the solution of (4.9) still equals (4.5). Computing (4.5) at $t = kT$ and $t = (k+1)T$ yields

$$\mathbf{x}[k] := \mathbf{x}(kT) = e^{\mathbf{A}kT}\mathbf{x}(0) + \int_0^{kT} e^{\mathbf{A}(kT-\tau)}\mathbf{B}\mathbf{u}(\tau)\,d\tau \tag{4.13}$$

and

$$\mathbf{x}[k+1] := \mathbf{x}((k+1)T) = e^{\mathbf{A}(k+1)T}\mathbf{x}(0) + \int_0^{(k+1)T} e^{\mathbf{A}((k+1)T-\tau)}\mathbf{B}\mathbf{u}(\tau)\,d\tau \tag{4.14}$$

Equation (4.14) can be written as

$$\mathbf{x}[k+1] = e^{\mathbf{A}T}\left[e^{\mathbf{A}kT}\mathbf{x}(0) + \int_0^{kT} e^{\mathbf{A}(kT-\tau)}\mathbf{B}\mathbf{u}(\tau)d\tau \right]$$

$$+ \int_{kT}^{(k+1)T} e^{\mathbf{A}(kT+T-\tau)}\mathbf{B}\mathbf{u}(\tau)\,d\tau$$

which becomes, after substituting (4.12) and (4.13) and introducing the new variable $\alpha :=kT + T - \tau$,

$$\mathbf{x}[k+1] = e^{\mathbf{A}T}\mathbf{x}[k] + \left(\int_0^T e^{\mathbf{A}\alpha}\,d\alpha \right)\mathbf{B}\mathbf{u}[k]$$

Thus, if an input changes value only at discrete-time instants kT and if we compute only the responses at $t = kT$, then (4.9) and (4.10) become

$$\mathbf{x}[k+1] = \mathbf{A}_d\mathbf{x}[k] + \mathbf{B}_d\mathbf{u}[k] \tag{4.15}$$

$$\mathbf{y}[k] = \mathbf{C}_d\mathbf{x}[k] + \mathbf{D}_d\mathbf{u}[k] \tag{4.16}$$

with

$$\mathbf{A}_d = e^{\mathbf{A}T} \qquad \mathbf{B}_d = \left(\int_0^T e^{\mathbf{A}\tau} d\tau \right) \mathbf{B} \qquad \mathbf{C}_d = \mathbf{C} \qquad \mathbf{D}_d = \mathbf{D} \qquad (4.17)$$

This is a discrete-time state-space equation. Note that there is no approximation involved in this derivation and (4.15) yields the exact solution of (4.9) at $t = kT$ if the input is piecewise constant.

We discuss the computation of $\mathbf{B}_d$. Using (3.51), we have

$$\int_0^T \left(\mathbf{I} + \mathbf{A}\tau + \mathbf{A}^2 \frac{\tau^2}{2!} + \cdots \right) d\tau$$

$$= T\mathbf{I} + \frac{T^2}{2!}\mathbf{A} + \frac{T^3}{3!}\mathbf{A}^2 + \frac{T^4}{4!}\mathbf{A}^3 + \cdots$$

This power series can be computed recursively as in computing (3.51). If $\mathbf{A}$ is nonsingular, then the series can be written as, using (3.51),

$$\mathbf{A}^{-1} \left(T\mathbf{A} + \frac{T^2}{2!}\mathbf{A}^2 + \frac{T^3}{3!}\mathbf{A}^3 + \cdots + \mathbf{I} - \mathbf{I} \right) = \mathbf{A}^{-1}(e^{\mathbf{A}T} - \mathbf{I})$$

Thus we have

$$\mathbf{B}_d = \mathbf{A}^{-1}(\mathbf{A}_d - \mathbf{I})\mathbf{B} \qquad \text{(if } \mathbf{A} \text{ is nonsingular)} \qquad (4.18)$$

Using this formula, we can avoid computing an infinite series.

The MATLAB function [ad,bd]=c2d(a,b,T) transforms the continuous-time state equation in (4.9) into the discrete-time state equation in (4.15).

4.2.2 Solution of Discrete-Time Equations

Consider the discrete-time state-space equation

$$\mathbf{x}[k+1] = \mathbf{A}\mathbf{x}[k] + \mathbf{B}\mathbf{u}[k]$$
$$\mathbf{y}[k] = \mathbf{C}\mathbf{x}[k] + \mathbf{D}\mathbf{u}[k] \qquad (4.19)$$

where the subscript d has been dropped. It is understood that if the equation is obtained from a continuous-time equation, then the four matrices must be computed from (4.17). The two equations in (4.19) are algebraic equations. Once $\mathbf{x}[0]$ and $\mathbf{u}[k]$, $k = 0, 1, \ldots$, are given, the response can be computed recursively from the equations.

The MATLAB function dstep computes unit-step responses of discrete-time state-space equations. It also computes unit-step responses of discrete transfer functions; internally, it first transforms the transfer function into a discrete-time state-space equation by calling tf2ss, which will be discussed later, and then uses dstep. The function dlsim, an acronym for discrete linear simulation, computes responses excited by any input. The function step computes unit-step responses of continuous-time state-space equations. Internally, it first uses the function c2d to transform a continuous-time state equation into a discrete-time equation and then carries out the computation. If the function step is applied to a continuous-time transfer function, then it first uses tf2ss to transform the transfer function into a continuous-time state equation and then discretizes it by using c2d and then uses dstep to compute the response. Similar remarks apply to `lsim`, which computes responses of continuous-time state equations or transfer functions excited by any input.

In order to discuss the general behavior of discrete-time state equations, we will develop a general form of solutions. We compute

$$\mathbf{x}[1] = \mathbf{Ax}[0] + \mathbf{Bu}[0]$$

$$\mathbf{x}[2] = \mathbf{Ax}[1] + \mathbf{Bu}[1] = \mathbf{A}^2\mathbf{x}[0] + \mathbf{ABu}[0] + \mathbf{Bu}[1]$$

Proceeding forward, we can readily obtain, for $k > 0$,

$$\mathbf{x}[k] = \mathbf{A}^k\mathbf{x}[0] + \sum_{m=0}^{k-1} \mathbf{A}^{k-1-m}\mathbf{Bu}[m] \tag{4.20}$$

$$\mathbf{y}[k] = \mathbf{CA}^k\mathbf{x}[0] + \sum_{m=0}^{k-1} \mathbf{CA}^{k-1-m}\mathbf{Bu}[m] + \mathbf{Du}[k] \tag{4.21}$$

They are the discrete counterparts of (4.5) and (4.7). Their derivations are considerably simpler than the continuous-time case.

We discuss a general property of the zero-input response $\mathbf{A}^k\mathbf{x}[0]$. Suppose $\mathbf{A}$ has eigenvalue λ_1 with multiplicity 4 and eigenvalue λ_2 with multiplicity 1 and suppose its Jordan form is as shown in the second matrix in (3.39). In other words, λ_1 has index 3 and λ_2 has index 1. Then we have

$$\mathbf{A}^k = \mathbf{Q} \begin{bmatrix} \lambda_1^k & k\lambda_1^{k-1} & k(k-1)\lambda_1^{k-2}/2 & 0 & 0 \\ 0 & \lambda_1^k & k\lambda_1^{k-1} & 0 & 0 \\ 0 & 0 & \lambda_1^k & 0 & 0 \\ 0 & 0 & 0 & \lambda_1^k & 0 \\ 0 & 0 & 0 & 0 & \lambda_2^k \end{bmatrix} \mathbf{Q}^{-1}$$

which implies that every entry of the zero-input response is a linear combination of $\{\lambda_1^k, k\lambda_1^{k-1}, k^2\lambda_1^{k-2}, \lambda_2^k\}$. These terms are dictated by the eigenvalues and their indices.

If every eigenvalue, simple or repeated, of $\mathbf{A}$ has magnitude less than 1, then every zero-input response will approach zero as $k \to \infty$. If $\mathbf{A}$ has an eigenvalue, simple or repeated, with magnitude larger than 1, then most zero-input responses will grow unbounded as $k \to \infty$. If $\mathbf{A}$ has some eigenvalues with magnitude 1 and all with index 1 and the remaining eigenvalues all have magnitudes less than 1, then no zero-input response will grow unbounded. However, if the index is 2 or higher, then some zero-state response may become unbounded. For example, if $\mathbf{A}$ has eigenvalue 1 with index 2, then $\mathbf{A}^k$ contains the terms $\{1, k\}$. If a zero-input response contains the term k, then it will grow unbounded as $k \to \infty$.

4.3 Equivalent State Equations

The example that follows provides a motivation for studying equivalent state equations.

EXAMPLE 4.3 Consider the network shown in Fig. 4.1. It consists of one capacitor, one inductor, one resistor, and one voltage source. First we select the inductor current x_1 and capacitor voltage x_2 as state variables as shown. The voltage across the inductor is $\dot{x}_1$ and the current through the capacitor is $\dot{x}_2$. The voltage across the resistor is x_2; thus its current is $x_2/1 = x_2$. Clearly we have $x_1 = x_2 + \dot{x}_2$ and $\dot{x}_1 + x_2 - u = 0$. Thus the network is described by the following state equation:

$$\begin{bmatrix} \dot{x}_1 \\ \dot{x}_2 \end{bmatrix} = \begin{bmatrix} 0 & -1 \\ 1 & -1 \end{bmatrix} \begin{bmatrix} x_1 \\ x_2 \end{bmatrix} + \begin{bmatrix} 1 \\ 0 \end{bmatrix} u$$

$$y = [0 \quad 1]\mathbf{x} \tag{4.22}$$

If, instead, the loop currents $\bar{x}_1$ and $\bar{x}_2$ are chosen as state variables as shown, then the voltage across the inductor is $\dot{\bar{x}}_1$ and the voltage across the resistor is $(\bar{x}_1 - \bar{x}_2) \cdot 1$. From the left-hand-side loop, we have

$$u = \dot{\bar{x}}_1 + \bar{x}_1 - \bar{x}_2 \quad \text{or} \quad \dot{\bar{x}}_1 = -\bar{x}_1 + \bar{x}_2 + u$$

The voltage across the capacitor is the same as the one across the resistor, which is $\bar{x}_1 - \bar{x}_2$. Thus the current through the capacitor is $\dot{\bar{x}}_1 - \dot{\bar{x}}_2$, which equals $\bar{x}_2$ or

$$\dot{\bar{x}}_2 = \dot{\bar{x}}_1 - \dot{\bar{x}}_2 = -\bar{x}_1 + u$$

Thus the network is also described by the state equation

$$\begin{bmatrix} \dot{\bar{x}}_1 \\ \dot{\bar{x}}_2 \end{bmatrix} = \begin{bmatrix} -1 & 1 \\ -1 & 0 \end{bmatrix} \begin{bmatrix} \bar{x}_1 \\ \bar{x}_2 \end{bmatrix} + \begin{bmatrix} 1 \\ 1 \end{bmatrix} u$$

$$y = [1 \quad -1]\bar{\mathbf{x}} \tag{4.23}$$

The state equations in (4.22) and (4.23) describe the same network; therefore they must be closely related. In fact, they are equivalent as will be established shortly.

Consider the n-dimensional state equation

$$\dot{\mathbf{x}}(t) = \mathbf{A}\mathbf{x}(t) + \mathbf{B}u(t)$$

$$\mathbf{y}(t) = \mathbf{C}\mathbf{x}(t) + \mathbf{D}u(t) \tag{4.24}$$

where $\mathbf{A}$ is an $n \times n$ constant matrix mapping an n-dimensional real space $\mathcal{R}^n$ into itself. The state $\mathbf{x}$ is a vector in $\mathcal{R}^n$ for all t; thus the real space is also called the state space. The state equation in (4.24) can be considered to be associated with the orthonormal basis in (3.8). Now we study the effect on the equation by choosing a different basis.

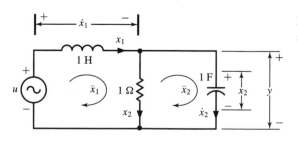

Figure 4.1   Network with two different sets of state variables.

> **Definition 4.1** Let **P** *be an* $n \times n$ *real nonsingular matrix and let* $\bar{\mathbf{x}} = \mathbf{Px}$. *Then the state equation,*
>
> $$\dot{\bar{\mathbf{x}}}(t) = \bar{\mathbf{A}}\bar{\mathbf{x}}(t) + \bar{\mathbf{B}}\mathbf{u}(t)$$
>
> $$\mathbf{y}(t) = \bar{\mathbf{C}}\bar{\mathbf{x}}(t) + \bar{\mathbf{D}}\mathbf{u}(t) \tag{4.25}$$
>
> *where*
>
> $$\bar{\mathbf{A}} = \mathbf{PAP}^{-1} \qquad \bar{\mathbf{B}} = \mathbf{PB} \qquad \bar{\mathbf{C}} = \mathbf{CP}^{-1} \qquad \bar{\mathbf{D}} = \mathbf{D} \tag{4.26}$$
>
> *is said to be* (algebraically) *equivalent to (4.24) and* $\bar{\mathbf{x}} = \mathbf{Px}$ *is called an* equivalence transformation.

Equation (4.26) is obtained from (4.24) by substituting $\mathbf{x}(t) = \mathbf{P}^{-1}\bar{\mathbf{x}}(t)$ and $\dot{\mathbf{x}}(t) = \mathbf{P}^{-1}\dot{\bar{\mathbf{x}}}(t)$. In this substitution, we have changed, as in Equation (3.7), the basis vectors of the state space from the orthonormal basis to the columns of $\mathbf{P}^{-1} =: \mathbf{Q}$. Clearly $\mathbf{A}$ and $\bar{\mathbf{A}}$ are similar and $\bar{\mathbf{A}}$ is simply a different representation of $\mathbf{A}$. To be precise, let $\mathbf{Q} = \mathbf{P}^{-1} = [\mathbf{q}_1 \ \mathbf{q}_2 \ \cdots \ \mathbf{q}_n]$. Then the ith column of $\bar{\mathbf{A}}$ is, as discussed in Section 3.4, the representation of $\mathbf{Aq}_i$ with respect to the basis $\{\mathbf{q}_1, \ \mathbf{q}_2, \ \cdots, \ \mathbf{q}_n\}$. From the equation $\bar{\mathbf{B}} = \mathbf{PB}$ or $\mathbf{B} = \mathbf{P}^{-1}\bar{\mathbf{B}} = [\mathbf{q}_1 \ \mathbf{q}_2 \ \cdots \ \mathbf{q}_n]\bar{\mathbf{B}}$, we see that the ith column of $\bar{\mathbf{B}}$ is the representation of the ith column of $\mathbf{B}$ with respect to the basis $\{\mathbf{q}_1, \ \mathbf{q}_2, \ \cdots, \ \mathbf{q}_n\}$. The matrix $\bar{\mathbf{C}}$ is to be computed from $\mathbf{CP}^{-1}$. The matrix $\mathbf{D}$, called the *direct transmission part* between the input and output, has nothing to do with the state space and is not affected by the equivalence transformation.

We show that (4.24) and (4.25) have the same set of eigenvalues and the same transfer matrix. Indeed, we have, using $\det(\mathbf{P})\det(\mathbf{P}^{-1}) = 1$,

$$\bar{\Delta}(\lambda) = \det(\lambda\mathbf{I} - \bar{\mathbf{A}}) = \det(\lambda\mathbf{PP}^{-1} - \mathbf{PAP}^{-1}) = \det[\mathbf{P}(\lambda\mathbf{I} - \mathbf{A})\mathbf{P}^{-1}]$$

$$= \det(\mathbf{P})\det(\lambda\mathbf{I} - \mathbf{A})\det(\mathbf{P}^{-1}) = \det(\lambda\mathbf{I} - \mathbf{A}) = \Delta(\lambda)$$

and

$$\hat{\bar{\mathbf{G}}}(s) = \bar{\mathbf{C}}(s\mathbf{I} - \bar{\mathbf{A}})^{-1}\bar{\mathbf{B}} + \bar{\mathbf{D}} = \mathbf{CP}^{-1}[\mathbf{P}(s\mathbf{I} - \mathbf{A})\mathbf{P}^{-1}]^{-1}\mathbf{PB} + \mathbf{D}$$

$$= \mathbf{CP}^{-1}\mathbf{P}(s\mathbf{I} - \mathbf{A})^{-1}\mathbf{P}^{-1}\mathbf{PB} + \mathbf{D} = \mathbf{C}(s\mathbf{I} - \mathbf{A})^{-1}\mathbf{B} + \mathbf{D} = \hat{\mathbf{G}}(s)$$

Thus equivalent state equations have the same characteristic polynomial and, consequently, the same set of eigenvalues and same transfer matrix. In fact, all properties of (4.24) are preserved or invariant under any equivalence transformation.

Consider again the network shown in Fig. 4.1, which can be described by (4.22) and (4.23). We show that the two equations are equivalent. From Fig. 4.1, we have $x_1 = \bar{x}_1$. Because the voltage across the resistor is x_2, its current is $x_2/1$ and equals $\bar{x}_1 - \bar{x}_2$. Thus we have

$$\begin{bmatrix} x_1 \\ x_2 \end{bmatrix} = \begin{bmatrix} 1 & 0 \\ 1 & -1 \end{bmatrix} \begin{bmatrix} \bar{x}_1 \\ \bar{x}_2 \end{bmatrix}$$

or

$$\begin{bmatrix} \bar{x}_1 \\ \bar{x}_2 \end{bmatrix} = \begin{bmatrix} 1 & 0 \\ 1 & -1 \end{bmatrix}^{-1} \begin{bmatrix} x_1 \\ x_2 \end{bmatrix} = \begin{bmatrix} 1 & 0 \\ 1 & -1 \end{bmatrix} \begin{bmatrix} x_1 \\ x_2 \end{bmatrix} \tag{4.27}$$

Note that, for this **P**, its inverse happens to equal itself. It is straightforward to verify that (4.22) and (4.23) are related by the equivalence transformation in (4.26).

The MATLAB function `[ab,bb,cb,db]=ss2ss(a,b,c,d,p)` carries out equivalence transformations.

Two state equations are said to be *zero-state equivalent* if they have the same transfer matrix or

$$\mathbf{D} + \mathbf{C}(s\mathbf{I} - \mathbf{A})^{-1}\mathbf{B} = \bar{\mathbf{D}} + \bar{\mathbf{C}}(s\mathbf{I} - \bar{\mathbf{A}})^{-1}\bar{\mathbf{B}}$$

This becomes, after substituting (3.57),

$$\mathbf{D} + \mathbf{C}\mathbf{B}s^{-1} + \mathbf{C}\mathbf{A}\mathbf{B}s^{-2} + \mathbf{C}\mathbf{A}^2\mathbf{B}s^{-3} + \cdots = \bar{\mathbf{D}} + \bar{\mathbf{C}}\bar{\mathbf{B}}s^{-1} + \bar{\mathbf{C}}\bar{\mathbf{A}}\bar{\mathbf{B}}s^{-2} + \bar{\mathbf{C}}\bar{\mathbf{A}}^2\bar{\mathbf{B}}s^{-3} + \cdots$$

Thus we have the theorem that follows.

▶ **Theorem 4.1**

Two linear time-invariant state equations {**A**, **B**, **C**, **D**} and {**Ā**, **B̄**, **C̄**, **D̄**} are zero-state equivalent or have the same transfer matrix if and only if $\mathbf{D} = \bar{\mathbf{D}}$ and

$$\mathbf{C}\mathbf{A}^m\mathbf{B} = \bar{\mathbf{C}}\bar{\mathbf{A}}^m\bar{\mathbf{B}} \qquad m = 0, 1, 2, \ldots$$

It is clear that (algebraic) equivalence implies zero-state equivalence. In order for two state equations to be equivalent, they must have the same dimension. This is, however, not the case for zero-state equivalence, as the next example shows.

EXAMPLE 4.4 Consider the two networks shown in Fig. 4.2. The capacitor is assumed to have capacitance -1 F. Such a negative capacitance can be realized using an op-amp circuit. For the circuit in Fig. 4.2(a), we have $y(t) = 0.5 \cdot u(t)$ or $\hat{y}(s) = 0.5\hat{u}(s)$. Thus its transfer function is 0.5. To compute the transfer function of the network in Fig. 4.2(b), we may assume the initial voltage across the capacitor to be zero. Because of the symmetry of the four resistors, half of the current will go through each resistor or $i(t) = 0.5u(t)$, where $i(t)$ denotes the right upper resistor's current. Consequently, $y(t) = i(t) \cdot 1 = 0.5u(t)$ and the transfer function also equals 0.5. Thus the two networks, or more precisely their state equations, are zero-state equivalent. This fact can also be verified by using Theorem 4.1. The network in Fig. 4.2(a) is described by the zero-dimensional state equation $y(t) = 0.5u(t)$ or $\mathbf{A} = \mathbf{B} = \mathbf{C} = 0$ and $\mathbf{D} = 0.5$. To

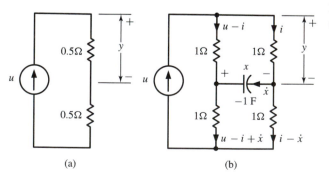

(a)   (b)

Figure 4.2 Two zero-state equivalent networks.

develop a state equation for the network in Fig. 4.2(b), we assign the capacitor voltage as state variable x with polarity shown. Its current is $\dot{x}$ flowing from the negative to positive polarity because of the negative capacitance. If we assign the right upper resistor's current as $i(t)$, then the right lower resistor's current is $i - \dot{x}$, the left upper resistor's current is $u - i$, and the left lower resistor's current is $u - i + \dot{x}$. The total voltage around the upper right-hand loop is 0:

$$i - x - (u - i) = 0 \quad \text{or} \quad i = 0.5(x + u)$$

which implies

$$y = 1 \cdot i = i = 0.5(x + u)$$

The total voltage around the lower right-hand loop is 0:

$$x + (i - \dot{x}) - (u - i + \dot{x}) = 0$$

which implies

$$2\dot{x} = 2i + x - u = x + u + x - u = 2x$$

Thus the network in Fig. 4.2(b) is described by the one-dimensional state equation

$$\dot{x}(t) = x(t)$$
$$y(t) = 0.5x(t) + 0.5u(t)$$

with $\bar{\mathbf{A}} = 1$, $\bar{\mathbf{B}} = 0$, $\bar{\mathbf{C}} = 0.5$, and $\bar{\mathbf{D}} = 0.5$. We see that $\mathbf{D} = \bar{\mathbf{D}} = 0.5$ and $\mathbf{C}\mathbf{A}^m\mathbf{B} = \bar{\mathbf{C}}\bar{\mathbf{A}}^m\bar{\mathbf{B}} = 0$ for $m = 0, 1, \ldots$. Thus the two equations are zero-state equivalent.

4.3.1 Canonical Forms

MATLAB contains the function [ab,bb,cb,db,P]=canon(a,b,c,d, 'type'). If type=companion, the function will generate an equivalent state equation with $\bar{\mathbf{A}}$ in the companion form in (3.24). This function works only if $\mathbf{Q} := [\mathbf{b}_1 \ \mathbf{Ab}_1 \ \cdots \ \mathbf{A}^{n-1}\mathbf{b}_1]$ is nonsingular, where $\mathbf{b}_1$ is the first column of $\mathbf{B}$. This condition is the same as $\{\mathbf{A}, \mathbf{b}_1\}$ controllable, as we will discuss in Chapter 6. The $\mathbf{P}$ that the function canon generates equals $\mathbf{Q}^{-1}$. See the discussion in Section 3.4.

We discuss a different canonical form. Suppose $\mathbf{A}$ has two real eigenvalues and two complex eigenvalues. Because $\mathbf{A}$ has only real coefficients, the two complex eigenvalues must be complex conjugate. Let $\lambda_1, \lambda_2, \alpha + j\beta$, and $\alpha - j\beta$ be the eigenvalues and $\mathbf{q}_1, \mathbf{q}_2, \mathbf{q}_3$, and $\mathbf{q}_4$ be the corresponding eigenvectors, where $\lambda_1, \lambda_2, \alpha, \beta, \mathbf{q}_1$, and $\mathbf{q}_2$ are all real and $\mathbf{q}_4$ equals the complex conjugate of $\mathbf{q}_3$. Define $\mathbf{Q} = [\mathbf{q}_1 \ \mathbf{q}_2 \ \mathbf{q}_3 \ \mathbf{q}_4]$. Then we have

$$\mathbf{J} := \begin{bmatrix} \lambda_1 & 0 & 0 & 0 \\ 0 & \lambda_2 & 0 & 0 \\ 0 & 0 & \alpha + j\beta & 0 \\ 0 & 0 & 0 & \alpha - j\beta \end{bmatrix} = \mathbf{Q}^{-1}\mathbf{A}\mathbf{Q}$$

Note that $\mathbf{Q}$ and $\mathbf{J}$ can be obtained from [q,j]=eig(a) in MATLAB as shown in Examples 3.5 and 3.6. This form is useless in practice but can be transformed into a real matrix by the following similarity transformation

$$\bar{\mathbf{Q}}^{-1}\mathbf{J}\bar{\mathbf{Q}} := \begin{bmatrix} 1 & 0 & 0 & 0 \\ 0 & 1 & 0 & 0 \\ 0 & 0 & 1 & 1 \\ 0 & 0 & j & -j \end{bmatrix} \begin{bmatrix} \lambda_1 & 0 & 0 & 0 \\ 0 & \lambda_2 & 0 & 0 \\ 0 & 0 & \alpha+j\beta & 0 \\ 0 & 0 & 0 & \alpha-j\beta \end{bmatrix}$$

$$\cdot \begin{bmatrix} 1 & 0 & 0 & 0 \\ 0 & 1 & 0 & 0 \\ 0 & 0 & 0.5 & -0.5j \\ 0 & 0 & 0.5 & 0.5j \end{bmatrix} = \begin{bmatrix} \lambda_1 & 0 & 0 & 0 \\ 0 & \lambda_2 & 0 & 0 \\ 0 & 0 & \alpha & \beta \\ 0 & 0 & -\beta & \alpha \end{bmatrix} =: \bar{\mathbf{A}}$$

We see that this transformation transforms the complex eigenvalues on the diagonal into a block with the real part of the eigenvalues on the diagonal and the imaginary part on the off-diagonal. This new A-matrix is said to be in *modal* form. The MATLAB function `[ab,bb,cb,db,P]=canon(a,b,c,d, 'modal')` or `canon(a,b,c,d)` with no type specified will yield an equivalent state equation with $\bar{\mathbf{A}}$ in modal form. Note that there is no need to transform **A** into a diagonal form and then to a modal form. The two transformations can be combined into one as

$$\mathbf{P}^{-1} = \mathbf{Q}\bar{\mathbf{Q}} = [\mathbf{q}_1 \ \mathbf{q}_2 \ \mathbf{q}_3 \ \mathbf{q}_4] \begin{bmatrix} 1 & 0 & 0 & 0 \\ 0 & 1 & 0 & 0 \\ 0 & 0 & 0.5 & -0.5j \\ 0 & 0 & 0.5 & 0.5j \end{bmatrix}$$

$$= [\mathbf{q}_1 \ \mathbf{q}_2 \ \mathrm{Re}(\mathbf{q}_3) \ \mathrm{Im}(\mathbf{q}_3)]$$

where Re and Im stand, respectively, for the real part and imaginary part and we have used in the last equality the fact that $\mathbf{q}_4$ is the complex conjugate of $\mathbf{q}_3$. We give one more example. The modal form of a matrix with real eigenvalue λ_1 and two pairs of distinct complex conjugate eigenvalues $\alpha_i \pm j\beta_i$, for $i = 1, 2$, is

$$\bar{\mathbf{A}} = \begin{bmatrix} \lambda_1 & 0 & 0 & 0 & 0 \\ 0 & \alpha_1 & \beta_1 & 0 & 0 \\ 0 & -\beta_1 & \alpha_1 & 0 & 0 \\ 0 & 0 & 0 & \alpha_2 & \beta_2 \\ 0 & 0 & 0 & -\beta_2 & \alpha_2 \end{bmatrix} \tag{4.28}$$

It is block diagonal and can be obtained by the similarity transformation

$$\mathbf{P}^{-1} = [\mathbf{q}_1 \ \mathrm{Re}(\mathbf{q}_2) \ \mathrm{Im}(\mathbf{q}_2) \ \mathrm{Re}(\mathbf{q}_4) \ \mathrm{Im}(\mathbf{q}_4)]$$

where $\mathbf{q}_1, \mathbf{q}_2$, and $\mathbf{q}_4$ are, respectively, eigenvectors associated with $\lambda_1, \alpha_1 + j\beta_1$, and $\alpha_2 + j\beta_2$. This form is useful in state-space design.

4.3.2 Magnitude Scaling in Op-Amp Circuits

As discussed in Section 2.3.1, every LTI state equation can be implemented using an op-amp circuit.[1] In actual op-amp circuits, all signals are limited by power supplies. If we use ±15-volt

---

1. This subsection may be skipped without loss of continuity.

power supplies, then all signals are roughly limited to ±13 volts. If any signal goes outside the range, the circuit will saturate and will not behave as the state equation dictates. Therefore saturation is an important issue in actual op-amp circuit implementation.

Consider an LTI state equation and suppose all signals must be limited to ±M. For linear systems, if the input magnitude increases by α, so do the magnitudes of all state variables and the output. Thus there must be a limit on input magnitude. Clearly it is desirable to have the admissible input magnitude as large as possible. One way to achieve this is to use an equivalence transformation so that

$$|x_i(t)| \leq |y(t)| \leq M$$

for all i and for all t. The equivalence transformation, however, will not alter the relationship between the input and output; therefore we can use the original state equation to find the input range to achieve $|y(t)| \leq M$. In addition, we can use the same transformation to amplify some state variables to increase visibility or accuracy. This is illustrated in the next example.

EXAMPLE 4.5 Consider the state equation

$$\dot{\mathbf{x}} = \begin{bmatrix} -0.1 & 2 \\ 0 & -1 \end{bmatrix} \mathbf{x} + \begin{bmatrix} 10 \\ 0.1 \end{bmatrix} u$$

$$y = [0.1 \quad -1]\mathbf{x}$$

Suppose the input is a step function of various magnitude and the equation is to be implemented using an op-amp circuit in which all signals must be limited to ±10. First we use MATLAB to find its unit-step response. We type

```
a=[-0.1 2;0 -1];b=[10;0.1];c=[0.2 -1];d=0;
[y,x,t]=step(a,b,c,d);
plot(t,y,t,x)
```

which yields the plot in Fig. 4.3(a). We see that $|x_1|_{max} = 100 > |y|_{max} = 20$ and $|x_2| << |y|_{max}$. The state variable x_2 is hardly visible and its largest magnitude is found to be 0.1 by plotting it separately (not shown). From the plot, we see that if $|u(t)| \leq 0.5$, then the output will not saturate but $x_1(t)$ will.

Let us introduce new state variables as

$$\bar{x}_1 = \frac{20}{100}x_1 = 0.2x_1 \qquad \bar{x}_2 = \frac{20}{0.1}x_2 = 200x_2$$

With this transformation, the maximum magnitudes of $\bar{x}_1(t)$ and $\bar{x}_2(t)$ will equal $|y|_{max}$. Thus if $y(t)$ does not saturate, neither will all the state variables $\bar{x}_i$. The transformation can be expressed as $\bar{\mathbf{x}} = \mathbf{P}\mathbf{x}$ with

$$\mathbf{P} = \begin{bmatrix} 0.2 & 0 \\ 0 & 200 \end{bmatrix} \qquad \mathbf{P}^{-1} = \begin{bmatrix} 5 & 0 \\ 0 & 0.005 \end{bmatrix}$$

Then its equivalent state equation can readily be computed from (4.26) as

$$\dot{\bar{\mathbf{x}}} = \begin{bmatrix} -0.1 & 0.002 \\ 0 & -1 \end{bmatrix} \bar{\mathbf{x}} + \begin{bmatrix} 2 \\ 20 \end{bmatrix} u$$

$$y = [1 \quad -0.005]\bar{\mathbf{x}}$$

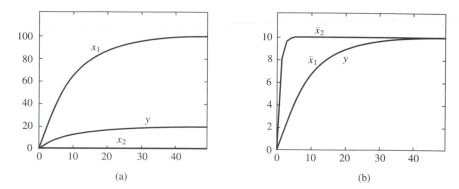

Figure 4.3   Time responses.

Its step responses due to $u(t) = 0.5$ are plotted in Fig. 4.3(b). We see that all signals lie inside the range ± 10 and occupy the full scale. Thus the equivalence state equation is better for op-amp circuit implementation or simulation.

The magnitude scaling is important in using op-amp circuits to implement or simulate continuous-time systems. Although we discuss only step inputs, the idea is applicable to any input. We mention that analog computers are essentially op-amp circuits. Before the advent of digital computers, magnitude scaling in analog computer simulation was carried out by trial and error. With the help of digital computer simulation, the magnitude scaling can now be carried out easily.

4.4 Realizations

Every linear time-invariant (LTI) system can be described by the input–output description

$$\hat{\mathbf{y}}(s) = \hat{\mathbf{G}}(s)\hat{\mathbf{u}}(s)$$

and, if the system is lumped as well, by the state-space equation description

$$\dot{\mathbf{x}}(t) = \mathbf{A}\mathbf{x}(t) + \mathbf{B}\mathbf{u}(t)$$
$$\mathbf{y}(t) = \mathbf{C}\mathbf{x}(t) + \mathbf{D}\mathbf{u}(t)$$

(4.29)

If the state equation is known, the transfer matrix can be computed as $\hat{\mathbf{G}}(s) = \mathbf{C}(s\mathbf{I}-\mathbf{A})^{-1}\mathbf{B}+\mathbf{D}$. The computed transfer matrix is unique. Now we study the converse problem, that is, to find a state-space equation from a given transfer matrix. This is called the *realization* problem. This terminology is justified by the fact that, by using the state equation, we can build an op-amp circuit for the transfer matrix.

A transfer matrix $\hat{\mathbf{G}}(s)$ is said to be *realizable* if there exists a finite-dimensional state equation (4.29) or, simply, $\{\mathbf{A}, \mathbf{B}, \mathbf{C}, \mathbf{D}\}$ such that

$$\hat{\mathbf{G}}(s) = \mathbf{C}(s\mathbf{I} - \mathbf{A})^{-1}\mathbf{B} + \mathbf{D}$$

and $\{\mathbf{A}, \mathbf{B}, \mathbf{C}, \mathbf{D}\}$ is called a *realization* of $\hat{\mathbf{G}}(s)$. An LTI distributed system can be described by a transfer matrix, but not by a finite-dimensional state equation. Thus not every $\hat{\mathbf{G}}(s)$ is realizable. If $\hat{\mathbf{G}}(s)$ is realizable, then it has infinitely many realizations, not necessarily of the same dimension. Thus the realization problem is fairly complex. We study here only the realizability condition. The other issues will be studied in later chapters.

▶ **Theorem 4.2**

A transfer matrix $\hat{\mathbf{G}}(s)$ is realizable if and only if $\hat{\mathbf{G}}(s)$ is a proper rational matrix.

We use (3.19) to write

$$\hat{\mathbf{G}}_{sp}(s) := \mathbf{C}(s\mathbf{I} - \mathbf{A})^{-1}\mathbf{B} = \frac{1}{\det(s\mathbf{I} - \mathbf{A})}\mathbf{C}[\text{Adj } (s\mathbf{I} - \mathbf{A})]\mathbf{B} \qquad (4.30)$$

If $\mathbf{A}$ is $n \times n$, then $\det (s\mathbf{I} - \mathbf{A})$ has degree n. Every entry of Adj $(s\mathbf{I} - \mathbf{A})$ is the determinant of an $(n - 1) \times (n - 1)$ submatrix of $(s\mathbf{I} - \mathbf{A})$; thus it has at most degree $(n - 1)$. Their linear combinations again have at most degree $(n - 1)$. Thus we conclude that $\mathbf{C}(s\mathbf{I} - \mathbf{A})^{-1}\mathbf{B}$ is a strictly proper rational matrix. If $\mathbf{D}$ is a nonzero matrix, then $\mathbf{C}(s\mathbf{I} - \mathbf{A})^{-1}\mathbf{B} + \mathbf{D}$ is proper. This shows that if $\hat{\mathbf{G}}(s)$ is realizable, then it is a proper rational matrix. Note that we have

$$\hat{\mathbf{G}}(\infty) = \mathbf{D}$$

Next we show the converse; that is, if $\hat{\mathbf{G}}(s)$ is a $q \times p$ proper rational matrix, then there exists a realization. First we decompose $\hat{\mathbf{G}}(s)$ as

$$\hat{\mathbf{G}}(s) = \hat{\mathbf{G}}(\infty) + \hat{\mathbf{G}}_{sp}(s) \qquad (4.31)$$

where $\hat{\mathbf{G}}_{sp}$ is the strictly proper part of $\hat{\mathbf{G}}(s)$. Let

$$d(s) = s^r + \alpha_1 s^{r-1} + \cdots + \alpha_{r-1} s + \alpha_r \qquad (4.32)$$

be the least common denominator of all entries of $\hat{\mathbf{G}}_{sp}(s)$. Here we require $d(s)$ to be monic; that is, its leading coefficient is 1. Then $\hat{\mathbf{G}}_{sp}(s)$ can be expressed as

$$\hat{\mathbf{G}}_{sp}(s) = \frac{1}{d(s)}[\mathbf{N}(s)] = \frac{1}{d(s)}\left[\mathbf{N}_1 s^{r-1} + \mathbf{N}_2 s^{r-2} + \cdots + \mathbf{N}_{r-1} s + \mathbf{N}_r\right] \qquad (4.33)$$

where $\mathbf{N}_i$ are $q \times p$ constant matrices. Now we claim that the set of equations

$$\dot{\mathbf{x}} = \begin{bmatrix} -\alpha_1 \mathbf{I}_p & -\alpha_2 \mathbf{I}_p & \cdots & -\alpha_{r-1} \mathbf{I}_p & -\alpha_r \mathbf{I}_p \\ \mathbf{I}_p & 0 & \cdots & 0 & 0 \\ 0 & \mathbf{I}_p & \cdots & 0 & 0 \\ \vdots & \vdots & & \vdots & \vdots \\ 0 & 0 & \cdots & \mathbf{I}_p & 0 \end{bmatrix} \mathbf{x} + \begin{bmatrix} \mathbf{I}_p \\ 0 \\ 0 \\ \vdots \\ 0 \end{bmatrix} \mathbf{u} \qquad (4.34)$$

$$\mathbf{y} = \begin{bmatrix} \mathbf{N}_1 & \mathbf{N}_2 & \cdots \mathbf{N}_{r-1} & \mathbf{N}_r \end{bmatrix} \mathbf{x} + \hat{\mathbf{G}}(\infty)\mathbf{u}$$

is a realization of $\hat{\mathbf{G}}(s)$. The matrix $\mathbf{I}_p$ is the $p \times p$ unit matrix and every $\mathbf{0}$ is a $p \times p$ zero matrix. The A-matrix is said to be in block companion form; it consists of r rows and r columns of $p \times p$ matrices; thus the A-matrix has order $rp \times rp$. The B-matrix has order $rp \times p$. Because the C-matrix consists of r number of $\mathbf{N}_i$, each of order $q \times p$, the C-matrix has order $q \times rp$. The realization has dimension rp and is said to be in *controllable canonical form*.

We show that (4.34) is a realization of $\hat{\mathbf{G}}(s)$ in (4.31) and (4.33). Let us define

$$
\mathbf{Z} := \begin{bmatrix} \mathbf{Z}_1 \\ \mathbf{Z}_2 \\ \vdots \\ \mathbf{Z}_r \end{bmatrix} := (s\mathbf{I} - \mathbf{A})^{-1}\mathbf{B}
\tag{4.35}
$$

where $\mathbf{Z}_i$ is $p \times p$ and $\mathbf{Z}$ is $rp \times p$. Then the transfer matrix of (4.34) equals

$$
\mathbf{C}(s\mathbf{I} - \mathbf{A})^{-1}\mathbf{B} + \hat{\mathbf{G}}(\infty) = \mathbf{N}_1\mathbf{Z}_1 + \mathbf{N}_2\mathbf{Z}_2 + \cdots + \mathbf{N}_r\mathbf{Z}_r + \hat{\mathbf{G}}(\infty)
\tag{4.36}
$$

We write (4.35) as $(s\mathbf{I} - \mathbf{A})\mathbf{Z} = \mathbf{B}$ or

$$
s\mathbf{Z} = \mathbf{A}\mathbf{Z} + \mathbf{B}
\tag{4.37}
$$

Using the shifting property of the companion form of $\mathbf{A}$, from the second to the last block of equations in (4.37), we can readily obtain

$$
s\mathbf{Z}_2 = \mathbf{Z}_1, \quad s\mathbf{Z}_3 = \mathbf{Z}_2, \quad \cdots, \quad s\mathbf{Z}_r = \mathbf{Z}_{r-1}
$$

which implies

$$
\mathbf{Z}_2 = \frac{1}{s}\mathbf{Z}_1, \quad \mathbf{Z}_3 = \frac{1}{s^2}\mathbf{Z}_1, \quad \cdots, \quad \mathbf{Z}_r = \frac{1}{s^{r-1}}\mathbf{Z}_1
$$

Substituting these into the first block of equations in (4.37) yields

$$
s\mathbf{Z}_1 = -\alpha_1\mathbf{Z}_1 - \alpha_2\mathbf{Z}_2 - \cdots - \alpha_r\mathbf{Z}_r + \mathbf{I}_p
$$
$$
= -\left(\alpha_1 + \frac{\alpha_2}{s} + \cdots + \frac{\alpha_r}{s^{r-1}}\right)\mathbf{Z}_1 + \mathbf{I}_p
$$

or, using (4.32),

$$
\left(s + \alpha_1 + \frac{\alpha_2}{s} + \cdots + \frac{\alpha_r}{s^{r-1}}\right)\mathbf{Z}_1 = \frac{d(s)}{s^{r-1}}\mathbf{Z}_1 = \mathbf{I}_p
$$

Thus we have

$$
\mathbf{Z}_1 = \frac{s^{r-1}}{d(s)}\mathbf{I}_p, \quad \mathbf{Z}_2 = \frac{s^{r-2}}{d(s)}\mathbf{I}_p, \quad \cdots, \quad \mathbf{Z}_r = \frac{1}{d(s)}\mathbf{I}_p
$$

Substituting these into (4.36) yields

$$
\mathbf{C}(s\mathbf{I} - \mathbf{A})^{-1}\mathbf{B} + \hat{\mathbf{G}}(\infty) = \frac{1}{d(s)}[\mathbf{N}_1 s^{r-1} + \mathbf{N}_2 s^{r-2} + \cdots + \mathbf{N}_r] + \hat{\mathbf{G}}(\infty)
$$

This equals $\hat{\mathbf{G}}(s)$ in (4.31) and (4.33). This shows that (4.34) is a realization of $\hat{\mathbf{G}}(s)$.

EXAMPLE 4.6 Consider the proper rational matrix

$$\hat{G}(s) = \begin{bmatrix} \dfrac{4s-10}{2s+1} & \dfrac{3}{s+2} \\ \dfrac{1}{(2s+1)(s+2)} & \dfrac{s+1}{(s+2)^2} \end{bmatrix}$$

$$= \begin{bmatrix} 2 & 0 \\ 0 & 0 \end{bmatrix} + \begin{bmatrix} \dfrac{-12}{2s+1} & \dfrac{3}{s+2} \\ \dfrac{1}{(2s+1)(s+2)} & \dfrac{s+1}{(s+2)^2} \end{bmatrix}$$

(4.38)

where we have decomposed $\hat{G}(s)$ into the sum of a constant matrix and a strictly proper rational matrix $\hat{G}_{sp}(s)$. The monic least common denominator of $\hat{G}_{sp}(s)$ is $d(s) = (s+0.5)(s+2)^2 = s^3 + 4.5s^2 + 6s + 2$. Thus we have

$$\hat{G}_{sp}(s) = \frac{1}{s^3+4.5s^2+6s+2}\begin{bmatrix} -6(s+2)^2 & 3(s+2)(s+0.5) \\ 0.5(s+2) & (s+1)(s+0.5) \end{bmatrix}$$

$$= \frac{1}{d(s)}\left(\begin{bmatrix} -6 & 3 \\ 0 & 1 \end{bmatrix}s^2 + \begin{bmatrix} -24 & 7.5 \\ 0.5 & 1.5 \end{bmatrix}s + \begin{bmatrix} -24 & 3 \\ 1 & 0.5 \end{bmatrix}\right)$$

and a realization of (4.38) is

$$\dot{\mathbf{x}} = \begin{bmatrix} -4.5 & 0 & \vdots & -6 & 0 & \vdots & -2 & 0 \\ 0 & -4.5 & \vdots & 0 & -6 & \vdots & 0 & -2 \\ \cdots & \cdots & \cdots & \cdots & \cdots & \cdots & \cdots & \cdots \\ 1 & 0 & \vdots & 0 & 0 & \vdots & 0 & 0 \\ 0 & 1 & \vdots & 0 & 0 & \vdots & 0 & 0 \\ \cdots & \cdots & \cdots & \cdots & \cdots & \cdots & \cdots & \cdots \\ 0 & 0 & \vdots & 1 & 0 & \vdots & 0 & 0 \\ 0 & 0 & \vdots & 0 & 1 & \vdots & 0 & 0 \end{bmatrix}\mathbf{x} + \begin{bmatrix} 1 & 0 \\ 0 & 1 \\ \cdots & \cdots \\ 0 & 0 \\ 0 & 0 \\ \cdots & \cdots \\ 0 & 0 \\ 0 & 0 \end{bmatrix}\begin{bmatrix} u_1 \\ u_2 \end{bmatrix}$$

$$\mathbf{y} = \begin{bmatrix} -6 & 3 & \vdots & -24 & 7.5 & \vdots & -24 & 3 \\ 0 & 1 & \vdots & 0.5 & 1.5 & \vdots & 1 & 0.5 \end{bmatrix}\mathbf{x} + \begin{bmatrix} 2 & 0 \\ 0 & 0 \end{bmatrix}\begin{bmatrix} u_1 \\ u_2 \end{bmatrix}$$

(4.39)

This is a six-dimensional realization.

We discuss a special case of (4.31) and (4.34) in which $p = 1$. To save space, we assume $r = 4$ and $q = 2$. However, the discussion applies to any positive integers r and q. Consider the 2×1 proper rational matrix

$$\hat{G}(s) = \begin{bmatrix} d_1 \\ d_2 \end{bmatrix} + \frac{1}{s^4 + \alpha_1 s^3 + \alpha_2 s^2 + \alpha_3 s + \alpha_4}$$

$$\cdot \begin{bmatrix} \beta_{11}s^3 + \beta_{12}s^2 + \beta_{13}s + \beta_{14} \\ \beta_{21}s^3 + \beta_{22}s^2 + \beta_{23}s + \beta_{24} \end{bmatrix}$$

(4.40)

Then its realization can be obtained directly from (4.34) as

$$\dot{\mathbf{x}} = \begin{bmatrix} -\alpha_1 & -\alpha_2 & -\alpha_3 & -\alpha_4 \\ 1 & 0 & 0 & 0 \\ 0 & 1 & 0 & 0 \\ 0 & 0 & 1 & 0 \end{bmatrix} \mathbf{x} + \begin{bmatrix} 1 \\ 0 \\ 0 \\ 0 \end{bmatrix} u$$

$$\mathbf{y} = \begin{bmatrix} \beta_{11} & \beta_{12} & \beta_{13} & \beta_{14} \\ \beta_{21} & \beta_{22} & \beta_{23} & \beta_{24} \end{bmatrix} \mathbf{x} + \begin{bmatrix} d_1 \\ d_2 \end{bmatrix} u$$

(4.41)

This controllable-canonical-form realization can be read out from the coefficients of $\hat{\mathbf{G}}(s)$ in (4.40).

There are many ways to realize a proper transfer matrix. For example, Problem 4.9 gives a different realization of (4.33) with dimension rq. Let $\hat{\mathbf{G}}_{ci}(s)$ be the ith column of $\hat{\mathbf{G}}(s)$ and let u_i be the ith component of the input vector $\mathbf{u}$. Then $\hat{\mathbf{y}}(s) = \hat{\mathbf{G}}(s)\hat{\mathbf{u}}(s)$ can be expressed as

$$\hat{\mathbf{y}}(s) = \hat{\mathbf{G}}_{c1}(s)\hat{u}_1(s) + \hat{\mathbf{G}}_{c2}(s)\hat{u}_2(s) + \cdots =: \hat{\mathbf{y}}_{c1}(s) + \hat{\mathbf{y}}_{c2}(s) + \cdots$$

as shown in Fig. 4.4(a). Thus we can realize each column of $\hat{\mathbf{G}}(s)$ and then combine them to yield a realization of $\hat{\mathbf{G}}(s)$. Let $\hat{\mathbf{G}}_{ri}(s)$ be the ith row of $\hat{\mathbf{G}}(s)$ and let y_i be the ith component of the output vector $\mathbf{y}$. Then $\hat{\mathbf{y}}(s) = \hat{\mathbf{G}}(s)\hat{\mathbf{u}}(s)$ can be expressed as

$$\hat{y}_i(s) = \hat{\mathbf{G}}_{ri}(s)\hat{\mathbf{u}}(s)$$

as shown in Fig. 4.4(b). Thus we can realize each row of $\hat{\mathbf{G}}(s)$ and then combine them to obtain a realization of $\hat{\mathbf{G}}(s)$. Clearly we can also realize each entry of $\hat{\mathbf{G}}(s)$ and then combine them to obtain a realization of $\hat{\mathbf{G}}(s)$. See Reference [6, pp. 158–160].

The MATLAB function `[a,b,c,d]=tf2ss(num,den)` generates the controllable-canonical-form realization shown in (4.41) for any single-input multiple-output transfer matrix $\hat{\mathbf{G}}(s)$. In its employment, there is no need to decompose $\hat{\mathbf{G}}(s)$ as in (4.31). But we must compute its least common denominator, not necessarily monic. The next example will apply `tf2ss` to each column of $\hat{\mathbf{G}}(s)$ in (4.38) and then combine them to form a realization of $\hat{\mathbf{G}}(s)$.

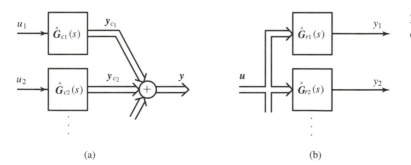

(a)   (b)

Figure 4.4   Realizations of $\hat{\mathbf{G}}(s)$ by columns and by rows.

EXAMPLE 4.7 Consider the proper rational matrix in (4.38). Its first column is

$$\hat{\mathbf{G}}_{c1}(s) = \begin{bmatrix} \dfrac{4s-10}{2s+1} \\ \dfrac{1}{(2s+1)(s+2)} \end{bmatrix} = \begin{bmatrix} \dfrac{(4s-10)(s+2)}{(2s+1)(s+2)} \\ \dfrac{1}{2s^2+5s+2} \end{bmatrix} = \begin{bmatrix} \dfrac{4s^2-2s-20}{2s^2+5s+2} \\ \dfrac{1}{2s^2+5s+2} \end{bmatrix}$$

Typing

```
n1=[4 -2 -20;0 0 1];d1=[2 5 2]; [a,b,c,d]=tf2ss(n1,d1)
```

yields the following realization for the first column of $\hat{\mathbf{G}}(s)$:

$$\dot{\mathbf{x}}_1 = \mathbf{A}_1\mathbf{x}_1 + \mathbf{b}_1 u_1 = \begin{bmatrix} -2.5 & -1 \\ 1 & 0 \end{bmatrix}\mathbf{x}_1 + \begin{bmatrix} 1 \\ 0 \end{bmatrix} u_1$$

$$\mathbf{y}_{c1} = \mathbf{C}_1\mathbf{x}_1 + \mathbf{d}_1 u_1 = \begin{bmatrix} -6 & -12 \\ 0 & 0.5 \end{bmatrix}\mathbf{x}_1 + \begin{bmatrix} 2 \\ 0 \end{bmatrix} u_1$$

(4.42)

Similarly, the function `tf2ss` can generate the following realization for the second column of $\hat{\mathbf{G}}(s)$:

$$\dot{\mathbf{x}}_2 = \mathbf{A}_2\mathbf{x}_2 + \mathbf{b}_2 u_2 = \begin{bmatrix} -4 & -4 \\ 1 & 0 \end{bmatrix}\mathbf{x}_2 + \begin{bmatrix} 1 \\ 0 \end{bmatrix} u_2$$

$$\mathbf{y}_{c2} = \mathbf{C}_2\mathbf{x}_2 + \mathbf{d}_2 u_2 = \begin{bmatrix} 3 & 6 \\ 1 & 1 \end{bmatrix}\mathbf{x}_2 + \begin{bmatrix} 0 \\ 0 \end{bmatrix} u_2$$

(4.43)

These two realizations can be combined as

$$\begin{bmatrix} \dot{\mathbf{x}}_1 \\ \dot{\mathbf{x}}_2 \end{bmatrix} = \begin{bmatrix} \mathbf{A}_1 & \mathbf{0} \\ \mathbf{0} & \mathbf{A}_2 \end{bmatrix}\begin{bmatrix} \mathbf{x}_1 \\ \mathbf{x}_2 \end{bmatrix} + \begin{bmatrix} \mathbf{b}_1 & \mathbf{0} \\ \mathbf{0} & \mathbf{b}_2 \end{bmatrix}\begin{bmatrix} u_1 \\ u_2 \end{bmatrix}$$
$$\mathbf{y} = \mathbf{y}_{c1} + \mathbf{y}_{c2} = [\mathbf{C}_1 \ \ \mathbf{C}_2]\mathbf{x} + [\mathbf{d}_1 \ \ \mathbf{d}_2]\mathbf{u}$$

or

$$\dot{\mathbf{x}} = \begin{bmatrix} -2.5 & -1 & 0 & 0 \\ 1 & 0 & 0 & 0 \\ 0 & 0 & -4 & -4 \\ 0 & 0 & 1 & 0 \end{bmatrix}\mathbf{x} + \begin{bmatrix} 1 & 0 \\ 0 & 0 \\ 0 & 1 \\ 0 & 0 \end{bmatrix}\mathbf{u}$$

$$\mathbf{y} = \begin{bmatrix} -6 & -12 & 3 & 6 \\ 0 & 0.5 & 1 & 1 \end{bmatrix}\mathbf{x} + \begin{bmatrix} 2 & 0 \\ 0 & 0 \end{bmatrix}\mathbf{u}$$

(4.44)

This is a different realization of the $\hat{\mathbf{G}}(s)$ in (4.38). This realization has dimension 4, two less than the one in (4.39).

The two state equations in (4.39) and (4.44) are zero-state equivalent because they have the same transfer matrix. They are, however, not algebraically equivalent. More will be said in Chapter 7 regarding realizations. We mention that all discussion in this section, including tf2ss, applies without any modification to the discrete-time case

4.5 Solution of Linear Time-Varying (LTV) Equations

Consider the linear time-varying (LTV) state equation

$$\dot{\mathbf{x}}(t) = \mathbf{A}(t)\mathbf{x}(t) + \mathbf{B}(t)\mathbf{u}(t) \tag{4.45}$$

$$\mathbf{y}(t) = \mathbf{C}(t)\mathbf{x}(t) + \mathbf{D}(t)\mathbf{u}(t) \tag{4.46}$$

It is assumed that, for every initial state $\mathbf{x}(t_0)$ and any input $\mathbf{u}(t)$, the state equation has a unique solution. A sufficient condition for such an assumption is that every entry of $\mathbf{A}(t)$ is a continuous function of t. Before considering the general case, we first discuss the solutions of $\dot{\mathbf{x}}(t) = \mathbf{A}(t)\mathbf{x}(t)$ and the reasons why the approach taken in the time-invariant case cannot be used here.

The solution of the time-invariant equation $\dot{\mathbf{x}} = \mathbf{A}\mathbf{x}$ can be extended from the scalar equation $\dot{x} = ax$. The solution of $\dot{x} = ax$ is $x(t) = e^{at}x(0)$ with $d(e^{at})/dt = ae^{at} = e^{at}a$. Similarly, the solution of $\dot{\mathbf{x}} = \mathbf{A}\mathbf{x}$ is $\mathbf{x}(t) = e^{\mathbf{A}t}\mathbf{x}(0)$ with

$$\frac{d}{dt}e^{\mathbf{A}t} = \mathbf{A}e^{\mathbf{A}t} = e^{\mathbf{A}t}\mathbf{A}$$

where the commutative property is crucial. Note that, in general, we have $\mathbf{AB} \neq \mathbf{BA}$ and $e^{(\mathbf{A}+\mathbf{B})t} \neq e^{\mathbf{A}t}e^{\mathbf{B}t}$.

The solution of the scalar time-varying equation $\dot{x} = a(t)x$ due to $x(0)$ is

$$x(t) = e^{\int_0^t a(\tau)d\tau}x(0)$$

with

$$\frac{d}{dt}e^{\int_0^t a(\tau)d\tau} = a(t)e^{\int_0^t a(\tau)d\tau} = e^{\int_0^t a(\tau)d\tau}a(t)$$

Extending this to the matrix case becomes

$$\mathbf{x}(t) = e^{\int_0^t \mathbf{A}(\tau)d\tau}\mathbf{x}(0) \tag{4.47}$$

with, using (3.51),

$$e^{\int_0^t \mathbf{A}(\tau)d\tau} = \mathbf{I} + \int_0^t \mathbf{A}(\tau)d\tau + \frac{1}{2}\left(\int_0^t \mathbf{A}(\tau)d\tau\right)\left(\int_0^t \mathbf{A}(s)ds\right) + \cdots$$

This extension, however, is not valid because

$$\frac{d}{dt}e^{\int_0^t \mathbf{A}(\tau)d\tau} = \mathbf{A}(t) + \frac{1}{2}\mathbf{A}(t)\left(\int_0^t \mathbf{A}(s)ds\right) + \frac{1}{2}\left(\int_0^t \mathbf{A}(\tau)d\tau\right)\mathbf{A}(t) + \cdots$$

$$\neq \mathbf{A}(t)e^{\int_0^t \mathbf{A}(\tau)d\tau} \tag{4.48}$$

Thus, in general, (4.47) is not a solution of $\dot{\mathbf{x}} = \mathbf{A}(t)\mathbf{x}$. In conclusion, we cannot extend the solution of scalar time-varying equations to the matrix case and must use a different approach to develop the solution.

Consider

$$\dot{\mathbf{x}} = \mathbf{A}(t)\mathbf{x} \tag{4.49}$$

where $\mathbf{A}$ is $n \times n$ with continuous functions of t as its entries. Then for every initial state $\mathbf{x}_i(t_0)$, there exists a unique solution $\mathbf{x}_i(t)$, for $i = 1, 2, \ldots, n$. We can arrange these n solutions as $\mathbf{X} = [\mathbf{x}_1 \ \mathbf{x}_2 \ \cdots \ \mathbf{x}_n]$, a square matrix of order n. Because every $\mathbf{x}_i$ satisfies (4.49), we have

$$\dot{\mathbf{X}}(t) = \mathbf{A}(t)\mathbf{X}(t) \tag{4.50}$$

If $\mathbf{X}(t_0)$ is nonsingular or the n initial states are linearly independent, then $\mathbf{X}(t)$ is called a *fundamental matrix* of (4.49). Because the initial states can arbitrarily be chosen, as long as they are linearly independent, the fundamental matrix is not unique.

EXAMPLE 4.8 Consider the homogeneous equation

$$\dot{\mathbf{x}}(t) = \begin{bmatrix} 0 & 0 \\ t & 0 \end{bmatrix} \mathbf{x}(t) \tag{4.51}$$

or

$$\dot{x}_1(t) = 0 \qquad \dot{x}_2(t) = tx_1(t)$$

The solution of $\dot{x}_1(t) = 0$ for $t_0 = 0$ is $x_1(t) = x_1(0)$; the solution of $\dot{x}_2(t) = tx_1(t) = tx_1(0)$ is

$$x_2(t) = \int_0^t \tau x_1(0)d\tau + x_2(0) = 0.5t^2 x_1(0) + x_2(0)$$

Thus we have

$$\mathbf{x}(0) = \begin{bmatrix} x_1(0) \\ x_2(0) \end{bmatrix} = \begin{bmatrix} 1 \\ 0 \end{bmatrix} \Rightarrow \mathbf{x}(t) = \begin{bmatrix} 1 \\ 0.5t^2 \end{bmatrix}$$

and

$$\mathbf{x}(0) = \begin{bmatrix} x_1(0) \\ x_2(0) \end{bmatrix} = \begin{bmatrix} 1 \\ 2 \end{bmatrix} \Rightarrow \mathbf{x}(t) = \begin{bmatrix} 1 \\ 0.5t^2 + 2 \end{bmatrix}$$

The two initial states are linearly independent; thus

$$\mathbf{X}(t) = \begin{bmatrix} 1 & 1 \\ 0.5t^2 & 0.5t^2 + 2 \end{bmatrix} \tag{4.52}$$

is a fundamental matrix.

A very important property of the fundamental matrix is that $\mathbf{X}(t)$ is nonsingular for all t. For example, $\mathbf{X}(t)$ in (4.52) has determinant $0.5t^2 + 2 - 0.5t^2 = 2$; thus it is nonsingular for all t. We argue intuitively why this is the case. If $\mathbf{X}(t)$ is singular at some t_1, then there exists a nonzero vector $\mathbf{v}$ such that $\mathbf{x}(t_1) := \mathbf{X}(t_1)\mathbf{v} = \mathbf{0}$, which, in turn, implies $\mathbf{x}(t) := \mathbf{X}(t)\mathbf{v} \equiv \mathbf{0}$ for all t, in particular, at $= t_0$. This is a contradiction. Thus $\mathbf{X}(t)$ is nonsingular for all t.

Definition 4.2 *Let* $\mathbf{X}(t)$ *be any fundamental matrix of* $\dot{\mathbf{x}} = \mathbf{A}(t)\mathbf{x}$. *Then*

$$\boldsymbol{\Phi}(t, t_0) := \mathbf{X}(t)\mathbf{X}^{-1}(t_0)$$

is called the state transition matrix *of* $\dot{\mathbf{x}} = \mathbf{A}(t)\mathbf{x}$. *The state transition matrix is also the unique solution of*

$$\frac{\partial}{\partial t}\boldsymbol{\Phi}(t, t_0) = \mathbf{A}(t)\boldsymbol{\Phi}(t, t_0) \tag{4.53}$$

with the initial condition $\boldsymbol{\Phi}(t_0, t_0) = \mathbf{I}$.

Because $\mathbf{X}(t)$ is nonsingular for all t, its inverse is well defined. Equation (4.53) follows directly from (4.50). From the definition, we have the following important properties of the state transition matrix:

$$\boldsymbol{\Phi}(t, t) = \mathbf{I} \tag{4.54}$$

$$\boldsymbol{\Phi}^{-1}(t, t_0) = [\mathbf{X}(t)\mathbf{X}^{-1}(t_0)]^{-1} = \mathbf{X}(t_0)\mathbf{X}^{-1}(t) = \boldsymbol{\Phi}(t_0, t) \tag{4.55}$$

$$\boldsymbol{\Phi}(t, t_0) = \boldsymbol{\Phi}(t, t_1)\boldsymbol{\Phi}(t_1, t_0) \tag{4.56}$$

for every t, t_0, and t_1.

EXAMPLE 4.9 Consider the homogeneous equation in Example 4.8. Its fundamental matrix was computed as

$$\mathbf{X}(t) = \begin{bmatrix} 1 & 1 \\ 0.5t^2 & 0.5t^2 + 2 \end{bmatrix}$$

Its inverse is, using (3.20),

$$\mathbf{X}^{-1}(t) = \begin{bmatrix} 0.25t^2 + 1 & -0.5 \\ -0.25t^2 & 0.5 \end{bmatrix}$$

Thus the state transition matrix is given by

$$\boldsymbol{\Phi}(t, t_0) = \begin{bmatrix} 1 & 1 \\ 0.5t^2 & 0.5t^2 + 2 \end{bmatrix} \begin{bmatrix} 0.25t_0^2 + 1 & -0.5 \\ -0.25t_0^2 & 0.5 \end{bmatrix}$$

$$= \begin{bmatrix} 1 & 0 \\ 0.5(t^2 - t_0^2) & 1 \end{bmatrix}$$

It is straightforward to verify that this transition matrix satisfies (4.53) and has the three properties listed in (4.54) through (4.56).

Now we claim that the solution of (4.45) excited by the initial state $\mathbf{x}(t_0) = \mathbf{x}_0$ and the input $\mathbf{u}(t)$ is given by

$$\mathbf{x}(t) = \boldsymbol{\Phi}(t, t_0)\mathbf{x}_0 + \int_{t_0}^{t} \boldsymbol{\Phi}(t, \tau)\mathbf{B}(\tau)\mathbf{u}(\tau)\, d\tau \tag{4.57}$$

$$= \boldsymbol{\Phi}(t, t_0)\left[\mathbf{x}_0 + \int_{t_0}^{t} \boldsymbol{\Phi}(t_0, \tau)\mathbf{B}(\tau)\mathbf{u}(\tau)\, d\tau\right] \tag{4.58}$$

where $\mathbf{\Phi}(t, \tau)$ is the state transition matrix of $\dot{\mathbf{x}} = \mathbf{A}(t)\mathbf{x}$. Equation (4.58) follows from (4.57) by using $\mathbf{\Phi}(t, \tau) = \mathbf{\Phi}(t, t_0)\mathbf{\Phi}(t_0, \tau)$. We show that (4.57) satisfies the initial condition and the state equaion. At $t = t_0$, we have

$$\mathbf{x}(t_0) = \mathbf{\Phi}(t_0, t_0)\mathbf{x}_0 + \int_{t_0}^{t_0} \mathbf{\Phi}(t, \tau)\mathbf{B}(\tau)\mathbf{u}(\tau)\, d\tau = \mathbf{I}\mathbf{x}_0 + \mathbf{0} = \mathbf{x}_0$$

Thus (4.57) satisfies the initial condition. Using (4.53) and (4.6), we have

$$\frac{d}{dt}\mathbf{x}(t) = \frac{\partial}{\partial t}\mathbf{\Phi}(t, t_0)\mathbf{x}_0 + \frac{\partial}{\partial t}\int_{t_0}^{t} \mathbf{\Phi}(t, \tau)\mathbf{B}(\tau)\mathbf{u}(\tau)\, d\tau$$

$$= \mathbf{A}(t)\mathbf{\Phi}(t, t_0)\mathbf{x}_0 + \int_{t_0}^{t} \left(\frac{\partial}{\partial t}\mathbf{\Phi}(t, \tau)\mathbf{B}(\tau) \right) d\tau + \mathbf{\Phi}(t, t)\mathbf{B}(t)\mathbf{u}(t)$$

$$= \mathbf{A}(t)\mathbf{\Phi}(t, t_0)\mathbf{x}_0 + \int_{t_0}^{t} \mathbf{A}(t)\mathbf{\Phi}(t, \tau)\mathbf{B}(\tau)\mathbf{u}(\tau)\, d\tau + \mathbf{B}(t)\mathbf{u}(t)$$

$$= \mathbf{A}(t)\left[\mathbf{\Phi}(t, t_0)\mathbf{x}_0 + \int_{t_0}^{t} \mathbf{\Phi}(t, \tau)\mathbf{B}(\tau)\mathbf{u}(\tau)\, d\tau \right] + \mathbf{B}(t)\mathbf{u}(t)$$

$$= \mathbf{A}(t)\mathbf{x}(t) + \mathbf{B}(t)\mathbf{u}(t)$$

Thus (4.57) is the solution. Substituting (4.57) into (4.46) yields

$$\mathbf{y}(t) = \mathbf{C}(t)\mathbf{\Phi}(t, t_0)\mathbf{x}_0 + \mathbf{C}(t)\int_{t_0}^{t} \mathbf{\Phi}(t, \tau)\mathbf{B}(\tau)\mathbf{u}(\tau)\, d\tau + \mathbf{D}(t)\mathbf{u}(t) \qquad (4.59)$$

If the input is identically zero, then Equation (4.57) reduces to

$$\mathbf{x}(t) = \mathbf{\Phi}(t, t_0)\mathbf{x}_0$$

This is the zero-input response. Thus the state transition matrix governs the unforced propagation of the state vector. If the initial state is zero, then (4.59) reduces to

$$\mathbf{y}(t) = \mathbf{C}(t)\int_{t_0}^{t} \mathbf{\Phi}(t, \tau)\mathbf{B}(\tau)\mathbf{u}(\tau)\, d\tau + \mathbf{D}(t)\mathbf{u}(t)$$

$$= \int_{t_0}^{t} [\mathbf{C}(t)\mathbf{\Phi}(t, \tau)\mathbf{B}(\tau) + \mathbf{D}\delta(t - \tau)]\,\mathbf{u}(\tau)\, d\tau \qquad (4.60)$$

This is the zero-state response. As discussed in (2.5), the zero-state response can be described by

$$\mathbf{y}(t) = \int_{t_0}^{t} \mathbf{G}(t, \tau)\mathbf{u}(\tau)\, d\tau \qquad (4.61)$$

where $\mathbf{G}(t, \tau)$ is the impulse response matrix and is the output at time t excited by an impulse input applied at time τ. Comparing (4.60) and (4.61) yields

$$\mathbf{G}(t, \tau) = \mathbf{C}(t)\mathbf{\Phi}(t, \tau)\mathbf{B}(\tau) + \mathbf{D}(t)\delta(t - \tau)$$

$$= \mathbf{C}(t)\mathbf{X}(t)\mathbf{X}^{-1}(\tau)\mathbf{B}(\tau) + \mathbf{D}(t)\delta(t - \tau) \qquad (4.62)$$

This relates the input–output and state-space descriptions.

The solutions in (4.57) and (4.59) hinge on solving (4.49) or (4.53). If $\mathbf{A}(t)$ is triangular such as

$$\begin{bmatrix} \dot{x}_1(t) \\ \dot{x}_2(t) \end{bmatrix} = \begin{bmatrix} a_{11}(t) & 0 \\ a_{21}(t) & a_{22}(t) \end{bmatrix} \begin{bmatrix} x_1(t) \\ x_2(t) \end{bmatrix}$$

we can solve the scalar equation $\dot{x}_1(t) = a_{11}(t)x_1(t)$ and then substitute it into

$$\dot{x}_2(t) = a_{22}(t)x_2(t) + a_{21}(t)x_1(t)$$

Because $x_1(t)$ has been solved, the preceding scalar equation can be solved for $x_2(t)$. This is what we did in Example 4.8. If $\mathbf{A}(t)$, such as $\mathbf{A}(t)$ diagonal or constant, has the commutative property

$$\mathbf{A}(t) \left(\int_{t_0}^{t} \mathbf{A}(\tau) \, d\tau \right) = \left(\int_{t_0}^{t} \mathbf{A}(\tau) \, d\tau \right) \mathbf{A}(t)$$

for all t_0 and t, then the solution of (4.53) can be shown to be

$$\mathbf{\Phi}(t, t_0) = e^{\int_{t_0}^{t} \mathbf{A}(\tau) \, d\tau} = \sum_{k=0}^{\infty} \frac{1}{k!} \left(\int_{t_0}^{t} \mathbf{A}(\tau) \, d\tau \right)^{k} \tag{4.63}$$

For $\mathbf{A}(t)$ constant, (4.63) reduces to

$$\mathbf{\Phi}(t, \tau) = e^{\mathbf{A}(t-\tau)} = \mathbf{\Phi}(t - \tau)$$

and $\mathbf{X}(t) = e^{\mathbf{A}t}$. Other than the preceding special cases, computing state transition matrices is generally difficult.

4.5.1 Discrete-Time Case

Consider the discrete-time state equation

$$\mathbf{x}[k + 1] = \mathbf{A}[k]\mathbf{x}[k] + \mathbf{B}[k]\mathbf{u}[k] \tag{4.64}$$

$$\mathbf{y}[k] = \mathbf{C}[k]\mathbf{x}[k] + \mathbf{D}[k]\mathbf{u}[k] \tag{4.65}$$

The set consists of algebraic equations and their solutions can be computed recursively once the initial state $\mathbf{x}[k_0]$ and the input $\mathbf{u}[k]$, for $k \geq k_0$, are given. The situation here is much simpler than the continuous-time case.

As in the continuous-time case, we can define the discrete state transition matrix as the solution of

$$\mathbf{\Phi}[k + 1, k_0] = \mathbf{A}[k]\mathbf{\Phi}[k, k_0] \qquad \text{with } \mathbf{\Phi}[k_0, k_0] = \mathbf{I}$$

for $k = k_0, k_0 + 1, \ldots$. This is the discrete counterpart of (4.53) and its solution can be obtained directly as

$$\mathbf{\Phi}[k, k_0] = \mathbf{A}[k - 1]\mathbf{A}[k - 2] \cdots \mathbf{A}[k_0] \tag{4.66}$$

for $k > k_0$ and $\mathbf{\Phi}[k_0, k_0] = \mathbf{I}$. We discuss a significant difference between the continuous- and discrete-time cases. Because the fundamental matrix in the continuous-time case is nonsingular for all t, the state transition matrix $\Phi(t, t_0)$ is defined for $t \geq t_0$ and $t < t_0$ and can govern the propagation of the state vector in the positive-time and negative-time directions. In the discrete-time case, the A-matrix can be singular; thus the inverse of $\Phi[k, k_0]$ may not be defined. Thus $\Phi[k, k_0]$ is defined only for $k \geq k_0$ and governs the propagation of the state vector in only the positive-time direction. Therefore the discrete counterpart of (4.56) or

$$\Phi[k, k_0] = \Phi[k, k_1]\Phi[k_1, k_0]$$

holds only for $k \geq k_1 \geq k_0$.

Using the discrete state transition matrix, we can express the solutions of (4.64) and (4.65) as, for $k > k_0$,

$$\mathbf{x}[k] = \Phi[k, k_0]\mathbf{x}_0 + \sum_{m=k_0}^{k-1} \Phi[k, m + 1]\mathbf{B}[m]\mathbf{u}[m]$$

$$(4.67)$$

$$\mathbf{y}[k] = \mathbf{C}[k]\Phi[k, k_0]\mathbf{x}_0 + \mathbf{C}[k] \sum_{m=k_0}^{k-1} \Phi[k, m + 1]\mathbf{B}[m]\mathbf{u}[m] + \mathbf{D}[k]\mathbf{u}[k]$$

Their derivations are similar to those of (4.20) and (4.21) and will not be repeated.

If the initial state is zero, Equation (4.67) reduces to

$$\mathbf{y}[k] = \mathbf{C}[k] \sum_{m=k_0}^{k-1} \Phi[k, m + 1]\mathbf{B}[m]\mathbf{u}[m] + \mathbf{D}[k]\mathbf{u}[k] \qquad (4.68)$$

for $k > k_0$. This describes the zero-state response of (4.65). If we define $\Phi[k, m] = \mathbf{0}$ for $k < m$, then (4.68) can be written as

$$\mathbf{y}[k] = \sum_{m=k_0}^{k} \left(\mathbf{C}[k]\Phi[k, m + 1]\mathbf{B}[m] + \mathbf{D}[m]\delta[k - m]\right)\mathbf{u}[m]$$

where the impulse sequence $\delta[k - m]$ equals 1 if $k = m$ and 0 if $k \neq m$. Comparing this with the multivariable version of (2.34), we have

$$\mathbf{G}[k, m] = \mathbf{C}[k]\Phi[k, m + 1]\mathbf{B}[m] + \mathbf{D}[m]\delta[k - m]$$

for $k \geq m$. This relates the impulse response sequence and the state equation and is the discrete counterpart of (4.62).

4.6 Equivalent Time-Varying Equations

This section extends the equivalent state equations discussed in Section 4.3 to the time-varying case. Consider the n-dimensional linear time-varying state equation

$$\dot{\mathbf{x}} = \mathbf{A}(t)\mathbf{x} + \mathbf{B}(t)\mathbf{u}$$

$$(4.69)$$

$$\mathbf{y} = \mathbf{C}(t)\mathbf{x} + \mathbf{D}(t)\mathbf{u}$$

Let $\mathbf{P}(t)$ be an $n \times n$ matrix. It is assumed that $\mathbf{P}(t)$ is nonsingular and both $\mathbf{P}(t)$ and $\dot{\mathbf{P}}(t)$ are continuous for all t. Let $\bar{\mathbf{x}} = \mathbf{P}(t)\mathbf{x}$. Then the state equation

$$\dot{\bar{\mathbf{x}}} = \bar{\mathbf{A}}(t)\bar{\mathbf{x}} + \bar{\mathbf{B}}(t)\mathbf{u}$$

$$\mathbf{y} = \bar{\mathbf{C}}(t)\bar{\mathbf{x}} + \bar{\mathbf{D}}(t)\mathbf{u} \tag{4.70}$$

where

$$\bar{\mathbf{A}}(t) = [\mathbf{P}(t)\mathbf{A}(t) + \dot{\mathbf{P}}(t)]\mathbf{P}^{-1}(t)$$

$$\bar{\mathbf{B}}(t) = \mathbf{P}(t)\mathbf{B}(t)$$

$$\bar{\mathbf{C}}(t) = \mathbf{C}(t)\mathbf{P}^{-1}(t)$$

$$\bar{\mathbf{D}}(t) = \mathbf{D}(t)$$

is said to be (algebraically) equivalent to (4.69) and $\mathbf{P}(t)$ is called an (*algebraic*) *equivalence* transformation.

Equation (4.70) is obtained from (4.69) by substituting $\bar{\mathbf{x}} = \mathbf{P}(t)\mathbf{x}$ and $\dot{\bar{\mathbf{x}}} = \dot{\mathbf{P}}(t)\mathbf{x} + \mathbf{P}(t)\dot{\mathbf{x}}$. Let $\mathbf{X}$ be a fundamental matrix of (4.69). Then we claim that

$$\bar{\mathbf{X}}(t) := \mathbf{P}(t)\mathbf{X}(t) \tag{4.71}$$

is a fundamental matrix of (4.70). By definition, $\dot{\mathbf{X}}(t) = \mathbf{A}(t)\mathbf{X}(t)$ and $\mathbf{X}(t)$ is nonsingular for all t. Because the rank of a matrix will not change by multiplying a nonsingular matrix, the matrix $\mathbf{P}(t)\mathbf{X}(t)$ is also nonsingular for all t. Now we show that $\mathbf{P}(t)\mathbf{X}(t)$ satisfies the equation $\dot{\bar{\mathbf{x}}} = \bar{\mathbf{A}}(t)\bar{\mathbf{x}}$. Indeed, we have

$$\frac{d}{dt}[\mathbf{P}(t)\mathbf{X}(t)] = \dot{\mathbf{P}}(t)\mathbf{X}(t) + \mathbf{P}(t)\dot{\mathbf{X}}(t) = \dot{\mathbf{P}}(t)\mathbf{X}(t) + \mathbf{P}(t)\mathbf{A}(t)\mathbf{X}(t)$$

$$= [\dot{\mathbf{P}}(t) + \mathbf{P}(t)\mathbf{A}(t)][\mathbf{P}^{-1}(t)\mathbf{P}(t)]\mathbf{X}(t) = \bar{\mathbf{A}}(t)[\mathbf{P}(t)\mathbf{X}(t)]$$

Thus $\mathbf{P}(t)\mathbf{X}(t)$ is a fundamental matrix of $\dot{\bar{\mathbf{x}}}(t) = \bar{\mathbf{A}}(t)\bar{\mathbf{x}}(t)$.

▷ **Theorem 4.3**

Let $\mathbf{A}_o$ be an arbitrary constant matrix. Then there exists an equivalence transformation that transforms (4.69) into (4.70) with $\bar{\mathbf{A}}(t) = \mathbf{A}_o$.

 Proof: Let $\mathbf{X}(t)$ be a fundamental matrix of $\dot{\mathbf{x}} = \mathbf{A}(t)\mathbf{x}$. The differentiation of $\mathbf{X}^{-1}(t)$ $\mathbf{X}(t) = \mathbf{I}$ yields

$$\dot{\mathbf{X}}^{-1}(t)\mathbf{X}(t) + \mathbf{X}^{-1}\dot{\mathbf{X}}(t) = \mathbf{0}$$

which implies

$$\dot{\mathbf{X}}^{-1}(t) = -\mathbf{X}^{-1}(t)\mathbf{A}(t)\mathbf{X}(t)\mathbf{X}^{-1}(t) = -\mathbf{X}^{-1}(t)\mathbf{A}(t) \tag{4.72}$$

Because $\bar{\mathbf{A}}(t) = \mathbf{A}_o$ is a constant matrix, $\bar{\mathbf{X}}(t) = e^{\mathbf{A}_o t}$ is a fundamental matrix of $\dot{\bar{\mathbf{x}}} = \bar{\mathbf{A}}(t)\bar{\mathbf{x}} = \mathbf{A}_o\bar{\mathbf{x}}$. Following (4.71), we define

$$\mathbf{P}(t) := \bar{\mathbf{X}}(t)\mathbf{X}^{-1}(t) = e^{\mathbf{A}_o t}\mathbf{X}^{-1}(t) \tag{4.73}$$

and compute

$$\bar{\mathbf{A}}(t) = [\mathbf{P}(t)\mathbf{A}(t) + \dot{\mathbf{P}}(t)]\mathbf{P}^{-1}(t)$$

$$= [e^{\mathbf{A}_o t}\mathbf{X}^{-1}(t)\mathbf{A}(t) + \mathbf{A}_o e^{\mathbf{A}_o t}\mathbf{X}^{-1}(t) + e^{\mathbf{A}_o t}\dot{\mathbf{X}}^{-1}(t)]\mathbf{X}(t)e^{-\mathbf{A}_o t}$$

which becomes, after substituting (4.72),

$$\bar{\mathbf{A}}(t) = \mathbf{A}_o e^{\mathbf{A}_o t}\mathbf{X}^{-1}(t)\mathbf{X}(t)e^{-\mathbf{A}_o t} = \mathbf{A}_o$$

This establishes the theorem.   Q.E.D.

If $\mathbf{A}_o$ is chosen as a zero matrix, then $\mathbf{P}(t) = \mathbf{X}^{-1}(t)$ and (4.70) reduces to

$$\bar{\mathbf{A}}(t) = \mathbf{0} \qquad \bar{\mathbf{B}}(t) = \mathbf{X}^{-1}(t)\mathbf{B}(t) \qquad \bar{\mathbf{C}}(t) = \mathbf{C}(t)\mathbf{X}(t) \qquad \bar{\mathbf{D}}(t) = \mathbf{D}(t) \qquad (4.74)$$

The block diagrams of (4.69) with $\mathbf{A}(t) \neq \mathbf{0}$ and $\mathbf{A}(t) = \mathbf{0}$ are plotted in Fig. 4.5. The block diagram with $\mathbf{A}(t) = \mathbf{0}$ has no feedback and is considerably simpler. Every time-varying state equation can be transformed into such a block diagram. However, in order to do so, we must know its fundamental matrix.

The impulse response matrix of (4.69) is given in (4.62). The impulse response matrix of (4.70) is, using (4.71) and (4.72),

$$\bar{\mathbf{G}}(t, \tau) = \bar{\mathbf{C}}(t)\bar{\mathbf{X}}(t)\bar{\mathbf{X}}^{-1}(\tau)\bar{\mathbf{B}}(\tau) + \bar{\mathbf{D}}(t)\delta(t - \tau)$$

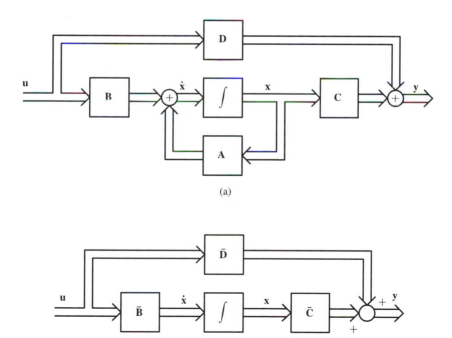

(a)

(b)

Figure 4.5   Block daigrams with feedback and without feedback.

$$= \mathbf{C}(t)\mathbf{P}^{-1}(t)\mathbf{P}(t)\mathbf{X}(t)\mathbf{X}^{-1}(\tau)\mathbf{P}^{-1}(\tau)\mathbf{P}(\tau)\mathbf{B}(\tau) + \mathbf{D}(t)\delta(t - \tau)$$

$$= \mathbf{C}(t)\mathbf{X}(t)\mathbf{X}^{-1}(\tau)\mathbf{B}(\tau) + \mathbf{D}(t)\delta(t - \tau) = \mathbf{G}(t, \tau)$$

Thus the impulse response matrix is invariant under any equivalence transformation. The property of the A-matrix, however, may not be preserved in equivalence transformations. For example, every A-matrix can be transformed, as shown in Theorem 4.3, into a constant or a zero matrix. Clearly the zero matrix does not have any property of $\mathbf{A}(t)$. In the time-invariant case, equivalence transformations will preserve all properties of the original state equation. Thus the equivalence transformation in the time-invariant case is not a special case of the time-varying case.

Definition 4.3 *A matrix $\mathbf{P}(t)$ is called a* Lyapunov transformation *if $\mathbf{P}(t)$ is nonsingular, $\mathbf{P}(t)$ and $\dot{\mathbf{P}}(t)$ are continuous, and $\mathbf{P}(t)$ and $\mathbf{P}^{-1}(t)$ are bounded for all t. Equations (4.69) and (4.70) are said to be* Lyapunov equivalent *if $\mathbf{P}(t)$ is a Lyapunov transformation.*

It is clear that if $\mathbf{P}(t) = \mathbf{P}$ is a constant matrix, then it is a Lyapunov transformation. Thus the (algebraic) transformation in the time-invariant case is a special case of the Lyapunov transformation. If $\mathbf{P}(t)$ is required to be a Lyapunov transformation, then Theorem 4.3 does not hold in general. In other words, not every time-varying state equation can be Lyapunov equivalent to a state equation with a constant A-matrix. However, this is true if $\mathbf{A}(t)$ is periodic.

Periodic state equations   Consider the linear time-varying state equation in (4.69). It is assumed that

$$\mathbf{A}(t + T) = \mathbf{A}(t)$$

for all t and for some positive constant T. That is, $\mathbf{A}(t)$ is periodic with period T. Let $\mathbf{X}(t)$ be a fundamental matrix of $\dot{\mathbf{x}} = \mathbf{A}(t)\mathbf{x}$ or $\dot{\mathbf{X}}(t) = \mathbf{A}(t)\mathbf{X}(t)$ with $\mathbf{X}(0)$ nonsingular. Then we have

$$\dot{\mathbf{X}}(t + T) = \mathbf{A}(t + T)\mathbf{X}(t + T) = \mathbf{A}(t)\mathbf{X}(t + T)$$

Thus $\mathbf{X}(t + T)$ is also a fundamental matrix. Furthermore, it can be expressed as

$$\mathbf{X}(t + T) = \mathbf{X}(t)\mathbf{X}^{-1}(0)\mathbf{X}(T) \tag{4.75}$$

This can be verified by direct substitution. Let us define $\mathbf{Q} = \mathbf{X}^{-1}(0)\mathbf{X}(T)$. It is a constant nonsingular matrix. For this $\mathbf{Q}$ there exists a constant matrix $\bar{\mathbf{A}}$ such that $e^{\bar{\mathbf{A}}T} = \mathbf{Q}$ (Problem 3.24). Thus (4.75) can be written as

$$\mathbf{X}(t + T) = \mathbf{X}(t)e^{\bar{\mathbf{A}}T} \tag{4.76}$$

Define

$$\mathbf{P}(t) := e^{\bar{\mathbf{A}}t}\mathbf{X}^{-1}(t) \tag{4.77}$$

We show that $\mathbf{P}(t)$ is periodic with period T:

$$\mathbf{P}(t + T) = e^{\bar{\mathbf{A}}(t+T)}\mathbf{X}^{-1}(t + T) = e^{\bar{\mathbf{A}}t}e^{\bar{\mathbf{A}}T}[e^{-\bar{\mathbf{A}}T}\mathbf{X}^{-1}(t)]$$

$$= e^{\bar{\mathbf{A}}t}\mathbf{X}^{-1}(t) = \mathbf{P}(t)$$

▶ **Theorem 4.4**

Consider (4.69) with $\mathbf{A}(t) = \mathbf{A}(t + T)$ for all t and some $T > 0$. Let $\mathbf{X}(t)$ be a fundamental matrix of $\dot{\mathbf{x}} = \mathbf{A}(t)\mathbf{x}$. Let $\bar{\mathbf{A}}$ be the constant matrix computed from $e^{\bar{\mathbf{A}}T} = \mathbf{X}^{-1}(0)\mathbf{X}(T)$. Then (4.69) is Lyapunov equivalent to

$$\dot{\bar{\mathbf{x}}}(t) = \bar{\mathbf{A}}\bar{\mathbf{x}}(t) + \mathbf{P}(t)\mathbf{B}(t)\mathbf{u}(t)$$
$$\bar{\mathbf{y}}(t) = \mathbf{C}(t)\mathbf{P}^{-1}(t)\bar{\mathbf{x}}(t) + \mathbf{D}(t)\mathbf{u}(t)$$

where $\mathbf{P}(t) = e^{\bar{\mathbf{A}}t}\mathbf{X}^{-1}(t)$.

The matrix $\mathbf{P}(t)$ in (4.77) satisfies all conditions in Definition 4.3; thus it is a Lyapunov transformation. The rest of the theorem follows directly from Theorem 4.3. The homogeneous part of Theorem 4.4 is the *theory of Floquet*. It states that if $\dot{\mathbf{x}} = \mathbf{A}(t)\mathbf{x}$ and if $\mathbf{A}(t + T) = \mathbf{A}(t)$ for all t, then its fundamental matrix is of the form $\mathbf{P}^{-1}(t)e^{\bar{\mathbf{A}}t}$, where $\mathbf{P}^{-1}(t)$ is a periodic function. Furthermore, $\dot{\mathbf{x}} = \mathbf{A}(t)\mathbf{x}$ is Lyapunov equivalent to $\dot{\bar{\mathbf{x}}} = \bar{\mathbf{A}}\bar{\mathbf{x}}$.

4.7 Time-Varying Realizations

We studied in Section 4.4 the realization problem for linear time-invariant systems. In this section, we study the corresponding problem for linear time-varying systems. The Laplace transform cannot be used here; therefore we study the problem directly in the time domain.

Every linear time-varying system can be described by the input–output description

$$\mathbf{y}(t) = \int_{t_0}^{t} \mathbf{G}(t, \tau)\mathbf{u}(\tau)\, d\tau$$

and, if the system is lumped as well, by the state equation

$$\dot{\mathbf{x}}(t) = \mathbf{A}(t)\mathbf{x}(t) + \mathbf{B}(t)\mathbf{u}(t)$$
$$\mathbf{y}(t) = \mathbf{C}(t)\mathbf{x}(t) + \mathbf{D}(t)\mathbf{u}(t) \tag{4.78}$$

If the state equation is available, the impulse response matrix can be computed from

$$\mathbf{G}(t, \tau) = \mathbf{C}(t)\mathbf{X}(t)\mathbf{X}^{-1}(\tau)\mathbf{B}(\tau) + \mathbf{D}(t)\delta(t - \tau) \qquad \text{for } t \geq \tau \tag{4.79}$$

where $\mathbf{X}(t)$ is a fundamental matrix of $\dot{\mathbf{x}} = \mathbf{A}(t)\mathbf{x}$. The converse problem is to find a state equation from a given impulse response matrix. An impulse response matrix $\mathbf{G}(t, \tau)$ is said to be *realizable* if there exists $\{\mathbf{A}(t),\ \mathbf{B}(t),\ \mathbf{C}(t),\ \mathbf{D}(t)\}$ to meet (4.79).

▶ **Theorem 4.5**

A $q \times p$ impulse response matrix $\mathbf{G}(t, \tau)$ is realizable if and only if $\mathbf{G}(t, \tau)$ can be decomposed as

$$\mathbf{G}(t, \tau) = \mathbf{M}(t)\mathbf{N}(\tau) + \mathbf{D}(t)\delta(t - \tau) \tag{4.80}$$

for all $t \geq \tau$, where $\mathbf{M}$, $\mathbf{N}$, and $\mathbf{D}$ are, respectively, $q \times n$, $n \times p$, and $q \times p$ matrices for some integer n.

Proof: If $\mathbf{G}(t, \tau)$ is realizable, there exists a realization that meets (4.79). Identifying $\mathbf{M}(t) = \mathbf{C}(t)\mathbf{X}(t)$ and $\mathbf{N}(\tau) = \mathbf{X}^{-1}(\tau)\mathbf{B}(\tau)$ establishes the necessary part of the theorem. If $\mathbf{G}(t, \tau)$ can be decomposed as in (4.80), then the n-dimensional state equation

$$\dot{\mathbf{x}}(t) = \mathbf{N}(t)\mathbf{u}(t)$$
$$\mathbf{y}(t) = \mathbf{M}(t)\mathbf{x}(t) + \mathbf{D}(t)\mathbf{u}(t)$$

(4.81)

is a realization. Indeed, a fundamental matrix of $\dot{\mathbf{x}} = \mathbf{0} \cdot \mathbf{x}$ is $\mathbf{X}(t) = \mathbf{I}$. Thus the impulse response matrix of (4.81) is

$$\mathbf{M}(t)\mathbf{I} \cdot \mathbf{I}^{-1}\mathbf{N}(\tau) + \mathbf{D}(t)\delta(t - \tau)$$

which equals $\mathbf{G}(t, \tau)$. This shows the sufficiency of the theorem.   Q.E.D.

Although Theorem 4.5 can also be applied to time-invariant systems, the result is not useful in practical implementation, as the next example illustrates.

EXAMPLE 4.10 Consider $g(t) = te^{\lambda t}$ or

$$g(t, \tau) = g(t - \tau) = (t - \tau)e^{\lambda(t-\tau)}$$

It is straightforward to verify

$$g(t - \tau) = [e^{\lambda t} \quad te^{\lambda t}]\begin{bmatrix} -\tau e^{-\lambda\tau} \\ e^{-\lambda\tau} \end{bmatrix}$$

Thus the two-dimensional time-varying state equation

$$\dot{\mathbf{x}}(t) = \begin{bmatrix} 0 & 0 \\ 0 & 0 \end{bmatrix}\mathbf{x} + \begin{bmatrix} -te^{-\lambda t} \\ e^{-\lambda t} \end{bmatrix}u(t)$$
$$y(t) = [e^{\lambda t} \quad te^{\lambda t}]\mathbf{x}(t)$$

(4.82)

is a realization of the impulse response $g(t) = te^{\lambda t}$.

The Laplace transform of the impulse response is

$$\hat{g}(s) = \mathcal{L}[te^{\lambda t}] = \frac{1}{(s - \lambda)^2} = \frac{1}{s^2 - 2\lambda s + \lambda^2}$$

Using (4.41), we can readily obtain

$$\dot{\mathbf{x}}(t) = \begin{bmatrix} 2\lambda & -\lambda^2 \\ 1 & 0 \end{bmatrix}\mathbf{x}(t) + \begin{bmatrix} 1 \\ 0 \end{bmatrix}u(t)$$
$$y(t) = [0 \quad 1]\mathbf{x}(t)$$

(4.83)

This LTI state equation is a different realization of the same impulse response. This realization is clearly more desirable because it can readily be implemented using an op-amp circuit. The implementation of (4.82) is much more difficult in practice.

PROBLEMS

4.1 An oscillation can be generated by

$$\dot{\mathbf{x}} = \begin{bmatrix} 0 & 1 \\ -1 & 0 \end{bmatrix} \mathbf{x}$$

Show that its solution is

$$\mathbf{x}(t) = \begin{bmatrix} \cos t & \sin t \\ -\sin t & \cos t \end{bmatrix} \mathbf{x}(0)$$

4.2 Use two different methods to find the unit-step response of

$$\dot{\mathbf{x}} = \begin{bmatrix} 0 & 1 \\ -2 & -2 \end{bmatrix} \mathbf{x} + \begin{bmatrix} 1 \\ 1 \end{bmatrix} u$$

$$y = [2 \ 3]\mathbf{x}$$

4.3 Discretize the state equation in Problem 4.2 for $T = 1$ and $T = \pi$.

4.4 Find the companion-form and modal-form equivalent equations of

$$\dot{\mathbf{x}} = \begin{bmatrix} -2 & 0 & 0 \\ 1 & 0 & 1 \\ 0 & -2 & -2 \end{bmatrix} \mathbf{x} + \begin{bmatrix} 1 \\ 0 \\ 1 \end{bmatrix} u$$

$$y = [1 \ -1 \ 0]\mathbf{x}$$

4.5 Find an equivalent state equation of the equation in Problem 4.4 so that all state variables have their largest magnitudes roughly equal to the largest magnitude of the output. If all signals are required to lie inside ± 10 volts and if the input is a step function with magnitude a, what is the permissible largest a?

4.6 Consider

$$\dot{\mathbf{x}} = \begin{bmatrix} \lambda & 0 \\ 0 & \bar{\lambda} \end{bmatrix} \mathbf{x} + \begin{bmatrix} b_1 \\ \bar{b}_1 \end{bmatrix} u \qquad y = [c_1 \ \bar{c}_1]\mathbf{x}$$

where the overbar denotes complex conjugate. Verify that the equation can be transformed into

$$\dot{\bar{\mathbf{x}}} = \bar{\mathbf{A}}\bar{\mathbf{x}} + \bar{\mathbf{b}}u \qquad y = \bar{\mathbf{c}}\bar{\mathbf{x}}$$

with

$$\bar{\mathbf{A}} = \begin{bmatrix} 0 & 1 \\ -\lambda\bar{\lambda} & \lambda + \bar{\lambda} \end{bmatrix} \qquad \bar{\mathbf{b}} = \begin{bmatrix} 0 \\ 1 \end{bmatrix} \qquad \bar{\mathbf{c}}_1 = [-2\mathrm{Re}(\bar{\lambda}b_1c_1) \ \ 2\mathrm{Re}(b_1c_1)]$$

by using the transformation $\mathbf{x} = \mathbf{Q}\bar{\mathbf{x}}$ with

$$\mathbf{Q}_1 = \begin{bmatrix} -\bar{\lambda}b_1 & b_1 \\ -\lambda\bar{b}_1 & \bar{b}_1 \end{bmatrix}$$

4.7 Verify that the Jordan-form equation

$$\dot{x} = \begin{bmatrix} \lambda & 1 & 0 & 0 & 0 & 0 \\ 0 & \lambda & 1 & 0 & 0 & 0 \\ 0 & 0 & \lambda & 0 & 0 & 0 \\ 0 & 0 & 0 & \bar{\lambda} & 1 & 0 \\ 0 & 0 & 0 & 0 & \bar{\lambda} & 1 \\ 0 & 0 & 0 & 0 & 0 & \bar{\lambda} \end{bmatrix} x + \begin{bmatrix} b_1 \\ b_2 \\ b_3 \\ \bar{b}_1 \\ \bar{b}_2 \\ \bar{b}_3 \end{bmatrix} u$$

$$y = [c_1 \ c_2 \ c_3 \ \bar{c}_1 \ \bar{c}_2 \ \bar{c}_3]x$$

can be transformed into

$$\dot{\bar{x}} = \begin{bmatrix} \bar{A} & I_2 & 0 \\ 0 & \bar{A} & I_2 \\ 0 & 0 & \bar{A} \end{bmatrix} \bar{x} + \begin{bmatrix} \bar{b} \\ \bar{b} \\ \bar{b} \end{bmatrix} u \qquad y = [\bar{c}_1 \ \bar{c}_2 \ \bar{c}_3]\bar{x}$$

where $\bar{A}$, $\bar{b}$, and $\bar{c}_i$ are defined in Problem 4.6 and I_2 is the unit matrix of order 2. [*Hint*: Change the order of the state variables from $[x_1 \ x_2 \ x_3 \ x_4 \ x_5 \ x_6]'$ to $[x_1 \ x_4 \ x_2 \ x_5 \ x_3 \ x_6]'$ and then apply the equivalence transformation $x = Q\bar{x}$ with $Q = \text{diag}(Q_1, Q_2, Q_3)$.]

4.8 Are the two sets of state equations

$$\dot{x} = \begin{bmatrix} 2 & 1 & 2 \\ 0 & 2 & 2 \\ 0 & 0 & 1 \end{bmatrix} x + \begin{bmatrix} 1 \\ 1 \\ 0 \end{bmatrix} u \qquad y = [1 \ -1 \ 0]x$$

and

$$\dot{x} = \begin{bmatrix} 2 & 1 & 1 \\ 0 & 2 & 1 \\ 0 & 0 & -1 \end{bmatrix} x + \begin{bmatrix} 1 \\ 1 \\ 0 \end{bmatrix} u \qquad y = [1 \ -1 \ 0]x$$

equivalent? Are they zero-state equivalent?

4.9 Verify that the transfer matrix in (4.33) has the following realization:

$$\dot{x} = \begin{bmatrix} -\alpha_1 I_q & I_q & 0 & \cdots & 0 \\ -\alpha_2 I_q & 0 & I_q & \cdots & 0 \\ \vdots & \vdots & \vdots & & \vdots \\ -\alpha_{r-1} I_q & 0 & 0 & \cdots & I_q \\ -\alpha_r I_q & 0 & 0 & \cdots & 0 \end{bmatrix} x + \begin{bmatrix} N_1 \\ N_2 \\ \vdots \\ N_{r-1} \\ N_r \end{bmatrix} u$$

$$y = [I_q \ 0 \ 0 \ \cdots \ 0]x$$

This is called the *observable canonical form realization* and has dimension rq. It is dual to (4.34).

4.10 Consider the 1×2 proper rational matrix

$$\hat{G}(s) = [d_1 \ d_2] + \frac{1}{s^4 + \alpha_1 s^3 + \alpha_2 s^2 + \alpha_3 s + \alpha_4}$$

$$\times \ [\beta_{11}s^3 + \beta_{21}s^2 + \beta_{31}s + \beta_{41} \quad \beta_{12}s^3 + \beta_{22}s^2 + \beta_{32}s + \beta_{42}]$$

Show that its observable canonical form realization can be reduced from Problem 4.9 as

$$\dot{\mathbf{x}} = \begin{bmatrix} -\alpha_1 & 1 & 0 & 0 \\ -\alpha_2 & 0 & 1 & 0 \\ -\alpha_3 & 0 & 0 & 1 \\ -\alpha_4 & 0 & 0 & 0 \end{bmatrix} \mathbf{x} + \begin{bmatrix} \beta_{11} & \beta_{12} \\ \beta_{21} & \beta_{22} \\ \beta_{31} & \beta_{32} \\ \beta_{41} & \beta_{42} \end{bmatrix} \mathbf{u}$$

$$y = [1 \ 0 \ 0 \ 0]\mathbf{x} + [d_1 \ d_2]\mathbf{u}$$

4.11 Find a realization for the proper rational matrix

$$\hat{\mathbf{G}}(s) = \begin{bmatrix} \dfrac{2}{s+1} & \dfrac{2s-3}{(s+1)(s+2)} \\ \dfrac{s-2}{s+1} & \dfrac{s}{s+2} \end{bmatrix}$$

4.12 Find a realization for each column of $\hat{\mathbf{G}}(s)$ in Problem 4.11 and then connect them, as shown in Fig. 4.4(a), to obtain a realization of $\hat{\mathbf{G}}(s)$. What is the dimension of this realization? Compare this dimension with the one in Problem 4.11.

4.13 Find a realization for each row of $\hat{\mathbf{G}}(s)$ in Problem 4.11 and then connect them, as shown in Fig. 4.4(b), to obtain a realization of $\hat{\mathbf{G}}(s)$. What is the dimension of this realization? Compare this dimension with the ones in Problems 4.11 and 4.12.

4.14 Find a realization for

$$\hat{\mathbf{G}}(s) = \begin{bmatrix} \dfrac{-(12s+6)}{3s+34} & \dfrac{22s+23}{3s+34} \end{bmatrix}$$

4.15 Consider the n-dimensional state equation

$$\dot{\mathbf{x}} = \mathbf{A}\mathbf{x} + \mathbf{b}u \qquad y = \mathbf{c}\mathbf{x}$$

Let $\hat{g}(s)$ be its transfer function. Show that $\hat{g}(s)$ has m zeros or, equivalently, the numerator of $\hat{g}(s)$ has degree m if and only if

$$\mathbf{c}\mathbf{A}^i\mathbf{b} = 0 \qquad \text{for } i = 0, 1, 2, \dots, n - m - 2$$

and $\mathbf{c}\mathbf{A}^{n-m-1}\mathbf{b} \neq 0$. Or, equivalently, the difference between the degrees of the denominator and numerator of $\hat{g}(s)$ is $\alpha = n - m$ if and only if

$$\mathbf{c}\mathbf{A}^{\alpha-1}\mathbf{b} \neq 0 \quad \text{and} \quad \mathbf{c}\mathbf{A}^i\mathbf{b} = 0$$

for $i = 0, 1, 2, \dots, \alpha - 2$.

4.16 Find fundamental matrices and state transition matrices for

$$\dot{\mathbf{x}} = \begin{bmatrix} 0 & 1 \\ 0 & t \end{bmatrix} \mathbf{x}$$

and

$$\dot{\mathbf{x}} = \begin{bmatrix} -1 & e^{2t} \\ 0 & -1 \end{bmatrix} \mathbf{x}$$

4.17 Show $\partial \boldsymbol{\Phi}(t_0, t)/\partial t = -\boldsymbol{\Phi}(t_0, t)\mathbf{A}(t)$.

4.18 Given

$$\mathbf{A}(t) = \begin{bmatrix} a_{11}(t) & a_{12}(t) \\ a_{21}(t) & a_{22}(t) \end{bmatrix}$$

show

$$\det \boldsymbol{\Phi}(t, t_0) = \exp \left[\int_{t_0}^{t} (a_{11}(\tau) + a_{22}(\tau)) \, d\tau \right]$$

4.19 Let

$$\boldsymbol{\Phi}(t, t_0) = \begin{bmatrix} \boldsymbol{\Phi}_{11}(t, t_0) & \boldsymbol{\Phi}_{12}(t, t_0) \\ \boldsymbol{\Phi}_{21}(t, t_0) & \boldsymbol{\Phi}_{22}(t, t_0) \end{bmatrix}$$

be the state transition matrix of

$$\dot{\mathbf{x}}(t) = \begin{bmatrix} \mathbf{A}_{11}(t) & \mathbf{A}_{12}(t) \\ \mathbf{0} & \mathbf{A}_{22}(t) \end{bmatrix} \mathbf{x}(t)$$

Show that $\boldsymbol{\Phi}_{21}(t, t_0) = \mathbf{0}$ for all t and t_0 and that $(\partial/\partial t)\boldsymbol{\Phi}_{ii}(t, t_0) = \mathbf{A}_{ii}\boldsymbol{\Phi}_{ii}(t, t_0)$, for $i = 1, 2$.

4.20 Find the state transition matrix of

$$\dot{\mathbf{x}} = \begin{bmatrix} -\sin t & 0 \\ 0 & -\cos t \end{bmatrix} \mathbf{x}$$

4.21 Verify that $\mathbf{X}(t) = e^{\mathbf{A}t} \mathbf{C} e^{\mathbf{B}t}$ is the solution of

$$\dot{\mathbf{X}} = \mathbf{A}\mathbf{X} + \mathbf{X}\mathbf{B} \qquad \mathbf{X}(0) = \mathbf{C}$$

4.22 Show that if $\dot{\mathbf{A}}(t) = \mathbf{A}_1 \mathbf{A}(t) - \mathbf{A}(t)\mathbf{A}_1$, then

$$\mathbf{A}(t) = e^{\mathbf{A}_1 t} \mathbf{A}(0) e^{-\mathbf{A}_1 t}$$

Show also that the eigenvalues of $\mathbf{A}(t)$ are independent of t.

4.23 Find an equivalent time-invariant state equation of the equation in Problem 4.20.

4.24 Transform a time-invariant $(\mathbf{A}, \mathbf{B}, \mathbf{C})$ into $(\mathbf{0}, \bar{\mathbf{B}}(t), \bar{\mathbf{C}}(t))$ by a time-varying equivalence transformation.

4.25 Find a time-varying realization and a time-invariant realization of the impulse response $g(t) = t^2 e^{\lambda t}$.

4.26 Find a realization of $g(t, \tau) = \sin t (e^{-(t-\tau)}) \cos \tau$. Is it possible to find a time-invariant state equation realization?

Chapter

Stability

5.1 Introduction

Systems are designed to perform some tasks or to process signals. If a system is not stable, the system may burn out, disintegrate, or saturate when a signal, no matter how small, is applied. Therefore an unstable system is useless in practice and stability is a basic requirement for all systems. In addition to stability, systems must meet other requirements, such as to track desired signals and to suppress noise, to be really useful in practice.

The response of linear systems can always be decomposed as the zero-state response and the zero-input response. It is customary to study the stabilities of these two responses separately. We will introduce the BIBO (bounded-input bounded-output) stability for the zero-state response and marginal and asymptotic stabilities for the zero-input response. We study first the time-invariant case and then the time-varying case.

5.2 Input–Output Stability of LTI Systems

Consider a SISO linear time-invariant (LTI) system described by

$$y(t) = \int_0^t g(t - \tau)u(\tau)\,d\tau = \int_0^t g(\tau)u(t - \tau)\,d\tau \tag{5.1}$$

where $g(t)$ is the impulse response or the output excited by an impulse input applied at $t = 0$. Recall that in order to be describable by (5.1), the system must be linear, time-invariant, and causal. In addition, the system must be initially relaxed at $t = 0$.

An input $u(t)$ is said to be *bounded* if $u(t)$ does not grow to positive or negative infinity or, equivalently, there exists a constant u_m such that

$$|u(t)| \leq u_m < \infty \qquad \text{for all } t \geq 0$$

A system is said to be *BIBO stable* (bounded-input bounded-output stable) if every bounded input excites a bounded output. This stability is defined for the zero-state response and is applicable only if the system is initially relaxed.

▶ **Theorem 5.1**

A SISO system described by (5.1) is BIBO stable if and only if $g(t)$ is absolutely integrable in $[0, \infty)$, or

$$\int_0^\infty |g(t)| dt \leq M < \infty$$

for some constant M.

Proof: First we show that if $g(t)$ is absolutely integrable, then every bounded input excites a bounded output. Let $u(t)$ be an arbitrary input with $|u(t)| \leq u_m < \infty$ for all $t \geq 0$. Then

$$|y(t)| = \left| \int_0^t g(\tau)u(t - \tau)\, d\tau \right| \leq \int_0^t |g(\tau)||u(t - \tau)|\, d\tau$$

$$\leq u_m \int_0^\infty |g(\tau)|\, d\tau \leq u_m M$$

Thus the output is bounded. Next we show intuitively that if $g(t)$ is not absolutely integrable, then the system is not BIBO stable. If $g(t)$ is not absolutely integrable, then there exists a t_1 such that

$$\int_0^{t_1} |g(\tau)|\, d\tau = \infty$$

Let us choose

$$u(t_1 - \tau) = \begin{cases} 1 & \text{if } g(\tau) \geq 0 \\ -1 & \text{if } g(\tau) < 0 \end{cases}$$

Clearly u is bounded. However, the output excited by this input equals

$$y(t_1) = \int_0^{t_1} g(\tau)u(t_1 - \tau)\, d\tau = \int_0^{t_1} |g(\tau)|\, d\tau = \infty$$

which is not bounded. Thus the system is not BIBO stable. This completes the proof of Theorem 5.1.   Q.E.D.

A function that is absolutely integrable may not be bounded or may not approach zero as $t \to \infty$. Indeed, consider the function defined by

$$f(t - n) = \begin{cases} n + (t - n)n^4 & \text{for } n - 1/n^3 \leq t \leq n \\ n - (t - n)n^4 & \text{for } n < t \leq n + 1/n^3 \end{cases}$$

for $n = 2, 3, \ldots$ and plotted in Fig. 5.1. The area under each triangle is $1/n^2$. Thus the absolute

Figure 5.1 Function.

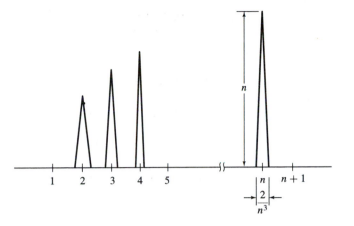

integration of the function equals $\sum_{n=2}^{\infty}(1/n^2) < \infty$. This function is absolutely integrable but is not bounded and does not approach zero as $t \to \infty$.

▶ **Theorem 5.2**

If a system with impulse response $g(t)$ is BIBO stable, then, as $t \to \infty$:

1. The output excited by $u(t) = a$, for $t \geq 0$, approaches $\hat{g}(0) \cdot a$.
2. The output excited by $u(t) = \sin \omega_o t$, for $t \geq 0$, approaches

$$|\hat{g}(j\omega_o)| \sin(\omega_o t + \measuredangle \hat{g}(j\omega_o))$$

where $\hat{g}(s)$ is the Laplace transform of $g(t)$ or

$$\hat{g}(s) = \int_0^\infty g(\tau)e^{-s\tau}\, d\tau \tag{5.2}$$

Proof: If $u(t) = a$ for all $t \geq 0$, then (5.1) becomes

$$y(t) = \int_0^t g(\tau)u(t-\tau)\, d\tau = a\int_0^t g(\tau)\, d\tau$$

which implies

$$y(t) \to a \int_0^\infty g(\tau)\, d\tau = a\hat{g}(0) \qquad \text{as } t \to \infty$$

where we have used (5.2) with $s = 0$. This establishes the first part of Theorem 5.2. If $u(t) = \sin \omega_o t$, then (5.1) becomes

$$y(t) = \int_0^t g(\tau) \sin \omega_o(t-\tau)\, d\tau$$

$$= \int_0^t g(\tau) \left[\sin \omega_o t \cos \omega_o \tau - \cos \omega_o t \sin \omega_o \tau \right]\, d\tau$$

$$= \sin \omega_o t \int_0^t g(\tau) \cos \omega_o \tau\, d\tau - \cos \omega_o t \int_0^t g(\tau) \sin \omega_o \tau\, d\tau$$

Thus we have, as $t \to \infty$,

$$y(t) \to \sin \omega_o t \int_0^\infty g(\tau) \cos \omega_o \tau \, d\tau - \cos \omega_o t \int_0^\infty g(\tau) \sin \omega_o \tau \, d\tau \qquad (5.3)$$

If $g(t)$ is absolutely integrable, we can replace s by $j\omega$ in (5.2) to yield

$$\hat{g}(j\omega) = \int_0^\infty g(\tau)[\cos \omega\tau - j \sin \omega\tau] \, d\tau$$

The impulse response $g(t)$ is assumed implicitly to be real; thus we have

$$\text{Re}[\hat{g}(j\omega)] = \int_0^\infty g(\tau) \cos \omega\tau \, d\tau$$

and

$$\text{Im}[\hat{g}(j\omega)] = -\int_0^\infty g(\tau) \sin \omega\tau \, d\tau$$

where Re and Im denote, respectively, the real part and imaginary part. Substituting these into (5.3) yields

$$y(t) \to \sin \omega_o t \, (\text{Re}[\hat{g}(j\omega_o)]) + \cos \omega_o t \, (\text{Im}[\hat{g}(j\omega_o)])$$
$$= |\hat{g}(j\omega_o)| \sin(\omega_o t + \angle\hat{g}(j\omega_o))$$

This completes the proof of Theorem 5.2.   Q.E.D.

Theorem 5.2 is a basic result; filtering of signals is based essentially on this theorem. Next we state the BIBO stability condition in terms of proper rational transfer functions.

▶ **Theorem 5.3**

A SISO system with proper rational transfer function $\hat{g}(s)$ is BIBO stable if and only if every pole of $\hat{g}(s)$ has a negative real part or, equivalently, lies inside the left-half s-plane.

If $\hat{g}(s)$ has pole p_i with multiplicity m_i, then its partial fraction expansion contains the factors

$$\frac{1}{s - p_i}, \quad \frac{1}{(s - p_i)^2}, \quad \cdots, \quad \frac{1}{(s - p_i)^{m_i}}$$

Thus the inverse Laplace transform of $\hat{g}(s)$ or the impulse response contains the factors

$$e^{p_i t}, \quad t e^{p_i t}, \quad \cdots, \quad t^{m_i - 1} e^{p_i t}$$

It is straightforward to verify that every such term is absolutely integrable if and only if p_i has a negative real part. Using this fact, we can establish Theorem 5.3.

EXAMPLE 5.1 Consider the positive feedback system shown in Fig. 2.5(a). Its impulse response was computed in (2.9) as

$$g(t) = \sum_{i=1}^\infty a^i \delta(t - i)$$

where the gain a can be positive or negative. The impulse is defined as the limit of the pulse in Fig. 2.3 and can be considered to be positive. Thus we have

$$|g(t)| = \sum_{i=1}^{\infty} |a|^i \delta(t - i)$$

and

$$\int_0^{\infty} |g(t)|\, dt = \sum_{i=1}^{\infty} |a|^i = \begin{cases} \infty & \text{if } |a| \geq 1 \\ |a|/(1 - |a|) < \infty & \text{if } |a| < 1 \end{cases}$$

Thus we conclude that the positive feedback system in Fig. 2.5(a) is BIBO stable if and only if the gain a has a magnitude less than 1.

The transfer function of the system was computed in (2.12) as

$$\hat{g}(s) = \frac{ae^{-s}}{1 - ae^{-s}}$$

It is an irrational function of s and Theorem 5.3 is not applicable. In this case, it is simpler to use Theorem 5.1 to check its stability.

For multivariable systems, we have the following results.

▶ **Theorem 5.M1**

A multivariable system with impulse response matrix $\mathbf{G}(t) = [g_{ij}(t)]$ is BIBO stable if and only if every $g_{ij}(t)$ is absolutely integrable in $[0, \infty)$.

▶ **Theorem 5.M3**

A multivariable system with proper rational transfer matrix $\hat{\mathbf{G}}(s) = [\hat{g}_{ij}(s)]$ is BIBO stable if and only if every pole of every $\hat{g}_{ij}(s)$ has a negative real part.

We now discuss the BIBO stability of state equations. Consider

$$\begin{aligned} \dot{\mathbf{x}}(t) &= \mathbf{A}\mathbf{x}(t) + \mathbf{B}\mathbf{u}(t) \\ \mathbf{y}(t) &= \mathbf{C}\mathbf{x}(t) + \mathbf{D}\mathbf{u}(t) \end{aligned} \tag{5.4}$$

Its transfer matrix is

$$\hat{\mathbf{G}}(s) = \mathbf{C}(s\mathbf{I} - \mathbf{A})^{-1}\mathbf{B} + \mathbf{D}$$

Thus Equation (5.4) or, to be more precise, the zero-state response of (5.4) is BIBO stable if and only if every pole of $\hat{\mathbf{G}}(s)$ has a negative real part. Recall that every pole of every entry of $\hat{\mathbf{G}}(s)$ is called a pole of $\hat{\mathbf{G}}(s)$.

We discuss the relationship between the poles of $\hat{\mathbf{G}}(s)$ and the eigenvalues of $\mathbf{A}$. Because of

$$\hat{\mathbf{G}}(s) = \frac{1}{\det(s\mathbf{I} - \mathbf{A})}\mathbf{C}[\text{Adj}(s\mathbf{I} - \mathbf{A})]\mathbf{B} + \mathbf{D} \tag{5.5}$$

every pole of $\hat{\mathbf{G}}(s)$ is an eigenvalue of $\mathbf{A}$. Thus if every eigenvalue of $\mathbf{A}$ has a negative real part, then every pole has a negative real part and (5.4) is BIBO stable. On the other hand, because of possible cancellation in (5.5), not every eigenvalue is a pole. Thus, even if $\mathbf{A}$ has some eigenvalues with zero or positive real part, (5.4) may still be BIBO stable, as the next example shows.

EXAMPLE 5.2 Consider the network shown in Fig. 4.2(b). Its state equation was derived in Example 4.4 as

$$\dot{x}(t) = x(t) + 0 \cdot u(t) \qquad y(t) = 0.5x(t) + 0.5u(t) \tag{5.6}$$

The A-matrix is 1 and its eigenvalue is 1. It has a positive real part. The transfer function of the equation is

$$\hat{g}(s) = 0.5(s-1)^{-1} \cdot 0 + 0.5 = 0.5$$

The transfer function equals 0.5. It has no pole and no condition to meet. Thus (5.6) is BIBO stable even though it has an eigenvalue with a positive real part. We mention that BIBO stability does not say anything about the zero-input response, which will be discussed later.

5.2.1 Discrete-Time Case

Consider a discrete-time SISO system described by

$$y[k] = \sum_{m=0}^{k} g[k-m]u[m] = \sum_{m=0}^{k} g[m]u[k-m] \tag{5.7}$$

where $g[k]$ is the impulse response sequence or the output sequence excited by an impulse sequence applied at $k = 0$. Recall that in order to be describable by (5.7), the discrete-time system must be linear, time-invariant, and causal. In addition, the system must be initially relaxed at $k = 0$.

An input sequence $u[k]$ is said to be *bounded* if $u[k]$ does not grow to positive or negative infinity or there exists a constant u_m such that

$$|u[k]| \leq u_m < \infty \qquad \text{for } k = 0, 1, 2, \ldots$$

A system is said to be *BIBO stable* (bounded-input bounded-output stable) if every bounded-input sequence excites a bounded-output sequence. This stability is defined for the zero-state response and is applicable only if the system is initially relaxed.

▶ **Theorem 5.D1**

A discrete-time SISO system described by (5.7) is BIBO stable if and only if $g[k]$ is absolutely summable in $[0, \infty)$ or

$$\sum_{k=0}^{\infty} |g[k]| \leq M < \infty$$

for some constant M.

Its proof is similar to the proof of Theorem 5.1 and will not be repeated. We give a discrete counterpart of Theorem 5.2 in the following.

▶ **Theorem 5.D2**

If a discrete-time system with impulse response sequence $g[k]$ is BIBO stable, then, as $k \to \infty$:

1. The output excited by $u[k] = a$, for $k \geq 0$, approaches $\hat{g}(1) \cdot a$.
2. The output excited by $u[k] = \sin \omega_o k$, for $k \geq 0$, approaches

$$|\hat{g}(e^{j\omega_o})| \sin(\omega_o k + \measuredangle \hat{g}(e^{j\omega_o}))$$

where $\hat{g}(z)$ is the z-transform of $g[k]$ or

$$\hat{g}(z) = \sum_{m=0}^{\infty} g[m]z^{-m} \tag{5.8}$$

⇒ ***Proof:*** If $u[k] = a$ for all $k \geq 0$, then (5.7) becomes

$$y[k] = \sum_{0}^{k} g[m]u[k-m] = a \sum_{k=0}^{k} g[m]$$

which implies

$$y[k] \to a \sum_{m=0}^{\infty} g[m] = a\hat{g}(1) \qquad \text{as } k \to \infty$$

where we have used (5.8) with $z = 1$. This establishes the first part of Theorem 5.D2. If $u[k] = \sin \omega_o k$, then (5.7) becomes

$$y[k] = \sum_{m=0}^{k} g[m] \sin \omega_o[k-m]$$

$$= \sum_{m=0}^{k} g[m](\sin \omega_o k \cos \omega_o m - \cos \omega_o k \sin \omega_o m)$$

$$= \sin \omega_o k \sum_{m=0}^{k} g[m] \cos \omega_o m - \cos \omega_o k \sum_{m=0}^{k} g[m] \sin \omega_o m$$

Thus we have, as $k \to \infty$,

$$y[k] \to \sin \omega_o k \sum_{m=0}^{\infty} g[m] \cos \omega_o m - \cos \omega_o k \sum_{m=0}^{\infty} g[m] \sin \omega_o m \tag{5.9}$$

If $g[k]$ is absolutely summable, we can replace z by $e^{j\omega}$ in (5.8) to yield

$$\hat{g}(e^{j\omega}) = \sum_{m=0}^{\infty} g[m]e^{-j\omega m} = \sum_{m=0}^{\infty} g[m][\cos \omega m - j \sin \omega m]$$

Thus (5.9) becomes

$$y[k] \rightarrow \sin \omega_o k (\text{Re}[\hat{g}(e^{j\omega_o})]) + \cos \omega_o k (\text{Im}[\hat{g}(e^{j\omega_o})])$$
$$= |\hat{g}(e^{j\omega_o})| \sin(\omega_o k + \measuredangle \hat{g}(e^{j\omega_o}))$$

This completes the proof of Theorem 5.D2.   Q.E.D.

Theorem 5.D2 is a basic result in digital signal processing. Next we state the BIBO stability in terms of discrete proper rational transfer functions.

▶ **Theorem 5.D3**

A discrete-time SISO system with proper rational transfer function $\hat{g}(z)$ is BIBO stable if and only if every pole of $\hat{g}(z)$ has a magnitude less than 1 or, equivalently, lies inside the unit circle on the z-plane.

If $\hat{g}(z)$ has pole p_i with multiplicity m_i, then its partial fraction expansion contains the factors

$$\frac{1}{z - p_i}, \quad \frac{1}{(z - p_i)^2}, \quad \cdots, \quad \frac{1}{(z - p_i)^{m_i}}$$

Thus the inverse z-transform of $\hat{g}(z)$ or the impulse response sequence contains the factors

$$p_i^k, \quad k p_i^k, \quad \cdots, \quad k^{m_i - 1} p_i^k$$

It is straightforward to verify that every such term is absolutely summable if and only if p_i has a magnitude less than 1. Using this fact, we can establish Theorem 5.D3.

In the continuous-time case, an absolutely integrable function $f(t)$, as shown in Fig. 5.1, may not be bounded and may not approach zero as $t \rightarrow \infty$. In the discrete-time case, if $g[k]$ is absolutely summable, then it must be bounded and approach zero as $k \rightarrow \infty$. However, the converse is not true as the next example shows.

EXAMPLE 5.3 Consider a discrete-time LTI system with impulse response sequence $g[k] = 1/k$, for $k = 1, 2, \ldots$, and $g[0] = 0$. We compute

$$S := \sum_{k=0}^{\infty} |g[k]| = \sum_{k=1}^{\infty} \frac{1}{k} = 1 + \frac{1}{2} + \frac{1}{3} + \frac{1}{4} + \cdots$$
$$= 1 + \frac{1}{2} + \left(\frac{1}{3} + \frac{1}{4} \right) + \left(\frac{1}{5} + \cdots + \frac{1}{8} \right) + \left(\frac{1}{9} + \cdots + \frac{1}{16} \right) + \cdots$$

There are two terms, each is $\frac{1}{4}$ or larger, in the first pair of parentheses; therefore their sum is larger than $\frac{1}{2}$. There are four terms, each is $\frac{1}{8}$ or larger, in the second pair of parentheses; therefore their sum is larger than $\frac{1}{2}$. Proceeding forward we conclude

$$S > 1 + \frac{1}{2} + \frac{1}{2} + \frac{1}{2} + \cdots = \infty$$

This impulse response sequence is bounded and approaches 0 as $k \rightarrow \infty$ but is not absolutely summable. Thus the discrete-time system is not BIBO stable according to Theorem 5.D1. The transfer function of the system can be shown to equal

$$\hat{g}(z) = -\ln(1 + z^{-1})$$

It is not a rational function of z and Theorem 5.D3 is not applicable.

For multivariable discrete-time systems, we have the following results.

▶ **Theorem 5.MD1**

A MIMO discrete-time system with impulse response sequence matrix $\mathbf{G}[k] = \left[g_{ij}[k]\right]$ is BIBO stable if and only if every $g_{ij}[k]$ is absolutely summable.

▶ **Theorem 5.MD3**

A MIMO discrete-time system with discrete proper rational transfer matrix $\hat{\mathbf{G}}(z) = \left[\hat{g}_{ij}(z)\right]$ is BIBO stable if and only if every pole of every $\hat{g}_{ij}(z)$ has a magnitude less than 1.

We now discuss the BIBO stability of discrete-time state equations. Consider

$$\mathbf{x}[k + 1] = \mathbf{Ax}[k] + \mathbf{Bu}[k]$$
$$\mathbf{y}[k] = \mathbf{Cx}[k] + \mathbf{Du}[k] \tag{5.10}$$

Its discrete transfer matrix is

$$\hat{\mathbf{G}}(z) = \mathbf{C}(z\mathbf{I} - \mathbf{A})^{-1}\mathbf{B} + \mathbf{D}$$

Thus Equation (5.10) or, to be more precise, the zero-state response of (5.10) is BIBO stable if and only if every pole of $\hat{\mathbf{G}}(z)$ has a magnitude less than 1.

We discuss the relationship between the poles of $\hat{\mathbf{G}}(z)$ and the eigenvalues of $\mathbf{A}$. Because of

$$\hat{\mathbf{G}}(z) = \frac{1}{\det(z\mathbf{I} - \mathbf{A})}\mathbf{C}[\mathrm{Adj}(z\mathbf{I} - \mathbf{A})]\mathbf{B} + \mathbf{D}$$

every pole of $\hat{\mathbf{G}}(z)$ is an eigenvalue of $\mathbf{A}$. Thus if every eigenvalue of $\mathbf{A}$ has a negative real part, then (5.10) is BIBO stable. On the other hand, even if $\mathbf{A}$ has some eigenvalues with zero or positive real part, (5.10) may, as in the continuous-time case, still be BIBO stable.

5.3 Internal Stability

The BIBO stability is defined for the zero-state response. Now we study the stability of the zero-input response or the response of

$$\dot{\mathbf{x}}(t) = \mathbf{Ax}(t) \tag{5.11}$$

excited by nonzero initial state $\mathbf{x}_o$. Clearly, the solution of (5.11) is

$$\mathbf{x}(t) = e^{\mathbf{A}t}\mathbf{x}_o \tag{5.12}$$

> **Definition 5.1** *The zero-input response of (5.4) or the equation* $\dot{\mathbf{x}} = \mathbf{A}\mathbf{x}$ *is marginally stable or stable in the sense of Lyapunov if every finite initial state* $\mathbf{x}_o$ *excites a bounded response. It is asymptotically stable if every finite initial state excites a bounded response, which, in addition, approaches* $\mathbf{0}$ *as* $t \rightarrow \infty$.

We mention that this definition is applicable only to linear systems. The definition that is applicable to both linear and nonlinear systems must be defined using the concept of equivalence states and can be found, for example, in Reference [6, pp. 401–403]. This text studies only linear systems; therefore we use the simplified Definition 5.1.

▶ **Theorem 5.4**

1. The equation $\dot{\mathbf{x}} = \mathbf{A}\mathbf{x}$ is marginally stable if and only if all eigenvalues of $\mathbf{A}$ have zero or negative real parts and those with zero real parts are simple roots of the minimal polynomial of $\mathbf{A}$.

2. The equation $\dot{\mathbf{x}} = \mathbf{A}\mathbf{x}$ is asymptotically stable if and only if all eigenvalues of $\mathbf{A}$ have negative real parts.

We first mention that any (algebraic) equivalence transformation will not alter the stability of a state equation. Consider $\bar{\mathbf{x}} = \mathbf{P}\mathbf{x}$, where $\mathbf{P}$ is a nonsingular matrix. Then $\dot{\mathbf{x}} = \mathbf{A}\mathbf{x}$ is equivalent to $\dot{\bar{\mathbf{x}}} = \bar{\mathbf{A}}\bar{\mathbf{x}} = \mathbf{P}\mathbf{A}\mathbf{P}^{-1}\bar{\mathbf{x}}$. Because $\mathbf{P}$ is nonsingular, if $\mathbf{x}$ is bounded, so is $\bar{\mathbf{x}}$; if $\mathbf{x}$ approaches $\mathbf{0}$ as $t \rightarrow \infty$, so does $\bar{\mathbf{x}}$. Thus we may study the stability of $\mathbf{A}$ by using $\bar{\mathbf{A}}$. Note that the eigenvalues of $\mathbf{A}$ and of $\bar{\mathbf{A}}$ are the same as discussed in Section 4.3.

The response of $\dot{\bar{\mathbf{x}}} = \bar{\mathbf{A}}\bar{\mathbf{x}}$ excited by $\bar{\mathbf{x}}(0)$ equals $\bar{\mathbf{x}}(t) = e^{\bar{\mathbf{A}}t}\mathbf{x}(0)$. It is clear that the response is bounded if and only if every entry of $e^{\bar{\mathbf{A}}t}$ is bounded for all $t \geq 0$. If $\bar{\mathbf{A}}$ is in Jordan form, then $e^{\bar{\mathbf{A}}t}$ is of the form shown in (3.48). Using (3.48), we can show that if an eigenvalue has a negative real part, then every entry of (3.48) is bounded and approaches 0 as $t \rightarrow \infty$. If an eigenvalue has zero real part and has no Jordan block of order 2 or higher, then the corresponding entry in (3.48) is a constant or is sinusoidal for all t and is, therefore, bounded. This establishes the sufficiency of the first part of Theorem 5.4. If $\bar{\mathbf{A}}$ has an eigenvalue with a positive real part, then every entry in (3.48) will grow without bound. If $\bar{\mathbf{A}}$ has an eigenvalue with zero real part and its Jordan block has order 2 or higher, then (3.48) has at least one entry that grows unbounded. This completes the proof of the first part. To be asymptotically stable, every entry of (3.48) must approach zero as $t \rightarrow \infty$. Thus no eigenvalue with zero real part is permitted. This establishes the second part of the theroem.

EXAMPLE 5.4 Consider

$$\dot{\mathbf{x}} = \begin{bmatrix} 0 & 0 & 0 \\ 0 & 0 & 0 \\ 0 & 0 & -1 \end{bmatrix} \mathbf{x}$$

Its characteristic polynomial is $\Delta(\lambda) = \lambda^2(\lambda + 1)$ and its mimimal polynomial is $\psi(\lambda) = \lambda(\lambda + 1)$. The matrix has eigenvalues 0, 0, and -1. The eigenvlaue 0 is a simple root of the minimal polynomial. Thus the equation is marginally stable. The equation

$$\dot{\mathbf{x}} = \begin{bmatrix} 0 & 1 & 0 \\ 0 & 0 & 0 \\ 0 & 0 & -1 \end{bmatrix} \mathbf{x}$$

is not marginally stable, however, because its minimal polynomial is $\lambda^2(\lambda + 1)$ and $\lambda = 0$ is not a simple root of the minimal polynomial.

As discussed earlier, every pole of the transfer matrix

$$\hat{\mathbf{G}}(s) = \mathbf{C}(s\mathbf{I} - \mathbf{A})^{-1}\mathbf{B} + \mathbf{D}$$

is an eigenvalue of $\mathbf{A}$. Thus asymptotic stability implies BIBO stability. Note that asymptotic stability is defined for the zero-input response, whereas BIBO stability is defined for the zero-state response. The system in Example 5.2 has eigenvalue 1 and is not asymptotically stable; however, it is BIBO stable. Thus BIBO stability, in general, does not imply asymptotic stability. We mention that marginal stability is useful only in the design of oscillators. Other than oscillators, every physical system is designed to be asymptotically stable or BIBO stable with some additional conditions, as we will discuss in Chapter 7.

5.3.1 Discrete-Time Case

This subsection studies the internal stability of discrete-time systems or the stability of

$$\mathbf{x}[k + 1] = \mathbf{A}\mathbf{x}[k] \tag{5.13}$$

excited by nonzero initial state $\mathbf{x}_o$. The solution of (5.13) is, as derived in (4.20),

$$\mathbf{x}[k] = \mathbf{A}^k\mathbf{x}_o \tag{5.14}$$

Equation (5.13) is said to be *marginally stable* or *stable in the sense of Lyapunov* if every finite initial state $\mathbf{x}_o$ excites a bounded response. It is *asymptotically stable* if every finite initial state excites a bounded response, which, in addition, approaches $\mathbf{0}$ as $k \to \infty$. These definitions are identical to the continuous-time case.

▶ **Theorem 5.D4**

1. The equation $\mathbf{x}[k + 1] = \mathbf{A}\mathbf{x}[k]$ is marginally stable if and only if all eigenvalues of $\mathbf{A}$ have magnitudes less than or equal to 1 and those equal to 1 are simple roots of the minimal polynomial of $\mathbf{A}$.

2. The equation $\mathbf{x}[k + 1] = \mathbf{A}\mathbf{x}[k]$ is asymptotically stable if and only if all eigenvalues of $\mathbf{A}$ have magnitudes less than 1.

As in the continuous-time case, any (algebraic) equivalence transformation will not alter the stability of a state equation. Thus we can use Jordan form to establish the theorem. The proof is similar to the continuous-time case and will not be repeated. Asymptotic stability implies BIBO stability but not the converse. We mention that marginal stability is useful only in the design of discrete-time oscillators. Other than oscillators, every discrete-time physical system is designed to be asymptotically stable or BIBO stable with some additional conditions, as we will discuss in Chapter 7.

5.4 Lyapunov Theorem

This section introduces a different method of checking asymptotic stability of $\dot{\mathbf{x}} = \mathbf{A}\mathbf{x}$. For convenience, we call $\mathbf{A}$ stable if every eigenvalue of $\mathbf{A}$ has a negative real part.

▷ **Theorem 5.5**

All eigenvalues of $\mathbf{A}$ have negative real parts if and only if for any given positive definite symmetric matrix $\mathbf{N}$, the *Lyapunov* equation

$$\mathbf{A}'\mathbf{M} + \mathbf{M}\mathbf{A} = -\mathbf{N} \tag{5.15}$$

has a unique symmetric solution $\mathbf{M}$ and $\mathbf{M}$ is positive definite.

▷ **Corollary 5.5**

All eigenvalues of an $n \times n$ matrix $\mathbf{A}$ have negative real parts if and only if for any given $m \times n$ matrix $\bar{\mathbf{N}}$ with $m < n$ and with the property

$$\text{rank } O := \text{rank} \begin{bmatrix} \bar{\mathbf{N}} \\ \bar{\mathbf{N}}\mathbf{A} \\ \vdots \\ \bar{\mathbf{N}}\mathbf{A}^{n-1} \end{bmatrix} = n \quad \text{(full column rank)} \tag{5.16}$$

where O is an $nm \times n$ matrix, the Lyapunov equation

$$\mathbf{A}'\mathbf{M} + \mathbf{M}\mathbf{A} = -\bar{\mathbf{N}}'\bar{\mathbf{N}} =: -\mathbf{N} \tag{5.17}$$

has a unique symmetric solution $\mathbf{M}$ and $\mathbf{M}$ is positive definite.

For any $\bar{\mathbf{N}}$, the matrix $\mathbf{N}$ in (5.17) is positive semidefinite (Theorem 3.7). Theorem 5.5 and its corollary are valid for any given $\mathbf{N}$; therefore we shall use the simplest possible $\mathbf{N}$. Even so, using them to check stability of $\mathbf{A}$ is not simple. It is much simpler to compute, using MATLAB, the eigenvalues of $\mathbf{A}$ and then check their real parts. Thus the importance of Theorem 5.5 and its corollary is not in checking the stability of $\mathbf{A}$ but rather in studying the stability of nonlinear systems. They are essential in using the so-called second method of Lyapunov. We mention that Corollary 5.5 can be used to prove the Routh–Hurwitz test. See Reference [6, pp. 417–419].

⟶ **Proof of Theorem 5.5** *Necessity:* Equation (5.15) is a special case of (3.59) with $\mathbf{A} = \mathbf{A}'$ and $\mathbf{B} = \mathbf{A}$. Because $\mathbf{A}$ and $\mathbf{A}'$ have the same set of eigenvalues, if $\mathbf{A}$ is stable, $\mathbf{A}$ has no two eigenvalues such that $\lambda_i + \lambda_j = 0$. Thus the Lyapunov equation is nonsingular and has a unique solution $\mathbf{M}$ for any $\mathbf{N}$. We claim that the solution can be expressed as

$$\mathbf{M} = \int_0^\infty e^{\mathbf{A}'t}\mathbf{N}e^{\mathbf{A}t}\,dt \tag{5.18}$$

Indeed, substituting (5.18) into (5.15) yields

$$\begin{aligned}
\mathbf{A}'\mathbf{M} + \mathbf{M}\mathbf{A} &= \int_0^\infty \mathbf{A}' e^{\mathbf{A}'t}\mathbf{N}e^{\mathbf{A}t}\,dt + \int_0^\infty e^{\mathbf{A}'t}\mathbf{N}e^{\mathbf{A}t}\mathbf{A}\,dt \\
&= \int_0^\infty \frac{d}{dt}\left(e^{\mathbf{A}'t}\mathbf{N}e^{\mathbf{A}t}\right)dt = e^{\mathbf{A}'t}\mathbf{N}e^{\mathbf{A}t}\Big|_{t=0}^\infty \\
&= \mathbf{0} - \mathbf{N} = -\mathbf{N} \tag{5.19}
\end{aligned}$$

where we have used the fact $e^{\mathbf{A}t} = \mathbf{0}$ at $t = \infty$ for stable $\mathbf{A}$. This shows that the $\mathbf{M}$ in (5.18) is the solution. It is clear that if $\mathbf{N}$ is symmetric, so is $\mathbf{M}$. Let us decompose $\mathbf{N}$ as $\mathbf{N} = \bar{\mathbf{N}}'\bar{\mathbf{N}}$, where $\bar{\mathbf{N}}$ is nonsingular (Theorem 3.7) and consider

$$\mathbf{x}'\mathbf{M}\mathbf{x} = \int_0^\infty \mathbf{x}' e^{\mathbf{A}'t}\bar{\mathbf{N}}'\bar{\mathbf{N}}e^{\mathbf{A}t}\mathbf{x}\,dt = \int_0^\infty ||\bar{\mathbf{N}}e^{\mathbf{A}t}\mathbf{x}||_2^2\,dt \tag{5.20}$$

Because both $\bar{\mathbf{N}}$ and $e^{\mathbf{A}t}$ are nonsingular, for any nonzero $\mathbf{x}$, the integrand of (5.20) is positive for every t. Thus $\mathbf{x}'\mathbf{M}\mathbf{x}$ is positive for any $\mathbf{x} \neq \mathbf{0}$. This shows the positive definiteness of $\mathbf{M}$.

Sufficiency: We show that if $\mathbf{N}$ and $\mathbf{M}$ are positive definite, then $\mathbf{A}$ is stable. Let λ be an eigenvalue of $\mathbf{A}$ and $\mathbf{v} \neq \mathbf{0}$ be a corresponding eigenvector; that is, $\mathbf{A}\mathbf{v} = \lambda\mathbf{v}$. Even though $\mathbf{A}$ is a real matrix, its eigenvalue and eigenvector can be complex, as shown in Example 3.6. Taking the complex-conjugate transpose of $\mathbf{A}\mathbf{v} = \lambda\mathbf{v}$ yields $\mathbf{v}^*\mathbf{A}^* = \mathbf{v}^*\mathbf{A}' = \lambda^*\mathbf{v}^*$, where the asterisk denotes complex-conjugate transpose. Premultiplying $\mathbf{v}^*$ and postmultiplying $\mathbf{v}$ to (5.15) yields

$$\begin{aligned}
-\mathbf{v}^*\mathbf{N}\mathbf{v} &= \mathbf{v}^*\mathbf{A}'\mathbf{M}\mathbf{v} + \mathbf{v}^*\mathbf{M}\mathbf{A}\mathbf{v} \\
&= (\lambda^* + \lambda)\mathbf{v}^*\mathbf{M}\mathbf{v} = 2\mathrm{Re}(\lambda)\mathbf{v}^*\mathbf{M}\mathbf{v} \tag{5.21}
\end{aligned}$$

Because $\mathbf{v}^*\mathbf{M}\mathbf{v}$ and $\mathbf{v}^*\mathbf{N}\mathbf{v}$ are, as discussed in Section 3.9, both real and positive, (5.21) implies $\mathrm{Re}(\lambda) < 0$. This shows that every eigenvalue of $\mathbf{A}$ has a negative real part. Q.E.D.

The proof of Corollary 5.5 follows the proof of Theorem 5.5 with some modification. We discuss only where the proof of Theorem 5.5 is not applicable. Consider (5.20). Now $\bar{\mathbf{N}}$ is $m \times n$ with $m < n$ and $\mathbf{N} = \bar{\mathbf{N}}'\bar{\mathbf{N}}$ is positive semidefinite. Even so, $\mathbf{M}$ in (5.18) can still be positive definite if the integrand of (5.20) is not identically zero for all t. Suppose the integrand of (5.20) is identically zero or $\bar{\mathbf{N}}e^{\mathbf{A}t}\mathbf{x} \equiv \mathbf{0}$. Then its derivative with respect to t yields $\bar{\mathbf{N}}\mathbf{A}e^{\mathbf{A}t}\mathbf{x} = \mathbf{0}$. Proceeding forward, we can obtain

$$\begin{bmatrix} \bar{\mathbf{N}} \\ \bar{\mathbf{N}}\mathbf{A} \\ \vdots \\ \bar{\mathbf{N}}\mathbf{A}^{n-1} \end{bmatrix} e^{\mathbf{A}t}\mathbf{x} = \mathbf{0} \tag{5.22}$$

This equation implies that, because of (5.16) and the nonsingularity of $e^{\mathbf{A}t}$, the only $\mathbf{x}$ meeting

(5.22) is $\mathbf{0}$. Thus the integrand of (5.20) cannot be identically zero for any $\mathbf{x} \neq \mathbf{0}$. Thus $\mathbf{M}$ is positive definite under the condition in (5.16). This shows the necessity of Corollary 5.5. Next we consider (5.21) with $\mathbf{N} = \bar{\mathbf{N}}'\bar{\mathbf{N}}$ or[1]

$$2\mathrm{Re}(\lambda)\mathbf{v}^*\mathbf{M}\mathbf{v} = -\mathbf{v}^*\bar{\mathbf{N}}'\bar{\mathbf{N}}\mathbf{v} = -||\bar{\mathbf{N}}\mathbf{v}||_2^2 \tag{5.23}$$

We show that $\bar{\mathbf{N}}\mathbf{v}$ is nonzero under (5.16). Because of $\mathbf{A}\mathbf{v} = \lambda\mathbf{v}$, we have $\mathbf{A}^2\mathbf{v} = \lambda\mathbf{A}\mathbf{v} = \lambda^2\mathbf{v}$, $\dots$, $\mathbf{A}^{n-1}\mathbf{v} = \lambda^{n-1}\mathbf{v}$. Consider

$$\begin{bmatrix} \bar{\mathbf{N}} \\ \bar{\mathbf{N}}\mathbf{A} \\ \vdots \\ \bar{\mathbf{N}}\mathbf{A}^{n-1} \end{bmatrix} \mathbf{v} = \begin{bmatrix} \bar{\mathbf{N}}\mathbf{v} \\ \bar{\mathbf{N}}\mathbf{A}\mathbf{v} \\ \vdots \\ \bar{\mathbf{N}}\mathbf{A}^{n-1}\mathbf{v} \end{bmatrix} = \begin{bmatrix} \bar{\mathbf{N}}\mathbf{v} \\ \lambda\bar{\mathbf{N}}\mathbf{v} \\ \vdots \\ \lambda^{n-1}\bar{\mathbf{N}}\mathbf{v} \end{bmatrix}$$

If $\bar{\mathbf{N}}\mathbf{v} = \mathbf{0}$, the rightmost matrix is zero; the leftmost matrix, however, is nonzero under the conditions of (5.16) and $\mathbf{v} \neq \mathbf{0}$. This is a contradiction. Thus $\bar{\mathbf{N}}\mathbf{v}$ is nonzero and (5.23) implies $\mathrm{Re}(\lambda) < 0$. This completes the proof of Corollary 5.5.

In the proof of Theorem of 5.5, we have established the following result. For easy reference, we state it as a theorem.

▶ **Theorem 5.6**

If all eigenvalues of $\mathbf{A}$ have negative real parts, then the Lyapunov equation

$$\mathbf{A}'\mathbf{M} + \mathbf{M}\mathbf{A} = -\mathbf{N}$$

has a unique solution for every $\mathbf{N}$, and the solution can be expressed as

$$\mathbf{M} = \int_0^\infty e^{\mathbf{A}'t}\mathbf{N}e^{\mathbf{A}t}\,dt \tag{5.24}$$

Because of the importance of this theorem, we give a different proof of the uniqueness of the solution. Suppose there are two solutions $\mathbf{M}_1$ and $\mathbf{M}_2$. Then we have

$$\mathbf{A}'(\mathbf{M}_1 - \mathbf{M}_2) + (\mathbf{M}_1 - \mathbf{M}_2)\mathbf{A} = \mathbf{0}$$

which implies

$$e^{\mathbf{A}'t}[\mathbf{A}'(\mathbf{M}_1 - \mathbf{M}_2) + (\mathbf{M}_1 - \mathbf{M}_2)\mathbf{A}]e^{\mathbf{A}t} = \frac{d}{dt}[e^{\mathbf{A}'t}(\mathbf{M}_1 - \mathbf{M}_2)e^{\mathbf{A}t}] = \mathbf{0}$$

Its integration from 0 to ∞ yields

$$[e^{\mathbf{A}'t}(\mathbf{M}_1 - \mathbf{M}_2)e^{\mathbf{A}t}]\Big|_0^\infty = \mathbf{0}$$

or, using $e^{\mathbf{A}t} \to \mathbf{0}$ as $t \to \infty$,

$$\mathbf{0} - (\mathbf{M}_1 - \mathbf{M}_2) = \mathbf{0}$$

---

1. Note that if $\mathbf{x}$ is a complex vector, then the Euclidean norm defined in Section 3.2 must be modified as $||\mathbf{x}||_2^2 = \mathbf{x}^*\mathbf{x}$, where $\mathbf{x}^*$ is the complex conjugate transpose of $\mathbf{x}$.

This shows the uniqueness of $\mathbf{M}$. Although the solution can be expressed as in (5.24), the integration is not used in computing the solution. It is simpler to arrange the Lyapunov equation, after some transformations, into a standard linear algebraic equation as in (3.60) and then solve the equation. Note that even if $\mathbf{A}$ is not stable, a unique solution still exists if $\mathbf{A}$ has no two eigenvalues such that $\lambda_i + \lambda_j = 0$. The solution, however, cannot be expressed as in (5.24); the integration will diverge and is meaningless. If $\mathbf{A}$ is singular or, equivalently, has at least one zero eigenvalue, then the Lyapunov equation is always singular and solutions may or may not exist depending on whether or not $\mathbf{N}$ lies in the range space of the equation.

5.4.1 Discrete-Time Case

Before discussing the discrete counterpart of Theorems 5.5 and 5.6, we discuss the discrete counterpart of the Lyapunov equation in (3.59). Consider

$$\mathbf{M} - \mathbf{AMB} = \mathbf{C} \tag{5.25}$$

where $\mathbf{A}$ and $\mathbf{B}$ are, respectively, $n \times n$ and $m \times m$ matrices, and $\mathbf{M}$ and $\mathbf{C}$ are $n \times m$ matrices. As (3.60), Equation (5.25) can be expressed as $\mathbf{Ym} = \mathbf{c}$, where $\mathbf{Y}$ is an $nm \times nm$ matrix; $\boldsymbol{m}$ and $\mathbf{c}$ are $nm \times 1$ column vectors with the m columns of $\mathbf{M}$ and $\mathbf{C}$ stacked in order. Thus (5.25) is essentially a set of linear algebraic equations. Let η_k be an eigenvalue of $\mathbf{Y}$ or of (5.25). Then we have

$$\eta_k = 1 - \lambda_i \mu_j \qquad \text{for } i = 1, 2, \ldots, n; \ j = 1, 2, \ldots, m$$

where λ_i and μ_j are, respectively, the eigenvalues of $\mathbf{A}$ and $\mathbf{B}$. This can be established intuitively as follows. Let us define $\mathcal{A}(\mathbf{M}) := \mathbf{M} - \mathbf{AMB}$. Then (5.25) can be written as $\mathcal{A}(\mathbf{M}) = \mathbf{C}$. A scalar η is an eigenvalue of $\mathcal{A}$ if there exists a nonzero $\mathbf{M}$ such that $\mathcal{A}(\mathbf{M}) = \eta\mathbf{M}$. Let $\mathbf{u}$ be an $n \times 1$ right eigenvector of $\mathbf{A}$ associated with λ_i; that is, $\mathbf{Au} = \lambda_i\mathbf{u}$. Let $\mathbf{v}$ be a $1 \times m$ left eigenvector of $\mathbf{B}$ associated with μ_j; that is, $\mathbf{vB} = \mathbf{v}\mu_j$. Applying $\mathcal{A}$ to the $n \times m$ nonzero matrix $\mathbf{uv}$ yields

$$\mathcal{A}(\mathbf{uv}) = \mathbf{uv} - \mathbf{AuvB} = (1 - \lambda_i \mu_j)\mathbf{uv}$$

Thus the eigenvalues of (5.25) are $1 - \lambda_i\mu_j$, for all i and j. If there are no i and j such that $\lambda_i\mu_j = 1$, then (5.25) is nonsingular and, for any $\mathbf{C}$, a unique solution $\mathbf{M}$ exists in (5.25). If $\lambda_i\mu_j = 1$ for some i and j, then (5.25) is singular and, for a given $\mathbf{C}$, solutions may or may not exist. The situation here is similar to what was discussed in Section 3.7.

▶ **Theorem 5.D5**

All eigenvalues of an $n \times n$ matrix $\mathbf{A}$ have magnitudes less than 1 if and only if for any given positive definite symmetric matrix $\mathbf{N}$ or for $\mathbf{N} = \bar{\mathbf{N}}'\bar{\mathbf{N}}$, where $\bar{\mathbf{N}}$ is any given $m \times n$ matrix with $m < n$ and with the property in (5.16), the discrete Lyapunov equation

$$\mathbf{M} - \mathbf{A}'\mathbf{MA} = \mathbf{N} \tag{5.26}$$

has a unique symmetric solution $\mathbf{M}$ and $\mathbf{M}$ is positive definite.

We sketch briefly its proof for $\mathbf{N} > 0$. If all eigenvalues of $\mathbf{A}$ and, consequently, of $\mathbf{A}'$ have magnitudes less than 1, then we have $|\lambda_i \lambda_j| < 1$ for all i and j. Thus $\lambda_i \lambda_j \neq 1$ and (5.26) is nonsingular. Therefore, for any $\mathbf{N}$, a unique solution exists in (5.26). We claim that the solution can be expressed as

$$\mathbf{M} = \sum_{m=0}^{\infty} (\mathbf{A}')^m \mathbf{N} \mathbf{A}^m \tag{5.27}$$

Because $|\lambda_i| < 1$ for all i, this infinite series converges and is well defined. Substituting (5.27) into (5.26) yields

$$\sum_{m=0}^{\infty} (\mathbf{A}')^m \mathbf{N} \mathbf{A}^m - \mathbf{A}' \left(\sum_{m=0}^{\infty} (\mathbf{A}')^m \mathbf{N} \mathbf{A}^m \right) \mathbf{A}$$

$$= \mathbf{N} + \sum_{m=1}^{\infty} (\mathbf{A}')^m \mathbf{N} \mathbf{A}^m - \sum_{m=1}^{\infty} (\mathbf{A}')^m \mathbf{N} \mathbf{A}^m = \mathbf{N}$$

Thus (5.27) is the solution. If $\mathbf{N}$ is symmetric, so is $\mathbf{M}$. If $\mathbf{N}$ is positive definite, so is $\mathbf{M}$. This establishes the necessity. To show sufficiency, let λ be an eigenvalue of $\mathbf{A}$ and $\mathbf{v} \neq \mathbf{0}$ be a corresponding eigenvector; that is, $\mathbf{A}\mathbf{v} = \lambda\mathbf{v}$. Then we have

$$\mathbf{v}^*\mathbf{N}\mathbf{v} = \mathbf{v}^*\mathbf{M}\mathbf{v} - \mathbf{v}^*\mathbf{A}'\mathbf{M}\mathbf{A}\mathbf{v}$$

$$= \mathbf{v}^*\mathbf{M}\mathbf{v} - \lambda^*\mathbf{v}^*\mathbf{M}\mathbf{v}\lambda = (1 - |\lambda|^2)\mathbf{v}^*\mathbf{M}\mathbf{v}$$

Because both $\mathbf{v}^*\mathbf{N}\mathbf{v}$ and $\mathbf{v}^*\mathbf{M}\mathbf{v}$ are real and positive, we conclude $(1 - |\lambda|^2) > 0$ or $|\lambda|^2 < 1$. This establishes the theorem for $\mathbf{N} > 0$. The case $\mathbf{N} \geq 0$ can similarly be established.

▷ **Theorem 5.D6**

If all eigenvalues of $\mathbf{A}$ have magnitudes less than 1, then the discrete Lyapunov equation

$$\mathbf{M} - \mathbf{A}'\mathbf{M}\mathbf{A} = \mathbf{N}$$

has a unique solution for every $\mathbf{N}$, and the solution can be expressed as

$$\mathbf{M} = \sum_{m=0}^{\infty} (\mathbf{A}')^m \mathbf{N} \mathbf{A}^m$$

It is important to mention that even if $\mathbf{A}$ has one or more eigenvalues with magnitudes larger than 1, a unique solution still exists in the discrete Lyapunov equation if $\lambda_i \lambda_j \neq 1$ for all i and j. In this case, the solution cannot be expressed as in (5.27) but can be computed from a set of linear algebraic equations.

Let us discuss the relationships between the continuous-time and discrete-time Lyapunov equations. The stability condition for continuous-time systems is that all eigenvalues lie inside the open left-half s-plane. The stability condition for discrete-time systems is that all eigenvalues lie inside the unit circle on the z-plane. These conditions can be related by the bilinear transformation

$$s = \frac{z - 1}{z + 1} \qquad z = \frac{1 + s}{1 - s} \tag{5.28}$$

which maps the left-half s-plane into the interior of the unit circle on the z-plane and vice versa. To differentiate the continuous-time and discrete-time cases, we write

$$\mathbf{A}'\mathbf{M} + \mathbf{M}\mathbf{A} = -\mathbf{N} \tag{5.29}$$

and

$$\mathbf{M}_d - \mathbf{A}'_d\mathbf{M}_d\mathbf{A}_d = \mathbf{N}_d \tag{5.30}$$

Following (5.28), these two equations can be related by

$$\mathbf{A} = (\mathbf{A}_d + \mathbf{I})^{-1}(\mathbf{A}_d - \mathbf{I}) \qquad \mathbf{A}_d = (\mathbf{I} + \mathbf{A})(\mathbf{I} - \mathbf{A})^{-1}$$

Substituting the right-hand-side equation into (5.30) and performing a simple manipulation, we find

$$\mathbf{A}'\mathbf{M}_d + \mathbf{M}_d\mathbf{A} = -0.5(\mathbf{I} - \mathbf{A}')\mathbf{N}_d(\mathbf{I} - \mathbf{A})$$

Comparing this with (5.29) yields

$$\mathbf{A} = (\mathbf{A}_d + \mathbf{I})^{-1}(\mathbf{A}_d - \mathbf{I}) \qquad \mathbf{M} = \mathbf{M}_d \qquad \mathbf{N} = 0.5(\mathbf{I} - \mathbf{A}')\mathbf{N}_d(\mathbf{I} - \mathbf{A}) \tag{5.31}$$

These relate (5.29) and (5.30).

The MATLAB function `lyap` computes the Lyapunov equation in (5.29) and `dlyap` computes the discrete Lyapunov equation in (5.30). The function `dlyap` transforms (5.30) into (5.29) by using (5.31) and then calls `lyap`. The result yields $\mathbf{M} = \mathbf{M}_d$.

5.5 Stability of LTV Systems

Consider a SISO linear time-varying (LTV) system described by

$$y(t) = \int_{t_0}^{t} g(t, \tau)u(\tau)\,d\tau \tag{5.32}$$

The system is said to be BIBO stable if every bounded input excites a bounded output. The condition for (5.32) to be BIBO stable is that there exists a finite constant M such that

$$\int_{t_0}^{t} |g(t, \tau)|\,d\tau \leq M < \infty \tag{5.33}$$

for all t and t_0 with $t \geq t_0$. The proof in the time-invariant case applies here with only minor modification.

For the multivariable case, (5.32) becomes

$$\mathbf{y}(t) = \int_{t_0}^{t} \mathbf{G}(t, \tau)\mathbf{u}(\tau)\,d\tau \tag{5.34}$$

The condition for (5.34) to be BIBO stable is that every entry of $\mathbf{G}(t, \tau)$ meets the condition in (5.33). For multivariable systems, we can also express the condition in terms of norms. Any norm discussed in Section 3.11 can be used. However, the infinite-norm

$$||\mathbf{u}||_\infty = \max_i |u_i| \qquad ||\mathbf{G}||_\infty = \text{ largest row absolute sum}$$

is probably the most convenient to use in stability study. For convenience, no subscript will be attached to any norm. The necessary and sufficient condition for (5.34) to be BIBO stable is that there exists a finite constant M such that

$$\int_{t_0}^{t} ||\mathbf{G}(t, \tau)|| \, d\tau \leq M < \infty$$

for all t and t_0 with $t \geq t_0$.

The impulse response matrix of

$$\begin{aligned} \dot{\mathbf{x}} &= \mathbf{A}(t)\mathbf{x} + \mathbf{B}(t)\mathbf{u} \\ \mathbf{y} &= \mathbf{C}(t)\mathbf{x} + \mathbf{D}(t)\mathbf{u} \end{aligned} \qquad (5.35)$$

is

$$\mathbf{G}(t, \tau) = \mathbf{C}(t)\boldsymbol{\Phi}(t, \tau)\mathbf{B}(\tau) + \mathbf{D}(t)\delta(t - \tau)$$

and the zero-state response is

$$\mathbf{y}(t) = \int_{t_0}^{t} \mathbf{C}(t)\boldsymbol{\Phi}(t, \tau)\mathbf{B}(\tau)\mathbf{u}(\tau) \, d\tau + \mathbf{D}(t)\mathbf{u}(t)$$

Thus (5.35) or, more precisely, the zero-state response of (5.35) is BIBO stable if and only if there exist constants M_1 and M_2 such that

$$||\mathbf{D}(t)|| \leq M_1 < \infty$$

and

$$\int_{t_0}^{t} ||\mathbf{G}(t, \tau)|| \, d\tau \leq M_2 < \infty$$

for all t and t_0 with $t \geq t_0$.

Next we study the stability of the zero-input response of (5.35). As in the time-invariant case, we define the zero-input response of (5.35) or the equation $\dot{\mathbf{x}} = \mathbf{A}(t)\mathbf{x}$ to be marginally stable if every finite initial state excites a bounded response. Because the response is governed by

$$\mathbf{x}(t) = \boldsymbol{\Phi}(t, t_0)\mathbf{x}(t_0) \qquad (5.36)$$

we conclude that the response is marginally stable if and only if there exists a finite constant M such that

$$||\boldsymbol{\Phi}(t, t_0)|| \leq M < \infty \qquad (5.37)$$

for all t_0 and for all $t \geq t_0$. The equation $\dot{\mathbf{x}} = \mathbf{A}(t)\mathbf{x}$ is asymptotically stable if the response excited by every finite initial state is bounded and approaches zero as $t \to \infty$. The asymptotic stability conditions are the boundedness condition in (5.37) and

$$||\boldsymbol{\Phi}(t, t_0)|| \to 0 \qquad \text{as } t \to \infty \qquad (5.38)$$

A great deal can be said regarding these definitions and conditions. Does the constant M in

(5.37) depend on t_0? What is the rate for the state transition matrix to approach 0 in (5.38)? The interested reader is referred to References [4, 15].

A time-invariant equation $\dot{\mathbf{x}} = \mathbf{A}\mathbf{x}$ is asymptotically stable if all eigenvalues of $\mathbf{A}$ have negative real parts. Is this also true for the time-varying case? The answer is negative as the next example shows.

EXAMPLE 5.5 Consider the linear time-varying equation

$$\dot{\mathbf{x}} = \mathbf{A}(t)\mathbf{x} = \begin{bmatrix} -1 & e^{2t} \\ 0 & -1 \end{bmatrix} \mathbf{x} \tag{5.39}$$

The characteristic polynomial of $\mathbf{A}(t)$ is

$$\det(\lambda \mathbf{I} - \mathbf{A}(t)) = \det \begin{bmatrix} \lambda + 1 & -e^{2t} \\ 0 & \lambda + 1 \end{bmatrix} = (\lambda + 1)^2$$

Thus $\mathbf{A}(t)$ has eigenvalues -1 and -1 for all t. It can be verified directly that

$$\Phi(t, 0) = \begin{bmatrix} e^{-t} & 0.5(e^t - e^{-t}) \\ 0 & e^{-t} \end{bmatrix}$$

meets (4.53) and is therefore the state transition matrix of (5.39). See also Problem 4.16. Because the (1,2)th entry of Φ grows without bound, the equation is neither asymptotically stable nor marginally stable. This example shows that even though the eigenvalues can be defined for $\mathbf{A}(t)$ at every t, the concept of eigenvalues is not useful in the time-varying case.

All stability properties in the time-invariant case are invariant under any equivalence transformation. In the time-varying case, this is so only for BIBO stability, because the impulse response matrix is preserved. An equivalence transformation can transform, as shown in Theorem 4.3, any $\dot{\mathbf{x}} = \mathbf{A}(t)\mathbf{x}$ into $\dot{\bar{\mathbf{x}}} = \mathbf{A}_o\bar{\mathbf{x}}$, where $\mathbf{A}_o$ is any constant matrix; therefore, in the time-varying case, marginal and asymptotic stabilities are not invariant under any equivalence transformation.

▶ **Theorem 5.7**

Marginal and asymptotic stabilities of $\dot{\mathbf{x}} = \mathbf{A}(t)\mathbf{x}$ are invariant under any Lyapunov transformation.

As discussed in Section 4.6, if $\mathbf{P}(t)$ and $\dot{\mathbf{P}}(t)$ are continuous, and $\mathbf{P}(t)$ is nonsingular for all t, then $\bar{\mathbf{x}} = \mathbf{P}(t)\mathbf{x}$ is an algebraic transformation. If, in addition, $\mathbf{P}(t)$ and $\mathbf{P}^{-1}(t)$ are bounded for all t, then $\bar{\mathbf{x}} = \mathbf{P}(t)\mathbf{x}$ is a Lyapunov transformation. The fundamental matrix $\mathbf{X}(t)$ of $\dot{\mathbf{x}} = \mathbf{A}(t)\mathbf{x}$ and the fundamental matrix $\bar{\mathbf{X}}(t)$ of $\dot{\bar{\mathbf{x}}} = \bar{\mathbf{A}}(t)\bar{\mathbf{x}}$ are related by, as derived in (4.71),

$$\bar{\mathbf{X}}(t) = \mathbf{P}(t)\mathbf{X}(t)$$

which implies

$$\bar{\Phi}(t, \tau) = \bar{\mathbf{X}}(t)\bar{\mathbf{X}}^{-1}(\tau) = \mathbf{P}(t)\mathbf{X}(t)\mathbf{X}^{-1}(\tau)\mathbf{P}^{-1}(\tau)$$
$$= \mathbf{P}(t)\Phi(t, \tau)\mathbf{P}^{-1}(\tau) \tag{5.40}$$

Because both $\mathbf{P}(t)$ and $\mathbf{P}^{-1}(t)$ are bounded, if $||\mathbf{\Phi}(t, \tau)||$ is bounded, so is $||\bar{\mathbf{\Phi}}(t, \tau)||$; if $||\mathbf{\Phi}(t, \tau)|| \to 0$ as $t \to \infty$, so is $||\bar{\mathbf{\Phi}}(t, \tau)||$. This establishes Theorem 5.7.

In the time-invariant case, asymptotic stability of zero-input responses always implies BIBO stability of zero-state responses. This is not necessarily so in the time-varying case. A time-varying equation is asymptotically stable if

$$||\mathbf{\Phi}(t, t_0)|| \to 0 \qquad \text{as } t \to \infty \tag{5.41}$$

for all t, t_0 with $t \geq t_0$. It is BIBO stable if

$$\int_{t_0}^{t} ||\mathbf{C}(t)\mathbf{\Phi}(t, \tau)\mathbf{B}(\tau)|| \, d\tau < \infty \tag{5.42}$$

for all t, t_0 with $t \geq t_0$. A function that approaches 0, as $t \to \infty$, may not be absolutely integrable. Thus asymptotic stability may not imply BIBO stability in the time-varying case. However, if $||\mathbf{\Phi}(t, \tau)||$ decreases to zero rapidly, in particular, exponentially, and if $\mathbf{C}(t)$ and $\mathbf{B}(t)$ are bounded for all t, then asymptotic stability does imply BIBO stability. See References [4, 6, 15].

PROBLEMS

5.1   Is the network shown in Fig. 5.2 BIBO stable? If not, find a bounded input that will excite an unbounded output.

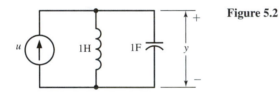

Figure 5.2

5.2   Consider a system with an irrational transfer function $\hat{g}(s)$. Show that a necessary condition for the system to be BIBO stable is that $|\hat{g}(s)|$ is finite for all Re $s \geq 0$.

5.3   Is a system with impulse response $g(t) = 1/(t+1)$ BIBO stable? How about $g(t) = te^{-t}$ for $t \geq 0$?

5.4   Is a system with transfer function $\hat{g}(s) = e^{-2s}/(s + 1)$ BIBO stable?

5.5   Show that the negative-feedback system shown in Fig. 2.5(b) is BIBO stable if and only if the gain a has a magnitude less than 1. For $a = 1$, find a bounded input $r(t)$ that will excite an unbounded output.

5.6   Consider a system with transfer function $\hat{g}(s) = (s-2)/(s+1)$. What are the steady-state responses excited by $u(t) = 3$, for $t \geq 0$, and by $u(t) = \sin 2t$, for $t \geq 0$?

5.7   Consider

$$\dot{\mathbf{x}} = \begin{bmatrix} -1 & 10 \\ 0 & 1 \end{bmatrix} \mathbf{x} + \begin{bmatrix} -2 \\ 0 \end{bmatrix} u$$

$$y = [-2 \quad 3]\mathbf{x} - 2u$$

Is it BIBO stable?

5.8 Consider a discrete-time system with impulse response sequence

$$g[k] = k(0.8)^k \qquad \text{for } k \geq 0$$

Is the system BIBO stable?

5.9 Is the state equation in Problem 5.7 marginally stable? Asymptotically stable?

5.10 Is the homogeneous state equation

$$\dot{\mathbf{x}} = \begin{bmatrix} -1 & 0 & 1 \\ 0 & 0 & 0 \\ 0 & 0 & 0 \end{bmatrix} \mathbf{x}$$

marginally stable? Asymptotically stable?

5.11 Is the homogeneous state equation

$$\dot{\mathbf{x}} = \begin{bmatrix} -1 & 0 & 1 \\ 0 & 0 & 1 \\ 0 & 0 & 0 \end{bmatrix} \mathbf{x}$$

marginally stable? Asymptotically stable?

5.12 Is the discrete-time homogeneous state equation

$$\mathbf{x}[k+1] = \begin{bmatrix} 0.9 & 0 & 1 \\ 0 & 1 & 0 \\ 0 & 0 & 1 \end{bmatrix} \mathbf{x}[k]$$

marginally stable? Asymptotically stable?

5.13 Is the discrete-time homogeneous state equation

$$\mathbf{x}[k+1] = \begin{bmatrix} 0.9 & 0 & 1 \\ 0 & 1 & 1 \\ 0 & 0 & 1 \end{bmatrix} \mathbf{x}[k]$$

marginally stable? Asymptotically stable?

5.14 Use Theorem 5.5 to show that all eigenvalues of

$$\mathbf{A} = \begin{bmatrix} 0 & 1 \\ -0.5 & -1 \end{bmatrix}$$

have negative real parts.

5.15 Use Theorem 5.D5 to show that all eigenvalues of the **A** in Problem 5.14 have magnitudes less than 1.

5.16 For any distinct negative real λ_i and any nonzero real a_i, show that the matrix

$$\mathbf{M} = \begin{bmatrix} -\dfrac{a_1^2}{2\lambda_1} & -\dfrac{a_1 a_2}{\lambda_1 + \lambda_2} & -\dfrac{a_1 a_3}{\lambda_1 + \lambda_3} \\[2ex] -\dfrac{a_2 a_1}{\lambda_2 + \lambda_1} & -\dfrac{a_2^2}{2\lambda_2} & -\dfrac{a_2 a_3}{\lambda_2 + \lambda_3} \\[2ex] -\dfrac{a_3 a_1}{\lambda_3 + \lambda_1} & -\dfrac{a_3 a_2}{\lambda_3 + \lambda_2} & -\dfrac{a_3^2}{2\lambda_3} \end{bmatrix}$$

is positive definite. [*Hint*: Use Corollary 5.5 and $\mathbf{A}$=diag$(\lambda_1, \lambda_2, \lambda_3)$.]

5.17 A real matrix $\mathbf{M}$ (not necessarily symmetric) is defined to be positive definite if $\mathbf{x}'\mathbf{Mx} > 0$ for any nonzero $\mathbf{x}$. Is it true that the matrix $\mathbf{M}$ is positive definite if all eigenvalues of $\mathbf{M}$ are real and positive or if all its leading principal minors are positive? If not, how do you check its positive definiteness? [*Hint*: Try

$$\begin{bmatrix} 0 & 1 \\ -2 & 3 \end{bmatrix} \qquad \begin{bmatrix} 2 & 1 \\ 1.9 & 1 \end{bmatrix}]$$

5.18 Show that all eigenvalues of $\mathbf{A}$ have real parts less than $-\mu < 0$ if and only if, for any given positive definite symmetric matrix $\mathbf{N}$, the equation

$$\mathbf{A}'\mathbf{M} + \mathbf{MA} + 2\mu\mathbf{M} = -\mathbf{N}$$

has a unique symmetric solution $\mathbf{M}$ and $\mathbf{M}$ is positive definite.

5.19 Show that all eigenvalues of $\mathbf{A}$ have magnitudes less than ρ if and only if, for any given positive definite symmetric matrix $\mathbf{N}$, the equation

$$\rho^2\mathbf{M} - \mathbf{A}'\mathbf{MA} = \rho^2\mathbf{N}$$

has a unique symmetric solution $\mathbf{M}$ and $\mathbf{M}$ is positive definite.

5.20 Is a system with impulse response $g(t, \tau) = e^{-2|t|-|\tau|}$, for $t \geq \tau$, BIBO stable? How about $g(t, \tau) = \sin t (e^{-(t-\tau)}) \cos \tau$?

5.21 Consider the time-varying equation

$$\dot{x} = 2tx + u \qquad y = e^{-t^2} x$$

Is the equation BIBO stable? Marginally stable? Asymptotically stable?

5.22 Show that the equation in Problem 5.21 can be transformed by using $\bar{x} = P(t)x$, with $P(t) = e^{-t^2}$, into

$$\dot{\bar{x}} = 0 \cdot \bar{x} + e^{-t^2} u \qquad y = \bar{x}$$

Is the equation BIBO stable? Marginally stable? Asymptotically stable? Is the transformation a Lyapunov transformation?

5.23 Is the homogeneous equation

$$\dot{\mathbf{x}} = \begin{bmatrix} -1 & 0 \\ -e^{-3t} & 0 \end{bmatrix} \mathbf{x}$$

for $t_0 \geq 0$, marginally stable? Asymptotically stable?

Chapter

6

Controllability
and Observability

6.1 Introduction

This chapter introduces the concepts of controllability and observability. Controllability deals with whether or not the state of a state-space equation can be controlled from the input, and observability deals with whether or not the initial state can be observed from the output. These concepts can be illustrated using the network shown in Fig. 6.1. The network has two state variables. Let x_i be the voltage across the capacitor with capacitance C_i, for $i = 1, 2$. The input u is a current source and the output y is the voltage shown. From the network, we see that, because of the open circuit across y, the input has no effect on x_2 or cannot control x_2. The current passing through the 2-Ω resistor always equals the current source u; therefore the response excited by the initial state x_1 will not appear in y. Thus the initial state x_1 cannot be observed from the output. Thus the equation describing the network cannot be controllable and observable.

These concepts are essential in discussing the internal structure of linear systems. They are also needed in studying control and filtering problems. We study first continuous-time

Figure 6.1 Network.

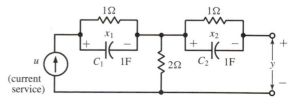

143

linear time-invariant (LTI) state equations and then discrete-time LTI state equations. Finally, we study the time-varying case.

6.2 Controllability

Consider the n-dimensional p-input state equation

$$\dot{\mathbf{x}} = \mathbf{A}\mathbf{x} + \mathbf{B}\mathbf{u} \tag{6.1}$$

where $\mathbf{A}$ and $\mathbf{B}$ are, respectively, $n \times n$ and $n \times p$ real constant matrices. Because the output does not play any role in controllability, we will disregard the output equation in this study.

> **Definition 6.1** *The state equation (6.1) or the pair* $(\mathbf{A}, \mathbf{B})$ *is said to be* controllable *if for any initial state* $\mathbf{x}(0) = \mathbf{x}_0$ *and any final state* $\mathbf{x}_1$, *there exists an input that transfers* $\mathbf{x}_0$ *to* $\mathbf{x}_1$ *in a finite time. Otherwise (6.1) or* $(\mathbf{A}, \mathbf{B})$ *is said to be* uncontrollable.

This definition requires only that the input be capable of moving any state in the state space to any other state in a finite time; what trajectory the state should take is not specified. Furthermore, there is no constraint imposed on the input; its magnitude can be as large as desired. We give an example to illustrate the concept.

EXAMPLE 6.1 Consider the network shown in Fig. 6.2(a). Its state variable x is the voltage across the capacitor. If $x(0) = 0$, then $x(t) = 0$ for all $t \geq 0$ no matter what input is applied. This is due to the symmetry of the network, and the input has no effect on the voltage across the capacitor. Thus the system or, more precisely, the state equation that describes the system is not controllable.

Next we consider the network shown in Fig. 6.2(b). It has two state variables x_1 and x_2 as shown. The input can transfer x_1 or x_2 to any value; but it cannot transfer x_1 *and* x_2 to any values. For example, if $x_1(0) = x_2(0) = 0$, then no matter what input is applied, $x_1(t)$ always equals $x_2(t)$ for all $t \geq 0$. Thus the equation that describes the network is not controllable.

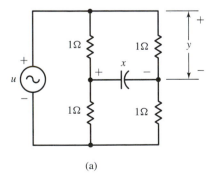

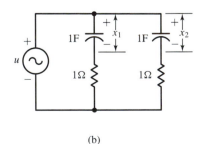

(a)   (b)

Figure 6.2   Uncontrollable networks.

▶ **Theorem 6.1**

The following statements are equivalent.

1. The n-dimensional pair $(\mathbf{A}, \mathbf{B})$ is controllable.
2. The $n \times n$ matrix

$$\mathbf{W}_c(t) = \int_0^t e^{\mathbf{A}\tau} \mathbf{B}\mathbf{B}' e^{\mathbf{A}'\tau} \, d\tau = \int_0^t e^{\mathbf{A}(t-\tau)} \mathbf{B}\mathbf{B}' e^{\mathbf{A}'(t-\tau)} \, d\tau \tag{6.2}$$

is nonsingular for any $t > 0$.

3. The $n \times np$ *controllability matrix*

$$C = [\mathbf{B} \ \mathbf{AB} \ \mathbf{A}^2\mathbf{B} \ \cdots \ \mathbf{A}^{n-1}\mathbf{B}] \tag{6.3}$$

has rank n (full row rank).

4. The $n \times (n + p)$ matrix $[\mathbf{A} - \lambda\mathbf{I} \ \mathbf{B}]$ has full row rank at every eigenvalue, λ, of $\mathbf{A}$.[1]
5. If, in addition, all eigenvalues of $\mathbf{A}$ have negative real parts, then the unique solution of

$$\mathbf{A}\mathbf{W}_c + \mathbf{W}_c\mathbf{A}' = -\mathbf{B}\mathbf{B}' \tag{6.4}$$

is positive definite. The solution is called the *controllability Gramian* and can be expressed as

$$\mathbf{W}_c = \int_0^\infty e^{\mathbf{A}\tau} \mathbf{B}\mathbf{B}' e^{\mathbf{A}'\tau} d\tau \tag{6.5}$$

➤ **Proof:** (1) ↔ (2): First we show the equivalence of the two forms in (6.2). Define $\bar{\tau} := t - \tau$. Then we have

$$\int_{\tau=0}^t e^{\mathbf{A}(t-\tau)} \mathbf{B}\mathbf{B}' e^{\mathbf{A}'(t-\tau)} \, d\tau = \int_{\bar{\tau}=t}^0 e^{\mathbf{A}\bar{\tau}} \mathbf{B}\mathbf{B}' e^{\mathbf{A}'\bar{\tau}} (-d\bar{\tau})$$

$$= \int_{\bar{\tau}=0}^t e^{\mathbf{A}\bar{\tau}} \mathbf{B}\mathbf{B}' e^{\mathbf{A}'\bar{\tau}} d\bar{\tau}$$

It becomes the first form of (6.2) after replacing $\bar{\tau}$ by τ. Because of the form of the integrand, $\mathbf{W}_c(t)$ is always positive semidefinite; it is positive definite if and only if it is nonsingular. See Section 3.9.

First we show that if $\mathbf{W}_c(t)$ is nonsingular, then (6.1) is controllable. The response of (6.1) at time t_1 was derived in (4.5) as

$$\mathbf{x}(t_1) = e^{\mathbf{A}t_1}\mathbf{x}(0) + \int_0^{t_1} e^{\mathbf{A}(t_1-\tau)}\mathbf{B}\mathbf{u}(\tau) \, d\tau \tag{6.6}$$

We claim that for any $\mathbf{x}(0) = \mathbf{x}_0$ and any $\mathbf{x}(t_1) = \mathbf{x}_1$, the input

$$\mathbf{u}(t) = -\mathbf{B}' e^{\mathbf{A}'(t_1-t)} \mathbf{W}_c^{-1}(t_1)[e^{\mathbf{A}t_1}\mathbf{x}_0 - \mathbf{x}_1] \tag{6.7}$$

will transfer $\mathbf{x}_0$ to $\mathbf{x}_1$ at time t_1. Indeed, substituting (6.7) into (6.6) yields

---

1. If λ is complex, then we must use complex numbers as scalars in checking the rank. See the discussion regarding (3.37).

$$\mathbf{x}(t_1) = e^{\mathbf{A}t_1}\mathbf{x}_0 - \left(\int_0^{t_1} e^{\mathbf{A}(t_1-\tau)}\mathbf{B}\mathbf{B}'e^{\mathbf{A}'(t_1-\tau)}d\tau \right) \mathbf{W}_c^{-1}(t_1)[e^{\mathbf{A}t_1}\mathbf{x}_0 - \mathbf{x}_1]$$

$$= e^{\mathbf{A}t_1}\mathbf{x}_0 - \mathbf{W}_c(t_1)\mathbf{W}_c^{-1}(t_1)[e^{\mathbf{A}t_1}\mathbf{x}_0 - \mathbf{x}_1] = \mathbf{x}_1$$

This shows that if $\mathbf{W}_c$ is nonsingular, then the pair $(\mathbf{A}, \mathbf{B})$ is controllable. We show the converse by contradiction. Suppose the pair is controllable but $\mathbf{W}_c(t_1)$ is not positive definite for some t_1. Then there exists an $n \times 1$ nonzero vector $\mathbf{v}$ such that

$$\mathbf{v}'\mathbf{W}_c(t_1)\mathbf{v} = \int_0^{t_1} \mathbf{v}'e^{\mathbf{A}(t_1-\tau)}\mathbf{B}\mathbf{B}'e^{\mathbf{A}'(t_1-\tau)}\mathbf{v}d\tau$$

$$= \int_0^{t_1} ||\mathbf{B}'e^{\mathbf{A}'(t_1-\tau)}\mathbf{v}||^2 d\tau = 0$$

which implies

$$\mathbf{B}'e^{\mathbf{A}'(t_1-\tau)}\mathbf{v} \equiv \mathbf{0} \quad \text{or} \quad \mathbf{v}'e^{\mathbf{A}(t_1-\tau)}\mathbf{B} \equiv \mathbf{0} \tag{6.8}$$

for all τ in $[0, t_1]$. If (6.1) is controllable, there exists an input that transfers the initial state $\mathbf{x}(0) = e^{-\mathbf{A}t_1}\mathbf{v}$ to $\mathbf{x}(t_1) = \mathbf{0}$ and (6.6) becomes

$$\mathbf{0} = \mathbf{v} + \int_0^{t_1} e^{\mathbf{A}(t_1-\tau)}\mathbf{B}\mathbf{u}(\tau)\,d\tau$$

Its premultiplication by $\mathbf{v}'$ yields

$$0 = \mathbf{v}'\mathbf{v} + \int_0^{t_1} \mathbf{v}'e^{\mathbf{A}(t_1-\tau)}\mathbf{B}\mathbf{u}(\tau)\,d\tau = ||\mathbf{v}||^2 + 0$$

which contradicts $\mathbf{v} \neq \mathbf{0}$. This establishes the equivalence of (1) and (2).

(2) $\leftrightarrow$ (3): Because every entry of $e^{\mathbf{A}t}\mathbf{B}$ is, as discussed at the end of Section 4.2, an analytical function of t, if $\mathbf{W}_c(t)$ is nonsingular for some t, then it is nonsingular for all t in $(-\infty, \infty)$. See Reference [6, p. 554]. Because of the equivalence of the two forms in (6.2), (6.8) implies that $\mathbf{W}_c(t)$ is nonsingular if and only if there exists no $n \times 1$ nonzero vector $\mathbf{v}$ such that

$$\mathbf{v}'e^{\mathbf{A}t}\mathbf{B} = \mathbf{0} \qquad \text{for all } t \tag{6.9}$$

Now we show that if $\mathbf{W}_c(t)$ is nonsingular, then the controllability matrix C has full row rank. Suppose C does not have full row rank, then there exists an $n \times 1$ nonzero vector $\mathbf{v}$ such that $\mathbf{v}'C = \mathbf{0}$ or

$$\mathbf{v}'\mathbf{A}^k\mathbf{B} = \mathbf{0} \qquad \text{for } k = 0, 1, 2, \ldots, n-1$$

Because $e^{\mathbf{A}t}\mathbf{B}$ can be expressed as a linear combination of $\{\mathbf{B}, \mathbf{A}\mathbf{B}, \ldots, \mathbf{A}^{n-1}\mathbf{B}\}$ (Theorem 3.5), we conclude $\mathbf{v}'e^{\mathbf{A}t}\mathbf{B} = \mathbf{0}$. This contradicts the nonsingularity assumption of $\mathbf{W}_c(t)$. Thus Condition (2) implies Condition (3). To show the converse, suppose C has full row rank but $\mathbf{W}_c(t)$ is singular. Then there exists a nonzero $\mathbf{v}$ such that (6.9) holds. Setting $t = 0$, we have $\mathbf{v}'\mathbf{B} = \mathbf{0}$. Differentiating (6.9) and then setting $t = 0$, we have $\mathbf{v}'\mathbf{A}\mathbf{B} = \mathbf{0}$. Proceeding forward yields $\mathbf{v}'\mathbf{A}^k\mathbf{B} = \mathbf{0}$ for $k = 0, 1, 2, \ldots$. They can be arranged as

$$\mathbf{v}'[\mathbf{B}\ \mathbf{A}\mathbf{B}\ \cdots\ \mathbf{A}^{n-1}\mathbf{B}] = \mathbf{v}'C = \mathbf{0}$$

This contradicts the hypothesis that C has full row rank. This shows the equivalence of (2) and (3).

(3) $\leftrightarrow$ (4): If C has full row rank, then the matrix $[\mathbf{A} - \lambda\mathbf{I} \ \mathbf{B}]$ has full row rank at every eigenvalue of $\mathbf{A}$. If not, there exists an eigenvalue λ_1 and a $1 \times n$ vector $\mathbf{q} \neq \mathbf{0}$ such that

$$\mathbf{q}[\mathbf{A} - \lambda_1\mathbf{I} \ \mathbf{B}] = \mathbf{0}$$

which implies $\mathbf{qA} = \lambda_1\mathbf{q}$ and $\mathbf{qB} = \mathbf{0}$. Thus $\mathbf{q}$ is a left eigenvector of $\mathbf{A}$. We compute

$$\mathbf{qA}^2 = (\mathbf{qA})\mathbf{A} = (\lambda_1\mathbf{q})\mathbf{A} = \lambda_1^2\mathbf{q}$$

Proceeding forward, we have $\mathbf{qA}^k = \lambda_1^k\mathbf{q}$. Thus we have

$$\mathbf{q}[\mathbf{B} \ \mathbf{AB} \ \cdots \ \mathbf{A}^{n-1}\mathbf{B}] = [\mathbf{qB} \ \lambda_1\mathbf{qB} \ \cdots \ \lambda_1^{n-1}\mathbf{qB}] = \mathbf{0}$$

This contradicts the hypothesis that C has full row rank.

In order to show that $\rho(C) < n$ implies $\rho([\mathbf{A} - \lambda\mathbf{I} \ \mathbf{B}]) < n$ at some eigenvalue λ_1 of $\mathbf{A}$, we need Theorems 6.2 and 6.6, which will be established later. Theorem 6.2 states that controllability is invariant under any equivalence transformation. Therefore we may show $\rho([\bar{\mathbf{A}} - \lambda\mathbf{I} \ \bar{\mathbf{B}}]) < n$ at some eigenvalue of $\bar{\mathbf{A}}$, where $(\bar{\mathbf{A}}, \bar{\mathbf{B}})$ is equivalent to $(\mathbf{A}, \mathbf{B})$. Theorem 6.6 states that if the rank of C is less than n or $\rho(C) = n - m$, for some integer $m \geq 1$, then there exists a nonsingular matrix $\mathbf{P}$ such that

$$\bar{\mathbf{A}} = \mathbf{PAP}^{-1} = \begin{bmatrix} \bar{\mathbf{A}}_c & \bar{\mathbf{A}}_{12} \\ \mathbf{0} & \bar{\mathbf{A}}_{\bar{c}} \end{bmatrix} \qquad \bar{\mathbf{B}} = \mathbf{PB} = \begin{bmatrix} \bar{\mathbf{B}}_c \\ \mathbf{0} \end{bmatrix}$$

where $\bar{\mathbf{A}}_{\bar{c}}$ is $m \times m$. Let λ_1 be an eigenvalue of $\bar{\mathbf{A}}_{\bar{c}}$ and $\mathbf{q}_1$ be a corresponding $1 \times m$ nonzero left eigenvector or $\mathbf{q}_1\bar{\mathbf{A}}_{\bar{c}} = \lambda_1\mathbf{q}_1$. Then we have $\mathbf{q}_1(\bar{\mathbf{A}}_{\bar{c}} - \lambda_1\mathbf{I}) = \mathbf{0}$. Now we form the $1 \times n$ vector $\mathbf{q} := [\mathbf{0} \ \mathbf{q}_1]$. We compute

$$\mathbf{q}[\bar{\mathbf{A}} - \lambda_1\mathbf{I} \ \bar{\mathbf{B}}] = [\mathbf{0} \ \mathbf{q}_1] \begin{bmatrix} \bar{\mathbf{A}}_c - \lambda_1\mathbf{I} & \bar{\mathbf{A}}_{12} & \bar{\mathbf{B}}_c \\ \mathbf{0} & \bar{\mathbf{A}}_{\bar{c}} - \lambda_1\mathbf{I} & \mathbf{0} \end{bmatrix} = \mathbf{0} \qquad (6.10)$$

which implies $\rho([\bar{\mathbf{A}} - \lambda\mathbf{I} \ \bar{\mathbf{B}}]) < n$ and, consequently, $\rho([\mathbf{A} - \lambda\mathbf{I} \ \mathbf{B}]) < n$ at some eigenvalue of $\mathbf{A}$. This establishes the equivalence of (3) and (4).

(2) $\leftrightarrow$ (5): If $\mathbf{A}$ is stable, then the unique solution of (6.4) can be expressed as in (6.5) (Theorem 5.6). The Gramian $\mathbf{W}_c$ is always positive semidefinite. It is positive definite if and only if $\mathbf{W}_c$ is nonsingular. This establishes the equivalence of (2) and (5).   Q.E.D.

EXAMPLE 6.2 Consider the inverted pendulum studied in Example 2.8. Its state equation was developed in (2.27). Suppose for a given pendulum, the equation becomes

$$\dot{\mathbf{x}} = \begin{bmatrix} 0 & 1 & 0 & 0 \\ 0 & 0 & -1 & 0 \\ 0 & 0 & 0 & 1 \\ 0 & 0 & 5 & 0 \end{bmatrix} \mathbf{x} + \begin{bmatrix} 0 \\ 1 \\ 0 \\ -2 \end{bmatrix} u$$

$$y = [1 \ 0 \ 0 \ 0]\mathbf{x}$$

(6.11)

We compute

$$C = [B \ AB \ A^2B \ A^3B] = \begin{bmatrix} 0 & 1 & 0 & 2 \\ 1 & 0 & 2 & 0 \\ 0 & -2 & 0 & -10 \\ -2 & 0 & -10 & 0 \end{bmatrix}$$

This matrix can be shown to have rank 4; thus the system is controllable. Therefore, if $x_3 = \theta$ deviates from zero slightly, we can find a control u to push it back to zero. In fact, a control exists to bring $x_1 = y$, x_3, and their derivatives back to zero. This is consistent with our experience of balancing a broom on our palm.

The MATLAB functions `ctrb` and `gram` will generate the controllability matrix and controllability Gramian. Note that the controllability Gramian is not computed from (6.5); it is obtained by solving a set of linear algebraic equations. Whether a state equation is controllable can then be determined by computing the rank of the controllability matrix or Gramian by using `rank` in MATLAB.

EXAMPLE 6.3 Consider the platform system shown in Fig. 6.3; it can be used to study suspension systems of automobiles. The system consists of one platform; both ends of the platform are supported on the ground by means of springs and dashpots, which provide viscous friction. The mass of the platform is assumed to be zero; thus the movements of the two spring systems are independent and half of the force is applied to each spring system. The spring constants of both springs are assumed to be 1 and the viscous friction coefficients are assumed to be 2 and 1 as shown. If the displacements of the two spring systems from equilibrium are chosen as state variables x_1 and x_2, then we have $x_1 + 2\dot{x}_1 = u$ and $x_2 + \dot{x}_2 = u$ or

$$\dot{\mathbf{x}} = \begin{bmatrix} -0.5 & 0 \\ 0 & -1 \end{bmatrix} \mathbf{x} + \begin{bmatrix} 0.5 \\ 1 \end{bmatrix} u \tag{6.12}$$

This state equation describes the system.

Now if the initial displacements are different from zero, and if no force is applied, the platform will return to zero exponentially. In theory, it will take an infinite time for x_i to equal 0 exactly. Now we pose the problem. If $x_1(0) = 10$ and $x_2(0) = -1$, can we apply a force to bring the platform to equilibrium in 2 seconds? The answer does not seem to be obvious because the *same* force is applied to the two spring systems.

Figure 6.3 Platform system.

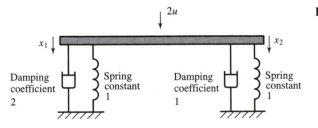

For Equation (6.12), we compute

$$\rho([\mathbf{B} \ \mathbf{AB}]) = \rho \begin{bmatrix} 0.5 & -0.25 \\ 1 & -1 \end{bmatrix} = 2$$

Thus the equation is controllable and, for any $\mathbf{x}(0)$, there exists an input that transfers $\mathbf{x}(0)$ to $\mathbf{0}$ in 2 seconds or in any finite time. We compute (6.2) and (6.7) for this system at $t_1 = 2$:

$$\mathbf{W}_c(2) = \int_0^2 \left(\begin{bmatrix} e^{-0.5\tau} & 0 \\ 0 & e^{-\tau} \end{bmatrix} \begin{bmatrix} 0.5 \\ 1 \end{bmatrix} [0.5 \ 1] \begin{bmatrix} e^{-0.5\tau} & 0 \\ 0 & e^{-\tau} \end{bmatrix} \right) d\tau$$

$$= \begin{bmatrix} 0.2162 & 0.3167 \\ 0.3167 & 0.4908 \end{bmatrix}$$

and

$$u_1(t) = -[0.5 \ 1] \begin{bmatrix} e^{-0.5(2-t)} & 0 \\ 0 & e^{-(2-t)} \end{bmatrix} \mathbf{W}_c^{-1}(2) \begin{bmatrix} e^{-1} & 0 \\ 0 & e^{-2} \end{bmatrix} \begin{bmatrix} 10 \\ -1 \end{bmatrix}$$

$$= -58.82e^{0.5t} + 27.96e^t$$

for t in $[0, \ 2]$. This input force will transfer $\mathbf{x}(0) = [10 \ -1]'$ to $[0 \ 0]'$ in 2 seconds as shown in Fig. 6.4(a) in which the input is also plotted. It is obtained by using the MATLAB function `lsim`, an acronym for *linear simulation*. The largest magnitude of the input is about 45.

Figure 6.4(b) plots the input u_2 that transfers $\mathbf{x}(0) = [10 \ -1]'$ to $\mathbf{0}$ in 4 seconds. We see that the smaller the time interval, the larger the input magnitude. If no restriction is imposed on the input, we can transfer $\mathbf{x}(0)$ to zero in an arbitrarily small time interval; however, the input magnitude may become very large. If some restriction is imposed on the input magnitude, then we cannot achieve the transfer as fast as desired. For example, if we require $|u(t)| < 9$, for all t, in Example 6.3, then we cannot transfer $\mathbf{x}(0)$ to $\mathbf{0}$ in less than 4 seconds. We remark that the input $\mathbf{u}(t)$ in (6.7) is called the *minimal energy control* in the sense that for any other input $\bar{\mathbf{u}}(t)$ that achieves the same transfer, we have

$$\int_{t_0}^{t_1} \bar{\mathbf{u}}'(t)\bar{\mathbf{u}}(t) \, dt \geq \int_{t_0}^{t_1} \mathbf{u}'(t)\mathbf{u}(t) \, dt$$

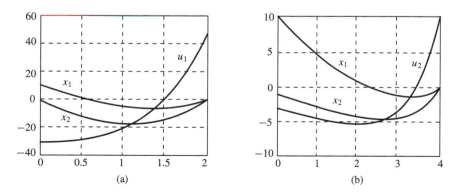

(a)   (b)

Figure 6.4   Transfer $\mathbf{x}(0) = [10 \ -1]'$ to $[0 \ 0]'$.

Its proof can be found in Reference [6, pp. 556–558].

EXAMPLE 6.4 Consider again the platform system shown in Fig. 6.3. We now assume that the viscous friction coefficients and spring constants of both spring systems all equal 1. Then the state equation that describes the system becomes

$$\dot{\mathbf{x}} = \begin{bmatrix} -1 & 0 \\ 0 & -1 \end{bmatrix} \mathbf{x} + \begin{bmatrix} 1 \\ 1 \end{bmatrix} u$$

Clearly we have

$$\rho(C) = \rho \begin{bmatrix} 1 & -1 \\ 1 & -1 \end{bmatrix} = 1$$

and the state equation is not controllable. If $x_1(0) \neq x_2(0)$, no input can transfer $\mathbf{x}(0)$ to zero in a finite time.

6.2.1 Controllability Indices

Let $\mathbf{A}$ and $\mathbf{B}$ be $n \times n$ and $n \times p$ constant matrices. We assume that $\mathbf{B}$ has rank p or full column rank. If $\mathbf{B}$ does not have full column rank, there is a redundancy in inputs. For example, if the second column of $\mathbf{B}$ equals the first column of $\mathbf{B}$, then the effect of the second input on the system can be generated from the first input. Thus the second input is redundant. In conclusion, deleting linearly dependent columns of $\mathbf{B}$ and the corresponding inputs will not affect the control of the system. Thus it is reasonable to assume that $\mathbf{B}$ has full column rank.

If $(\mathbf{A}, \mathbf{B})$ is controllable, its controllability matrix C has rank n and, consequently, n linearly independent columns. Note that there are np columns in C; therefore it is possible to find many sets of n linearly independent columns in C. We discuss in the following the most important way of searching these columns; the search also happens to be most natural. Let $\mathbf{b}_i$ be the ith column of $\mathbf{B}$. Then C can be written explicitly as

$$C = [\mathbf{b}_1 \ \cdots \ \mathbf{b}_p \vdots \mathbf{A}\mathbf{b}_1 \ \cdots \ \mathbf{A}\mathbf{b}_p \vdots \ \cdots \ \vdots \mathbf{A}^{n-1}\mathbf{b}_1 \ \cdots \ \mathbf{A}^{n-1}\mathbf{b}_p] \qquad (6.13)$$

Let us search linearly independent columns of C from left to right. Because of the pattern of C, if $\mathbf{A}^i\mathbf{b}_m$ depends on its left-hand-side (LHS) columns, then $\mathbf{A}^{i+1}\mathbf{b}_m$ will also depend on its LHS columns. It means that once a column associated with $\mathbf{b}_m$ becomes linearly dependent, then all columns associated with $\mathbf{b}_m$ thereafter are linearly dependent. Let μ_m be the number of the linearly independent columns associated with $\mathbf{b}_m$ in C. That is, the columns

$$\mathbf{b}_m, \ \mathbf{A}\mathbf{b}_m, \ \ldots, \ \mathbf{A}^{\mu_m-1}\mathbf{b}_m$$

are linearly independent in C and $\mathbf{A}^{\mu_m+i}\mathbf{b}_m$ are linearly dependent for $i = 0, 1, \ldots$. It is clear that if C has rank n, then

$$\mu_1 + \mu_2 + \cdots + \mu_p = n \qquad (6.14)$$

The set $\{\mu_1, \mu_2, \ldots, \mu_p\}$ is called the *controllability indices* and

$$\mu = \max (\mu_1, \ \mu_2, \ \ldots, \ \mu_p)$$

is called the *controllability index* of $(\mathbf{A}, \mathbf{B})$. Or, equivalently, if $(\mathbf{A}, \mathbf{B})$ is controllable, the controllability index μ is the least integer such that

$$\rho(C_\mu) = \rho([\mathbf{B} \ \mathbf{AB} \ \cdots \ \mathbf{A}^{\mu-1}\mathbf{B}]) = n \tag{6.15}$$

Now we give a range of μ. If $\mu_1 = \mu_2 = \cdots = \mu_p$, then $n/p \leq \mu$. If all μ_m, except one, equal 1, then $\mu = n - (p - 1)$; this is the largest possible μ. Let $\bar{n}$ be the degree of the minimal polynomial of $\mathbf{A}$. Then, by definition, there exist α_i such that

$$\mathbf{A}^{\bar{n}} = \alpha_1\mathbf{A}^{\bar{n}-1} + \alpha_2\mathbf{A}^{\bar{n}-2} + \cdots + \alpha_{\bar{n}}\mathbf{I}$$

which implies that $\mathbf{A}^{\bar{n}}\mathbf{B}$ can be written as a linear combination of $\{\mathbf{B}, \mathbf{AB}, \ldots, \mathbf{A}^{\bar{n}-1}\mathbf{B}\}$. Thus we conclude

$$n/p \leq \mu \leq \min(\bar{n}, n - p + 1) \tag{6.16}$$

where $\rho(\mathbf{B}) = p$. Because of (6.16), in checking controllability, it is unnecessary to check the $n \times np$ matrix C. It is sufficient to check a matrix of lesser columns. Because the degree of the minimal polynomial is generally not available—whereas the rank of $\mathbf{B}$ can readily be computed—we can use the following corollary to check controllability. The second part of the corollary follows Theorem 3.8.

▶ **Corollary 6.1**

The n-dimensional pair $(\mathbf{A}, \mathbf{B})$ is controllable if and only if the matrix

$$C_{n-p+1} := [\mathbf{B} \ \mathbf{AB} \ \cdots \ \mathbf{A}^{n-p}\mathbf{B}] \tag{6.17}$$

where $\rho(\mathbf{B}) = p$, has rank n or the $n \times n$ matrix $C_{n-p+1}C'_{n-p+1}$ is nonsingular.

EXAMPLE 6.5 Consider the satellite system studied in Fig. 2.13. Its linearized state equation was developed in (2.29). From the equation, we can see that the control of the first four state variables by the first two inputs and the control of the last two state variables by the last input are decoupled; therefore we can consider only the following subequation of (2.29):

$$\dot{\mathbf{x}} = \begin{bmatrix} 0 & 1 & 0 & 0 \\ 3 & 0 & 0 & 2 \\ 0 & 0 & 0 & 1 \\ 0 & -2 & 0 & 0 \end{bmatrix} \mathbf{x} + \begin{bmatrix} 0 & 0 \\ 1 & 0 \\ 0 & 0 \\ 0 & 1 \end{bmatrix} \mathbf{u} \tag{6.18}$$

$$\mathbf{y} = \begin{bmatrix} 1 & 0 & 0 & 0 \\ 0 & 0 & 1 & 0 \end{bmatrix} \mathbf{x}$$

where we have assumed, for simplicity, $\omega_o = m = r_o = 1$. The controllability matrix of (6.18) is of order 4×8. If we use Corollary 6.1, then we can check its controllability by using the following 4×6 matrix:

$$[\mathbf{B} \ \mathbf{AB} \ \mathbf{A}^2\mathbf{B}] = \begin{bmatrix} 0 & 0 & 1 & 0 & 0 & 2 \\ 1 & 0 & 0 & 2 & -1 & 0 \\ 0 & 0 & 0 & 1 & -2 & 0 \\ 0 & 1 & -2 & 0 & 0 & -4 \end{bmatrix} \tag{6.19}$$

It has rank 4. Thus (6.18) is controllable. From (6.19), we can readily verify that the controllability indices are 2 and 2, and the controllability index is 2.

▶ **Theorem 6.2**

The controllability property is invariant under any equivalence transformation.

 Proof: Consider the pair $(\mathbf{A}, \mathbf{B})$ with controllability matrix

$$C = [\mathbf{B} \quad \mathbf{AB} \quad \cdots \quad \mathbf{A}^{n-1}\mathbf{B}]$$

and its equivalent pair $(\bar{\mathbf{A}}, \bar{\mathbf{B}})$ with $\bar{\mathbf{A}} = \mathbf{PAP}^{-1}$ and $\bar{\mathbf{B}} = \mathbf{PB}$, where $\mathbf{P}$ is a nonsingular matrix. The controllability matrix of $(\bar{\mathbf{A}}, \bar{\mathbf{B}})$ is

$$
\begin{aligned}
\bar{C} &= [\bar{\mathbf{B}} \quad \bar{\mathbf{A}}\bar{\mathbf{B}} \quad \cdots \quad \bar{\mathbf{A}}^{n-1}\bar{\mathbf{B}}] \\
&= [\mathbf{PB} \quad \mathbf{PAP}^{-1}\mathbf{PB} \quad \cdots \quad \mathbf{PA}^{n-1}\mathbf{P}^{-1}\mathbf{PB}] \\
&= [\mathbf{PB} \quad \mathbf{PAB} \quad \cdots \quad \mathbf{PA}^{n-1}\mathbf{B}] \\
&= \mathbf{P}[\mathbf{B} \quad \mathbf{AB} \quad \cdots \quad \mathbf{A}^{n-1}\mathbf{B}] = \mathbf{P}C
\end{aligned}
\tag{6.20}
$$

Because $\mathbf{P}$ is nonsingular, we have $\rho(C) = \rho(\bar{C})$ (see Equation (3.62)). This establishes Theorem 6.2.   Q.E.D.

▶ **Theorem 6.3**

The set of the controllability indices of $(\mathbf{A}, \mathbf{B})$ is invariant under any equivalence transformation and any reordering of the columns of $\mathbf{B}$.

⟹ **Proof:** Let us define

$$C_k = [\mathbf{B} \quad \mathbf{AB} \quad \cdots \quad \mathbf{A}^{k-1}\mathbf{B}] \tag{6.21}$$

Then we have, following the proof of Theorem 6.2,

$$\rho(C_k) = \rho(\bar{C}_k)$$

for $k = 0, 1, 2, \ldots$. Thus the set of controllability indices is invariant under any equivalence transformation.

The rearrangement of the columns of $\mathbf{B}$ can be achieved by

$$\hat{\mathbf{B}} = \mathbf{BM}$$

where $\mathbf{M}$ is a $p \times p$ nonsingular permutation matrix. It is straightforward to verify

$$\hat{C}_k := [\hat{\mathbf{B}} \quad \mathbf{A}\hat{\mathbf{B}} \quad \cdots \quad \mathbf{A}^{k-1}\hat{\mathbf{B}}] = C_k \mathrm{diag}(\mathbf{M}, \mathbf{M}, \ldots, \mathbf{M})$$

Because diag $(\mathbf{M}, \mathbf{M}, \ldots, \mathbf{M})$ is nonsingular, we have $\rho(\hat{C}_k) = \rho(C_k)$ for $k = 0, 1, \ldots$. Thus the set of controllability indices is invariant under any reordering of the columns of $\mathbf{B}$.   Q.E.D.

Because the set of the controllability indices is invariant under any equivalence transformation and any rearrangement of the inputs, it is an intrinsic property of the system that the state equation describes. The physical significance of the controllability index is not transparent here; but it becomes obvious in the discrete-time case. As we will discuss in later chapters, the controllability index can also be computed from transfer matrices and dictates the minimum degree required to achieve pole placement and model matching.

6.3 Observability

The concept of observability is dual to that of controllability. Roughly speaking, controllability studies the possibility of steering the state from the input; observability studies the possibility of estimating the state from the output. These two concepts are defined under the assumption that the state equation or, equivalently, all $\mathbf{A}$, $\mathbf{B}$, $\mathbf{C}$, and $\mathbf{D}$ are known. Thus the problem of observability is different from the problem of realization or identification, which is to determine or estimate $\mathbf{A}$, $\mathbf{B}$, $\mathbf{C}$, and $\mathbf{D}$ from the information collected at the input and output terminals.

Consider the n-dimensional p-input q-output state equation

$$\dot{\mathbf{x}} = \mathbf{A}\mathbf{x} + \mathbf{B}\mathbf{u}$$
$$\mathbf{y} = \mathbf{C}\mathbf{x} + \mathbf{D}\mathbf{u}$$

(6.22)

where $\mathbf{A}$, $\mathbf{B}$, $\mathbf{C}$, and $\mathbf{D}$ are, respectively, $n \times n$, $n \times p$, $q \times n$, and $q \times p$ constant matrices.

> **Definition 6.O1** *The state equation (6.22) is said to be* observable *if for any unknown initial state* $\mathbf{x}(0)$, *there exists a finite* $t_1 > 0$ *such that the knowledge of the input* $\mathbf{u}$ *and the output* $\mathbf{y}$ *over* $[0, \ t_1]$ *suffices to determine uniquely the initial state* $\mathbf{x}(0)$. *Otherwise, the equation is said to be* unobservable.

EXAMPLE 6.6 Consider the network shown in Fig. 6.5. If the input is zero, no matter what the initial voltage across the capacitor is, the output is identically zero because of the symmetry of the four resistors. We know the input and output (both are identically zero), but we cannot determine uniquely the initial state. Thus the network or, more precisely, the state equation that describes the network is not observable.

EXAMPLE 6.7 Consider the network shown in Fig. 6.6(a). The network has two state variables: the current x_1 through the inductor and the voltage x_2 across the capacitor. The input u is a

Figure 6.5 Unobservable network.

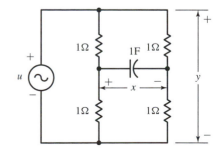

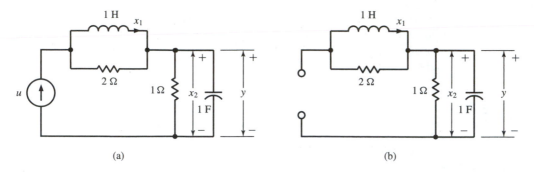

(a)   (b)

Figure 6.6 Unobservable network.

current source. If $u = 0$, the network reduces to the one shown in Fig. 6.6(b). If $x_1(0) = a \neq 0$ and $x_2 = 0$, then the output is identically zero. Any $\mathbf{x}(0) = [a \ \ 0]'$ and $u(t) \equiv 0$ yield the same output $y(t) \equiv 0$. Thus there is no way to determine the initial state $[a \ \ 0]'$ uniquely and the equation that describes the network is not observable.

The response of (6.22) excited by the initial state $\mathbf{x}(0)$ and the input $\mathbf{u}(t)$ was derived in (4.7) as

$$\mathbf{y}(t) = \mathbf{C}e^{\mathbf{A}t}\mathbf{x}(0) + \mathbf{C}\int_0^t e^{\mathbf{A}(t-\tau)}\mathbf{B}\mathbf{u}(\tau)\,d\tau + \mathbf{D}\mathbf{u}(t) \tag{6.23}$$

In the study of observability, the output $\mathbf{y}$ and the input $\mathbf{u}$ are assumed to be known; the inital state $\mathbf{x}(0)$ is the only unknown. Thus we can write (6.23) as

$$\mathbf{C}e^{\mathbf{A}t}\mathbf{x}(0) = \bar{\mathbf{y}}(t) \tag{6.24}$$

where

$$\bar{\mathbf{y}}(t) := \mathbf{y}(t) - \mathbf{C}\int_0^t e^{\mathbf{A}(t-\tau)}\mathbf{B}\mathbf{u}(\tau)\,d\tau - \mathbf{D}\mathbf{u}(t)$$

is a known function. Thus the observability problem reduces to solving $\mathbf{x}(0)$ from (6.24). If $\mathbf{u} \equiv \mathbf{0}$, then $\bar{\mathbf{y}}(t)$ reduces to the zero-input response $\mathbf{C}e^{\mathbf{A}t}\mathbf{x}(0)$. Thus Definition 6.O1 can be modified as follows: Equation (6.22) is observable if and only if the initial state $\mathbf{x}(0)$ can be determined uniquely from its zero-input response over a finite time interval.

Next we discuss how to solve $\mathbf{x}(0)$ from (6.24). For a fixed t, $\mathbf{C}e^{\mathbf{A}t}$ is a $q \times n$ constant matrix, and $\bar{\mathbf{y}}(t)$ is a $q \times 1$ constant vector. Thus (6.24) is a set of linear algebraic equations with n unknowns. Because of the way it is developed, for every fixed t, $\bar{\mathbf{y}}(t)$ is in the range space of $\mathbf{C}e^{\mathbf{A}t}$ and solutions always exist in (6.24). The only question is whether the solution is unique. If $q < n$, as is the case in general, the $q \times n$ matrix $\mathbf{C}e^{\mathbf{A}t}$ has rank at most q and, consequently, has nullity $n - q$ or larger. Thus solutions are not unique (Theorem 3.2). In conclusion, we cannot find a unique $\mathbf{x}(0)$ from (6.24) at an isolated t. In order to determine $\mathbf{x}(0)$ uniquely from (6.24), we must use the knowledge of $\mathbf{u}(t)$ and $\mathbf{y}(t)$ over a nonzero time interval as stated in the next theorem.

▶ **Theorem 6.4**

The state equation (6.22) is observable if and only if the $n \times n$ matrix

$$\mathbf{W}_o(t) = \int_0^t e^{\mathbf{A}'\tau}\mathbf{C}'\mathbf{C}e^{\mathbf{A}\tau}d\tau \tag{6.25}$$

is nonsingular for any $t > 0$.

Proof: We premultiply (6.24) by $e^{\mathbf{A}'t}\mathbf{C}'$ and then integrate it over $[0,\ t_1]$ to yield

$$\left(\int_0^{t_1} e^{\mathbf{A}'t}\mathbf{C}'\mathbf{C}e^{\mathbf{A}t}dt\right)\mathbf{x}(0) = \int_0^{t_1} e^{\mathbf{A}'t}\mathbf{C}'\bar{\mathbf{y}}(t)\,dt$$

If $\mathbf{W}_o(t_1)$ is nonsingular, then

$$\mathbf{x}(0) = \mathbf{W}_o^{-1}(t_1)\int_0^{t_1} e^{\mathbf{A}'t}\mathbf{C}'\bar{\mathbf{y}}(t)\,dt \tag{6.26}$$

This yields a unique $\mathbf{x}(0)$. This shows that if $\mathbf{W}_o(t)$, for any $t > 0$, is nonsingular, then (6.22) is observable. Next we show that if $\mathbf{W}_o(t_1)$ is singular or, equivalently, positive semidefinite for all t_1, then (6.22) is not observable. If $\mathbf{W}_o(t_1)$ is positive semidefinite, there exists an $n \times 1$ nonzero constant vector $\mathbf{v}$ such that

$$\mathbf{v}'\mathbf{W}_o(t_1)\mathbf{v} = \int_0^{t_1} \mathbf{v}'e^{\mathbf{A}'\tau}\mathbf{C}'\mathbf{C}e^{\mathbf{A}\tau}\mathbf{v}\,d\tau$$

$$= \int_0^{t_1} ||\mathbf{C}e^{\mathbf{A}\tau}\mathbf{v}||^2 d\tau = 0$$

which implies

$$\mathbf{C}e^{\mathbf{A}\tau}\mathbf{v} \equiv \mathbf{0} \tag{6.27}$$

for all t in $[0,\ t_1]$. If $\mathbf{u} \equiv \mathbf{0}$, then $\mathbf{x}_1(0) = \mathbf{v} \neq \mathbf{0}$ and $\mathbf{x}_2(0) = \mathbf{0}$ both yield the same

$$\mathbf{y}(t) = \mathbf{C}e^{\mathbf{A}t}\mathbf{x}_i(0) \equiv \mathbf{0}$$

Two different initial states yield the same zero-input response; therefore we cannot uniquely determine $\mathbf{x}(0)$. Thus (6.22) is not observable. This completes the proof of Theorem 6.4.   Q.E.D.

We see from this theorem that observability depends only on $\mathbf{A}$ and $\mathbf{C}$. This can also be deduced from Definition 6.O1 by choosing $\mathbf{u}(t) \equiv \mathbf{0}$. Thus observability is a property of the pair $(\mathbf{A}, \mathbf{C})$ and is independent of $\mathbf{B}$ and $\mathbf{D}$. As in the controllability part, if $\mathbf{W}_o(t)$ is nonsingular for some t, then it is nonsingular for every t and the initial state can be computed from (6.26) by using any nonzero time interval.

▶ **Theorem 6.5 (Theorem of duality)**

The pair $(\mathbf{A}, \mathbf{B})$ is controllable if and only if the pair $(\mathbf{A}', \mathbf{B}')$ is observable.

Proof: The pair $(\mathbf{A}, \mathbf{B})$ is controllable if and only if

$$\mathbf{W}_c(t) = \int_0^t e^{\mathbf{A}\tau} \mathbf{B}\mathbf{B}' e^{\mathbf{A}'\tau} d\tau$$

is nonsingular for any t. The pair $(\mathbf{A}', \mathbf{B}')$ is observable if and only if, by replacing $\mathbf{A}$ by $\mathbf{A}'$ and $\mathbf{C}$ by $\mathbf{B}'$ in (6.25),

$$\mathbf{W}_o(t) = \int_0^t e^{\mathbf{A}\tau} \mathbf{B}\mathbf{B}' e^{\mathbf{A}'\tau} d\tau$$

is nonsingular for any t. The two conditions are identical and the theorem follows. Q.E.D.

We list in the following the observability counterpart of Theorem 6.1. It can be proved either directly or by applying the theorem of duality.

▶ **Theorem 6.O1**

The following statements are equivalent.

1. The n-dimensional pair $(\mathbf{A}, \mathbf{C})$ is observable.
2. The $n \times n$ matrix

$$\mathbf{W}_o(t) = \int_0^t e^{\mathbf{A}'\tau} \mathbf{C}' \mathbf{C} e^{\mathbf{A}\tau} d\tau \tag{6.28}$$

is nonsingular for any $t > 0$.

3. The $nq \times n$ *observability matrix*

$$O = \begin{bmatrix} \mathbf{C} \\ \mathbf{C}\mathbf{A} \\ \vdots \\ \mathbf{C}\mathbf{A}^{n-1} \end{bmatrix} \tag{6.29}$$

has rank n (full column rank). This matrix can be generated by calling `obsv` in MATLAB.

4. The $(n + q) \times n$ matrix

$$\begin{bmatrix} \mathbf{A} - \lambda\mathbf{I} \\ \mathbf{C} \end{bmatrix}$$

has full column rank at every eigenvalue, λ, of $\mathbf{A}$.

5. If, in addition, all eigenvalues of $\mathbf{A}$ have negative real parts, then the unique solution of

$$\mathbf{A}'\mathbf{W}_o + \mathbf{W}_o\mathbf{A} = -\mathbf{C}'\mathbf{C} \tag{6.30}$$

is positive definite. The solution is called the *observability Gramian* and can be expressed as

$$\mathbf{W}_o = \int_0^\infty e^{\mathbf{A}'\tau} \mathbf{C}' \mathbf{C} e^{\mathbf{A}\tau} d\tau \tag{6.31}$$

6.3.1 Observability Indices

Let $\mathbf{A}$ and $\mathbf{C}$ be $n \times n$ and $q \times n$ constant matrices. We assume that $\mathbf{C}$ has rank q (full row rank). If $\mathbf{C}$ does not have full row rank, then the output at some output terminal can be expressed as a linear combination of other outputs. Thus the output does not offer any new information regarding the system and the terminal can be eliminated. By deleting the corresponding row, the reduced $\mathbf{C}$ will then have full row rank.

If $(\mathbf{A}, \mathbf{C})$ is observable, its observability matrix O has rank n and, consequently, n linearly independent rows. Let $\mathbf{c}_i$ be the ith row of $\mathbf{C}$. Let us search linearly independent rows of O in order from top to bottom. Dual to the controllability part, if a row associated with $\mathbf{c}_m$ becomes linearly dependent on its upper rows, then all rows associated with $\mathbf{c}_m$ thereafter will also be dependent. Let ν_m be the number of the linearly independent rows associated with $\mathbf{c}_m$. It is clear that if O has rank n, then

$$\nu_1 + \nu_2 + \cdots + \nu_q = n \tag{6.32}$$

The set $\{\nu_1, \nu_2, \ldots, \nu_q\}$ is called the *observability indices* and

$$\nu = \max(\nu_1, \nu_2, \ldots, \nu_q) \tag{6.33}$$

is called the *observability index* of $(\mathbf{A}, \mathbf{C})$. If $(\mathbf{A}, \mathbf{C})$ is observable, it is the least integer such that

$$\rho(O_\nu) := \begin{bmatrix} \mathbf{C} \\ \mathbf{CA} \\ \mathbf{CA}^2 \\ \vdots \\ \mathbf{CA}^{\nu-1} \end{bmatrix} = n$$

Dual to the controllability part, we have

$$n/q \leq \nu \leq \min(\bar{n}, n - q + 1) \tag{6.34}$$

where $\rho(\mathbf{C}) = q$ and $\bar{n}$ is the degree of the minimal polynomial of A.

▶ **Corollary 6.01**

The n-dimensional pair $(\mathbf{A}, \mathbf{C})$ is observable if and only if the matrix

$$O_{n-q+1} = \begin{bmatrix} \mathbf{C} \\ \mathbf{CA} \\ \vdots \\ \mathbf{CA}^{n-q} \end{bmatrix} \tag{6.35}$$

where $\rho(\mathbf{C}) = q$, has rank n or the $n \times n$ matrix $O'_{n-q+1} O_{n-q+1}$ is nonsingular.

▶ **Theorem 6.02**

The observability property is invariant under any equivalence transformation.

▶ **Theorem 6.03**

The set of the observability indices of $(\mathbf{A}, \mathbf{C})$ is invariant under any equivalence transformation and any reordering of the rows of $\mathbf{C}$.

Before concluding this section, we discuss a different way of solving (6.24). Differentiating (6.24) repeatedly and setting $t = 0$, we can obtain

$$
\begin{bmatrix} \mathbf{C} \\ \mathbf{CA} \\ \vdots \\ \mathbf{CA}^{\nu-1} \end{bmatrix} \mathbf{x}(0) = \begin{bmatrix} \bar{\mathbf{y}}(0) \\ \dot{\bar{\mathbf{y}}}(0) \\ \vdots \\ \bar{\mathbf{y}}^{(\nu-1)}(0) \end{bmatrix}
$$

or

$$
O_\nu \mathbf{x}(0) = \tilde{\mathbf{y}}(0) \tag{6.36}
$$

where $\bar{\mathbf{y}}^{(i)}(t)$ is the ith derivative of $\bar{\mathbf{y}}(t)$, and $\tilde{\mathbf{y}}(0) := [\bar{\mathbf{y}}'(0) \ \dot{\bar{\mathbf{y}}}'(0) \ \cdots \ (\bar{\mathbf{y}}^{(\nu-1)})']'$. Equation (6.36) is a set of linear algebraic equations. Because of the way it is developed, $\tilde{\mathbf{y}}(0)$ must lie in the range space of O_ν. Thus a solution $\mathbf{x}(0)$ exists in (6.36). If $(\mathbf{A}, \mathbf{C})$ is observable, then O_ν has full column rank and, following Theorem 3.2, the solution is unique. Premultiplying (6.36) by O'_ν and then using Theorem 3.8, we can obtain the solution as

$$
\mathbf{x}(0) = \left[O'_\nu O_\nu \right]^{-1} O'_\nu \tilde{\mathbf{y}}(0) \tag{6.37}
$$

We mention that in order to obtain $\dot{\bar{\mathbf{y}}}(0), \ddot{\bar{\mathbf{y}}}(0), \ldots$, we need knowledge of $\bar{\mathbf{y}}(t)$ in the neighborhood of $t = 0$. This is consistent with the earlier assertion that we need knowledge of $\bar{\mathbf{y}}(t)$ over a nonzero time interval in order to determine $\mathbf{x}(0)$ uniquely from (6.24). In conclusion, the initial state can be computed using (6.26) or (6.37).

The output $y(t)$ measured in practice is often corrupted by high-frequency noise. Because

- differentiation will amplify high-frequency noise and
- integration will suppress or smooth high-frequency noise,

the result obtained from (6.36) or (6.37) may differ greatly from the actual initial state. Thus (6.26) is preferable to (6.36) in computing initial states.

The physical significance of the observability index can be seen from (6.36). It is the smallest integer in order to determine $\mathbf{x}(0)$ uniquely from (6.36) or (6.37). It also dictates the minimum degree required to achieve pole placement and model matching, as we will discuss in Chapter 9.

6.4 Canonical Decomposition

This section discusses canonical decomposition of state equations. This fundamental result will be used to establish the relationship between the state-space description and the transfer-matrix description. Consider

$$\dot{\mathbf{x}} = \mathbf{Ax} + \mathbf{Bu}$$

$$\mathbf{y} = \mathbf{Cx} + \mathbf{Du} \tag{6.38}$$

Let $\bar{\mathbf{x}} = \mathbf{Px}$, where $\mathbf{P}$ is a nonsingular matrix. Then the state equation

$$\dot{\bar{\mathbf{x}}} = \bar{\mathbf{A}}\bar{\mathbf{x}} + \bar{\mathbf{B}}\mathbf{u}$$

$$\mathbf{y} = \bar{\mathbf{C}}\bar{\mathbf{x}} + \bar{\mathbf{D}}\mathbf{u} \tag{6.39}$$

with $\bar{\mathbf{A}} = \mathbf{PAP}^{-1}$, $\bar{\mathbf{B}} = \mathbf{PB}$, $\bar{\mathbf{C}} = \mathbf{CP}^{-1}$, and $\bar{\mathbf{D}} = \mathbf{D}$ is equivalent to (6.38). All properties of (6.38), including stability, controllability, and observability, are preserved in (6.39). We also have

$$\bar{C} = PC \qquad \bar{O} = OP^{-1}$$

▶ **Theorem 6.6**

Consider the n-dimensional state equation in (6.38) with

$$\rho(C) = \rho([\mathbf{B} \ \mathbf{AB} \ \cdots \ \mathbf{A}^{n-1}\mathbf{B}]) = n_1 < n$$

We form the $n \times n$ matrix

$$\mathbf{P}^{-1} := [\mathbf{q}_1 \ \cdots \ \mathbf{q}_{n_1} \cdots \ \mathbf{q}_n]$$

where the first n_1 columns are any n_1 linearly independent columns of C, and the remaining columns can arbitrarily be chosen as long as $\mathbf{P}$ is nonsingular. Then the equivalence transformation $\bar{\mathbf{x}} = \mathbf{Px}$ or $\mathbf{x} = \mathbf{P}^{-1}\bar{\mathbf{x}}$ will transform (6.38) into

$$\begin{bmatrix} \dot{\bar{\mathbf{x}}}_c \\ \dot{\bar{\mathbf{x}}}_{\bar{c}} \end{bmatrix} = \begin{bmatrix} \bar{\mathbf{A}}_c & \bar{\mathbf{A}}_{12} \\ \mathbf{0} & \bar{\mathbf{A}}_{\bar{c}} \end{bmatrix} \begin{bmatrix} \bar{\mathbf{x}}_c \\ \bar{\mathbf{x}}_{\bar{c}} \end{bmatrix} + \begin{bmatrix} \bar{\mathbf{B}}_c \\ \mathbf{0} \end{bmatrix} \mathbf{u}$$

$$\mathbf{y} = [\bar{\mathbf{C}}_c \ \bar{\mathbf{C}}_{\bar{c}}] \begin{bmatrix} \bar{\mathbf{x}}_c \\ \bar{\mathbf{x}}_{\bar{c}} \end{bmatrix} + \mathbf{Du} \tag{6.40}$$

where $\bar{\mathbf{A}}_c$ is $n_1 \times n_1$ and $\bar{\mathbf{A}}_{\bar{c}}$ is $(n - n_1) \times (n - n_1)$, and the n_1-dimensional subequation of (6.40),

$$\dot{\bar{\mathbf{x}}}_c = \bar{\mathbf{A}}_c\bar{\mathbf{x}}_c + \bar{\mathbf{B}}_c\mathbf{u}$$

$$\bar{\mathbf{y}} = \bar{\mathbf{C}}_c\bar{\mathbf{x}}_c + \mathbf{Du} \tag{6.41}$$

is controllable and has the same transfer matrix as (6.38).

Proof: As discussed in Section 4.3, the transformation $\mathbf{x} = \mathbf{P}^{-1}\bar{\mathbf{x}}$ changes the basis of the state space from the orthonormal basis in (3.8) to the columns of $\mathbf{Q} := \mathbf{P}^{-1}$ or $\{\mathbf{q}_1, \ldots, \mathbf{q}_{n_1}, \ldots, \mathbf{q}_n\}$. The ith column of $\bar{\mathbf{A}}$ is the representation of $\mathbf{Aq}_i$ with respect to $\{\mathbf{q}_1, \ldots, \mathbf{q}_{n_1}, \ldots, \mathbf{q}_n\}$. Now the vector $\mathbf{Aq}_i$, for $i = 1, 2, \ldots, n_1$, are linearly dependent on the set $\{\mathbf{q}_1, \ldots, \mathbf{q}_{n_1}\}$; they are linearly independent of $\{\mathbf{q}_{n_1+1}, \ldots, \mathbf{q}_n\}$. Thus the matrix $\bar{\mathbf{A}}$ has the form shown in (6.40). The columns of $\bar{\mathbf{B}}$ are the representation of the columns of $\mathbf{B}$ with respect to $\{\mathbf{q}_1, \ldots, \mathbf{q}_{n_1}, \ldots, \mathbf{q}_n\}$. Now the columns of $\mathbf{B}$ depend only on $\{\mathbf{q}_1, \ldots, \mathbf{q}_{n_1}\}$; thus $\bar{\mathbf{B}}$ has the form shown in (6.40). We mention that if the $n \times p$

matrix $\mathbf{B}$ has rank p and if its columns are chosen as the first p columns of $\mathbf{P}^{-1}$, then the upper part of $\bar{\mathbf{B}}$ is the unit matrix of order p.

Let $\bar{C}$ be the controllability matrix of (6.40). Then we have $\rho(C) = \rho(\bar{C}) = n_1$. It is straightforward to verify

$$\bar{C} = \begin{bmatrix} \bar{\mathbf{B}}_c & \bar{\mathbf{A}}_c\bar{\mathbf{B}}_c & \cdots & \bar{\mathbf{A}}_c^{n_1}\bar{\mathbf{B}}_c & \cdots & \bar{\mathbf{A}}_c^{n-1}\bar{\mathbf{B}}_c \\ \mathbf{0} & \mathbf{0} & \cdots & \mathbf{0} & \cdots & \mathbf{0} \end{bmatrix}$$

$$= \begin{bmatrix} \bar{C}_c & \bar{\mathbf{A}}_c^{n_1}\bar{\mathbf{B}}_c & \cdots & \bar{\mathbf{A}}_c^{n-1}\bar{\mathbf{B}}_c \\ \mathbf{0} & \mathbf{0} & \cdots & \mathbf{0} \end{bmatrix}$$

where $\bar{C}_c$ is the controllability matrix of $(\bar{\mathbf{A}}_c, \bar{\mathbf{B}}_c)$. Because the columns of $\bar{\mathbf{A}}_c^k\bar{\mathbf{B}}_c$, for $k \geq n_1$, are linearly dependent on the columns of $\bar{C}_c$, the condition $\rho(C) = n_1$ implies $\rho(\bar{C}_c) = n_1$. Thus the n_1-dimensional state equation in (6.41) is controllable.

Next we show that (6.41) has the same transfer matrix as (6.38). Because (6.38) and (6.40) have the same transfer matrix, we need to show only that (6.40) and (6.41) have the same transfer matrix. By direct verification, we can show

$$\begin{bmatrix} s\mathbf{I} - \bar{\mathbf{A}}_c & -\bar{\mathbf{A}}_{12} \\ \mathbf{0} & s\mathbf{I} - \bar{\mathbf{A}}_{\bar{c}} \end{bmatrix}^{-1} = \begin{bmatrix} (s\mathbf{I} - \bar{\mathbf{A}}_c)^{-1} & \mathbf{M} \\ \mathbf{0} & (s\mathbf{I} - \bar{\mathbf{A}}_{\bar{c}})^{-1} \end{bmatrix} \tag{6.42}$$

where

$$\mathbf{M} = (s\mathbf{I} - \bar{\mathbf{A}}_c)^{-1}\bar{\mathbf{A}}_{12}(s\mathbf{I} - \bar{\mathbf{A}}_{\bar{c}})^{-1}$$

Thus the transfer matrix of (6.40) is

$$\begin{bmatrix} \bar{\mathbf{C}}_c & \bar{\mathbf{C}}_{\bar{c}} \end{bmatrix} \begin{bmatrix} s\mathbf{I} - \bar{\mathbf{A}}_c & -\bar{\mathbf{A}}_{12} \\ \mathbf{0} & s\mathbf{I} - \bar{\mathbf{A}}_{\bar{c}} \end{bmatrix}^{-1} \begin{bmatrix} \bar{\mathbf{B}}_c \\ \mathbf{0} \end{bmatrix} + \mathbf{D}$$

$$= \begin{bmatrix} \bar{\mathbf{C}}_c & \bar{\mathbf{C}}_{\bar{c}} \end{bmatrix} \begin{bmatrix} (s\mathbf{I} - \bar{\mathbf{A}}_c)^{-1} & \mathbf{M} \\ \mathbf{0} & (s\mathbf{I} - \bar{\mathbf{A}}_{\bar{c}})^{-1} \end{bmatrix} \begin{bmatrix} \bar{\mathbf{B}}_c \\ \mathbf{0} \end{bmatrix} + \mathbf{D}$$

$$= \bar{\mathbf{C}}_c(s\mathbf{I} - \bar{\mathbf{A}}_c)^{-1}\bar{\mathbf{B}}_c + \mathbf{D}$$

which is the transfer matrix of (6.41). This completes the proof of Theorem 6.6.   Q.E.D.

In the equivalence transformation $\bar{\mathbf{x}} = \mathbf{Px}$, the n-dimensional state space is divided into two subspaces. One is the n_1-dimensional subspace that consists of all vectors of the form $[\bar{\mathbf{x}}_c'\ \mathbf{0}']'$; the other is the $(n - n_1)$-dimensional subspace that consists of all vectors of the form $[\mathbf{0}'\ \bar{\mathbf{x}}_{\bar{c}}']'$. Because (6.41) is controllable, the input $\mathbf{u}$ can transfer $\bar{\mathbf{x}}_c$ from any state to any other state. However, the input $\mathbf{u}$ cannot control $\bar{\mathbf{x}}_{\bar{c}}$ because, as we can see from (6.40), $\mathbf{u}$ does not affect $\bar{\mathbf{x}}_{\bar{c}}$ directly, nor indirectly through the state $\bar{\mathbf{x}}_c$. By dropping the uncontrollable state vector, we obtain a controllable state equation of lesser dimension that is zero-state equivalent to the original equation.

EXAMPLE 6.8 Consider the three-dimensional state equation

$$\dot{\mathbf{x}} = \begin{bmatrix} 1 & 1 & 0 \\ 0 & 1 & 0 \\ 0 & 1 & 1 \end{bmatrix} \mathbf{x} + \begin{bmatrix} 0 & 1 \\ 1 & 0 \\ 0 & 1 \end{bmatrix} u \qquad y = [1\ 1\ 1]\mathbf{x} \tag{6.43}$$

The rank of **B** is 2; therefore we can use $C_2 = [\textbf{B} \quad \textbf{AB}]$, instead of $C = [\textbf{B} \quad \textbf{AB} \quad \textbf{A}^2\textbf{B}]$, to check the controllability of (6.43) (Corollary 6.1). Because

$$\rho(C_2) = \rho([\textbf{B} \quad \textbf{AB}]) = \rho \begin{bmatrix} 0 & 1 & 1 & 1 \\ 1 & 0 & 1 & 0 \\ 0 & 1 & 1 & 1 \end{bmatrix} = 2 < 3$$

the state equation in (6.43) is not controllable. Let us choose

$$\textbf{P}^{-1} = \textbf{Q} := \begin{bmatrix} 0 & 1 & 1 \\ 1 & 0 & 0 \\ 0 & 1 & 0 \end{bmatrix}$$

The first two columns of **Q** are the first two linearly independent columns of C_2; the last column is chosen arbitrarily to make **Q** nonsingular. Let $\bar{\textbf{x}} = \textbf{Px}$. We compute

$$\bar{\textbf{A}} = \textbf{PAP}^{-1} = \begin{bmatrix} 0 & 1 & 0 \\ 0 & 0 & 1 \\ 1 & 0 & -1 \end{bmatrix} \begin{bmatrix} 1 & 1 & 0 \\ 0 & 1 & 0 \\ 0 & 1 & 1 \end{bmatrix} \begin{bmatrix} 0 & 1 & 1 \\ 1 & 0 & 0 \\ 0 & 1 & 0 \end{bmatrix}$$

$$= \begin{bmatrix} 1 & 0 & \vdots & 0 \\ 1 & 1 & \vdots & 0 \\ \cdots & \cdots & \cdots & \cdots \\ 0 & 0 & \vdots & 1 \end{bmatrix}$$

$$\bar{\textbf{B}} = \textbf{PB} = \begin{bmatrix} 0 & 1 & 0 \\ 0 & 0 & 1 \\ 1 & 0 & -1 \end{bmatrix} \begin{bmatrix} 0 & 1 \\ 1 & 0 \\ 0 & 1 \end{bmatrix} = \begin{bmatrix} 1 & 0 \\ 0 & 1 \\ \cdots & \cdots \\ 0 & 0 \end{bmatrix}$$

$$\bar{\textbf{C}} = \textbf{CP}^{-1} = [1 \quad 1 \quad 1] \begin{bmatrix} 0 & 1 & 1 \\ 1 & 0 & 0 \\ 0 & 1 & 0 \end{bmatrix} = [1 \quad 2 \quad \vdots \quad 1]$$

Note that the 1×2 submatrix $\bar{\textbf{A}}_{21}$ of $\bar{\textbf{A}}$ and $\bar{\textbf{B}}_{\bar{c}}$ are zero as expected. The 2×1 submatrix $\bar{\textbf{A}}_{12}$ happens to be zero; it could be nonzero. The upper part of $\bar{\textbf{B}}$ is a unit matrix because the columns of **B** are the first two columns of **Q**. Thus (6.43) can be reduced to

$$\dot{\bar{\textbf{x}}}_c = \begin{bmatrix} 1 & 0 \\ 1 & 1 \end{bmatrix} \bar{\textbf{x}}_c + \begin{bmatrix} 1 & 0 \\ 0 & 1 \end{bmatrix} \textbf{u} \qquad y = [1 \quad 2]\bar{\textbf{x}}_c$$

This equation is controllable and has the same transfer matrix as (6.43).

The MATLAB function `ctrbf` transforms (6.38) into (6.40) except that the order of the columns in $\textbf{P}^{-1}$ is reversed. Thus the resulting equation has the form

$$\begin{bmatrix} \bar{\textbf{A}}_{\bar{c}} & \textbf{0} \\ \bar{\textbf{A}}_{21} & \bar{\textbf{A}}_c \end{bmatrix} \qquad \begin{bmatrix} \textbf{0} \\ \bar{\textbf{B}}_c \end{bmatrix}$$

Theorem 6.6 is established from the controllability matrix. In actual computation, it is unnecessary to form the controllability matrix. The result can be obtained by carrying out a sequence of similarity transformations to transform $[\mathbf{B} \ \mathbf{A}]$ into a Hessenberg form. See Reference [6, pp. 220–222]. This procedure is efficient and numerically stable and should be used in actual computation.

Dual to Theorem 6.6, we have the following theorem for unobservable state equations.

▶ **Theorem 6.O6**

Consider the n-dimensional state equation in (6.38) with

$$\rho(O) = \rho \begin{bmatrix} \mathbf{C} \\ \mathbf{CA} \\ \vdots \\ \mathbf{CA}^{n-1} \end{bmatrix} = n_2 < n$$

We form the $n \times n$ matrix

$$\mathbf{P} = \begin{bmatrix} \mathbf{p}_1 \\ \vdots \\ \mathbf{p}_{n_2} \\ \vdots \\ \mathbf{p}_n \end{bmatrix}$$

where the first n_2 rows are any n_2 linearly independent rows of O, and the remaining rows can be chosen arbitrarily as long as $\mathbf{P}$ is nonsingular. Then the equivalence transformation $\bar{\mathbf{x}} = \mathbf{Px}$ will transform (6.38) into

$$\begin{bmatrix} \dot{\bar{\mathbf{x}}}_o \\ \dot{\bar{\mathbf{x}}}_{\bar{o}} \end{bmatrix} = \begin{bmatrix} \bar{\mathbf{A}}_o & \mathbf{0} \\ \bar{\mathbf{A}}_{21} & \bar{\mathbf{A}}_{\bar{o}} \end{bmatrix} \begin{bmatrix} \bar{\mathbf{x}}_o \\ \bar{\mathbf{x}}_{\bar{o}} \end{bmatrix} + \begin{bmatrix} \bar{\mathbf{B}}_o \\ \bar{\mathbf{B}}_{\bar{o}} \end{bmatrix} \mathbf{u}$$

$$\mathbf{y} = [\bar{\mathbf{C}}_o \ \ \mathbf{0}] \begin{bmatrix} \bar{\mathbf{x}}_o \\ \bar{\mathbf{x}}_{\bar{o}} \end{bmatrix} + \mathbf{Du} \tag{6.44}$$

where $\bar{\mathbf{A}}_o$ is $n_2 \times n_2$ and $\bar{\mathbf{A}}_{\bar{o}}$ is $(n - n_2) \times (n - n_2)$, and the n_2-dimensional subequation of (6.44),

$$\dot{\bar{\mathbf{x}}}_o = \bar{\mathbf{A}}_c \bar{\mathbf{x}}_o + \bar{\mathbf{B}}_o \mathbf{u}$$

$$\bar{\mathbf{y}} = \bar{\mathbf{C}}_o \bar{\mathbf{x}}_o + \mathbf{Du}$$

is observable and has the same transfer matrix as (6.38).

In the equivalence transformation $\bar{\mathbf{x}} = \mathbf{Px}$, the n-dimensional state space is divided into two subspaces. One is the n_2-dimensional subspace that consists of all vectors of the form $[\bar{\mathbf{x}}_o' \ \mathbf{0}']'$; the other is the $(n - n_2)$-dimensional subspace consisting of all vectors of the form $[\mathbf{0}' \ \bar{\mathbf{x}}_{\bar{o}}']'$. The state $\bar{\mathbf{x}}_o$ can be detected from the output. However, $\bar{\mathbf{x}}_{\bar{o}}$ cannot be detected from the output because, as we can see from (6.44), it is not connected to the output either directly, or indirectly through the state $\bar{\mathbf{x}}_o$. By dropping the unobservable state vector, we obtain an observable state equation of lesser dimension that is zero-state equivalent to the original equation. The MATLAB function obsvf is the counterpart of ctrbf. Combining Theorems 6.6 and 6.06, we have the following *Kalman decomposition theorem*.

▶ **Theorem 6.7**

Every state-space equation can be transformed, by an equivalence transformation, into the following canonical form

$$
\begin{bmatrix} \dot{\bar{\mathbf{x}}}_{co} \\ \dot{\bar{\mathbf{x}}}_{c\bar{o}} \\ \dot{\bar{\mathbf{x}}}_{\bar{c}o} \\ \dot{\bar{\mathbf{x}}}_{\bar{c}\bar{o}} \end{bmatrix} = \begin{bmatrix} \bar{\mathbf{A}}_{co} & 0 & \bar{\mathbf{A}}_{13} & 0 \\ \bar{\mathbf{A}}_{21} & \bar{\mathbf{A}}_{c\bar{o}} & \bar{\mathbf{A}}_{23} & \bar{\mathbf{A}}_{24} \\ 0 & 0 & \bar{\mathbf{A}}_{\bar{c}o} & 0 \\ 0 & 0 & \bar{\mathbf{A}}_{43} & \bar{\mathbf{A}}_{\bar{c}\bar{o}} \end{bmatrix} \begin{bmatrix} \bar{\mathbf{x}}_{co} \\ \bar{\mathbf{x}}_{c\bar{o}} \\ \bar{\mathbf{x}}_{\bar{c}o} \\ \bar{\mathbf{x}}_{\bar{c}\bar{o}} \end{bmatrix} + \begin{bmatrix} \bar{\mathbf{B}}_{co} \\ \bar{\mathbf{B}}_{c\bar{o}} \\ 0 \\ 0 \end{bmatrix} \mathbf{u} \tag{6.45}
$$

$$
\mathbf{y} = [\bar{\mathbf{C}}_{co} \ \ 0 \ \ \bar{\mathbf{C}}_{\bar{c}o} \ \ 0]\bar{\mathbf{x}} + \mathbf{Du}
$$

where the vector $\bar{\mathbf{x}}_{co}$ is controllable and observable, $\bar{\mathbf{x}}_{c\bar{o}}$ is controllable but not observable, $\bar{\mathbf{x}}_{\bar{c}o}$ is observable but not controllable, and $\bar{\mathbf{x}}_{\bar{c}\bar{o}}$ is neither controllable nor observable. Furthermore, the state equation is zero-state equivalent to the controllable and observable state equation

$$
\dot{\bar{\mathbf{x}}}_{co} = \bar{\mathbf{A}}_{co}\bar{\mathbf{x}}_{co} + \bar{\mathbf{B}}_{co}\mathbf{u}
$$
$$
\mathbf{y} = \bar{\mathbf{C}}_{co}\bar{\mathbf{x}}_{co} + \mathbf{Du} \tag{6.46}
$$

and has the transfer matrix

$$
\hat{\mathbf{G}}(s) = \bar{\mathbf{C}}_{co}(s\mathbf{I} - \bar{\mathbf{A}}_{co})^{-1}\bar{\mathbf{B}}_{co} + \mathbf{D}
$$

This theorem can be illustrated symbolically as shown in Fig. 6.7. The equation is first decomposed, using Theorem 6.6, into controllable and uncontrollable subequations. We then decompose each subequation, using Theorem 6.O6, into observable and unobservable parts. From the figure, we see that only the controllable and observable part is connected to both the input and output terminals. Thus the transfer matrix describes only this part of the system. This is the reason that the transfer-function description and the state-space description are not necessarily equivalent. For example, if any A-matrix other than $\bar{\mathbf{A}}_{co}$ has an eigenvalue with a

Figure 6.7 Kalman decomposition.

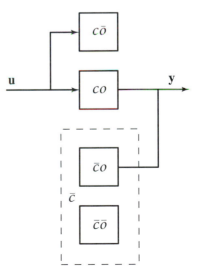

positive real part, then some state variable may grow without bound and the system may burn out. This phenomenon, however, cannot be detected from the transfer matrix.

The MATLAB function `minreal`, an acronym for *minimal realization*, can reduce any state equation to (6.46). The reason for calling it minimal realization will be given in the next chapter.

EXAMPLE 6.9 Consider the network shown in Fig. 6.8(a). Because the input is a current source, responses due to the initial conditions in C_1 and L_1 will not appear at the output. Thus the state variables associated with C_1 and L_1 are not observable; whether or not they are controllable is immaterial in subsequent discussion. Similarly, the state variable associated with L_2 is not controllable. Because of the symmetry of the four 1-Ω resistors, the state variable associated with C_2 is neither controllable nor observable. By dropping the state variables that are either uncontrollable or unobservable, the network in Fig. 6.8(a) can be reduced to the one in Fig. 6.8(b). The current in each branch is $u/2$; thus the output y equals $2 \cdot (u/2)$ or $y = u$. Thus the transfer function of the network in Fig. 6.8(a) is $\hat{g}(s) = 1$.

If we assign state variables as shown, then the network can be described by

$$\dot{\mathbf{x}} = \begin{bmatrix} 0 & -0.5 & 0 & 0 \\ 1 & 0 & 0 & 0 \\ 0 & 0 & -0.5 & 0 \\ 0 & 0 & 0 & -1 \end{bmatrix} \mathbf{x} + \begin{bmatrix} 0.5 \\ 0 \\ 0 \\ 0 \end{bmatrix} u$$

$$y = [0\ 0\ 0\ 1]\mathbf{x} + u$$

Because the equation is already of the form shown in (6.40), it can be reduced to the following controllable state equation

$$\dot{\mathbf{x}}_c = \begin{bmatrix} 0 & -0.5 \\ 1 & 0 \end{bmatrix} \mathbf{x}_c + \begin{bmatrix} 0.5 \\ 0 \end{bmatrix} u$$

$$y = [0\ 0]\mathbf{x}_c + u$$

The output is independent of $\mathbf{x}_c$; thus the equation can be further reduced to $y = u$. This is what we will obtain by using the MATLAB function `minreal`.

6.5 Conditions in Jordan-Form Equations

Controllability and observability are invariant under any equivalence transformation. If a state equation is transformed into Jordan form, then the controllability and observability conditions become very simple and can often be checked by inspection. Consider the state equation

$$\dot{\mathbf{x}} = \mathbf{Jx} + \mathbf{Bx}$$

$$\mathbf{y} = \mathbf{Cx}$$

(6.47)

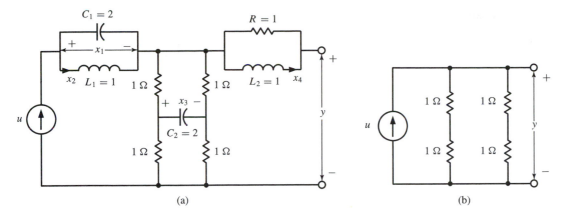

Figure 6.8 Networks.

where $\mathbf{J}$ is in Jordan form. To simplify discussion, we assume that $\mathbf{J}$ has only two distinct eigenvalues λ_1 and λ_2 and can be written as

$$\mathbf{J} = \text{diag } (\mathbf{J}_1, \mathbf{J}_2)$$

where $\mathbf{J}_1$ consists of all Jordan blocks associated with λ_1 and $\mathbf{J}_2$ consists of all Jordan blocks associated with λ_2. Again to simplify discussion, we assume that $\mathbf{J}_1$ has three Jordan blocks and $\mathbf{J}_2$ has two Jordan blocks or

$$\mathbf{J}_1 = \text{diag } (\mathbf{J}_{11}, \mathbf{J}_{12}, \mathbf{J}_{13}) \qquad \mathbf{J}_2 = \text{diag } (\mathbf{J}_{21}, \mathbf{J}_{22})$$

The row of $\mathbf{B}$ corresponding to the *last row* of $\mathbf{J}_{ij}$ is denoted by $\mathbf{b}_{lij}$. The column of $\mathbf{C}$ corresponding to the *first column* of $\mathbf{J}_{ij}$ is denoted by $\mathbf{c}_{fij}$.

▶ **Theorem 6.8**

1. The state equation in (6.47) is controllable if and only if the three row vectors $\{\mathbf{b}_{l11}, \mathbf{b}_{l12}, \mathbf{b}_{l13}\}$ are linearly independent and the two row vectors $\{\mathbf{b}_{l21}, \mathbf{b}_{l22}\}$ are linearly independent.

2. The state equation in (6.47) is observable if and only if the three column vectors $\{\mathbf{c}_{f11}, \mathbf{c}_{f12}, \mathbf{c}_{f13}\}$ are linearly independent and the two column vectors $\{\mathbf{c}_{f21}, \mathbf{c}_{f22}\}$ are linearly independent.

We discuss first the implications of this theorem. If a state equation is in Jordan form, then the controllability of the state variables associated with one eigenvalue can be checked independently from those associated with different eigenvalues. The controllability of the state variables associated with the same eigenvalue depends only on the rows of $\mathbf{B}$ corresponding to the last row of all Jordan blocks associated with the eigenvalue. All other rows of $\mathbf{B}$ play no role in determining the controllability. Similar remarks apply to the observability part except that the columns of $\mathbf{C}$ corresponding to the first column of all Jordan blocks determine the observability. We use an example to illustrate the use of Theorem 6.8.

EXAMPLE 6.10 Consider the Jordan-form state equation

$$\dot{\mathbf{x}} = \begin{bmatrix} \lambda_1 & 1 & 0 & 0 & 0 & 0 & 0 \\ 0 & \lambda_1 & 0 & 0 & 0 & 0 & 0 \\ 0 & 0 & \lambda_1 & 0 & 0 & 0 & 0 \\ 0 & 0 & 0 & \lambda_1 & 0 & 0 & 0 \\ 0 & 0 & 0 & 0 & \lambda_2 & 1 & 0 \\ 0 & 0 & 0 & 0 & 0 & \lambda_2 & 1 \\ 0 & 0 & 0 & 0 & 0 & 0 & \lambda_2 \end{bmatrix} \mathbf{x} + \begin{bmatrix} 0 & 0 & 0 \\ 1 & 0 & 0 \\ 0 & 1 & 0 \\ 1 & 1 & 1 \\ 1 & 2 & 3 \\ 0 & 1 & 0 \\ 1 & 1 & 1 \end{bmatrix} \mathbf{u}$$

(6.48)

$$\mathbf{y} = \begin{bmatrix} 1 & 1 & 2 & 0 & 0 & 2 & 1 \\ 1 & 0 & 1 & 2 & 0 & 1 & 1 \\ 1 & 0 & 2 & 3 & 0 & 2 & 0 \end{bmatrix} \mathbf{x}$$

The matrix $\mathbf{J}$ has two distinct eigenvalues λ_1 and λ_2. There are three Jordan blocks, with order 2, 1, and 1, associated with λ_1. The rows of $\mathbf{B}$ corresponding to the last row of the three Jordan blocks are [1 0 0], [0 1 0], and [1 1 1]. The three rows are linearly independent. There is only one Jordan block, with order 3, associated with λ_2. The row of $\mathbf{B}$ corresponding to the last row of the Jordan block is [1 1 1], which is nonzero and is therefore linearly independent. Thus we conclude that the state equation in (6.48) is controllable.

The conditions for (6.48) to be observable are that the three columns [1 1 1]′, [2 1 2]′, and [0 2 3]′ are linearly independent (they are) and the one column [0 0 0]′ is linearly independent (it is not). Therefore the state equation is not observable.

Before proving Theorem 6.8, we draw a block diagram to show how the conditions in the theorem arise. The inverse of $(s\mathbf{I} - \mathbf{J})$ is of the form shown in (3.49), whose entries consist of only $1/(s - \lambda_i)^k$. Using (3.49), we can draw a block diagram for (6.48) as shown in Fig. 6.9. Each chain of blocks corresponds to one Jordan block in the equation. Because (6.48) has four Jordan blocks, the figure has four chains. The output of each block can be assigned as a state variable as shown in Fig. 6.10. Let us consider the last chain in Fig. 6.9. If $\mathbf{b}_{l21} = \mathbf{0}$, the state variable x_{l21} is not connected to the input and is not controllable no matter what values $\mathbf{b}_{221}$ and $\mathbf{b}_{121}$ assume. On the other hand, if $\mathbf{b}_{l21}$ is nonzero, then all state variables in the chain are controllable. If there are two or more chains associated with the same eigenvalue, then we require the linear independence of the first gain vectors of those chains. The chains associated with different eigenvalues can be checked separately. All discussion applies to the observability part except that the column vector $\mathbf{c}_{fij}$ plays the role of the row vector $\mathbf{b}_{lij}$.

Proof of Theorem 6.8   We prove the theorem by using the condition that the matrix $[\mathbf{A} - s\mathbf{I} \ \mathbf{B}]$ or $[s\mathbf{I} - \mathbf{A} \ \mathbf{B}]$ has full row rank at every eigenvalue of $\mathbf{A}$. In order not to be overwhelmed by notation, we assume $[s\mathbf{I} - \mathbf{J} \ \mathbf{B}]$ to be of the form

$$\begin{bmatrix} s - \lambda_1 & -1 & 0 & 0 & 0 & 0 & 0 & \mathbf{b}_{111} \\ 0 & s - \lambda_1 & -1 & 0 & 0 & 0 & 0 & \mathbf{b}_{211} \\ 0 & 0 & s - \lambda_1 & 0 & 0 & 0 & 0 & \mathbf{b}_{l11} \\ 0 & 0 & 0 & s - \lambda_1 & -1 & 0 & 0 & \mathbf{b}_{112} \\ 0 & 0 & 0 & 0 & s - \lambda_1 & 0 & 0 & \mathbf{b}_{l12} \\ 0 & 0 & 0 & 0 & 0 & s - \lambda_2 & -1 & \mathbf{b}_{121} \\ 0 & 0 & 0 & 0 & 0 & 0 & s - \lambda_2 & \mathbf{b}_{l21} \end{bmatrix}$$

(6.49)

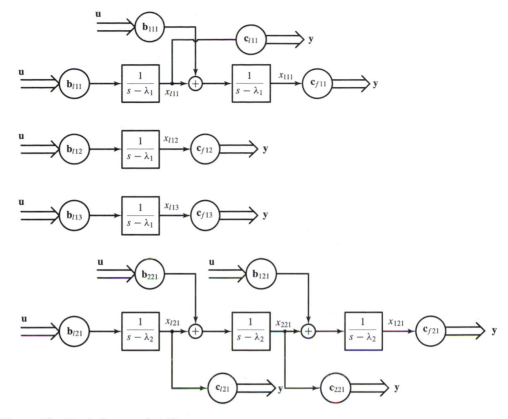

Figure 6.9   Block diagram of (6.48).

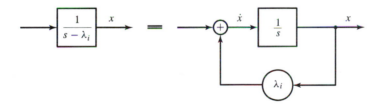

Figure 6.10   Internal structure of $1/(s - \lambda_i)$.

The Jordan-form matrix **J** has two distinct eigenvalues λ_1 and λ_2. There are two Jordan blocks associated with λ_1 and one associated with λ_2. If $s = \lambda_1$, (6.49) becomes

$$
\begin{bmatrix}
0 & -1 & 0 & 0 & 0 & 0 & 0 & \mathbf{b}_{111} \\
0 & 0 & -1 & 0 & 0 & 0 & 0 & \mathbf{b}_{211} \\
0 & 0 & 0 & 0 & 0 & 0 & 0 & \mathbf{b}_{l11} \\
0 & 0 & 0 & 0 & -1 & 0 & 0 & \mathbf{b}_{112} \\
0 & 0 & 0 & 0 & 0 & 0 & 0 & \mathbf{b}_{l12} \\
0 & 0 & 0 & 0 & 0 & \lambda_1 - \lambda_2 & -1 & \mathbf{b}_{121} \\
0 & 0 & 0 & 0 & 0 & 0 & \lambda_1 - \lambda_2 & \mathbf{b}_{l21}
\end{bmatrix}
$$

(6.50)

The rank of the matrix will not change by elementary column operations. We add the product of the second column of (6.50) by $\mathbf{b}_{l11}$ to the last block column. Repeating the process for the third and fifth columns, we can obtain

$$
\begin{bmatrix}
0 & -1 & 0 & 0 & 0 & 0 & 0 & \mathbf{0} \\
0 & 0 & -1 & 0 & 0 & 0 & 0 & \mathbf{0} \\
0 & 0 & 0 & 0 & 0 & 0 & 0 & \mathbf{b}_{l11} \\
0 & 0 & 0 & 0 & -1 & 0 & 0 & \mathbf{0} \\
0 & 0 & 0 & 0 & 0 & 0 & 0 & \mathbf{b}_{l12} \\
0 & 0 & 0 & 0 & 0 & \lambda_1 - \lambda_2 & -1 & \mathbf{b}_{l21} \\
0 & 0 & 0 & 0 & 0 & 0 & \lambda_1 - \lambda_2 & \mathbf{b}_{l21}
\end{bmatrix}
$$

Because λ_1 and λ_2 are distinct, $\lambda_1 - \lambda_2$ is nonzero. We add the product of the seventh column and $-\mathbf{b}_{l21}/(\lambda_1 - \lambda_2)$ to the last column and then use the sixth column to eliminate its right-hand-side entries to yield

$$
\begin{bmatrix}
0 & -1 & 0 & 0 & 0 & 0 & 0 & \mathbf{0} \\
0 & 0 & -1 & 0 & 0 & 0 & 0 & \mathbf{0} \\
0 & 0 & 0 & 0 & 0 & 0 & 0 & \mathbf{b}_{l11} \\
0 & 0 & 0 & 0 & -1 & 0 & 0 & \mathbf{0} \\
0 & 0 & 0 & 0 & 0 & 0 & 0 & \mathbf{b}_{l12} \\
0 & 0 & 0 & 0 & 0 & \lambda_1 - \lambda_2 & 0 & \mathbf{0} \\
0 & 0 & 0 & 0 & 0 & 0 & \lambda_1 - \lambda_2 & \mathbf{0}
\end{bmatrix} \qquad (6.51)
$$

It is clear that the matrix in (6.51) has full row rank if and only if $\mathbf{b}_{l11}$ and $\mathbf{b}_{l12}$ are linearly independent. Proceeding similarly for each eigenvalue, we can establish Theorem 6.8. Q.E.D.

Consider an n-dimensional Jordan-form state equation with p inputs and q outputs. If there are m, with $m > p$, Jordan blocks associated with the same eigenvalue, then m number of $1 \times p$ row vectors can never be linearly independent and the state equation can never be controllable. Thus a necessary condition for the state equation to be controllable is $m \leq p$. Similarly, a necessary condition for the state equation to be observable is $m \leq q$. For the single-input or single-output case, we then have the following corollaries.

▶ **Corollary 6.8**

A single-input Jordan-form state equation is controllable if and only if there is only one Jordan block associated with each distinct eigenvalue and every entry of $\mathbf{B}$ corresponding to the last row of each Jordan block is different from zero.

▶ **Corollary 6.O8**

A single-output Jordan-form state equation is observable if and only if there is only one Jordan block associated with each distinct eigenvalue and every entry of $\mathbf{C}$ corresponding to the first column of each Jordan block is different from zero.

EXAMPLE 6.11 Consider the state equation

$$\dot{\mathbf{x}} = \begin{bmatrix} 0 & 1 & 0 & 0 \\ 0 & 0 & 1 & 0 \\ 0 & 0 & 0 & 0 \\ 0 & 0 & 0 & -2 \end{bmatrix} \mathbf{x} + \begin{bmatrix} 10 \\ 9 \\ 0 \\ 1 \end{bmatrix} u \tag{6.52}$$

$$y = [1 \ 0 \ 0 \ 2]\mathbf{x}$$

There are two Jordan blocks, one with order 3 and associated with eigenvalue 0, the other with order 1 and associated with eigenvalue -2. The entry of **B** corresponding to the last row of the first Jordan block is zero; thus the state equation is not controllable. The two entries of **C** corresponding to the first column of both Jordan blocks are different from zero; thus the state equation is observable.

6.6 Discrete-Time State Equations

Consider the n-dimensional p-input q-output state equation

$$\mathbf{x}[k + 1] = \mathbf{Ax}[k] + \mathbf{Bu}[k]$$
$$\mathbf{y}[k] = \mathbf{Cx}[k] \tag{6.53}$$

where **A**, **B**, and **C** are, respectively, $n \times n$, $n \times p$, and $q \times n$ real constant matrices.

> **Definition 6.D1** *The discrete-time state equation (6.53) or the pair* (**A**, **B**) *is said to be* controllable *if for any initial state* $\mathbf{x}(0) = \mathbf{x}_0$ *and any final state* $\mathbf{x}_1$, *there exists an input sequence of finite length that transfers* $\mathbf{x}_0$ *to* $\mathbf{x}_1$. *Otherwise the equation or* (**A**, **B**) *is said to be* uncontrollable.

▶ **Theorem 6.D1**

The following statements are equivalent:

1. The n-dimensional pair (**A**, **B**) is controllable.

2. The $n \times n$ matrix

$$\mathbf{W}_{dc}[n - 1] = \sum_{m=0}^{n-1} (\mathbf{A})^m \mathbf{BB}'(\mathbf{A}')^m \tag{6.54}$$

is nonsingular.

3. The $n \times np$ *controllability matrix*

$$C_d = [\mathbf{B} \ \mathbf{AB} \ \mathbf{A}^2\mathbf{B} \ \cdots \ \mathbf{A}^{n-1}\mathbf{B}] \tag{6.55}$$

has rank n (full row rank). The matrix can be generated by calling `ctrb` in MATLAB.

4. The $n \times (n + p)$ matrix $[\mathbf{A} - \lambda\mathbf{I} \ \mathbf{B}]$ has full row rank at every eigenvalue, λ, of **A**.

5. If, in addition, all eigenvalues of **A** have magnitudes less than 1, then the unique solution of

$$\mathbf{W}_{dc} - \mathbf{A}\mathbf{W}_{dc}\mathbf{A}' = \mathbf{B}\mathbf{B}' \tag{6.56}$$

is positive definite. The solution is called the discrete *controllability Gramian* and can be obtained by using the MATLAB function `dgram`. The discrete Gramian can be expressed as

$$\mathbf{W}_{dc} = \sum_{m=0}^{\infty} \mathbf{A}^m \mathbf{B}\mathbf{B}'(\mathbf{A}')^m \tag{6.57}$$

The solution of (6.53) at $k = n$ was derived in (4.20) as

$$x[n] = \mathbf{A}^n \mathbf{x}[0] + \sum_{m=0}^{n-1} \mathbf{A}^{n-1-m} \mathbf{B}\mathbf{u}[m]$$

which can be written as

$$\mathbf{x}[n] - \mathbf{A}^n \mathbf{x}[0] = [\mathbf{B} \ \mathbf{AB} \ \cdots \ \mathbf{A}^{n-1}\mathbf{B}] \begin{bmatrix} \mathbf{u}[n-1] \\ \mathbf{u}[n-2] \\ \vdots \\ \mathbf{u}[0] \end{bmatrix} \tag{6.58}$$

It follows from Theorem 3.1 that for any $\mathbf{x}[0]$ and $\mathbf{x}[n]$, an input sequence exists if and only if the controllability matrix has full row rank. This shows the equivalence of (1) and (3). The matrix $\mathbf{W}_{dc}[n-1]$ can be written as

$$\mathbf{W}_{dc}[n-1] = [\mathbf{B} \ \mathbf{AB} \ \cdots \ \mathbf{A}^{n-1}\mathbf{B}] \begin{bmatrix} \mathbf{B}' \\ \mathbf{B}'\mathbf{A}' \\ \vdots \\ \mathbf{B}'(\mathbf{A}')^{n-1} \end{bmatrix}$$

The equivalence of (2) and (3) then follows Theorem 3.8. Note that $\mathbf{W}_{dc}[m]$ is always positive semidefinite. If it is nonsingular or, equivalently, positive definite, then (6.53) is controllable. The proof of the equivalence of (3) and (4) is identical to the continuous-time case. Condition (5) follows Condition (2) and Theorem 5.D6. We see that establishing Theorem 6.D1 is considerably simpler than establishing Theorem 6.1.

There is one important difference between the continuous- and discrete-time cases. If a continuous-time state equation is controllable, the input can transfer any state to any other state in any nonzero time interval, no matter how small. If a discrete-time state equation is controllable, an input sequence of length n can transfer any state to any other state. If we compute the controllability index μ as defined in (6.15), then the transfer can be achieved using an input sequence of length μ. If an input sequence is shorter than μ, it is not possible to transfer any state to any other state.

Definition 6.D2 *The discrete-time state equation (6.53) or the pair* $(\mathbf{A}, \mathbf{C})$ *is said to be* observable *if for any unknown initial state* $\mathbf{x}[0]$, *there exists a finite integer* $k_1 > 0$ *such that the knowledge of the input sequence* $\mathbf{u}[k]$ *and output sequence* $\mathbf{y}[k]$ *from* $k = 0$ *to* k_1 *suffices to determine uniquely the initial state* $\mathbf{x}[0]$. *Otherwise, the equation is said to be* unobservable.

▶ **Theorem 6.DO1**

The following statements are equivalent:

1. The n-dimensional pair $(\mathbf{A}, \mathbf{C})$ is observable.
2. The $n \times n$ matrix

$$\mathbf{W}_{do}[n-1] = \sum_{m=0}^{n-1} (\mathbf{A}')^m \mathbf{C}'\mathbf{C}\mathbf{A}^m \tag{6.59}$$

   is nonsingular or, equivalently, positive definite.
3. The $nq \times n$ *observability matrix*

$$O_d = \begin{bmatrix} \mathbf{C} \\ \mathbf{C}\mathbf{A} \\ \vdots \\ \mathbf{C}\mathbf{A}^{n-1} \end{bmatrix} \tag{6.60}$$

   has rank n (full column rank). The matrix can be generated by calling `obsv` in MATLAB.
4. The $(n+q) \times n$ matrix

$$\begin{bmatrix} \mathbf{A} - \lambda\mathbf{I} \\ \mathbf{B} \end{bmatrix}$$

   has full column rank at every eigenvalue, λ, of $\mathbf{A}$.
5. If, in addition, all eigenvalues of $\mathbf{A}$ have magnitudes less than 1, then the unique solution of

$$\mathbf{W}_{do} - \mathbf{A}'\mathbf{W}_{do}\mathbf{A} = \mathbf{C}'\mathbf{C} \tag{6.61}$$

   is positive definite. The solution is called the discrete *observability Gramian* and can be expressed as

$$\mathbf{W}_{do} = \sum_{m=0}^{\infty} (\mathbf{A}')^m \mathbf{C}'\mathbf{C}\mathbf{A}^m \tag{6.62}$$

This can be proved directly or indirectly using the duality theorem. We mention that all other properties—such as controllability and observability indices, Kalman decomposition, and Jordan-form controllability and observability conditions—discussed for the continuous-time case apply to the discrete-time case without any modification. The controllability index and observability index, however, have simple interpretations in the discrete-time case. The controllability index is the shortest input sequence that can transfer any state to any other state. The observability index is the shortest input and output sequences needed to determine the initial state uniquely.

6.6.1 Controllability to the Origin and Reachability

In the literature, there are three different controllability definitions:

1. Transfer any state to any other state as adopted in Definition 6.D1.
2. Transfer any state to the zero state, called controllability to the origin.

3. Transfer the zero state to any state, called controllability from the origin or, more often, *reachability*.

In the continuous-time case, because $e^{\mathbf{A}t}$ is nonsingular, the three definitions are equivalent. In the discrete-time case, if $\mathbf{A}$ is nonsingular, the three definitions are again equivalent. But if $\mathbf{A}$ is singular, then (1) and (3) are equivalent, but not (2) and (3). The equivalence of (1) and (3) can easily be seen from (6.58). We use examples to discuss the difference between (2) and (3). Consider

$$\mathbf{x}[k+1] = \begin{bmatrix} 0 & 1 & 0 \\ 0 & 0 & 1 \\ 0 & 0 & 0 \end{bmatrix} \mathbf{x}[k] + \begin{bmatrix} 0 \\ 0 \\ 0 \end{bmatrix} u[k] \tag{6.63}$$

Its controllability matrix has rank 0 and the equation is not controllable as defined in (1) or not reachable as defined in (3). The matrix $\mathbf{A}$ has the form shown in (3.40) and has the property $\mathbf{A}^k = \mathbf{0}$ for $k \geq 3$. Thus we have

$$\mathbf{x}[3] = \mathbf{A}^3\mathbf{x}[0] = \mathbf{0}$$

for any initial state $\mathbf{x}[0]$. Thus every state propagates to the zero state whether or not an input sequence is applied. Thus the equation is controllable to the origin. A different example follows. Consider

$$\mathbf{x}[k+1] = \begin{bmatrix} 2 & 1 \\ 0 & 0 \end{bmatrix} \mathbf{x}[k] + \begin{bmatrix} -1 \\ 0 \end{bmatrix} u[k] \tag{6.64}$$

Its controllability matrix

$$\begin{bmatrix} -1 & -2 \\ 0 & 0 \end{bmatrix}$$

has rank 1 and the equation is not reachable. However, for any $x_1[0] = \alpha$ and $x_2[0] = \beta$, the input $u[0] = 2\alpha + \beta$ transfers $\mathbf{x}[0]$ to $\mathbf{x}[1] = \mathbf{0}$. Thus the equation is controllable to the origin. Note that the A-matrices in (6.63) and (6.64) are both singular. The definition adopted in Definition 6.D1 encompasses the other two definitions and makes the discussion simple. For a thorough discussion of the three definitions, see Reference [4].

6.7 Controllability After Sampling

Consider a continuous-time state equation

$$\dot{\mathbf{x}}(t) = \mathbf{A}\mathbf{x}(t) + \mathbf{B}u(t) \tag{6.65}$$

If the input is piecewise constant or

$$u[k] := u(kT) = u(t) \qquad \text{for } kT \leq t < (k+1)T$$

then the equation can be described, as developed in (4.17), by

$$\bar{\mathbf{x}}[k+1] = \bar{\mathbf{A}}\bar{\mathbf{x}}[k] + \bar{\mathbf{B}}u[k] \tag{6.66}$$

with

$$\bar{\mathbf{A}} = e^{\mathbf{A}T} \qquad \bar{\mathbf{B}} = \left(\int_0^T e^{\mathbf{A}t} dt \right) \mathbf{B} =: \mathbf{M}\mathbf{B} \tag{6.67}$$

The question is: If (6.65) is controllable, will its sampled equation in (6.66) be controllable? This problem is important in designing so-called dead-beat sampled-data systems and in computer control of continuous-time systems. The answer to the question depends on the sampling period T and the location of the eigenvalues of $\mathbf{A}$. Let λ_i and $\bar{\lambda}_i$ be, respectively, the eigenvalues of $\mathbf{A}$ and $\bar{\mathbf{A}}$. We use Re and Im to denote the real part and imaginary part. Then we have the following theorem.

▶ **Theorem 6.9**

Suppose (6.65) is controllable. A sufficient condition for its discretized equation in (6.66), with sampling period T, to be controllable is that $|\text{Im}[\lambda_i - \lambda_j]| \neq 2\pi m / T$ for $m = 1, 2, \ldots$, whenever $\text{Re}[\lambda_i - \lambda_j] = 0$. For the single-input case, the condition is necessary as well.

First we remark on the conditions. If $\mathbf{A}$ has only real eigenvalues, then the discretized equation with any sampling period $T > 0$ is always controllable. Suppose $\mathbf{A}$ has complex conjugate eigenvalues $\alpha \pm j\beta$. If the sampling period T does not equal any integer multiple of π/β, then the discretized state equation is controllable. If $T = m\pi/\beta$ for some integer m, then the discretized equation *may not* be controllable. The reason is as follows. Because $\bar{\mathbf{A}} = e^{\mathbf{A}T}$, if λ_i is an eigenvalue of $\mathbf{A}$, then $\bar{\lambda}_i := e^{\lambda_i T}$ is an eigenvalue of $\bar{\mathbf{A}}$ (Problem 3.19). If $T = m\pi/\beta$, the two distinct eigenvalues $\lambda_1 = \alpha + j\beta$ and $\lambda_2 = \alpha - j\beta$ of $\mathbf{A}$ become a repeated eigenvalue $-e^{\alpha T}$ or $e^{\alpha T}$ of $\bar{\mathbf{A}}$. This will cause the discretized equation to be uncontrollable, as we will see in the proof. We show Theorem 6.9 by assuming $\mathbf{A}$ to be in Jordan form. This is permitted because controllability is invariant under any equivalence transformation.

Proof of Theorem 6.9 To simplify the discussion, we assume $\mathbf{A}$ to be of the form

$$\mathbf{A} = \text{diag}(\mathbf{A}_{11}, \mathbf{A}_{12}, \mathbf{A}_{21}) = \begin{bmatrix} \lambda_1 & 1 & 0 & 0 & 0 & 0 \\ 0 & \lambda_1 & 1 & 0 & 0 & 0 \\ 0 & 0 & \lambda_1 & 0 & 0 & 0 \\ 0 & 0 & 0 & \lambda_1 & 0 & 0 \\ 0 & 0 & 0 & 0 & \lambda_2 & 1 \\ 0 & 0 & 0 & 0 & 0 & \lambda_2 \end{bmatrix} \tag{6.68}$$

In other words, $\mathbf{A}$ has two distinct eigenvalues λ_1 and λ_2. There are two Jordan blocks, one with order 3 and one with order 1, associated with λ_1 and only one Jordan block of order 2 associated with λ_2. Using (3.48), we have

$$\bar{\mathbf{A}} = \text{diag}(\bar{\mathbf{A}}_{11}, \bar{\mathbf{A}}_{12}, \bar{\mathbf{A}}_{21})$$
$$= \begin{bmatrix} e^{\lambda_1 T} & Te^{\lambda_1 T} & T^2 e^{\lambda_1 T}/2 & 0 & 0 & 0 \\ 0 & e^{\lambda_1 T} & Te^{\lambda_1 T} & 0 & 0 & 0 \\ 0 & 0 & e^{\lambda_1 T} & 0 & 0 & 0 \\ 0 & 0 & 0 & e^{\lambda_1 T} & 0 & 0 \\ 0 & 0 & 0 & 0 & e^{\lambda_2 T} & Te^{\lambda_2 T} \\ 0 & 0 & 0 & 0 & 0 & e^{\lambda_2 T} \end{bmatrix} \tag{6.69}$$

This is not in Jordan form. Because we will use Theorem 6.8, which is also applicable to the discrete-time case without any modification, to prove Theorem 6.9, we must transform $\bar{\mathbf{A}}$ in (6.69) into Jordan form. It turns out that the Jordan form of $\bar{\mathbf{A}}$ equals the one in (6.68) if λ_i is replaced by $\bar{\lambda}_i := e^{\lambda_i T}$ (Problem 3.17). In other words, there exists a nonsingular triangular matrix $\mathbf{P}$ such that the transformation $\tilde{\mathbf{x}} = \mathbf{P}\bar{\mathbf{x}}$ will transform (6.66) into

$$\tilde{\mathbf{x}}[k+1] = \mathbf{P}\bar{\mathbf{A}}\mathbf{P}^{-1}\tilde{\mathbf{x}}[k] + \mathbf{PMB}\mathbf{u}[k] \tag{6.70}$$

with $\mathbf{P}\bar{\mathbf{A}}\mathbf{P}^{-1}$ in the Jordan form in (6.68) with λ_i replaced by $\bar{\lambda}_i$. Now we are ready to establish Theorem 6.9.

First we show that $\mathbf{M}$ in (6.67) is nonsingular. If $\mathbf{A}$ is of the form shown in (6.68), then $\mathbf{M}$ is block diagonal and triangular. Its diagonal entry is of form

$$m_{ii} := \int_0^T e^{\lambda_i \tau} d\tau = \begin{cases} (e^{\lambda_i T} - 1)/\lambda_i & \text{if } \lambda_i \neq 0 \\ T & \text{if } \lambda_i = 0 \end{cases} \tag{6.71}$$

Let $\lambda_i = \alpha_i + j\beta_i$. The only way for $m_{ii} = 0$ is $\alpha_i = 0$ and $\beta_i T = 2\pi m$. In this case, $-j\beta_i$ is also an eigenvalue and the theorem requires that $2\beta_i T \neq 2\pi m$. Thus we conclude $m_{ii} \neq 0$ and $\mathbf{M}$ is nonsingular and triangular.

If $\mathbf{A}$ is of the form shown in (6.68), then it is controllable if and only if the third and fourth rows of $\mathbf{B}$ are linearly independent and the last row of $\mathbf{B}$ is nonzero (Theorem 6.8). Under the condition in Theorem 6.9, the two eigenvalues $\bar{\lambda}_1 = e^{\lambda_1 T}$ and $\bar{\lambda}_2 = e^{\lambda_2 T}$ of $\bar{\mathbf{A}}$ are distinct. Thus (6.70) is controllable if and only if the third and fourth rows of $\mathbf{PMB}$ are linearly independent and the last row of $\mathbf{PMB}$ is nonzero. Because $\mathbf{P}$ and $\mathbf{M}$ are both triangular and nonsingular, $\mathbf{PMB}$ and $\mathbf{B}$ have the same properties on the linear independence of their rows. This shows the sufficiency of the theorem. If the condition in Theorem 6.9 is not met, then $\bar{\lambda}_1 = \bar{\lambda}_2$. In this case, (6.70) is controllable if the third, fourth, and last rows of $\mathbf{PMB}$ are linearly independent. This is still possible if $\mathbf{B}$ has three or more columns. Thus the condition is not necessary. In the single-input case, if $\bar{\lambda}_1 = \bar{\lambda}_2$, then (6.70) has two or more Jordan blocks associated with the same eigenvalue and (6.70) is, following Corollary 6.8, not controllable. This establishes the theorem.   Q.E.D.

In the proof of Theorem 6.9, we have essentially established the theorem that follows.

▶ **Theorem 6.10**

If a continuous-time linear time-invariant state equation is not controllable, then its discretized state equation, with any sampling period, is not controllable.

This theorem is intuitively obvious. If a state equation is not controllable using any input, it is certainly not controllable using only piecewise constant input.

EXAMPLE 6.12 Consider the system shown in Fig. 6.11. Its input is sampled every T seconds and then kept constant using a hold circuit. The transfer function of the system is given as

$$\hat{g}(s) = \frac{s+2}{s^3 + 3s^2 + 7s + 5} = \frac{s+2}{(s+1)(s+1+j2)(s+1-j2)} \tag{6.72}$$

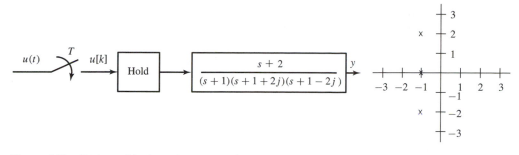

Figure 6.11 System with piecewise constant input.

Using (4.41), we can readily obtain the state equation

$$\dot{\mathbf{x}} = \begin{bmatrix} -3 & -7 & -5 \\ 1 & 0 & 0 \\ 0 & 1 & 0 \end{bmatrix} \mathbf{x} + \begin{bmatrix} 1 \\ 0 \\ 0 \end{bmatrix} u$$

$$y = [0 \quad 1 \quad 2]\mathbf{x}$$

(6.73)

to describe the system. It is a controllable-form realization and is clearly controllable. The eigenvalues of **A** are -1, $-1 \pm j2$ and are plotted in Fig. 6.11. The three eigenvalues have the same real part; their differences in imaginary parts are 2 and 4. Thus the discretized state equation is controllable if and only if

$$T \neq \frac{2\pi m}{2} = \pi m \quad \text{and} \quad T \neq \frac{2\pi m}{4} = 0.5\pi m$$

for $m = 1, 2, \ldots$. The second condition includes the first condition. Thus we conclude that the discretized equation of (6.73) is controllable if and only if $T \neq 0.5m\pi$ for any positive integer m.

We use MATLAB to check the result for $m = 1$ or $T = 0.5\pi$. Typing

```
a=[-3 -7 -5;1 0 0;0 1 0];b=[1;0;0];
[ad,bd]=c2d(a,b,pi/2)
```

yields the discretized state equation as

$$\bar{\mathbf{x}}[k+1] = \begin{bmatrix} -0.1039 & 0.2079 & 0.5197 \\ -0.1390 & -0.4158 & -0.5197 \\ 0.1039 & 0.2079 & 0.3118 \end{bmatrix} \bar{\mathbf{x}}[k] + \begin{bmatrix} -0.1039 \\ 0.1039 \\ 0.1376 \end{bmatrix} u[k]$$

(6.74)

Its controllability matrix can be obtained by typing `ctrb(ad,bd)`, which yields

$$C_d = \begin{bmatrix} -0.1039 & 0.1039 & -0.0045 \\ 0.1039 & -0.1039 & 0.0045 \\ 0.1376 & 0.0539 & 0.0059 \end{bmatrix}$$

Its first two rows are clearly linearly dependent. Thus C_d does not have full row rank and (6.74) is not controllable as predicted by Theorem 6.9. We mention that if we type rank(ctrb(ad,bd)), the result is 3 and (6.74) is controllable. This is incorrect and is due to roundoff errors. We see once again that the rank is very sensitive to roundoff errors.

What has been discussed is also applicable to the observability part. In other words, under the conditions in Theorem 6.9, if a continuous-time state equation is observable, its discretized equation is also observable.

6.8 LTV State Equations

Consider the n-dimensional p-input q-output state equation

$$\dot{\mathbf{x}} = \mathbf{A}(t)\mathbf{x} + \mathbf{B}(t)\mathbf{u}$$
$$\mathbf{y} = \mathbf{C}(t)\mathbf{x}$$

(6.75)

The state equation is said to be controllable at t_0, if there exists a finite $t_1 > t_0$ such that for any $\mathbf{x}(t_0) = \mathbf{x}_0$ and any $\mathbf{x}_1$, there exists an input that transfers $\mathbf{x}_0$ to $\mathbf{x}_1$ at time t_1. Otherwise the state equation is uncontrollable at t_0. In the time-invariant case, if a state equation is controllable, then it is controllable at every t_0 and for every $t_1 > t_0$; thus there is no need to specify t_0 and t_1. In the time-varying case, the specification of t_0 and t_1 is crucial.

▶ **Theorem 6.11**

The n-dimensional pair $(\mathbf{A}(t), \mathbf{B}(t))$ is controllable at time t_0 if and only if there exists a finite $t_1 > t_0$ such that the $n \times n$ matrix

$$\mathbf{W}_c(t_0, t_1) = \int_{t_0}^{t_1} \mathbf{\Phi}(t_1, \tau)\mathbf{B}(\tau)\mathbf{B}'(\tau)\mathbf{\Phi}'(t_1, \tau)\, d\tau$$

(6.76)

where $\mathbf{\Phi}(t, \tau)$ is the state transition matrix of $\dot{\mathbf{x}} = \mathbf{A}(t)\mathbf{x}$, is nonsingular.

 Proof: We first show that if $\mathbf{W}_c(t_0, t_1)$ is nonsingular, then (6.75) is controllable. The response of (6.75) at t_1 was computed in (4.57) as

$$\mathbf{x}(t_1) = \mathbf{\Phi}(t_1, t_0)\mathbf{x}_0 + \int_{t_0}^{t_1} \mathbf{\Phi}(t_1, \tau)\mathbf{B}(\tau)\mathbf{u}(\tau)\, d\tau$$

(6.77)

We claim that the input

$$\mathbf{u}(t) = -\mathbf{B}'(t)\mathbf{\Phi}'(t_1, t)\mathbf{W}_c^{-1}(t_0, t_1)[\mathbf{\Phi}(t_1, t_0)\mathbf{x}_0 - \mathbf{x}_1]$$

(6.78)

will transfer $\mathbf{x}_0$ at time t_0 to $\mathbf{x}_1$ at time t_1. Indeed, substituting (6.78) into (6.77) yields

$$\mathbf{x}(t_1) = \mathbf{\Phi}(t_1, t_0)\mathbf{x}_0 - \int_{t_0}^{t_1} \mathbf{\Phi}(t_1, \tau)\mathbf{B}(\tau)\mathbf{B}'(\tau)\mathbf{\Phi}'(t_1, \tau)\, d\tau$$

$$\cdot \mathbf{W}_c^{-1}(t_0, t_1)[\mathbf{\Phi}(t_1, t_0)\mathbf{x}_0 - \mathbf{x}_1]$$

$$= \mathbf{\Phi}(t_1, t_0)\mathbf{x}_0 - \mathbf{W}_c(t_0, t_1)\mathbf{W}_c^{-1}(t_0, t_1)[\mathbf{\Phi}(t_1, t_0)\mathbf{x}_0 - \mathbf{x}_1] = \mathbf{x}_1$$

Thus the equation is controllable at t_0. We show the converse by contradiction. Suppose (6.75) is controllable at t_0 but $\mathbf{W}_c(t_0, t)$ is singular or, positive semidefinite, for all $t_1 > t_0$. Then there exists an $n \times 1$ nonzero constant vector $\mathbf{v}$ such that

$$\mathbf{v}'\mathbf{W}_c(t_0, t_1)\mathbf{v} = \int_{t_0}^{t_1} \mathbf{v}'\mathbf{\Phi}(t_1, \tau)\mathbf{B}(\tau)\mathbf{B}'(\tau)\mathbf{\Phi}'(t_1, \tau)\mathbf{v}\, d\tau$$

$$= \int_{t_0}^{t_1} ||\mathbf{B}'(\tau)\mathbf{\Phi}'(t_1, \tau)\mathbf{v}||^2 d\tau = 0$$

which implies

$$\mathbf{B}'(\tau)\mathbf{\Phi}'(t_1, \tau)\mathbf{v} \equiv \mathbf{0} \quad \text{or} \quad \mathbf{v}'\mathbf{\Phi}(t_1, \tau)\mathbf{B}(\tau) \equiv \mathbf{0} \qquad (6.79)$$

for all τ in $[t_0, \ t_1]$. If (6.75) is controllable, there exists an input that transfers the initial state $\mathbf{x}_0 = \mathbf{\Phi}(t_0, t_1)\mathbf{v}$ at t_0 to $\mathbf{x}(t_1) = \mathbf{0}$. Then (6.77) becomes

$$\mathbf{0} = \mathbf{\Phi}(t_1, t_0)\mathbf{\Phi}(t_0, t_1)\mathbf{v} + \int_{t_0}^{t_1} \mathbf{\Phi}(t_1, \tau)\mathbf{B}(\tau)\mathbf{u}(\tau)\, d\tau \qquad (6.80)$$

Its premultiplication by $\mathbf{v}'$ yields

$$0 = \mathbf{v}'\mathbf{v} + \mathbf{v}' \int_{t_0}^{t_1} \mathbf{\Phi}(t_1, \tau)\mathbf{B}(\tau)\mathbf{u}(\tau)\, d\tau = ||\mathbf{v}||^2 + 0$$

This contradicts the hypothesis $\mathbf{v} \neq \mathbf{0}$. Thus if $(\mathbf{A}(t), \mathbf{B}(t))$ is controllable at t_0, $\mathbf{W}_c(t_0, t_1)$ must be nonsingular for some finite $t_1 > t_0$. This establishes Theorem 6.11.   Q.E.D.

In order to apply Theorem 6.11, we need knowledge of the state transition matrix, which, however, may not be available. Therefore it is desirable to develop a controllability condition without involving $\mathbf{\Phi}(t, \tau)$. This is possible if we have additional conditions on $\mathbf{A}(t)$ and $\mathbf{B}(t)$. Recall that we have assumed $\mathbf{A}(t)$ and $\mathbf{B}(t)$ to be continuous. Now we require them to be $(n-1)$ times continuously differentiable. Define $\mathbf{M}_0(t) = \mathbf{B}(t)$. We then define recursively a sequence of $n \times p$ matrices $\mathbf{M}_m(t)$ as

$$\mathbf{M}_{m+1}(t) := -\mathbf{A}(t)\mathbf{M}_m(t) + \frac{d}{dt}\mathbf{M}_m(t) \qquad (6.81)$$

for $m = 0, \ 1, \ \ldots, \ n - 1$. Clearly, we have

$$\mathbf{\Phi}(t_2, t)\mathbf{B}(t) = \mathbf{\Phi}(t_2, t)\mathbf{M}_0(t)$$

for any fixed t_2. Using

$$\frac{\partial}{\partial t}\mathbf{\Phi}(t_2, t) = -\mathbf{\Phi}(t_2, t)\mathbf{A}(t)$$

(Problem 4.17), we compute

$$\frac{\partial}{\partial t}[\mathbf{\Phi}(t_2, t)\mathbf{B}(t)] = \frac{\partial}{\partial t}[\mathbf{\Phi}(t_2, t)]\mathbf{B}(t) + \mathbf{\Phi}(t_2, t)\frac{d}{dt}\mathbf{B}(t)$$

$$= \mathbf{\Phi}(t_2, t)[-\mathbf{A}(t)\mathbf{M}_0(t) + \frac{d}{dt}\mathbf{M}_0(t)] = \mathbf{\Phi}(t_2, t)\mathbf{M}_1(t)$$

Proceeding forward, we have

$$\frac{\partial^m}{\partial t^m}\mathbf{\Phi}(t_2, t)\mathbf{B}(t) = \mathbf{\Phi}(t_2, t)\mathbf{M}_m(t) \tag{6.82}$$

for $m = 0, 1, 2, \ldots$. The following theorem is sufficient but not necessary for (6.75) to be controllable.

▶ **Theorem 6.12**

Let $\mathbf{A}(t)$ and $\mathbf{B}(t)$ be $n - 1$ times continuously differentiable. Then the n-dimensional pair $(\mathbf{A}(t), \mathbf{B}(t))$ is controllable at t_0 if there exists a finite $t_1 > t_0$ such that

$$\text{rank } [\mathbf{M}_0(t_1) \ \ \mathbf{M}_1(t_1) \ \ \cdots \ \ \mathbf{M}_{n-1}(t_1)] = n \tag{6.83}$$

⟹ *Proof:* We show that if (6.83) holds, then $\mathbf{W}_c(t_0, t)$ is nonsingular for all $t \geq t_1$. Suppose not, that is, $\mathbf{W}_c(t_0, t)$ is singular or positive semidefinite for some $t_2 \geq t_1$. Then there exists an $n \times 1$ nonzero constant vector $\mathbf{v}$ such that

$$\mathbf{v}'\mathbf{W}_c(t_0, t_2)\mathbf{v} = \int_{t_0}^{t_2} \mathbf{v}'\mathbf{\Phi}(t_2, \tau)\mathbf{B}(\tau)\mathbf{B}'(\tau)\mathbf{\Phi}'(t_2, \tau)\mathbf{v}\, d\tau$$

$$= \int_{t_0}^{t_2} ||\mathbf{B}'(\tau)\mathbf{\Phi}'(t_2, \tau)\mathbf{v}||^2\, d\tau = 0$$

which implies

$$\mathbf{B}'(\tau)\mathbf{\Phi}'(t_2, \tau)\mathbf{v} \equiv \mathbf{0} \quad \text{or} \quad \mathbf{v}'\mathbf{\Phi}(t_2, \tau)\mathbf{B}(\tau) \equiv \mathbf{0} \tag{6.84}$$

for all τ in $[t_0, \ t_2]$. Its differentiations with respect to τ yield, as derived in (6.82),

$$\mathbf{v}'\mathbf{\Phi}(t_2, \tau)\mathbf{M}_m(\tau) \equiv \mathbf{0}$$

for $m = 0, 1, 2, \ldots, n - 1$, and all τ in $[t_0, t_2]$, in particular, at t_1. They can be arranged as

$$\mathbf{v}'\mathbf{\Phi}(t_2, t_1)[\mathbf{M}_0(t_1) \ \ \mathbf{M}_1(t_1) \ \ \cdots \ \ \mathbf{M}_{n-1}(t_1)] = \mathbf{0} \tag{6.85}$$

Because $\mathbf{\Phi}(t_2, t_1)$ is nonsingular, $\mathbf{v}'\mathbf{\Phi}(t_2, t_1)$ is nonzero. Thus (6.85) contradicts (6.83). Therefore, under the condition in (6.83), $\mathbf{W}_c(t_0, t_2)$, for any $t_2 \geq t_1$, is nonsingular and $(\mathbf{A}(t), \mathbf{B}(t))$ is, following Theorem 6.11, controllable at t_0.   Q.E.D.

EXAMPLE 6.13 Consider

$$\dot{\mathbf{x}} = \begin{bmatrix} t & -1 & 0 \\ 0 & -t & t \\ 0 & 0 & t \end{bmatrix}\mathbf{x} + \begin{bmatrix} 0 \\ 1 \\ 1 \end{bmatrix}u \tag{6.86}$$

We have $\mathbf{M}_0 = [0 \ 1 \ 1]'$ and compute

$$\mathbf{M}_1 = -\mathbf{A}(t)\mathbf{M}_0 + \frac{d}{dt}\mathbf{M}_0 = \begin{bmatrix} 1 \\ 0 \\ -t \end{bmatrix}$$

$$\mathbf{M}_2 = -\mathbf{A}(t)\mathbf{M}_1 + \frac{d}{dt}\mathbf{M}_1 = \begin{bmatrix} -t \\ t^2 \\ t^2 - 1 \end{bmatrix}$$

The determinant of the matrix

$$[\mathbf{M}_0 \ \mathbf{M}_1 \ \mathbf{M}_2] = \begin{bmatrix} 0 & 1 & -t \\ 1 & 0 & t^2 \\ 1 & -t & t^2 - 1 \end{bmatrix}$$

is $t^2 + 1$, which is nonzero for all t. Thus the state equation in (6.86) is controllable at every t.

EXAMPLE 6.14 Consider

$$\dot{\mathbf{x}} = \begin{bmatrix} 1 & 0 \\ 0 & 2 \end{bmatrix} \mathbf{x} + \begin{bmatrix} 1 \\ 1 \end{bmatrix} u \tag{6.87}$$

and

$$\dot{\mathbf{x}} = \begin{bmatrix} 1 & 0 \\ 0 & 2 \end{bmatrix} \mathbf{x} + \begin{bmatrix} e^t \\ e^{2t} \end{bmatrix} u \tag{6.88}$$

Equation (6.87) is a time-invariant equation and is controllable according to Corollary 6.8. Equation (6.88) is a time-varying equation; the two entries of its B-matrix are nonzero for all t and one might be tempted to conclude that (6.88) is controllable. Let us check this by using Theorem 6.11. Its state transition matrix is

$$\boldsymbol{\Phi}(t, \tau) = \begin{bmatrix} e^{t-\tau} & 0 \\ 0 & e^{2(t-\tau)} \end{bmatrix}$$

and

$$\boldsymbol{\Phi}(t, \tau)\mathbf{B}(\tau) = \begin{bmatrix} e^{t-\tau} & 0 \\ 0 & e^{2(t-\tau)} \end{bmatrix} \begin{bmatrix} e^{\tau} \\ e^{2\tau} \end{bmatrix} = \begin{bmatrix} e^t \\ e^{2t} \end{bmatrix}$$

We compute

$$\mathbf{W}_c(t_0, t) = \int_{t_0}^{t} \begin{bmatrix} e^t \\ e^{2t} \end{bmatrix} [e^t \ \ e^{2t}] d\tau = \begin{bmatrix} \int_{t_0}^{t} e^{2t} d\tau & \int_{t_0}^{t} e^{3t} d\tau \\ \int_{t_0}^{t} e^{3t} d\tau & \int_{t_0}^{t} e^{4t} d\tau \end{bmatrix}$$

$$= \begin{bmatrix} e^{2t}(t - t_0) & e^{3t}(t - t_0) \\ e^{3t}(t - t_0) & e^{4t}(t - t_0) \end{bmatrix}$$

Its determinant is identically zero for all t_0 and t. Thus (6.88) is not controllable at any t_0. From this example, we see that, in applying a theorem, every condition should be checked carefully; otherwise, we might obtain an erroneous conclusion.

We now discuss the observability part. The linear time-varying state equation in (6.75) is observable at t_0 if there exists a finite t_1 such that for any state $\mathbf{x}(t_0) = \mathbf{x}_0$, the knowledge of the input and output over the time interval $[t_0, t_1]$ suffices to determine uniquely the initial state $\mathbf{x}_0$. Otherwise, the state equation is said to be unobservable at t_0.

▶ **Theorem 6.O11**

The pair $(\mathbf{A}(t), \mathbf{C}(t))$ is observable at time t_0 if and only if there exists a finite $t_1 > t_0$ such that the $n \times n$ matrix

$$\mathbf{W}_o(t_0, t_1) = \int_{t_0}^{t_1} \mathbf{\Phi}'(\tau, t_0)\mathbf{C}'(\tau)\mathbf{C}(\tau)\mathbf{\Phi}(\tau, t_0) \, d\tau \qquad (6.89)$$

where $\mathbf{\Phi}(t, \tau)$ is the state transition matrix of $\dot{\mathbf{x}} = \mathbf{A}(t)\mathbf{x}$, is nonsingular.

▶ **Theorem 6.O12**

Let $\mathbf{A}(t)$ and $\mathbf{C}(t)$ be $n-1$ times continuously differentiable. Then the n-dimensional pair $(\mathbf{A}(t), \mathbf{C}(t))$ is observable at t_0 if there exists a finite $t_1 > t_0$ such that

$$\text{rank} \begin{bmatrix} \mathbf{N}_0(t_1) \\ \mathbf{N}_1(t_1) \\ \vdots \\ \mathbf{N}_{n-1}(t_1) \end{bmatrix} = n \qquad (6.90)$$

where

$$\mathbf{N}_{m+1}(t) = \mathbf{N}_m(t)\mathbf{A}(t) + \frac{d}{dt}\mathbf{N}_m(t) \qquad m = 0, 1, \ldots, n-1$$

with

$$\mathbf{N}_0 = \mathbf{C}(t)$$

We mention that the duality theorem in Theorem 6.5 for time-invariant systems is not applicable to time-varying systems. It must be modified. See Problems 6.22 and 6.23.

PROBLEMS

6.1 Is the state equation

$$\dot{\mathbf{x}} = \begin{bmatrix} 0 & 1 & 0 \\ 0 & 0 & 1 \\ -1 & -3 & -3 \end{bmatrix} \mathbf{x} + \begin{bmatrix} 1 \\ 0 \\ 0 \end{bmatrix} u$$

$$y = [1 \ 2 \ 1]\mathbf{x}$$

controllable? Observable?

6.2 Is the state equation

$$\dot{\mathbf{x}} = \begin{bmatrix} 0 & 1 & 0 \\ 0 & 0 & 1 \\ 0 & 2 & -1 \end{bmatrix} \mathbf{x} + \begin{bmatrix} 0 & 1 \\ 1 & 0 \\ 0 & 0 \end{bmatrix} \mathbf{u}$$

$$y = [1 \ 0 \ 1]\mathbf{x}$$

controllable? Observable?

6.3 Is it true that the rank of $[\mathbf{B} \ \mathbf{AB} \ \cdots \ \mathbf{A}^{n-1}\mathbf{B}]$ equals the rank of $[\mathbf{AB} \ \mathbf{A}^2\mathbf{B} \ \cdots \ \mathbf{A}^n\mathbf{B}]$? If not, under what conditon will it be true?

6.4 Show that the state equation

$$\dot{\mathbf{x}} = \begin{bmatrix} \mathbf{A}_{11} & \mathbf{A}_{12} \\ \mathbf{A}_{21} & \mathbf{A}_{22} \end{bmatrix} \mathbf{x} + \begin{bmatrix} \mathbf{B}_1 \\ \mathbf{0} \end{bmatrix} \mathbf{u}$$

is controllable if and only if the pair $(\mathbf{A}_{22}, \mathbf{A}_{21})$ is controllable.

6.5 Find a state equation to describe the network shown in Fig. 6.1, and then check its controllability and observability.

6.6 Find the controllability index and observability index of the state equations in Problems 6.1 and 6.2.

6.7 What is the controllability index of the state equation

$$\dot{\mathbf{x}} = \mathbf{A}\mathbf{x} + \mathbf{I}\mathbf{u}$$

where $\mathbf{I}$ is the unit matrix?

6.8 Reduce the state equation

$$\dot{\mathbf{x}} = \begin{bmatrix} -1 & 4 \\ 4 & -1 \end{bmatrix} \mathbf{x} + \begin{bmatrix} 1 \\ 1 \end{bmatrix} u \qquad y = [1 \ 1]\mathbf{x}$$

to a controllable one. Is the reduced equation observable?

6.9 Reduce the state equation in Problem 6.5 to a controllable and observable equation.

6.10 Reduce the state equation

$$\dot{\mathbf{x}} = \begin{bmatrix} \lambda_1 & 1 & 0 & 0 & 0 \\ 0 & \lambda_1 & 1 & 0 & 0 \\ 0 & 0 & \lambda_1 & 0 & 0 \\ 0 & 0 & 0 & \lambda_2 & 1 \\ 0 & 0 & 0 & 0 & \lambda_2 \end{bmatrix} \mathbf{x} + \begin{bmatrix} 0 \\ 1 \\ 0 \\ 0 \\ 1 \end{bmatrix} u$$

$$y = [0 \ 1 \ 1 \ 1 \ 0 \ 1]\mathbf{x}$$

to a controllable and observable equation.

6.11 Consider the n-dimensional state equation

$$\dot{\mathbf{x}} = \mathbf{A}\mathbf{x} + \mathbf{B}\mathbf{u}$$
$$y = \mathbf{C}\mathbf{x} + \mathbf{D}\mathbf{u}$$

The rank of its controllability matrix is assumed to be $n_1 < n$. Let $\mathbf{Q}_1$ be an $n \times n_1$ matrix whose columns are any n_1 linearly independent columns of the controllability matrix. Let $\mathbf{P}_1$ be an $n_1 \times n$ matrix such that $\mathbf{P}_1\mathbf{Q}_1 = \mathbf{I}_{n_1}$, where $\mathbf{I}_{n_1}$ is the unit matrix of order n_1. Show that the following n_1-dimensional state equation

$$\dot{\bar{\mathbf{x}}}_1 = \mathbf{P}_1\mathbf{A}\mathbf{Q}_1\bar{\mathbf{x}}_1 + \mathbf{P}_1\mathbf{B}\mathbf{u}$$
$$\bar{y} = \mathbf{C}\mathbf{Q}_1\bar{\mathbf{x}}_1 + \mathbf{D}\mathbf{u}$$

is controllable and has the same transfer matrix as the original state equation.

6.12 In Problem 6.11, the reduction procedure reduces to solving for $\mathbf{P}_1$ in $\mathbf{P}_1\mathbf{Q}_1 = \mathbf{I}$. How do you solve $\mathbf{P}_1$?

6.13 Develop a similar statement as in Problem 6.11 for an unobservable state equation.

6.14 Is the Jordan-form state equation controllable and observable?

$$\dot{\mathbf{x}} = \begin{bmatrix} 2 & 1 & 0 & 0 & 0 & 0 & 0 \\ 0 & 2 & 0 & 0 & 0 & 0 & 0 \\ 0 & 0 & 2 & 0 & 0 & 0 & 0 \\ 0 & 0 & 0 & 2 & 0 & 0 & 0 \\ 0 & 0 & 0 & 0 & 1 & 1 & 0 \\ 0 & 0 & 0 & 0 & 0 & 1 & 0 \\ 0 & 0 & 0 & 0 & 0 & 0 & 1 \end{bmatrix} \mathbf{x} + \begin{bmatrix} 2 & 1 & 0 \\ 2 & 1 & 1 \\ 1 & 1 & 1 \\ 3 & 2 & 1 \\ -1 & 0 & 1 \\ 1 & 0 & 1 \\ 1 & 0 & 0 \end{bmatrix} \mathbf{u}$$

$$\mathbf{y} = \begin{bmatrix} 2 & 2 & 1 & 3 & -1 & 1 & 1 \\ 1 & 1 & 1 & 2 & 0 & 0 & 0 \\ 0 & 1 & 1 & 1 & 1 & 1 & 0 \end{bmatrix} \mathbf{x}$$

6.15 Is it possible to find a set of b_{ij} and a set of c_{ij} such that the state equation

$$\dot{\mathbf{x}} = \begin{bmatrix} 1 & 1 & 0 & 0 & 0 \\ 0 & 1 & 0 & 0 & 0 \\ 0 & 0 & 1 & 1 & 0 \\ 0 & 0 & 0 & 1 & 0 \\ 0 & 0 & 0 & 0 & 1 \end{bmatrix} \mathbf{x} + \begin{bmatrix} b_{11} & b_{12} \\ b_{21} & b_{22} \\ b_{31} & b_{32} \\ b_{41} & b_{42} \\ b_{51} & b_{52} \end{bmatrix} \mathbf{u}$$

$$\mathbf{y} = \begin{bmatrix} c_{11} & c_{12} & c_{13} & c_{14} & c_{15} \\ c_{21} & c_{22} & c_{23} & c_{24} & c_{25} \\ c_{31} & c_{32} & c_{33} & c_{34} & c_{35} \end{bmatrix}$$

is controllable? Observable?

6.16 Consider the state equation

$$\dot{\mathbf{x}} = \begin{bmatrix} \lambda_1 & 0 & 0 & 0 & 0 \\ 0 & \alpha_1 & \beta_1 & 0 & 0 \\ 0 & -\beta_1 & \alpha_1 & 0 & 0 \\ 0 & 0 & 0 & \alpha_2 & \beta_2 \\ 0 & 0 & 0 & -\beta_2 & \alpha_2 \end{bmatrix} \mathbf{x} + \begin{bmatrix} b_1 \\ b_{11} \\ b_{12} \\ b_{21} \\ b_{22} \end{bmatrix} \mathbf{u}$$

$$\mathbf{y} = \begin{bmatrix} c_1 & c_{11} & c_{12} & c_{21} & c_{22} \end{bmatrix}$$

It is the modal form discussed in (4.28). It has one real eigenvalue and two pairs of complex conjugate eigenvalues. It is assumed that they are distinct. Show that the state equation is controllable if and only if $b_1 \neq 0$; $b_{i1} \neq 0$ or $b_{i2} \neq 0$ for $i = 1, 2$. It is observable if and only if $c_1 \neq 0$; $c_{i1} \neq 0$ or $c_{i2} \neq 0$ for $i = 1, 2$.

6.17 Find two- and three-dimensional state equations to describe the network shown in Fig. 6.12. Discuss their controllability and observability.

6.18 Check controllability and observability of the state equation obtained in Problem 2.19. Can you give a physical interpretation directly from the network?

Figure 6.12

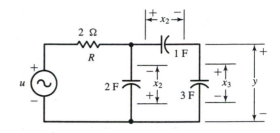

6.19 Consider the continuous-time state equation in Problem 4.2 and its discretized equations in Problem 4.3 with sampling period $T = 1$ and π. Discuss controllability and observability of the discretized equations.

6.20 Check controllability and observability of

$$\dot{\mathbf{x}} = \begin{bmatrix} 0 & 1 \\ 0 & t \end{bmatrix} \mathbf{x} + \begin{bmatrix} 0 \\ 1 \end{bmatrix} u \qquad y = [0 \ 1]\mathbf{x}$$

6.21 Check controllability and observability of

$$\dot{\mathbf{x}} = \begin{bmatrix} 0 & 0 \\ 0 & -1 \end{bmatrix} \mathbf{x} + \begin{bmatrix} 1 \\ e^{-t} \end{bmatrix} u \qquad y = [0 \ e^{-t}]\mathbf{x}$$

6.22 Show that $(\mathbf{A}(t), \mathbf{B}(t))$ is controllable at t_0 if and only if $(-\mathbf{A}'(t), \mathbf{B}'(t))$ is observable at t_0.

6.23 For time-invariant systems, show that $(\mathbf{A}, \mathbf{B})$ is controllable if and only if $(-\mathbf{A}, \mathbf{B})$ is controllable. Is this true for time-varying systems?

Chapter

7

Minimal Realizations and Coprime Fractions

7.1 Introduction

This chapter studies further the realization problem discussed in Section 4.4. Recall that a transfer matrix $\hat{\mathbf{G}}(s)$ is said to be realizable if there exists a state-space equation

$$\dot{\mathbf{x}} = \mathbf{A}\mathbf{x} + \mathbf{B}\mathbf{u}$$

$$\mathbf{y} = \mathbf{C}\mathbf{x} + \mathbf{D}\mathbf{u}$$

that has $\hat{\mathbf{G}}(s)$ as its transfer matrix. This is an important problem for the following reasons. First, many design methods and computational algorithms are developed for state equations. In order to apply these methods and algorithms, transfer matrices must be realized into state equations. As example, computing the response of a transfer function in MATLAB is achieved by first transforming the transfer function into a state equation. Second, once a transfer function is realized into a state equation, the transfer function can be implemented using op-amp circuits, as discussed in Section 2.3.1.

If a transfer function is realizable, then it has infinitely many realizations, not necessarily of the same dimension, as shown in Examples 4.6 and 4.7. An important question is then raised: What is the smallest possible dimension? Realizations with the smallest possible dimension are called *minimal-dimensional* or *minimal* realizations. If we use a minimal realization to implement a transfer function, then the number of integrators used in an op-amp circuit will be minimum. Thus minimal realizations are of practical importance.

In this chapter, we show how to obtain minimal realizations. We will show that a realization of $\hat{g}(s) = N(s)/D(s)$ is minimal if and only if it is controllable and observable, or if and only if its dimension equals the degree of $\hat{g}(s)$. The degree of $\hat{g}(s)$ is defined as the degree of $D(s)$ if the two polynomials $D(s)$ and $N(s)$ are coprime or have no common factors. Thus the concept of coprimeness is essential here. In fact, coprimeness in the fraction $N(s)/D(s)$ plays the same role of controllability and observability in state-space equations.

This chapter studies only linear time-invariant systems. We study first SISO systems and then MIMO systems.

7.2 Implications of Coprimeness

Consider a system with proper transfer function $\hat{g}(s)$. We decompose it as

$$\hat{g}(s) = \hat{g}(\infty) + \hat{g}_{sp}(s)$$

where $\hat{g}_{sp}(s)$ is strictly proper and $\hat{g}(\infty)$ is a constant. The constant $\hat{g}(\infty)$ yields the D-matrix in every realization and will not play any role in what will be discussed. Therefore we consider in this section only strictly proper rational functions. Consider

$$\hat{g}(s) = \frac{N(s)}{D(s)} = \frac{\beta_1 s^3 + \beta_2 s^2 + \beta_3 s + \beta_4}{s^4 + \alpha_1 s^3 + \alpha_2 s^2 + \alpha_3 s + \alpha_4} \tag{7.1}$$

To simplify the discussion, we have assumed that the denominator $D(s)$ has degree 4 and is monic (has 1 as its leading coefficient). In Section 4.4, we introduced for (7.1) the realization in (4.41) without any discussion of its state variables. Now we will redevelop (4.41) by first defining a set of state variables and then discussing the implications of the coprimeness of $D(s)$ and $N(s)$.

Consider

$$\hat{y}(s) = N(s)D^{-1}(s)\hat{u}(s) \tag{7.2}$$

Let us introduce a new variable $v(t)$ defined by $\hat{v}(s) = D^{-1}(s)\hat{u}(s)$. Then we have

$$D(s)\hat{v}(s) = \hat{u}(s) \tag{7.3}$$

$$\hat{y}(s) = N(s)\hat{v}(s) \tag{7.4}$$

Define state variables as

$$\mathbf{x}(t) := \begin{bmatrix} x_1(t) \\ x_2(t) \\ x_3(t) \\ x_4(t) \end{bmatrix} := \begin{bmatrix} v^{(3)}(t) \\ \ddot{v}(t) \\ \dot{v}(t) \\ v(t) \end{bmatrix} \quad \text{or} \quad \hat{\mathbf{x}}(s) = \begin{bmatrix} \hat{x}_1(s) \\ \hat{x}_2(s) \\ \hat{x}_3(s) \\ \hat{x}_4(s) \end{bmatrix} = \begin{bmatrix} s^3 \\ s^2 \\ s \\ 1 \end{bmatrix} \hat{v}(s) \tag{7.5}$$

Then we have

$$\dot{x}_2 = x_1, \quad \dot{x}_3 = x_2, \quad \text{and} \quad \dot{x}_4 = x_3 \tag{7.6}$$

They are independent of (7.1) and follow directly from the definition in (7.5). In order to develop an equation for $\dot{x}_1$, we substitute (7.5) into (7.3) or

$$(s^4 + \alpha_1 s^3 + \alpha_2 s^2 + \alpha_3 s + \alpha_4)\hat{v}(s) = \hat{u}(s)$$

to yield

$$s\hat{x}_1(s) = -\alpha_1 \hat{x}_1(s) - \alpha_2 \hat{x}_2(s) - \alpha_3 \hat{x}_3(s) - \alpha_4 \hat{x}_4(s) + \hat{u}(s)$$

which becomes, in the time domain,

$$\dot{x}_1(t) = [-\alpha_1 \ -\alpha_2 \ -\alpha_3 \ -\alpha_4]\mathbf{x}(t) + 1 \cdot u(t) \tag{7.7}$$

Substituting (7.5) into (7.4) yields

$$\hat{y}(s) = (\beta_1 s^3 + \beta_2 s^2 + \beta_3 s + \beta_4)\hat{v}(s)$$
$$= \beta_1 \hat{x}_1(s) + \beta_2 \hat{x}_2(s) + \beta_3 \hat{x}_3(s) + \beta_4 \hat{x}_4(s)$$
$$= [\beta_1 \ \beta_2 \ \beta_3 \ \beta_4]\hat{\mathbf{x}}(s)$$

which becomes, in the time domain,

$$y(t) = [\beta_4 \ \beta_3 \ \beta_2 \ \beta_1]\mathbf{x}(t) \tag{7.8}$$

Equations (7.6), (7.7), and (7.8) can be combined as

$$\dot{\mathbf{x}} = \mathbf{A}\mathbf{x} + \mathbf{b}u = \begin{bmatrix} -\alpha_1 & -\alpha_2 & -\alpha_3 & -\alpha_4 \\ 1 & 0 & 0 & 0 \\ 0 & 1 & 0 & 0 \\ 0 & 0 & 1 & 0 \end{bmatrix} \mathbf{x} + \begin{bmatrix} 1 \\ 0 \\ 0 \\ 0 \end{bmatrix} u \tag{7.9}$$

$$y = \mathbf{c}\mathbf{x} = [\beta_1 \ \beta_2 \ \beta_3 \ \beta_4]\mathbf{x}$$

This is a realization of (7.1) and was developed in (4.41) by direct verification.

Before proceeding, we mention that if $N(s)$ in (7.1) is 1, then $y(t) = v(t)$ and the output $y(t)$ and its derivatives can be chosen as state variables. However, if $N(s)$ is a polynomial of degree 1 or higher and if we choose the output and its derivatives as state variables, then its realization will be of the form

$$\dot{\mathbf{x}} = \mathbf{A}\mathbf{x} + \mathbf{b}u$$

$$y = \mathbf{c}\mathbf{x} + du + d_1\dot{u} + d_2\ddot{u} + \cdots$$

This equation requires differentiations of u and is not used. Therefore, in general, we cannot select the output y and its derivatives as state variables.[1] We must define state variables by using $v(t)$. Thus $v(t)$ is called a *pseudo state*.

Now we check the controllability and observability of (7.9). Its controllability matrix can readily be computed as

$$C = \begin{bmatrix} 1 & -\alpha_1 & \alpha_1^2 - \alpha_2 & -\alpha_1^3 + 2\alpha_1\alpha_2 - \alpha_3 \\ 0 & 1 & -\alpha_1 & \alpha_1^2 - \alpha_2 \\ 0 & 0 & 1 & -\alpha_1 \\ 0 & 0 & 0 & 1 \end{bmatrix} \tag{7.10}$$

---

1. See also Example 2.16, in particular, (2.47).

Its determinant is 1 for any α_i. Thus the controllability matrix C has full row rank and the state equation is always controllable. This is the reason that (7.9) is called a *controllable canonical form.*

Next we check its observability. It turns out that it depends on whether or not $N(s)$ and $D(s)$ are *coprime*. Two polynomials are said to be *coprime* if they have no common factor of degree at least 1. More specifically, a polynomial $R(s)$ is called a common factor or a common divisor of $D(s)$ and $N(s)$ if they can be expressed as $D(s) = \bar{D}(s)R(s)$ and $N(s) = \bar{N}(s)R(s)$, where $\bar{D}(s)$ and $\bar{N}(s)$ are polynomials. A polynomial $R(s)$ is called a *greatest common divisor* (gcd) of $D(s)$ and $N(s)$ if (1) it is a common divisor of $D(s)$ and $N(s)$ and (2) it can be divided without remainder by every other common divisor of $D(s)$ and $N(s)$. Note that if $R(s)$ is a gcd, so is $\alpha R(s)$ for any nonzero constant α. Thus greatest common divisors are not unique.[2] In terms of the gcd, the polynomials $D(s)$ and $N(s)$ are coprime if their gcd $R(s)$ is a nonzero constant, a polynomial of degree 0; they are not coprime if their gcd has degree 1 or higher.

▶ **Theorem 7.1**

The controllable canonical form in (7.9) is observable if and only if $D(s)$ and $N(s)$ in (7.1) are coprime.

➡ **Proof:** We first show that if (7.9) is observable, then $D(s)$ and $N(s)$ are coprime. We show this by contradiction. If $D(s)$ and $N(s)$ are not coprime, then there exists a λ_1 such that

$$N(\lambda_1) = \beta_1\lambda_1^3 + \beta_2\lambda_1^2 + \beta_3\lambda_1 + \beta_4 = 0 \tag{7.11}$$

$$D(\lambda_1) = \lambda_1^4 + \alpha_1\lambda_1^3 + \alpha_2\lambda_1^2 + \alpha_3\lambda_1 + \alpha_4 = 0 \tag{7.12}$$

Let us define $\mathbf{v} := [\lambda_1^3 \ \lambda_1^2 \ \lambda_1 \ 1]'$; it is a 4×1 nonzero vector. Then (7.11) can be written as $N(\lambda_1) = \mathbf{cv} = 0$, where $\mathbf{c}$ is defined in (7.9). Using (7.12) and the shifting property of the companion form, we can readily verify

$$\mathbf{Av} = \begin{bmatrix} -\alpha_1 & -\alpha_2 & -\alpha_3 & -\alpha_4 \\ 1 & 0 & 0 & 0 \\ 0 & 1 & 0 & 0 \\ 0 & 0 & 1 & 0 \end{bmatrix} \begin{bmatrix} \lambda_1^3 \\ \lambda_1^2 \\ \lambda_1 \\ 1 \end{bmatrix} = \begin{bmatrix} \lambda_1^4 \\ \lambda_1^3 \\ \lambda_1^2 \\ \lambda_1 \end{bmatrix} = \lambda_1 \mathbf{v} \tag{7.13}$$

Thus we have $\mathbf{A}^2\mathbf{v} = \mathbf{A}(\mathbf{Av}) = \lambda_1\mathbf{Av} = \lambda_1^2\mathbf{v}$ and $\mathbf{A}^3\mathbf{v} = \lambda_1^3\mathbf{v}$. We compute, using $\mathbf{cv} = 0$,

$$\mathbf{Ov} = \begin{bmatrix} \mathbf{c} \\ \mathbf{cA} \\ \mathbf{cA}^2 \\ \mathbf{cA}^3 \end{bmatrix} \mathbf{v} = \begin{bmatrix} \mathbf{cv} \\ \mathbf{cAv} \\ \mathbf{cA}^2\mathbf{v} \\ \mathbf{cA}^3\mathbf{v} \end{bmatrix} = \begin{bmatrix} \mathbf{cv} \\ \lambda_1\mathbf{cv} \\ \lambda_1^2\mathbf{cv} \\ \lambda_1^3\mathbf{cv} \end{bmatrix} = \mathbf{0}$$

which implies that the observability matrix does not have full column rank. This contradicts the hypothesis that (7.9) is observable. Thus if (7.9) is observable, then $D(s)$ and $N(s)$ are coprime.

Next we show the converse; that is, if $D(s)$ and $N(s)$ are coprime, then (7.9) is observable. We show this by contradiction. Suppose (7.9) is not observable, then Theorem 6.O1 implies that there exists an eigenvalue λ_1 of $\mathbf{A}$ and a nonzero vector $\mathbf{v}$ such that

---

2. If we require $R(s)$ to be monic, then the gcd is unique.

$$\begin{bmatrix} \mathbf{A} - \lambda_1 \mathbf{I} \\ \mathbf{c} \end{bmatrix} \mathbf{v} = \mathbf{0}$$

or

$$\mathbf{A}\mathbf{v} = \lambda_1 \mathbf{v} \quad \text{and} \quad \mathbf{c}\mathbf{v} = 0$$

Thus $\mathbf{v}$ is an eigenvector of $\mathbf{A}$ associated with eigenvalue λ_1. From (7.13), we see that $\mathbf{v} = [\lambda_1^3 \ \lambda_1^2 \ \lambda_1 \ 1]'$ is an eigenvector. Substituting this $\mathbf{v}$ into $\mathbf{c}\mathbf{v} = 0$ yields

$$N(\lambda_1) = \beta_1 \lambda_1^3 + \beta_2 \lambda_1^2 + \beta_3 \lambda_1 + \beta_4 = 0$$

Thus λ_1 is a root of $N(s)$. The eigenvalue of $\mathbf{A}$ is a root of its characteristic polynomial, which, because of the companion form of $\mathbf{A}$, equals $D(s)$. Thus we also have $D(\lambda_1) = 0$, and $D(s)$ and $N(s)$ have the same factor $s - \lambda_1$. This contradicts the hypothesis that $D(s)$ and $N(s)$ are coprime. Thus if $D(s)$ and $N(s)$ are coprime, then (7.9) is observable. This establishes the theorem.   Q.E.D.

If (7.9) is a realization of $\hat{g}(s)$, then we have, by definition,

$$\hat{g}(s) = \mathbf{c}(s\mathbf{I} - \mathbf{A})^{-1}\mathbf{b}$$

Taking its transpose yields

$$\hat{g}'(s) = \hat{g}(s) = [\mathbf{c}(s\mathbf{I} - \mathbf{A})^{-1}\mathbf{b}]' = \mathbf{b}'(s\mathbf{I} - \mathbf{A}')^{-1}\mathbf{c}'$$

Thus the state equation

$$\dot{\mathbf{x}} = \mathbf{A}'\mathbf{x} + \mathbf{c}'u = \begin{bmatrix} -\alpha_1 & 1 & 0 & 0 \\ -\alpha_2 & 0 & 1 & 0 \\ -\alpha_3 & 0 & 0 & 1 \\ -\alpha_4 & 0 & 0 & 0 \end{bmatrix} \mathbf{x} + \begin{bmatrix} \beta_1 \\ \beta_2 \\ \beta_3 \\ \beta_4 \end{bmatrix} u \tag{7.14}$$

$$y = \mathbf{b}'\mathbf{x} = [1 \ 0 \ 0 \ 0]\mathbf{x}$$

is a different realization of (7.1). This state equation is always observable and is called an *observable canonical form*. Dual to Theorem 7.1, Equation (7.14) is controllable if and only if $D(s)$ and $N(s)$ are coprime.

We mention that the equivalence transformation $\bar{\mathbf{x}} = \mathbf{P}\mathbf{x}$ with

$$\mathbf{P} = \begin{bmatrix} 0 & 0 & 0 & 1 \\ 0 & 0 & 1 & 0 \\ 0 & 1 & 0 & 0 \\ 1 & 0 & 0 & 0 \end{bmatrix} \tag{7.15}$$

will transform (7.9) into

$$\dot{\mathbf{x}} = \begin{bmatrix} 0 & 1 & 0 & 0 \\ 0 & 0 & 1 & 0 \\ 0 & 0 & 0 & 1 \\ -\alpha_4 & -\alpha_3 & -\alpha_2 & -\alpha_1 \end{bmatrix} \mathbf{x} + \begin{bmatrix} 0 \\ 0 \\ 0 \\ 1 \end{bmatrix} u$$

$$y = [\beta_4 \ \beta_3 \ \beta_2 \ \beta_1]\mathbf{x}$$

This is also called a controllable canonical form. Similarly, (7.15) will transform (7.14) into

$$\dot{\mathbf{x}} = \begin{bmatrix} 0 & 0 & 0 & -\alpha_4 \\ 1 & 0 & 0 & -\alpha_3 \\ 0 & 1 & 0 & -\alpha_2 \\ 0 & 0 & 1 & -\alpha_1 \end{bmatrix} \mathbf{x} + \begin{bmatrix} \beta_4 \\ \beta_3 \\ \beta_2 \\ \beta_1 \end{bmatrix} u$$

$$y = \mathbf{cx} = [0 \ 0 \ 0 \ 1]\mathbf{x}$$

This is a different observable canonical form.

7.2.1 Minimal Realizations

We first define a degree for proper rational functions. We call $N(s)/D(s)$ a *polynomial fraction* or, simply, a *fraction*. Because

$$\hat{g}(s) = \frac{N(s)}{D(s)} = \frac{N(s)Q(s)}{D(s)Q(s)}$$

for any polynomial $Q(s)$, fractions are not unique. Let $R(s)$ be a greatest common divisor (gcd) of $N(s)$ and $D(s)$. That is, if we write $N(s) = \bar{N}(s)R(s)$ and $D(s) = \bar{D}(s)R(s)$, then the polynomials $\bar{N}(s)$ and $\bar{D}(s)$ are coprime. Clearly every rational function $\hat{g}(s)$ can be reduced to $\hat{g}(s) = \bar{N}(s)/\bar{D}(s)$. Such an expression is called a *coprime fraction*. We call $\bar{D}(s)$ a *characteristic polynomial* of $\hat{g}(s)$. The degree of the characteristic polynomial is defined as the *degree* of $\hat{g}(s)$. Note that characteristic polynomials are not unique; they may differ by a nonzero constant. If we require the polynomial to be monic, then it is unique.

Consider the rational function

$$\hat{g}(s) = \frac{s^2 - 1}{4(s^3 - 1)}$$

Its numerator and denominator contain the common factor $s - 1$. Thus its coprime fraction is $\hat{g}(s) = (s+1)/4(s^2+s+1)$ and its characteristic polynomial is $4s^2+4s+4$. Thus the rational function has degree 2. Given a proper rational function, if its numerator and denominator are coprime—as is often the case—then its denominator is a characteristic polynomial and the degree of the denominator is the degree of the rational function.

▶ **Theorem 7.2**

A state equation $(\mathbf{A}, \mathbf{b}, \mathbf{c}, d)$ is a minimal realization of a proper rational function $\hat{g}(s)$ if and only if $(\mathbf{A}, \mathbf{b})$ is controllable and $(\mathbf{A}, \mathbf{c})$ is observable or if and only if

$$\dim \mathbf{A} = \deg \ \hat{g}(s)$$

Proof: If $(\mathbf{A}, \mathbf{b})$ is not controllable or if $(\mathbf{A}, \mathbf{c})$ is not observable, then the state equation can be reduced to a lesser dimensional state equation that has the same transfer function (Theorems 6.6 and 6.O6). Thus $(\mathbf{A}, \mathbf{b}, \mathbf{c}, d)$ is not a minimal realization. This shows the necessity of the theorem.

To show the sufficiency, consider the n-dimensional controllable and observable state equation

$$\dot{\mathbf{x}} = \mathbf{A}\mathbf{x} + \mathbf{b}u$$
$$y = \mathbf{c}\mathbf{x} + du \tag{7.16}$$

Clearly its $n \times n$ controllability matrix

$$C = [\mathbf{b} \ \mathbf{Ab} \ \cdots \ \mathbf{A}^{n-1}\mathbf{b}] \tag{7.17}$$

and its $n \times n$ observability matrix

$$O = \begin{bmatrix} \mathbf{c} \\ \mathbf{c}\mathbf{A} \\ \vdots \\ \mathbf{c}\mathbf{A}^{n-1} \end{bmatrix} \tag{7.18}$$

both have rank n. We show that (7.16) is a minimal realization by contradiction. Suppose the $\bar{n}$-dimensional state equation, with $\bar{n} < n$,

$$\dot{\bar{\mathbf{x}}} = \bar{\mathbf{A}}\bar{\mathbf{x}} + \bar{\mathbf{b}}u$$
$$y = \bar{\mathbf{c}}\bar{\mathbf{x}} + \bar{d}u \tag{7.19}$$

is a realization of $\hat{g}(s)$. Then Theorem 4.1 implies $d = \bar{d}$ and

$$\mathbf{c}\mathbf{A}^m\mathbf{b} = \bar{\mathbf{c}}\bar{\mathbf{A}}^m\bar{\mathbf{b}} \qquad \text{for } m = 0, \ 1, \ 2, \ \ldots \tag{7.20}$$

Let us consider the product

$$OC = \begin{bmatrix} \mathbf{c} \\ \mathbf{c}\mathbf{A} \\ \vdots \\ \mathbf{c}\mathbf{A}^{n-1} \end{bmatrix} [\mathbf{b} \ \mathbf{Ab} \ \cdots \mathbf{A}^{n-1}\mathbf{b}]$$

$$= \begin{bmatrix} \mathbf{cb} & \mathbf{cAb} & \mathbf{cA}^2\mathbf{b} & \cdots & \mathbf{cA}^{n-1}\mathbf{b} \\ \mathbf{cAb} & \mathbf{cA}^2\mathbf{b} & \mathbf{cA}^3\mathbf{b} & \cdots & \mathbf{cA}^n\mathbf{b} \\ \mathbf{cA}^2\mathbf{b} & \mathbf{cA}^3\mathbf{b} & \mathbf{cA}^4\mathbf{b} & \cdots & \mathbf{cA}^{n+1}\mathbf{b} \\ \vdots & \vdots & \vdots & \ddots & \vdots \\ \mathbf{cA}^{n-1}\mathbf{b} & \mathbf{cA}^n\mathbf{b} & \mathbf{cA}^{n+1}\mathbf{b} & \cdots & \mathbf{cA}^{2(n-1)}\mathbf{b} \end{bmatrix} \tag{7.21}$$

Using (7.20), we can replace every $\mathbf{cA}^m\mathbf{b}$ by $\bar{\mathbf{c}}\bar{\mathbf{A}}^m\bar{\mathbf{b}}$. Thus we have

$$OC = \bar{O}_n\bar{C}_n \tag{7.22}$$

where $\bar{O}_n$ is defined as in (6.21) for the $\bar{n}$-dimensional state equation in (7.19) and $\bar{C}_n$ is defined similarly. Because (7.16) is controllable and observable, we have $\rho(O) = n$ and $\rho(C) = n$. Thus (3.62) implies $\rho(OC) = n$. Now $\bar{O}_n$ and $\bar{C}_n$ are, respectively, $n \times \bar{n}$ and $\bar{n} \times n$; thus (3.61) implies that the matrix $\bar{O}_n\bar{C}_n$ has rank at most $\bar{n}$. This contradicts $\rho(\bar{O}_n\bar{C}_n) = \rho(OC) = n$. Thus $(\mathbf{A}, \mathbf{b}, \mathbf{c}, d)$ is minimal. This establishes the first part of the theorem.

The realization in (7.9) is controllable and observable if and only if $\hat{g}(s) = N(s)/D(s)$ is a coprime fraction (Theorem 7.1). In this case, we have dim $\mathbf{A}$ = deg $D(s)$ = deg $\hat{g}(s)$. Because all minimal realizations are equivalent, as will be established immediately, we conclude that every realization is minimal if and only if dim $\mathbf{A}$ = deg $\hat{g}(s)$. This establishes the theorem.   Q.E.D.

To complete the proof of Theorem 7.2, we need the following theorem.

▶ **Theorem 7.3**

All minimal realizations of $\hat{g}(s)$ are equivalent.

Proof: Let $(\mathbf{A}, \mathbf{b}, \mathbf{c}, d)$ and $(\bar{\mathbf{A}}, \bar{\mathbf{b}}, \bar{\mathbf{c}}, \bar{d})$ be minimal realizations of $\hat{g}(s)$. Then we have $\bar{d} = d$ and, following (7.22),

$$OC = \bar{O}\bar{C} \tag{7.23}$$

Multiplying $O\mathbf{A}C$ out explicitly and then using (7.20), we can show

$$O\mathbf{A}C = \bar{O}\bar{\mathbf{A}}\bar{C} \tag{7.24}$$

Note that the controllability and observability matrices are all nonsingular square matrices. Let us define

$$\mathbf{P} := \bar{O}^{-1}O$$

Then (7.23) implies

$$\mathbf{P} = \bar{O}^{-1}O = \bar{C}C^{-1} \quad \text{and} \quad \mathbf{P}^{-1} = O^{-1}\bar{O} = C\bar{C}^{-1} \tag{7.25}$$

From (7.23), we have $\bar{C} = \bar{O}^{-1}OC = \mathbf{P}C$. The first columns on both side of the equality yield $\bar{\mathbf{b}} = \mathbf{P}\mathbf{b}$. Again from (7.23), we have $\bar{O} = OC\bar{C}^{-1} = O\mathbf{P}^{-1}$. The first rows on both sides of the equality yield $\bar{\mathbf{c}} = \mathbf{c}\mathbf{P}^{-1}$. Equation (7.24) implies

$$\bar{\mathbf{A}} = \bar{O}^{-1}O\mathbf{A}C\bar{C}^{-1} = \mathbf{P}\mathbf{A}\mathbf{P}^{-1}$$

Thus $(\mathbf{A}, \mathbf{b}, \mathbf{c}\, d)$ and $(\bar{\mathbf{A}}, \bar{\mathbf{b}}, \bar{\mathbf{c}}, \bar{d})$ meet the conditions in (4.26) and, consequently, are equivalent. This establishes the theorem.   Q.E.D.

Theorem 7.2 has many important implications. Given a state equation, if we compute its transfer function and degree, then the minimality of the state equation can readily be determined without checking its controllability and observability. Thus the theorem provides an alternative way of checking controllability and observability. Conversely, given a rational function, if we compute first its common factors and reduce it to a coprime fraction, then the state equations obtained by using its coefficients as shown in (7.9) and (7.14) will automatically be controllable and observable.

Consider a proper rational function $\hat{g}(s) = N(s)/D(s)$. If the fraction is coprime, then every root of $D(s)$ is a pole of $\hat{g}(s)$ and vice versa. This is not true if $N(s)$ and $D(s)$ are not coprime. Let $(\mathbf{A}, \mathbf{b}.\mathbf{c}.d)$ be a minimal realization of $\hat{g}(s) = N(s)/D(s)$. Then we have

$$\frac{N(s)}{D(s)} = \mathbf{c}(s\mathbf{I} - \mathbf{A})^{-1}\mathbf{b} + d = \frac{1}{\det(s\mathbf{I} - \mathbf{A})}\mathbf{c}\left[\text{Adj}(s\mathbf{I} - \mathbf{A})\right]\mathbf{b} + d$$

If $N(s)$ and $D(s)$ are coprime, then deg $D(s) = $ deg $\hat{g}(s) = $ dim $\mathbf{A}$. Thus we have

$$D(s) = k \det(s\mathbf{I} - \mathbf{A})$$

for some nonzero constant k. Note that $k = 1$ if $D(s)$ is monic. This shows that if a state equation is controllable and observable, then every eigenvalue of $\mathbf{A}$ is a pole of $\hat{g}(s)$ and every pole of $\hat{g}(s)$ is an eigenvalue of $\mathbf{A}$. Thus we conclude that if $(\mathbf{A}, \mathbf{b}, \mathbf{c}, d)$ is controllable and observable, then we have

$$\text{Asymptotic stability} \iff \text{BIBO stability}$$

More generally, *controllable and observable state equations and coprime fractions contain essentially the same information and either description can be used to carry out analysis and design.*

7.3 Computing Coprime Fractions

The importance of coprime fractions and degrees was demonstrated in the preceding section. In this section, we discuss how to compute them. Consider a proper rational function

$$\hat{g}(s) = \frac{N(s)}{D(s)}$$

where $N(s)$ and $D(s)$ are polynomials. If we use the MATLAB function `roots` to compute their roots and then to cancel their common factors, we will obtain a coprime fraction. The MATLAB function `minreal` can also be used to obtain coprime fractions. In this section, we introduce a different method by solving a set of linear algebraic equations. The method does not offer any advantages over the aforementioned methods for scalar rational functions. However, it can readily be extended to the matrix case. More importantly, the method will be used to carry out design in Chapter 9.

Consider $N(s)/D(s)$. To simplify the discussion, we assume deg $N(s) \leq $ deg $D(s) = n = 4$. Let us write

$$\frac{N(s)}{D(s)} = \frac{\bar{N}(s)}{\bar{D}(s)}$$

which implies

$$D(s)(-\bar{N}(s)) + N(s)\bar{D}(s) = 0 \tag{7.26}$$

It is clear that $D(s)$ and $N(s)$ are not coprime if and only if there exist polynomials $\bar{N}(s)$ and $\bar{D}(s)$ with deg $\bar{N}(s) \leq $ deg $\bar{D}(s) < n = 4$ to meet (7.26). The condition deg $\bar{D}(s) < n$ is crucial; otherwise, (7.26) has infinitely many solutions $\bar{N}(s) = N(s)R(s)$ and $\bar{D}(s) = D(s)R(s)$ for any polynomial $R(s)$. Thus the coprimeness problem can be reduced to solving the polynomial equation in (7.26).

Instead of solving (7.26) directly, we will change it into solving a set of linear algebraic equations. We write

$$D(s) = D_0 + D_1 s + D_2 s^2 + D_3 s^3 + D_4 s^4$$

$$N(s) = N_0 + N_1 s + N_2 s^2 + N_3 s^3 + N_4 s^4$$

$$\bar{D}(s) = \bar{D}_0 + \bar{D}_1 s + \bar{D}_2 s^2 + \bar{D}_3 s^3$$

$$\bar{N}(s) = \bar{N}_0 + \bar{N}_1 s + \bar{N}_2 s^2 + \bar{N}_3 s^3 \qquad (7.27)$$

where $D_4 \neq 0$ and the remaining D_i, N_i, $\bar{D}_i$, and $\bar{N}_i$ can be zero or nonzero. Substituting these into (7.26) and equating to zero the coefficients associated with s^k, for $k = 0, 1, \ldots, 7$, we obtain

$$
\mathbf{Sm} :=
\begin{bmatrix}
D_0 & N_0 & \vdots & 0 & 0 & \vdots & 0 & 0 & \vdots & 0 & 0 \\
D_1 & N_1 & \vdots & D_0 & N_0 & \vdots & 0 & 0 & \vdots & 0 & 0 \\
D_2 & N_2 & \vdots & D_1 & N_1 & \vdots & D_0 & N_0 & \vdots & 0 & 0 \\
D_3 & N_3 & \vdots & D_2 & N_2 & \vdots & D_1 & N_1 & \vdots & D_0 & N_0 \\
D_4 & N_4 & \vdots & D_3 & N_3 & \vdots & D_2 & N_2 & \vdots & D_1 & N_1 \\
0 & 0 & \vdots & D_4 & N_4 & \vdots & D_3 & N_3 & \vdots & D_2 & N_2 \\
0 & 0 & \vdots & 0 & 0 & \vdots & D_4 & N_4 & \vdots & D_3 & N_3 \\
0 & 0 & \vdots & 0 & 0 & \vdots & 0 & 0 & \vdots & D_4 & N_4
\end{bmatrix}
\begin{bmatrix}
-\bar{N}_0 \\ \bar{D}_0 \\ \cdots \\ -\bar{N}_1 \\ \bar{D}_1 \\ \cdots \\ -\bar{N}_2 \\ \bar{D}_2 \\ \cdots \\ -\bar{N}_3 \\ \bar{D}_3
\end{bmatrix}
= \mathbf{0} \quad (7.28)
$$

This is a homogeneous linear algebraic equation. The first block column of $\mathbf{S}$ consists of two columns formed from the coefficients of $D(s)$ and $N(s)$ arranged in ascending powers of s. The second block column is the first block column shifted down one position. Repeating the process until $\mathbf{S}$ is a square matrix of order $2n = 8$. The square matrix $\mathbf{S}$ is called the *Sylvester resultant*. If the Sylvester resultant is singular, nonzero solutions exist in (7.28) (Theorem 3.3). This means that polynomials $\bar{N}(s)$ and $\bar{D}(s)$ of degree 3 or less exist to meet (7.26). Thus $D(s)$ and $N(s)$ are not coprime. If the Sylvester resultant is nonsingular, no nonzero solutions exist in (7.28) or, equivalently, no polynomials $\bar{N}(s)$ and $\bar{D}(s)$ of degree 3 or less exist to meet (7.26). Thus $D(s)$ and $N(s)$ are coprime. In conclusion, $D(s)$ *and* $N(s)$ *are coprime if and only if the Sylvester resultant is nonsingular.*

If the Sylvester resultant is singular, then $N(s)/D(s)$ can be reduced to

$$\frac{N(s)}{D(s)} = \frac{\bar{N}(s)}{\bar{D}(s)}$$

where $\bar{N}(s)$ and $\bar{D}(s)$ are coprime. We discuss how to obtain a coprime fraction directly from (7.28). Let us search linearly independent columns of $\mathbf{S}$ in order from left to right. We call columns formed from D_i D-columns and formed from N_i N-columns. Then every D-column is linearly independent of its left-hand-side (LHS) columns. Indeed, because $D_4 \neq 0$, the first D-column is linearly independent. The second D-column is also linearly independent of its LHS columns because the LHS entries of D_4 are all zero. Proceeding forward, we conclude that all D-columns are linearly independent of their LHS columns. On the other hand, an N-column can be dependent or independent of its LHS columns. Because of the repetitive pattern of **S**, if an N-column becomes linearly dependent on its LHS columns, then all subsequent N-columns are linearly dependent of their LHS columns. Let μ denote the number of linearly independent N-columns in **S**. Then the $(\mu + 1)$th N-column is the first N-column to become linearly dependent on its LHS columns and will be called the *primary dependent N-column*. Let us use $\mathbf{S}_1$ to denote the submatrix of **S** that consists of the primary dependent N-column and all its LHS columns. That is, $\mathbf{S}_1$ consists of $\mu + 1$ D-columns (all of them are linearly independent) and $\mu + 1$ N-columns (the last one is dependent). Thus $\mathbf{S}_1$ has $2(\mu + 1)$ columns but rank $2\mu + 1$. In other words, $\mathbf{S}_1$ has nullity 1 and, consequently, has one independent null vector. Note that if $\bar{\mathbf{n}}$ is a null vector, so is $\alpha \bar{\mathbf{n}}$ for any nonzero α. Although any null vector can be used, we will use exclusively the null vector with 1 as its last entry to develop $\bar{N}(s)$ and $\bar{D}(s)$. For convenience, we call such a null vector a *monic null vector*. If we use the MATLAB function `null` to generate a null vector, then the null vector must be divided by its last entry to yield a monic null vector. This is illustrated in the next example.

EXAMPLE 7.1 Consider

$$\frac{N(s)}{D(s)} = \frac{6s^3 + s^2 + 3s - 20}{2s^4 + 7s^3 + 15s^2 + 16s + 10} \tag{7.29}$$

We have $n = 4$ and its Sylvester resultant **S** is 8×8. The fraction is coprime if and only if **S** is nonsingular or has rank 8. We use MATLAB to check the rank of **S**. Because it is simpler to key in the transpose of **S**, we type

```
d=[10 16 15 7 2];n=[-20 3 1 6 0];
s=[d 0 0 0;n 0 0 0;0 d 0 0;0 n 0 0;...
   0 0 d 0;0 0 n 0;0 0 0 d;0 0 0 n]';
m=rank(s)
```

The answer is 6; thus $D(s)$ and $N(s)$ are not coprime. Because all four D-columns of **S** are linearly independent, we conclude that **S** has only two linearly independent N-columns and $\mu = 2$. The third N-column is the primary dependent N-column and all its LHS columns are linearly independent. Let $\mathbf{S}_1$ denote the first six columns of **S**, an 8×6 matrix. The submatrix $\mathbf{S}_1$ has three D-column (all linearly independent) and two linearly independent N-columns, thus it has rank 5 and nullity 1. Because all entries of the last row of $\mathbf{S}_1$ are zero, they can be skipped in forming $\mathbf{S}_1$. We type

```
s1=[d 0 0;n 0 0;0 d 0;0 n 0;0 0 d;0 0 n]';
z=null(s1)
```

which yields

```
ans z= [ 0.6860  0.3430  -0.5145  0.3430  0.0000  0.1715 ]'
```

This null vector does not have 1 as its last entry. We divide it by the last entry or the sixth entry of z by typing

```
zb=z/z(6)
```

which yields

```
ans zb= [4 2   -3 2 0 1]'
```

This monic null vector equals $[-\bar{N}_0 \quad \bar{D}_0 \quad -\bar{N}_1 \quad \bar{D}_1 \quad -\bar{N}_2 \quad \bar{D}_2]'$. Thus we have

$$\bar{N}(s) = -4 + 3s + 0 \cdot s^2 \qquad \bar{D}(s) = 2 + 2s + s^2$$

and

$$\frac{6s^3 + s^2 + 3s - 20}{2s^4 + 7s^3 + 15s^2 + 16s + 10} = \frac{3s - 4}{s^2 + 2s + 2}$$

Because the null vector is computed from the first linearly dependent N-column, the computed $\bar{N}(s)$ and $\bar{D}(s)$ have the smallest possible degrees to meet (7.26) and, therefore, are coprime. This completes the reduction of $N(s)/D(s)$ to a coprime fraction.

The preceding procedure can be summerized as a theorem.

▶ **Theroem 7.4**

Consider $\hat{g}(s) = N(s)/D(s)$. We use the coefficients of $D(s)$ and $N(s)$ to form the Sylvester resultant **S** in (7.28) and search its linearly independent columns in order from left to right. Then we have

$$\deg \hat{g}(s) = \text{number of linearly independent } N\text{-columns} =: \mu$$

and the coefficients of a coprime fraction $\hat{g}(s) = \bar{N}(s)/\bar{D}(s)$ or

$$[-\bar{N}_0 \quad \bar{D}_0 \quad -\bar{N}_1 \quad \bar{D}_1 \cdots -\bar{N}_\mu \quad \bar{D}_\mu]'$$

equals the monic null vector of the submatrix that consists of the primary dependent N-column and all its LHS linearly independent columns of **S**.

We mention that if D- and N-columns in **S** are arranged in descending powers of s, then it is not true that all D-columns are linearly independent of their LHS columns and that the degree of $\hat{g}(s)$ equals the number of linearly independent N-columns. See Problem 7.6. Thus it is essential to arrange the D- and N-columns in ascending powers of s in **S**.

7.3.1 QR Decomposition

As discussed in the preceding section, a coprime fraction can be obtained by searching linearly independent columns of the Sylvester resultant in order from left to right. It turns out the widely available QR decomposition can be used to achieve this searching.

Consider an $n \times m$ matrix **M**. Then there exists an $n \times n$ orthogonal matrix $\bar{\mathbf{Q}}$ such that

$$\bar{\mathbf{Q}}\mathbf{M} = \mathbf{R}$$

where **R** is an upper triangular matrix of the same dimensions as **M**. Because $\bar{\mathbf{Q}}$ operates on the rows of **M**, the linear independence of the columns of **M** is preserved in the columns of **R**. In other words, if a column of **R** is linearly dependent on its left-hand-side (LHS) columns, so is the corresponding column of $\mathbf{M}$. Now because $\mathbf{R}$ is in upper triangular form, its mth column is linearly independent of its LHS columns if and only if its mth entry at the diagonal position is nonzero. Thus using $\mathbf{R}$, the linearly independent columns of $\mathbf{M}$, in order from left to right, can be obtained by inspection. Because $\bar{\mathbf{Q}}$ is orthogonal, we have $\bar{\mathbf{Q}}^{-1} = \bar{\mathbf{Q}}' =: \mathbf{Q}$ and $\bar{\mathbf{Q}}\mathbf{M} = \mathbf{R}$ becomes $\mathbf{M} = \mathbf{QR}$. This is called *QR decomposition*. In MATLAB, $\mathbf{Q}$ and $\mathbf{R}$ can be obtained by typing $[q,r]=qr(m)$.

Let us apply QR decomposition to the resultant in Example 7.1. We type

```
d=[10 16 15 7 2];n=[-20 3 1 6 0];
s=[d 0 0 0;n 0 0 0;0 d 0 0;0 n 0 0;...
   0 0 d 0;0 0 n 0;0 0 0 d;0 0 0 n]';
[q,r]=qr(s)
```

Because $\mathbf{Q}$ is not needed, we show only $\mathbf{R}$:

$$
r = \begin{bmatrix}
-25.1 & 3.7 & -20.6 & 10.1 & -11.6 & 11.0 & -4.1 & 5.3 \\
0 & -20.7 & -10.3 & 4.3 & -7.2 & 2.1 & -3.6 & 6.7 \\
0 & 0 & -10.2 & -15.6 & -20.3 & 0.8 & -16.8 & 9.6 \\
0 & 0 & 0 & 8.9 & -3.5 & -17.9 & -11.2 & 7.3 \\
0 & 0 & 0 & 0 & -5.0 & 0 & -12.0 & -15.0 \\
0 & 0 & 0 & 0 & 0 & 0 & -2.0 & 0 \\
0 & 0 & 0 & 0 & 0 & 0 & -4.6 & 0 \\
0 & 0 & 0 & 0 & 0 & 0 & 0 & 0
\end{bmatrix}
$$

We see that the matrix is upper triangular. Because the sixth column has 0 as its sixth entry (diagonal position), it is linearly dependent on its LHS columns. So is the last column. To determine whether a column is linearly dependent, we need to know only whether the diagonal entry is zero or not. Thus the matrix can be simplified as

$$
r = \begin{bmatrix}
d & x & x & x & x & x & x & x \\
0 & n & x & x & x & x & x & x \\
0 & 0 & d & x & x & x & x & x \\
0 & 0 & 0 & n & x & x & x & x \\
0 & 0 & 0 & 0 & d & 0 & x & x \\
0 & 0 & 0 & 0 & 0 & 0 & x & 0 \\
0 & 0 & 0 & 0 & 0 & 0 & d & 0 \\
0 & 0 & 0 & 0 & 0 & 0 & 0 & 0
\end{bmatrix}
$$

where d, n, and x denote nonzero entries and d also denotes D-column and n denotes N-column. We see that every D-column is linearly independent of its LHS columns and there are only two linearly independent N-columns. Thus by employing QR decomposition, we obtain immediately μ and the primary dependent N-column. In scalar transfer functions, we can use either rank or qr to find μ. In the matrix case, using rank is very inconvenient; we will use QR decomposition.

7.4 Balanced Realization[3]

Every transfer function has infinitely many minimal realizations. Among these realizations, it is of interest to see which realizations are more suitable for practical implementation. If we use the controllable or observable canonical form, then the A-matrix and b- or c-vector have many zero entries, and its implementation will use a small number of components. However, either canonical form is very sensitive to parameter variations; therefore both forms should be avoided if sensitivity is an important issue. If all eigenvalues of **A** are distinct, we can transform **A**, using an equivalence transformation, into a diagonal form (if all eigenvalues are real) or into the modal form discussed in Section 4.3.1 (if some eigenvalues are complex). The diagonal or modal form has many zero entries in **A** and will use a small number of components in its implementation. More importantly, the diagonal and modal forms are least sensitive to parameter variations among all realizations; thus they are good candidates for practical implementation.

We discuss next a different minimal realization, called a balanced realization. However, the discussion is applicable only to stable **A**. Consider

$$\dot{\mathbf{x}} = \mathbf{A}\mathbf{x} + \mathbf{b}u$$
$$y = \mathbf{c}\mathbf{x} \tag{7.30}$$

It is assumed that **A** is stable or all its eigenvalues have negative real parts. Then the controllability Gramian $\mathbf{W}_c$ and the observability Gramian $\mathbf{W}_o$ are, respectively, the unique solutions of

$$\mathbf{A}\mathbf{W}_c + \mathbf{W}_c\mathbf{A}' = -\mathbf{b}\mathbf{b}' \tag{7.31}$$

and

$$\mathbf{A}'\mathbf{W}_o + \mathbf{W}_o\mathbf{A} = -\mathbf{c}'\mathbf{c} \tag{7.32}$$

They are positive definite if (7.30) is controllable and observable.

Different minimal realizations of the same transfer function have different controllability and observability Gramians. For example, the state equation, taken from Reference [23],

$$\dot{\mathbf{x}} = \begin{bmatrix} -1 & -4/\alpha \\ 4\alpha & -2 \end{bmatrix} \mathbf{x} + \begin{bmatrix} 1 \\ 2\alpha \end{bmatrix} u$$
$$y = [-1 \ \ 2/\alpha]\mathbf{x} \tag{7.33}$$

for any nonzero α, has transfer function $\hat{g}(s) = (3s + 18)/(s^2 + 3s + 18)$, and is controllable and observable. Its controllability and observability Gramians can be computed as

$$\mathbf{W}_c = \begin{bmatrix} 0.5 & 0 \\ 0 & \alpha^2 \end{bmatrix} \quad \text{and} \quad \mathbf{W}_o = \begin{bmatrix} 0.5 & 0 \\ 0 & 1/\alpha^2 \end{bmatrix} \tag{7.34}$$

We see that different α yields different minimal realization and different controllability and observability Gramians. Even though the controllability and observability Gramians will change, their product remains the same as diag (0.25, 1) for all α.

---

3. This section may be skipped without loss of continuity.

▷ **Theorem 7.5**

Let $(\mathbf{A}, \mathbf{b}, \mathbf{c})$ and $(\bar{\mathbf{A}}, \bar{\mathbf{b}}, \bar{\mathbf{c}})$ be minimal and equivalent and let $\mathbf{W}_c\mathbf{W}_o$ and $\bar{\mathbf{W}}_c\bar{\mathbf{W}}_o$ be the products of their controllability and observability Gramians. Then $\mathbf{W}_c\mathbf{W}_o$ and $\bar{\mathbf{W}}_c\bar{\mathbf{W}}_o$ are similar and their eigenvalues are all real and positive.

Proof: Let $\bar{\mathbf{x}} = \mathbf{Px}$, where $\mathbf{P}$ is a nonsingular constant matrix. Then we have

$$\bar{\mathbf{A}} = \mathbf{PAP}^{-1} \qquad \bar{\mathbf{b}} = \mathbf{Pb} \qquad \bar{\mathbf{c}} = \mathbf{cP}^{-1} \tag{7.35}$$

The controllability Gramian $\bar{\mathbf{W}}_c$ and observability Gramian $\bar{\mathbf{W}}_o$ of $(\bar{\mathbf{A}}, \bar{\mathbf{b}}, \bar{\mathbf{c}})$ are, respectively, the unique solutions of

$$\bar{\mathbf{A}}\bar{\mathbf{W}}_c + \bar{\mathbf{W}}_c\bar{\mathbf{A}}' = -\bar{\mathbf{b}}\bar{\mathbf{b}}' \tag{7.36}$$

and

$$\bar{\mathbf{A}}'\bar{\mathbf{W}}_o + \bar{\mathbf{W}}_o\bar{\mathbf{A}} = -\bar{\mathbf{c}}'\bar{\mathbf{c}} \tag{7.37}$$

Substituting $\bar{\mathbf{A}} = \mathbf{PAP}^{-1}$ and $\bar{\mathbf{b}} = \mathbf{Pb}$ into (7.36) yields

$$\mathbf{PAP}^{-1}\bar{\mathbf{W}}_c + \bar{\mathbf{W}}_c(\mathbf{P}')^{-1}\mathbf{A}'\mathbf{P}' = -\mathbf{Pbb}'\mathbf{P}'$$

which implies

$$\mathbf{AP}^{-1}\bar{\mathbf{W}}_c(\mathbf{P}')^{-1} + \mathbf{P}^{-1}\bar{\mathbf{W}}_c(\mathbf{P}')^{-1}\mathbf{A}' = -\mathbf{bb}'$$

Comparing this with (7.31) yields

$$\mathbf{W}_c = \mathbf{P}^{-1}\bar{\mathbf{W}}_c(\mathbf{P}')^{-1} \quad \text{or} \quad \bar{\mathbf{W}}_c = \mathbf{PW}_c\mathbf{P}' \tag{7.38}$$

Similarly, we can show

$$\mathbf{W}_o = \mathbf{P}'\bar{\mathbf{W}}_o\mathbf{P} \quad \text{or} \quad \bar{\mathbf{W}}_o = (\mathbf{P}')^{-1}\mathbf{W}_o\mathbf{P}^{-1} \tag{7.39}$$

Thus we have

$$\mathbf{W}_c\mathbf{W}_o = \mathbf{P}^{-1}\bar{\mathbf{W}}_c(\mathbf{P}')^{-1}\mathbf{P}'\bar{\mathbf{W}}_o\mathbf{P} = \mathbf{P}^{-1}\bar{\mathbf{W}}_c\bar{\mathbf{W}}_o\mathbf{P}$$

This shows that all $\mathbf{W}_c\mathbf{W}_o$ are similar and, consequently, have the same set of eigenvalues.

Next we show that all eigenvalues of $\mathbf{W}_c\mathbf{W}_o$ are real and positive. Note that both $\mathbf{W}_c$ and $\mathbf{W}_o$ are symmetric, but their product may not be. Therefore Theorem 3.6 is not directly applicable to $\mathbf{W}_c\mathbf{W}_o$. Now we apply Theorem 3.6 to $\mathbf{W}_c$:

$$\mathbf{W}_c = \mathbf{Q}'\mathbf{DQ} = \mathbf{Q}'\mathbf{D}^{1/2}\mathbf{D}^{1/2}\mathbf{Q} =: \mathbf{R}'\mathbf{R} \tag{7.40}$$

where $\mathbf{D}$ is a diagonal matrix with the eigenvalues of $\mathbf{W}_c$ on the diagonal. Because $\mathbf{W}_c$ is symmetric and positive definite, all its eigenvalues are real and positive. Thus we can express $\mathbf{D}$ as $\mathbf{D}^{1/2}\mathbf{D}^{1/2}$, where $\mathbf{D}^{1/2}$ is diagonal with positive square roots of the diagonal entries of $\mathbf{D}$ as its diagonal entries. Note that $\mathbf{Q}$ is orthogonal or $\mathbf{Q}^{-1} = \mathbf{Q}'$. The matrix $\mathbf{R} = \mathbf{D}^{1/2}\mathbf{Q}$ is not orthogonal but is nonsingular.

Consider $\mathbf{RW}_o\mathbf{R}'$; it is clearly symmetric and positive definite. Thus its eigenvalues are all real and positive. Using (7.40) and (3.66), we have

$$\det(\sigma^2\mathbf{I} - \mathbf{W}_c\mathbf{W}_o) = \det(\sigma^2\mathbf{I} - \mathbf{R}'\mathbf{R}\mathbf{W}_o) = \det(\sigma^2\mathbf{I} - \mathbf{R}\mathbf{W}_o\mathbf{R}') \tag{7.41}$$

which implies that $\mathbf{W}_c\mathbf{W}_o$ and $\mathbf{R}\mathbf{W}_o\mathbf{R}'$ have the same set of eigenvalues. Thus we conclude that all eigenvalues of $\mathbf{W}_c\mathbf{W}_o$ are real and positive. Q.E.D.

Let us define

$$\boldsymbol{\Sigma} = \mathrm{diag}(\sigma_1, \sigma_2, \ldots, \sigma_n) \tag{7.42}$$

where σ_i are positive square roots of the eigenvalues of $\mathbf{W}_c\mathbf{W}_o$. For convenience, we arrange them in descending order in magnitude or

$$\sigma_1 \geq \sigma_2 \geq \cdots \geq \sigma_n > 0$$

These eigenvalues are called the *Hankel singular values*. The product $\mathbf{W}_c\mathbf{W}_o$ of any minimal realization is similar to $\boldsymbol{\Sigma}^2$.

▶ **Theorem 7.6**

For any n-dimensional minimal state equation $(\mathbf{A}, \mathbf{b}, \mathbf{c})$, there exists an equivalence transformation $\bar{\mathbf{x}} = \mathbf{P}\mathbf{x}$ such that the controllability Gramian $\bar{\mathbf{W}}_c$ and observability Gramian $\bar{\mathbf{W}}_o$ of its equivalent state equation have the property

$$\bar{\mathbf{W}}_c = \bar{\mathbf{W}}_o = \boldsymbol{\Sigma} \tag{7.43}$$

This is called a *balanced realization*.

➡ *Proof:* We first compute $\mathbf{W}_c = \mathbf{R}'\mathbf{R}$ as in (7.40). We then apply Theorem 3.6 to the real and symmetric matrix $\mathbf{R}\mathbf{W}_o\mathbf{R}'$ to yield

$$\mathbf{R}\mathbf{W}_o\mathbf{R}' = \mathbf{U}\boldsymbol{\Sigma}^2\mathbf{U}'$$

where $\mathbf{U}$ is orthogonal or $\mathbf{U}'\mathbf{U} = \mathbf{I}$. Let

$$\mathbf{P}^{-1} = \mathbf{R}'\mathbf{U}\boldsymbol{\Sigma}^{-1/2} \quad \text{or} \quad \mathbf{P} = \boldsymbol{\Sigma}^{1/2}\mathbf{U}'(\mathbf{R}')^{-1}$$

Then (7.38) and $\mathbf{W}_c = \mathbf{R}'\mathbf{R}$ imply

$$\bar{\mathbf{W}}_c = \boldsymbol{\Sigma}^{1/2}\mathbf{U}'(\mathbf{R}')^{-1}\mathbf{W}_c\mathbf{R}^{-1}\mathbf{U}\boldsymbol{\Sigma}^{1/2} = \boldsymbol{\Sigma}$$

and (7.39) and $\mathbf{R}\mathbf{W}_o\mathbf{R}' = \mathbf{U}\boldsymbol{\Sigma}^2\mathbf{U}'$ imply

$$\bar{\mathbf{W}}_o = \boldsymbol{\Sigma}^{-1/2}\mathbf{U}'\mathbf{R}\mathbf{W}_o\mathbf{R}'\mathbf{U}\boldsymbol{\Sigma}^{-1/2} = \boldsymbol{\Sigma}$$

This establishes the theorem. Q.E.D.

By selecting a different $\mathbf{P}$, it is possible to find an equivalent state equation with $\bar{\mathbf{W}}_c = \mathbf{I}$ and $\bar{\mathbf{W}}_o = \boldsymbol{\Sigma}^2$. Such a state equation is called the *input-normal* realization. Similarly, we can have a state equation with $\bar{\mathbf{W}}_c = \boldsymbol{\Sigma}^2$ and $\bar{\mathbf{W}}_o = \mathbf{I}$, which is called the *output-normal* realization. The balanced realization in Theorem 7.5 can be used in system reduction. More specifically, suppose

$$\begin{bmatrix} \dot{\mathbf{x}}_1 \\ \dot{\mathbf{x}}_2 \end{bmatrix} = \begin{bmatrix} \mathbf{A}_{11} & \mathbf{A}_{12} \\ \mathbf{A}_{21} & \mathbf{A}_{22} \end{bmatrix} \begin{bmatrix} \mathbf{x}_1 \\ \mathbf{x}_2 \end{bmatrix} + \begin{bmatrix} \mathbf{b}_1 \\ \mathbf{b}_2 \end{bmatrix} u$$

$$y = [\mathbf{c}_1 \quad \mathbf{c}_2]\mathbf{x} \tag{7.44}$$

is a balanced minimal realization of a stable $\hat{g}(s)$ with

$$\mathbf{W}_c = \mathbf{W}_o = \text{diag}(\mathbf{\Sigma}_1, \mathbf{\Sigma}_2)$$

where the A-, b-, and c-matrices are partitioned according to the order of $\mathbf{\Sigma}_i$. If the Hankel singular values of $\mathbf{\Sigma}_1$ and $\mathbf{\Sigma}_2$ are disjoint, then the reduced state equation

$$\dot{\mathbf{x}}_1 = \mathbf{A}_{11}\mathbf{x}_1 + \mathbf{b}_1 u$$

$$y = \mathbf{c}_1 \mathbf{x}_1 \tag{7.45}$$

is balanced and $\mathbf{A}_{11}$ is stable. If the singular values of $\mathbf{\Sigma}_2$ are much smaller than those of $\mathbf{\Sigma}_1$, then the transfer function of (7.45) will be close to $\hat{g}(s)$. See Reference [23].

The MATLAB function `balreal` will transform $(\mathbf{A}, \mathbf{b}, \mathbf{c})$ into a balanced state equation. The reduced equation in (7.45) can be obtained by using `balred`. The results in this section are based on the controllability and observability Gramians. Because the Gramians in the MIMO case are square as in the SISO case; all results in this section apply to the MIMO case without any modification.

7.5 Realizations from Markov Parameters[4]

Consider the strictly proper rational function

$$\hat{g}(s) = \frac{\beta_1 s^{n-1} + \beta_2 s^{n-2} + \cdots + \beta_{n-1} s + \beta_n}{s^n + \alpha_1 s^{n-1} + \alpha_2 s^{s-2} + \cdots + \alpha_{n-1} s + \alpha_n} \tag{7.46}$$

We expand it into an infinite power series as

$$\hat{g}(s) = h(0) + h(1)s^{-1} + h(2)s^{-2} + \cdots \tag{7.47}$$

If $\hat{g}(s)$ is strictly proper as assumed in (7.46), then $h(0) = 0$. The coefficients $h(m)$, $m = 1, 2, \ldots$, are called *Markov parameters*. Let $g(t)$ be the inverse Laplace transform of $\hat{g}(s)$ or, equivalently, the impulse response of the system. Then we have

$$h(m) = \left. \frac{d^{m-1}}{dt^{m-1}} g(t) \right|_{t=0}$$

for $m = 1, 2, 3, \ldots$. This method of computing Markov parameters is impractical because it requires repetitive differentiations, and differentiations are susceptible to noise.[5] Equating (7.46) and (7.47) yields

---

4. This section may be skipped without loss of continuity.

5. In the discrete-time case, if we apply an impulse sequence to a system, then the output sequence directly yields Markov parameters. Thus Markov parameters can easily be generated in discrete-time systems.

$$\beta_1 s^{n-1} + \beta_2 s^{n-2} + \cdots + \beta_n$$
$$= (s^n + \alpha_1 s^{n-1} + \alpha_2 s^{n-2} + \cdots + \alpha_n)(h(1)s^{-1} + h(2)s^{-2} + \cdots)$$

From this equation, we can obtain the Markov parameters recursively as

$$h(1) = \beta_1$$
$$h(2) = -\alpha_1 h(1) + \beta_2$$
$$h(3) = -\alpha_1 h(2) - \alpha_2 h(1) + \beta_3$$
$$\vdots$$
$$h(n) = -\alpha_1 h(n-1) - \alpha_2 h(n-2) - \cdots - \alpha_{n-1}h(1) + \beta_n \tag{7.48}$$
$$h(m) = -\alpha_1 h(m-1) - \alpha_2 h(m-2) - \cdots - \alpha_{n-1}h(m-n+1)$$
$$- \alpha_n h(m-n) \tag{7.49}$$

for $m = n + 1, \ n + 2, \ \ldots$.

Next we use the Markov parameters to form the $\alpha \times \beta$ matrix

$$\mathbf{T}(\alpha, \beta) = \begin{bmatrix} h(1) & h(2) & h(3) & \cdots & h(\beta) \\ h(2) & h(3) & h(4) & \cdots & h(\beta+1) \\ h(3) & h(4) & h(5) & \cdots & h(\beta+2) \\ \vdots & \vdots & \vdots & \ddots & \vdots \\ h(\alpha) & h(\alpha+1) & h(\alpha+2) & \cdots & h(\alpha+\beta-1) \end{bmatrix} \tag{7.50}$$

It is called a *Hankel matrix*. It is important to mention that even if $h(0) \neq 0$, $h(0)$ does not appear in the Hankel matrix.

▶ **Theorem 7.7**

A strictly proper rational function $\hat{g}(s)$ has degree n if and only if

$$\rho \mathbf{T}(n, n) = \rho \mathbf{T}(n+k, n+l) = n \quad \text{for every } k, l = 1, 2, \ldots \tag{7.51}$$

where ρ denotes the rank.

⇒   ***Proof:*** We first show that if $\deg \hat{g}(s) = n$, then $\rho \mathbf{T}(n, n) = \rho \mathbf{T}(n+1, n) = \rho \mathbf{T}(\infty, n)$. If $\deg \hat{g}(s) = n$, then (7.49) holds, and n is the smallest integer having the property. Because of (7.49), the $(n+1)$th row of $\mathbf{T}(n+1, n)$ can be written as a linear combination of the first n rows. Thus we have $\rho \mathbf{T}(n, n) = \rho \mathbf{T}(n+1, n)$. Again, because of (7.49), the $(n+2)$th row of $\mathbf{T}(n+2, n)$ depends on its previous n rows and, consequently, on the first n rows. Proceeding forward, we can establish $\rho \mathbf{T}(n, n) = \rho \mathbf{T}(\infty, n)$. Now we claim $\rho \mathbf{T}(\infty, n) = n$. If not, there would be an integer $\bar{n} < n$ having the property (7.49). This contradicts the hypothesis that $\deg \hat{g}(s) = n$. Thus we have $\rho \mathbf{T}(n, n) = \rho \mathbf{T}(\infty, n) = n$. Applying (7.49) to the columns of $\mathbf{T}$ yields (7.51).

Now we show that if (7.51) holds, then $\hat{g}(s) = h(1)s^{-1} + h(2)s^{-2} + \cdots$ can be expressed as a strictly proper rational function of degree n. From the condition

$\rho \mathbf{T}(n+1, \infty) = \rho \mathbf{T}(n, \infty) = n$, we can compute $\{\alpha_i, \ i = 1, 2, \ldots, n\}$ to meet (7.49). We then use (7.48) to compute $\{\beta_i, \ i = 1, 2, \ldots, n\}$. Hence we have

$$\hat{g}(s) = h(1)s^{-1} + h(2)s^{-2} + h(3)s^{-3} \cdots$$
$$= \frac{\beta_1 s^{n-1} + \beta_2 s^{n-2} + \cdots + \beta_{n-1}s + \beta_n}{s^n + \alpha_1 s^{n-1} + \alpha_2 s^{s-2} + \cdots + \alpha_{n-1}s + \alpha_n}$$

Because the n is the smallest integer having the property in (7.51), we have deg $\hat{g}(s) = n$. This completes the proof of the theorem.   Q.E.D.

With this preliminary, we are ready to discuss the realization problem. Consider a strictly proper transfer function $\hat{g}(s)$ expressed as

$$\hat{g}(s) = h(1)s^{-1} + h(2)s^{-2} + h(3)s^{-3} + \cdots$$

If the triplet $(\mathbf{A}, \mathbf{b}, \mathbf{c})$ is a realization of $\hat{g}(s)$, then

$$\hat{g}(s) = \mathbf{c}(s\mathbf{I} - \mathbf{A})^{-1}\mathbf{b} = \mathbf{c}[s(\mathbf{I} - s^{-1}\mathbf{A})]^{-1}\mathbf{b}$$

which becomes, using (3.57),

$$\hat{g}(s) = \mathbf{cb}s^{-1} + \mathbf{cAb}s^{-2} + \mathbf{cA}^2\mathbf{b}s^{-3} + \cdots$$

Thus we conclude that $(\mathbf{A}, \mathbf{b}, \mathbf{c})$ is a realization of $\hat{g}(s)$ if and only if

$$h(m) = \mathbf{cA}^{m-1}\mathbf{b} \qquad \text{for } m = 1, \ 2, \ \ldots \tag{7.52}$$

Substituting (7.52) into the Hankel matrix $\mathbf{T}(n, n)$ yields

$$\mathbf{T}(n, n) = \begin{bmatrix} \mathbf{cb} & \mathbf{cAb} & \mathbf{cA}^2\mathbf{b} & \cdots & \mathbf{cA}^{n-1}\mathbf{b} \\ \mathbf{cAb} & \mathbf{cA}^2\mathbf{b} & \mathbf{cA}^3\mathbf{b} & \cdots & \mathbf{cA}^n\mathbf{b} \\ \mathbf{cA}^2\mathbf{b} & \mathbf{cA}^3\mathbf{b} & \mathbf{cA}^4\mathbf{b} & \cdots & \mathbf{cA}^{n+1}\mathbf{b} \\ \vdots & \vdots & \vdots & \ddots & \vdots \\ \mathbf{cA}^{n-1}\mathbf{b} & \mathbf{cA}^n\mathbf{b} & \mathbf{cA}^{n+1}\mathbf{b} & \cdots & \mathbf{cA}^{2(n-1)}\mathbf{b} \end{bmatrix}$$

which implies, as shown in (7.21),

$$\mathbf{T}(n, n) = OC \tag{7.53}$$

where O and C are, respectively, the $n \times n$ observability and controllability matrices of $(\mathbf{A}, \mathbf{b}, \mathbf{c})$. Define

$$\tilde{\mathbf{T}}(n, n) = \begin{bmatrix} h(2) & h(3) & h(4) & \cdots & h(n+1) \\ h(3) & h(4) & h(5) & \cdots & h(n+2) \\ h(4) & h(5) & h(6) & \cdots & h(n+3) \\ \vdots & \vdots & \vdots & \ddots & \vdots \\ h(n+1) & h(n+2) & h(n+3) & \cdots & h(2n) \end{bmatrix} \tag{7.54}$$

It is the submatrix of $\mathbf{T}(n+1, n)$ by deleting the first row or the submatrix of $\mathbf{T}(n, n+1)$ by deleting the first column. Then as with (7.53), we can readily show

$$\tilde{\mathbf{T}}(n, n) = OAC \tag{7.55}$$

Using (7.53) and (7.55), we can obtain many different realizations. We discuss here only a companion-form and a balanced-form realization.

Companion form There are many ways to decompose $\mathbf{T}(n, n)$ into OC. The simplest is to select $O = \mathbf{I}$ or $C = \mathbf{I}$. If we select $O = \mathbf{I}$, then (7.53) and (7.55) imply $C = \mathbf{T}(n, n)$ and $\mathbf{A} = \tilde{\mathbf{T}}(n, n)\mathbf{T}^{-1}(n, n)$. The state equation corresponding to $O = \mathbf{I}$, $C = \mathbf{T}(n, n)$, and $\mathbf{A} = \tilde{\mathbf{T}}(n, n)\mathbf{T}^{-1}(n, n)$ is

$$\dot{\mathbf{x}} = \begin{bmatrix} 0 & 1 & 0 & \cdots & 0 & 0 \\ 0 & 0 & 1 & \cdots & 0 & 0 \\ \vdots & \vdots & \vdots & \ddots & \vdots & \vdots \\ 0 & 0 & 0 & \cdots & 0 & 1 \\ -\alpha_n & -\alpha_{n-1} & -\alpha_{n-2} & \cdots & -\alpha_2 & -\alpha_1 \end{bmatrix} \mathbf{x} + \begin{bmatrix} h(1) \\ h(2) \\ \vdots \\ h(n-1) \\ h(n) \end{bmatrix} u$$

$$y = [1 \quad 0 \quad 0 \quad \cdots \quad 0 \quad 0]\mathbf{x} \tag{7.56}$$

Indeed, the first row of $O = \mathbf{I}$ and the first column of $C = \mathbf{T}(n, n)$ yield the $\mathbf{c}$ and $\mathbf{b}$ in (7.56). Instead of showing $\mathbf{A} = \tilde{\mathbf{T}}(n, n)\mathbf{T}^{-1}(n, n)$, we show

$$\mathbf{A}\mathbf{T}(n, n) = \tilde{\mathbf{T}}(n, n) \tag{7.57}$$

Using the shifting property of the companion-form matrix in (7.56), we can readily verify

$$\mathbf{A}\begin{bmatrix} h(1) \\ h(2) \\ \vdots \\ h(n) \end{bmatrix} = \begin{bmatrix} h(2) \\ h(3) \\ \vdots \\ h(n+1) \end{bmatrix}, \quad \mathbf{A}\begin{bmatrix} h(2) \\ h(3) \\ \vdots \\ h(n+1) \end{bmatrix} = \begin{bmatrix} h(3) \\ h(4) \\ \vdots \\ h(n+2) \end{bmatrix}, \quad \cdots \tag{7.58}$$

We see that the Markov parameters of a column are shifted up one position if the column is premultiplied by $\mathbf{A}$. Using this property, we can readily establish (7.57). Thus $O = \mathbf{I}$, $C = \mathbf{T}(n, n)$, and $\mathbf{A} = \tilde{\mathbf{T}}(n, n)\mathbf{T}^{-1}(n, n)$ generate the realization in (7.56). It is a companion-form realization. Now we use (7.52) to show that (7.56) is indeed a realization. Because of the form of $\mathbf{c}$, $\mathbf{c}\mathbf{A}^m\mathbf{b}$ equals simply the top entry of $\mathbf{A}^m\mathbf{b}$ or

$$\mathbf{c}\mathbf{b} = h(1), \quad \mathbf{c}\mathbf{A}\mathbf{b} = h(2), \quad \mathbf{c}\mathbf{A}^2\mathbf{b} = h(3), \quad \cdots$$

Thus (7.56) is a realization of $\hat{g}(s)$. The state equation is always observable because $O = \mathbf{I}$ has full rank. It is controllable if $C = \mathbf{T}(n, n)$ has rank n.

EXAMPLE 7.2 Consider

$$\hat{g}(s) = \frac{4s^2 - 2s - 6}{2s^4 + 2s^3 + 2s^2 + 3s + 1}$$

$$= 0 \cdot s^{-1} + 2s^{-2} - 3s^{-3} - 2s^{-4} + 2s^{-5} + 3.5s^{-6} + \cdots \tag{7.59}$$

We form $\mathbf{T}(4, 4)$ and compute its rank. The rank is 3; thus $\hat{g}(s)$ in (7.59) has degree 3 and its numerator and denominator have a common factor of degree 1. There is no need to cancel first the common factor in the expansion in (7.59). From the preceding derivation, we have

$$
\mathbf{A} = \begin{bmatrix} 2 & -3 & -2 \\ -3 & -2 & 2 \\ -2 & 2 & 3.5 \end{bmatrix} \begin{bmatrix} 0 & 2 & -3 \\ 2 & -3 & -2 \\ -3 & -2 & 2 \end{bmatrix}^{-1} = \begin{bmatrix} 0 & 1 & 0 \\ 0 & 0 & 1 \\ -0.5 & -1 & 0 \end{bmatrix}
\tag{7.60}
$$

and

$$
\mathbf{b} = [0 \ 2 \ -3]' \qquad \mathbf{c} = [1 \ 0 \ 0]
$$

The triplet $(\mathbf{A}, \mathbf{b}, \mathbf{c})$ is a minimal realization of $\hat{g}(s)$ in (7.59).

We mention that the matrix $\mathbf{A}$ in (7.60) can be obtained without computing $\tilde{\mathbf{T}}(n, n)$ $\mathbf{T}^{-1}(n, n)$. Using (7.49) we can show

$$
\mathbf{T}(3, 4)\mathbf{a} := \begin{bmatrix} 0 & 2 & -3 & -2 \\ 2 & -3 & -2 & 2 \\ -3 & -2 & 2 & 3.5 \end{bmatrix} \begin{bmatrix} \alpha_3 \\ \alpha_2 \\ \alpha_1 \\ 1 \end{bmatrix} = \mathbf{0}
$$

Thus $\mathbf{a}$ is a null vector of $\mathbf{T}(3, 4)$. The MATLAB function

```
t=[0 2 -3 -2;2 -3 -2 2;-3 -2 2 3.5];a=null(t)
```

yields a=$[-0.3333 \ -0.6667 \ 0.0000 \ -0.6667]'$. We normalize the last entry of a to 1 by typing a/a(4), which yields $[0.5 \ 1 \ 0 \ 1]'$. The first three entries, with sign reversed, are the last row of $\mathbf{A}$.

Balanced form   Next we discuss a different decomposition of $\mathbf{T}(n, n) = OC$, which will yield a realization with the property

$$
CC' = O'O
$$

First we use singular-value decomposition to express $\mathbf{T}(n, n)$ as

$$
\mathbf{T}(n, n) = \mathbf{K}\Lambda\mathbf{L}' = \mathbf{K}\Lambda^{1/2}\Lambda^{1/2}\mathbf{L}'
\tag{7.61}
$$

where $\mathbf{K}$ and $\mathbf{L}$ are orthogonal matrices and $\Lambda^{1/2}$ is diagonal with the singular values of $\mathbf{T}(n, n)$ on the diagonal. Let us select

$$
O = \mathbf{K}\Lambda^{1/2} \quad \text{and} \quad C = \Lambda^{1/2}\mathbf{L}'
\tag{7.62}
$$

Then we have

$$
O^{-1} = \Lambda^{-1/2}\mathbf{K}' \quad \text{and} \quad C^{-1} = \mathbf{L}\Lambda^{-1/2}
\tag{7.63}
$$

For this selection of C and O, the triplet

$$
\mathbf{A} = O^{-1}\tilde{\mathbf{T}}(n, n)C^{-1}
\tag{7.64}
$$

$$
\mathbf{b} = \text{first column of } C
\tag{7.65}
$$

$$
\mathbf{c} = \text{first row of } O
\tag{7.66}
$$

is a minimal realization of $\hat{g}(s)$. For this realization, we have

$$CC' = \Lambda^{1/2}\mathbf{L}'\mathbf{L}\Lambda^{1/2} = \Lambda$$

and

$$O'O = \Lambda^{1/2}\mathbf{K}'\mathbf{K}\Lambda^{1/2} = \Lambda = CC'$$

Thus it is called a *balanced realization*. This balanced realization is different from the balanced realization discussed in Section 7.4. It is not clear what the relationships between them are.

EXAMPLE 7.3 Consider the transfer function in Example 7.2. Now we will find a balanced realization from Hankel matrices. We type

```
t=[0 2 -3;2 -3 -2;-3 -2 2];tt=[2 -3 -2;-3 -2 2;-2 2 3.5];
[k,s,l]=svd(t);
s1=sqrt(s);
O=k*s1;C=s1*l';
a=inv(O)*tt*inv(C),
b=[C(1,1);C(2,1);C(3,1)],c=[O(1,1) O(1,2) O(1,3)]
```

This yields the following balanced realization:

$$\dot{\mathbf{x}} = \begin{bmatrix} 0.4003 & -1.0024 & -0.4805 \\ 1.0024 & -0.3121 & 0.3209 \\ 0.4805 & 0.3209 & -0.0882 \end{bmatrix}\mathbf{x} + \begin{bmatrix} 1.2883 \\ -0.7303 \\ 1.0614 \end{bmatrix}u$$

$$y = [1.2883 \; 0.7303 \; -1.0614]\mathbf{x} + 0 \cdot u$$

To check the correctness of this result, we type `[n,d]=ss2tf(a,b,c,0)`, which yields

$$\hat{g}(s) = \frac{2s - 3}{s^3 + s + 0.5}$$

This equals $\hat{g}(s)$ in (7.59) after canceling the common factor $2(s + 1)$.

7.6 Degree of Transfer Matrices

This section will extend the concept of degree for scalar rational functions to rational matrices. Given a proper rational matrix $\hat{\mathbf{G}}(s)$, it is assumed that every entry of $\hat{\mathbf{G}}(s)$ is a coprime fraction; that is, its numerator and denominator have no common factors. This will be a standing assumption throughout the remainder of this text.

> **Definition 7.1** The characteristic polynomial *of a proper rational matrix* $\hat{\mathbf{G}}(s)$ *is defined as the least common denominator of all minors of* $\hat{\mathbf{G}}(s)$. *The degree of the characteristic polynomial is defined as the* McMillan degree *or, simply, the* degree *of* $\hat{\mathbf{G}}(s)$ *and is denoted by* $\delta\hat{\mathbf{G}}(s)$.

EXAMPLE 7.4 Consider the rational matrices

$$\hat{\mathbf{G}}_1(s) = \begin{bmatrix} \frac{1}{s+1} & \frac{1}{s+1} \\ \frac{1}{s+1} & \frac{1}{s+1} \end{bmatrix} \qquad \hat{\mathbf{G}}_2(s) = \begin{bmatrix} \frac{2}{s+1} & \frac{1}{s+1} \\ \frac{1}{s+1} & \frac{1}{s+1} \end{bmatrix}$$

The matrix $\hat{\mathbf{G}}_1(s)$ has $1/(s+1)$, $1/(s+1)$, $1/(s+1)$, and $1/(s+1)$ as its minors of order 1, and $\det \hat{\mathbf{G}}_1(s) = 0$ as its minor of order 2. Thus the characteristic polynomial of $\hat{\mathbf{G}}_1(s)$ is $s+1$ and $\delta\hat{\mathbf{G}}_1(s) = 1$. The matrix $\hat{\mathbf{G}}_2(s)$ has $2/(s+1)$, $1/(s+1)$, $1/(s+1)$, and $1/(s+1)$ as its minors of order 1, and $\det \hat{\mathbf{G}}_2(s) = 1/(s+1)^2$ as its minor of order 2. Thus the characteristic polynomial of $\hat{\mathbf{G}}_2(s)$ is $(s+1)^2$ and $\delta\hat{\mathbf{G}}_2(s) = 2$.

From this example, we see that the characteristic polynomial of $\hat{\mathbf{G}}(s)$ is, in general, different from the denominator of the determinant of $\hat{\mathbf{G}}(s)$ [if $\hat{\mathbf{G}}(s)$ is square] and different from the least common denominator of all entries of $\hat{\mathbf{G}}(s)$.

EXAMPLE 7.5 Consider the 2×3 rational matrix

$$\hat{\mathbf{G}}(s) = \begin{bmatrix} \dfrac{s}{s+1} & \dfrac{1}{(s+1)(s+2)} & \dfrac{1}{s+3} \\ \dfrac{-1}{s+1} & \dfrac{1}{(s+1)(s+2)} & \dfrac{1}{s} \end{bmatrix}$$

Its minors of order 1 are the six entries of $\hat{\mathbf{G}}(s)$. The matrix has the following three minors of order 2:

$$\frac{s}{(s+1)^2(s+2)} + \frac{1}{(s+1)^2(s+2)} = \frac{s+1}{(s+1)^2(s+2)} = \frac{1}{(s+1)(s+2)}$$

$$\frac{s}{s+1} \cdot \frac{1}{s} + \frac{1}{(s+1)(s+3)} = \frac{s+4}{(s+1)(s+3)}$$

$$\frac{1}{(s+1)(s+2)s} - \frac{1}{(s+1)(s+2)(s+3)} = \frac{3}{s(s+1)(s+2)(s+3)}$$

The least common denominator of all these minors is $s(s+1)(s+2)(s+3)$. Thus the degree of $\hat{\mathbf{G}}(s)$ is 4. Note that $\hat{\mathbf{G}}(s)$ has no minors of order 3 or higher.

In computing the characteristic polynomial, every minor must be reduced to a coprime fraction as we did in the preceding example; otherwise, we will obtain an erroneous result. We discuss two special cases. If $\hat{\mathbf{G}}(s)$ is $1 \times p$ or $q \times 1$, then there are no minors of order 2 or higher. Thus the characteristic polynomial equals the least common denominator of all entries of $\hat{\mathbf{G}}(s)$. In particular, if $\hat{\mathbf{G}}(s)$ is scalar, then the characteristic polynomial equals its denominator. If every entry of $q \times p$ $\hat{\mathbf{G}}(s)$ has poles that differ from those of all other entries, such as

$$\hat{\mathbf{G}}(s) = \begin{bmatrix} \dfrac{1}{(s+1)^2(s+2)} & \dfrac{s+2}{s^2} \\ \dfrac{s-2}{s+3} & \dfrac{s}{(s+5)(s-3)} \end{bmatrix}$$

then its minors contain no poles with multiplicities higher than those of each entry. Thus the characteristic polynomial equals the product of the denominators of all entries of $\hat{\mathbf{G}}(s)$.

To conclude this section, we mention two important properties. Let $(\mathbf{A}, \mathbf{B}, \mathbf{C}, \mathbf{D})$ be a controllable and observable realization of $\hat{\mathbf{G}}(s)$. Then we have the following:

- Monic least common denominator of all *minors* of $\hat{\mathbf{G}}(s)$ = characteristic polynomial of $\mathbf{A}$.
- Monic least common denominator of all *entries* of $\hat{\mathbf{G}}(s)$ = minimal polynomial of $\mathbf{A}$.

For their proofs, see Reference [4, pp. 302–304].

7.7 Minimal Realizations—Matrix Case

We introduced in Section 7.2.1 minimal realizations for scalar transfer functions. Now we discuss the matrix case.

▶ **Theorem 7.M2**

A state equation $(\mathbf{A}, \mathbf{B}, \mathbf{C}, \mathbf{D})$ is a minimal realization of a proper rational matrix $\hat{\mathbf{G}}(s)$ if and only if $(\mathbf{A}, \mathbf{B})$ is controllable and $(\mathbf{A}, \mathbf{C})$ is observable or if and only if

$$\dim \mathbf{A} = \deg \hat{\mathbf{G}}(s)$$

 Proof: The proof of the first part is similar to the proof of Theorem 7.2. If $(\mathbf{A}, \mathbf{B})$ is not controllable or if $(\mathbf{A}, \mathbf{C})$ is not observable, then the state equation is zero-state equivalent to a lesser-dimensional state equation and is not minimal. If $(\mathbf{A}, \mathbf{B}, \mathbf{C}, \mathbf{D})$ is of dimension n and is controllable and observable, and if the $\bar{n}$-dimensional state equation $(\bar{\mathbf{A}}, \bar{\mathbf{B}}, \bar{\mathbf{C}}, \bar{\mathbf{D}})$, with $\bar{n} < n$, is a realization of $\hat{\mathbf{G}}(s)$, then Theorem 4.1 implies $\mathbf{D} = \bar{\mathbf{D}}$ and

$$\mathbf{C}\mathbf{A}^m\mathbf{B} = \bar{\mathbf{C}}\bar{\mathbf{A}}^m\bar{\mathbf{B}} \qquad \text{for } m = 0, 1, 2, \ldots$$

Thus, as in (7.22), we have

$$OC = \bar{O}_n \bar{C}_n$$

Note that O, C, $\bar{O}_n$, and $\bar{C}_n$ are, respectively, $nq \times n$, $n \times np$, $nq \times \bar{n}$, and $\bar{n} \times np$ matrices. Using the Sylvester inequality

$$\rho(O) + \rho(C) - n \leq \rho(OC) \leq \min(\rho(O), \rho(C))$$

which is proved in Reference [6, p. 31], and $\rho(O) = \rho(C) = n$, we have $\rho(OC) = n$. Similarly, we have $\rho(\bar{O}_n \bar{C}_n) = \bar{n} < n$. This contradicts $\rho(OC) = \rho(\bar{O}_n C_n)$. Thus every controllable and observable state equation is a minimal realization.

To show that $(\mathbf{A}, \mathbf{B}, \mathbf{C}, \mathbf{D})$ is minimal if and only if $\dim \mathbf{A} = \deg \hat{\mathbf{G}}(s)$ is much more complex and will be established in the remainder of this chapter.   Q.E.D.

▶ **Theorem 7.M3**

All minimal realizations of $\hat{\mathbf{G}}(s)$ are equivalent.

Proof: The proof follows closely the proof of Theorem 7.3. Let $(\mathbf{A}, \mathbf{B}, \mathbf{C}, \mathbf{D})$ and $(\bar{\mathbf{A}}, \bar{\mathbf{B}}, \bar{\mathbf{C}}, \bar{\mathbf{D}})$ be any two n-dimensional minimal realizations of a $q \times p$ proper rational matrix $\hat{\mathbf{G}}(s)$. As in (7.23) and (7.24), we have

$$OC = \bar{O}\bar{C} \tag{7.67}$$

and

$$OAC = \bar{O}\bar{A}\bar{C} \tag{7.68}$$

In the scalar case, O, C, $\bar{O}$, and $\bar{C}$ are all $n \times n$ nonsingular matrices and their inverses are well defined. Here O and $\bar{O}$ are $nq \times n$ matrices of rank n; C and $\bar{C}$ are $n \times np$ matrices of rank n. They are not square and their inverses are not defined. Let us define the $n \times nq$ matrix

$$O^+ := (O'O)^{-1}O' \tag{7.69}$$

Because O' is $n \times nq$ and O is $nq \times n$, the matrix $O'O$ is $n \times n$ and is, following Theorem 3.8, nonsingular. Clearly, we have

$$O^+O = (O'O)^{-1}O'O = \mathbf{I}$$

Thus O^+ is called the *pseudoinverse* or *left inverse* of O. Note that OO^+ is $nq \times nq$ and does not equal a unit matrix. Similarly, we define

$$C^+ := C'(CC')^{-1} \tag{7.70}$$

It is an $np \times n$ matrix and has the property

$$CC^+ = CC'(CC')^{-1} = \mathbf{I}$$

Thus C^+ is called the *pseudoinverse* or *right inverse* of C. In the scalar case, the equivalence transformation is defined in (7.25) as $\mathbf{P} = \bar{O}^{-1}O = \bar{C}C^{-1}$. Now we replace inverses by pseudoinverses to yield

$$\mathbf{P} := \bar{O}^+O = (\bar{O}'\bar{O})^{-1}\bar{O}'O \tag{7.71}$$

$$= \bar{C}C^+ = \bar{C}C'(CC')^{-1} \tag{7.72}$$

This equality can be verified directly by premultiplying $(\bar{O}'\bar{O})$ and postmultiplying (CC') and then using (7.67). The inverse of $\mathbf{P}$ in the scalar case is $\mathbf{P}^{-1} = O^{-1}\bar{O} = C\bar{C}^{-1}$. In the matrix case, it becomes

$$\mathbf{P}^{-1} := O^+\bar{O} = (O'O)^{-1}O'\bar{O} \tag{7.73}$$

$$= C\bar{C}^+ = C\bar{C}'(\bar{C}\bar{C}')^{-1} \tag{7.74}$$

This again can be verified using (7.67). From $\bar{O}\bar{C} = OC$, we have

$$\bar{C} = (\bar{O}'\bar{O})^{-1}\bar{O}'OC = \mathbf{P}C$$

$$\bar{O} = OC\bar{C}'(\bar{C}\bar{C}')^{-1} = O\mathbf{P}^{-1}$$

Their first p columns and first q rows are $\bar{\mathbf{B}} = \mathbf{P}\mathbf{B}$ and $\bar{\mathbf{C}} = \mathbf{C}\mathbf{P}^{-1}$. The equation $\bar{O}\bar{A}\bar{C} = OAC$ implies

$$\bar{A} = (\bar{O}'\bar{O})^{-1}\bar{O}'OAC\bar{C}'(\bar{C}\bar{C}')^{-1} = PAP^{-1}$$

This shows that all minimal realizations of the same transfer matrix are equivalent. Q.E.D.

We see from the proof of Theorem 7.M3 that the results in the scalar case can be extended directly to the matrix case if inverses are replaced by pseudoinverses. In MATLAB, the function `pinv` generates the pseudoinverse. We mention that minimal realization can be obtained from nonminimal realizations by applying Theorems 6.6 and 6.O6 or by calling the MATLAB function `minreal`, as the next example illustrates.

EXAMPLE 7.6 Consider the transfer matrix in Example 4.6 or

$$\hat{G}(s) = \begin{bmatrix} \dfrac{4s-10}{2s+1} & \dfrac{3}{s+2} \\ \dfrac{1}{(2s+1)(s+2)} & \dfrac{1}{(s+2)^2} \end{bmatrix} \tag{7.75}$$

Its characteristic polynomial can be computed as $(2s+1)(s+2)^2$. Thus the rational matrix has degree 3. Its six-dimensional realization in (4.39) and four-dimensional realization in (4.44) are clearly not minimal realizations. They can be reduced to minimal realizations by calling the MATLAB function `minreal`. For example, for the realization in (4.39) typing

```
a=[-4.5 0 -6 0 -2 0;0 -4.5 0 -6 0 -2;1 0 0 0 0 0;...
   0 1 0 0 0 0;0 0 1 0 0 0;0 0 0 1 0 0];
b=[1 0;0 1;0 0;0 0;0 0;0 0];
c=[-6 3 -24 7.5 -24 3;0 1 0.5 1.5 1 0.5];d=[2 0;0 0];
[am,bm,cm,dm]=minreal(a,b,c,d)
```

yields

$$\dot{x} = \begin{bmatrix} -0.8625 & -4.0897 & 3.2544 \\ 0.2921 & -3.0508 & 1.2709 \\ -0.0944 & 0.3377 & -0.5867 \end{bmatrix} x + \begin{bmatrix} 0.3218 & -0.5305 \\ 0.0459 & -0.4983 \\ -0.1688 & 0.0840 \end{bmatrix} u$$

$$y = \begin{bmatrix} 0 & -0.0339 & 35.5281 \\ 0 & -2.1031 & -0.5720 \end{bmatrix} x + \begin{bmatrix} 2 & 0 \\ 0 & 0 \end{bmatrix} u$$

Its dimension equals the degree of $\hat{G}(s)$; thus it is controllable and observable and is a minimal realization of $\hat{G}(s)$ in (7.75).

7.8 Matrix Polynomial Fractions

The degree of the scalar transfer function

$$\hat{g}(s) = \frac{N(s)}{D(s)} = N(s)D^{-1}(s) = D^{-1}(s)N(s)$$

is defined as the degree of $D(s)$ if $N(s)$ and $D(s)$ are coprime. It can also be defined as the smallest possible denominator degree. In this section, we shall develop similar results for transfer matrices. Because matrices are not commutative, their orders cannot be altered in our discussion.

Every $q \times p$ proper rational matrix $\hat{\mathbf{G}}(s)$ can be expressed as

$$\hat{\mathbf{G}}(s) = \mathbf{N}(s)\mathbf{D}^{-1}(s) \tag{7.76}$$

where $\mathbf{N}(s)$ and $\mathbf{D}(s)$ are $q \times p$ and $p \times p$ polynomial matrices. For example, the 2×3 rational matrix in Example 7.5 can be expressed as

$$\hat{\mathbf{G}}(s) = \begin{bmatrix} s & 1 & s \\ -1 & 1 & s+3 \end{bmatrix} \begin{bmatrix} s+1 & 0 & 0 \\ 0 & (s+1)(s+2) & 0 \\ 0 & 0 & s(s+3) \end{bmatrix}^{-1} \tag{7.77}$$

The three diagonal entries of $\mathbf{D}(s)$ in (7.77) are the least common denominators of the three columns of $\hat{\mathbf{G}}(s)$. The fraction in (7.76) or (7.77) is called a *right polynomial fraction* or, simply, *right fraction*. Dual to (7.76), the expression

$$\hat{\mathbf{G}}(s) = \bar{\mathbf{D}}^{-1}(s)\bar{\mathbf{N}}(s)$$

where $\bar{\mathbf{D}}(s)$ and $\bar{\mathbf{N}}(s)$ are $q \times q$ and $q \times p$ polynomial matrices, is called a *left polynomial fraction* or, simply, *left fraction*.

Let $\mathbf{R}(s)$ be any $p \times p$ nonsingular polynomial matrix. Then we have

$$\hat{\mathbf{G}}(s) = [\mathbf{N}(s)\mathbf{R}(s)][\mathbf{D}(s)\mathbf{R}(s)]^{-1}$$
$$= \mathbf{N}(s)\mathbf{R}(s)\mathbf{R}^{-1}(s)\mathbf{D}^{-1}(s) = \mathbf{N}(s)\mathbf{D}^{-1}(s)$$

Thus right fractions are not unique. The same holds for left fractions. We introduce in the following right coprime fractions.

Consider $\mathbf{A}(s) = \mathbf{B}(s)\mathbf{C}(s)$, where $\mathbf{A}(s)$, $\mathbf{B}(s)$, and $\mathbf{C}(s)$ are polynomials of compatible orders. We call $\mathbf{C}(s)$ a *right divisor* of $\mathbf{A}(s)$ and $\mathbf{A}(s)$ a *left multiple* of $\mathbf{C}(s)$. Similarly, we call $\mathbf{B}(s)$ a *left divisor* of $\mathbf{A}(s)$ and $\mathbf{A}(s)$ a *right multiple* of $\mathbf{B}(s)$.

Consider two polynomial matrices $\mathbf{D}(s)$ and $\mathbf{N}(s)$ with the same number of columns p. A $p \times p$ square polynomial matrix $\mathbf{R}(s)$ is called a common *right* divisor of $\mathbf{D}(s)$ and $\mathbf{N}(s)$ if there exist polynomial matrices $\hat{\mathbf{D}}(s)$ and $\hat{\mathbf{N}}(s)$ such that

$$\mathbf{D}(s) = \hat{\mathbf{D}}(s)\mathbf{R}(s) \quad \text{and} \quad \mathbf{N}(s) = \hat{\mathbf{N}}(s)\mathbf{R}(s)$$

Note that $\mathbf{D}(s)$ and $\mathbf{N}(s)$ can have different numbers of rows.

Definition 7.2 *A square polynomial matrix* $\mathbf{M}(s)$ *is called a* unimodular matrix *if its determinant is nonzero and independent of s.*

The following polynomial matrices are all unimodular matrices:

$$\begin{bmatrix} 2s & s^2+s+1 \\ 2 & s+1 \end{bmatrix}, \quad \begin{bmatrix} -2 & s^{10}+s+1 \\ 0 & 3 \end{bmatrix}, \quad \begin{bmatrix} s & s+1 \\ s-1 & s \end{bmatrix}$$

Products of unimodular matrices are clearly unimodular matrices. Consider

$$\det \mathbf{M}(s)\det \mathbf{M}^{-1}(s) = \det [\mathbf{M}(s)\mathbf{M}^{-1}(s)] = \det \mathbf{I} = 1$$

which implies that if the determinant of $\mathbf{M}(s)$ is a nonzero constant, so is the determinant of $\mathbf{M}^{-1}(s)$. Thus the inverse of a unimodular matrix $\mathbf{M}(s)$ is again a unimodular matrix.

Definition 7.3 *A square polynomial matrix* $\mathbf{R}(s)$ *is a* greatest common right divisor *(gcrd) of* $\mathbf{D}(s)$ *and* $\mathbf{N}(s)$ *if (i)* $\mathbf{R}(s)$ *is a common right divisor of* $\mathbf{D}(s)$ *and* $\mathbf{N}(s)$ *and (ii)* $\mathbf{R}(s)$ *is a left multiple of every common right divisor of* $\mathbf{D}(s)$ *and* $\mathbf{N}(s)$. *If a gcrd is a unimodular matrix, then* $\mathbf{D}(s)$ *and* $\mathbf{N}(s)$ *are said to be* right coprime.

Dual to this definition, a square polynomial matrix $\bar{\mathbf{R}}(s)$ is a *greatest common left divisor* (gcld) of $\bar{\mathbf{D}}(s)$ and $\bar{\mathbf{N}}(s)$ if (i) $\bar{\mathbf{R}}(s)$ is a common left divisor of $\bar{\mathbf{D}}(s)$ and $\bar{\mathbf{N}}(s)$ and (ii) $\bar{\mathbf{R}}(s)$ is a right multiple of every common left divisor of $\bar{\mathbf{D}}(s)$ and $\bar{\mathbf{N}}(s)$. If a gcld is a unimodular matrix, then $\bar{\mathbf{D}}(s)$ and $\bar{\mathbf{N}}(s)$ are said to be *left coprime*.

Definition 7.4 *Consider a proper rational matrix* $\hat{\mathbf{G}}(s)$ *factored as*

$$\hat{\mathbf{G}}(s) = \mathbf{N}(s)\mathbf{D}^{-1}(s) = \bar{\mathbf{D}}^{-1}(s)\bar{\mathbf{N}}(s)$$

where $\mathbf{N}(s)$ *and* $\mathbf{D}(s)$ *are right coprime, and* $\bar{\mathbf{N}}(s)$ *and* $\bar{\mathbf{D}}(s)$ *are left coprime. Then the characteristic polynomial of* $\hat{\mathbf{G}}(s)$ *is defined as*

$$\det \mathbf{D}(s) \quad \text{or} \quad \det \bar{\mathbf{D}}(s)$$

and the degree of $\hat{\mathbf{G}}(s)$ *is defined as*

$$\deg \hat{\mathbf{G}}(s) = \deg \det \mathbf{D}(s) = \deg \det \bar{\mathbf{D}}(s)$$

Consider

$$\hat{\mathbf{G}}(s) = \mathbf{N}(s)\mathbf{D}^{-1}(s) = [\mathbf{N}(s)\mathbf{R}(s)][\mathbf{D}(s)\mathbf{R}(s)]^{-1} \tag{7.78}$$

where $\mathbf{N}(s)$ and $\mathbf{D}(s)$ are right coprime. Define $\mathbf{D}_1(s) = \mathbf{D}(s)\mathbf{R}(s)$ and $\mathbf{N}_1(s) = \mathbf{N}(s)\mathbf{R}(s)$. Then we have

$$\det \mathbf{D}_1(s) = \det[\mathbf{D}(s)\mathbf{R}(s)] = \det \mathbf{D}(s) \det \mathbf{R}(s)$$

which implies

$$\deg \det \mathbf{D}_1(s) = \deg \det \mathbf{D}(s) + \deg \det \mathbf{R}(s)$$

Clearly we have $\deg \det \mathbf{D}_1(s) \geq \deg \det \mathbf{D}(s)$ and the equality holds if and only if $\mathbf{R}(s)$ is unimodular or, equivalently, $\mathbf{N}_1(s)$ anc $\mathbf{D}_1(s)$ are right coprime. Thus we conclude that if $\mathbf{N}(s)\mathbf{D}^{-1}(s)$ is a coprime fraction, then $\mathbf{D}(s)$ has the smallest possible determinantal degree and the degree is defined as the degree of the transfer matrix. Therefore a coprime fraction can also be defined as a polynomial matrix fraction with the smallest denominator's determinantal degree. From (7.78), we can see that coprime fractions are not unique; they can differ by unimodular matrices.

We have introduced Definitions 7.1 and 7.4 to define the degree of rational matrices. Their equivalence can be established by using the Smith–McMillan form and will not be discussed here. The interested reader is referred to, for example, Reference [3].

7.8.1 Column and Row Reducedness

In order to apply Definition 7.4 to determine the degree of $\hat{\mathbf{G}}(s) = \mathbf{N}(s)\mathbf{D}^{-1}(s)$, we must compute the determinant of $\mathbf{D}(s)$. This can be avoided if the coprime fraction has some additional property as we will discuss next.

The degree of a polynomial vector is defined as the highest power of s in all entries of the vector. Consider a polynomial matrix $\mathbf{M}(s)$. We define

$$\delta_{ci}\mathbf{M}(s) = \text{degree of } i\text{th column of } \mathbf{M}(s)$$

$$\delta_{ri}\mathbf{M}(s) = \text{degree of } i\text{th row of } \mathbf{M}(s)$$

and call δ_{ci} the *column degree* and δ_{ri} the *row degree*. For example, the matrix

$$\mathbf{M}(s) = \begin{bmatrix} s+1 & s^3 - 2s + 5 & -1 \\ s-1 & s^2 & 0 \end{bmatrix}$$

has $\delta_{c1} = 1$, $\delta_{c2} = 3$, $\delta_{c3} = 0$, $\delta_{r1} = 3$, and $\delta_{r2} = 2$.

Definition 7.5 *A nonsingular polynomial matrix* $\mathbf{M}(s)$ *is* column reduced *if*

$$\deg \det \mathbf{M}(s) = \text{sum of all column degrees}$$

It is row reduced *if*

$$\deg \det \mathbf{M}(s) = \text{sum of all row degrees}$$

A matrix can be column reduced but not row reduced and vice versa. For example, the matrix

$$\mathbf{M}(s) = \begin{bmatrix} 3s^2 + 2s & 2s+1 \\ s^2 + s - 3 & s \end{bmatrix} \tag{7.79}$$

has determinant $s^3 - s^2 + 5s + 3$. Its degree equals the sum of its column degrees 2 and 1. Thus $\mathbf{M}(s)$ in (7.79) is column reduced. The row degrees of $\mathbf{M}(s)$ are 2 and 2; their sum is larger than 3. Thus $\mathbf{M}(s)$ is not row reduced. A diagonal polynomial matrix is always both column and row reduced. If a square polynomial matrix is not column reduced, then the degree of its determinant is less than the sum of its column degrees. Every nonsingular polynomial matrix can be changed to a column- or row-reduced matrix by pre- or postmultiplying a unimodular matrix. See Reference [6, p. 603].

Let $\delta_{ci}\mathbf{M}(s) = k_{ci}$ and define $\mathbf{H}_c(s) = \text{diag}(s^{k_{c1}}, s^{k_{c2}}, \dots)$. Then the polynomial matrix $\mathbf{M}(s)$ can be expressed as

$$\mathbf{M}(s) = \mathbf{M}_{hc}\mathbf{H}_c(s) + \mathbf{M}_{lc}(s) \tag{7.80}$$

For example, the $\mathbf{M}(s)$ in (7.79) has column degrees 2 and 1 and can be expressed as

$$\mathbf{M}(s) = \begin{bmatrix} 3 & 2 \\ 1 & 1 \end{bmatrix} \begin{bmatrix} s^2 & 0 \\ 0 & s \end{bmatrix} + \begin{bmatrix} 2s & 1 \\ s-3 & 0 \end{bmatrix}$$

The constant matrix $\mathbf{M}_{hc}$ is called the *column-degree coefficient matrix*; its ith column is the coefficients of the ith column of $\mathbf{M}(s)$ associated with $s^{k_{ci}}$. The polynomial matrix $\mathbf{M}_{lc}(s)$ contains the remaining terms and its ith column has degree less than k_{ci}. If $\mathbf{M}(s)$ is expressed as in (7.80), then it can be verified that $\mathbf{M}(s)$ is column reduced if and only if its column-degree coefficient matrix $\mathbf{M}_{hc}$ is nonsingular. Dual to (7.80), we can express $\mathbf{M}(s)$ as

$$\mathbf{M}(s) = \mathbf{H}_r(s)\mathbf{M}_{hr} + \mathbf{M}_{lr}(s)$$

where $\delta_{ri}\mathbf{M}(s) = k_{ri}$ and $\mathbf{H}_r(s) = \text{diag}(s^{k_{r1}}, s^{k_{r2}}, \ldots)$. The matrix $\mathbf{M}_{hr}$ is called the *row-degree coefficient matrix*. Then $\mathbf{M}(s)$ is row reduced if and only if $\mathbf{M}_{hr}$ is nonsingular.

Using the concept of reducedness, we now can state the degree of a proper rational matrix as follows. Consider $\hat{\mathbf{G}}(s) = \mathbf{N}(s)\mathbf{D}^{-1}(s) = \bar{\mathbf{D}}^{-1}(s)\bar{\mathbf{N}}(s)$, where $\mathbf{N}(s)$ and $\mathbf{D}(s)$ are right coprime, $\bar{\mathbf{N}}(s)$ and $\bar{\mathbf{D}}(s)$ are left coprime, $\mathbf{D}(s)$ is column reduced, and $\bar{\mathbf{D}}(s)$ is row reduced. Then we have

$$\deg \hat{\mathbf{G}}(s) = \text{sum of column degrees of } \mathbf{D}(s)$$

$$= \text{sum of row degrees of } \bar{\mathbf{D}}(s)$$

We discuss another implication of reducedness. Consider $\hat{\mathbf{G}}(s) = \mathbf{N}(s)\mathbf{D}^{-1}(s)$. If $\hat{\mathbf{G}}(s)$ is strictly proper, then $\delta_{ci}\mathbf{N}(s) < \delta_{ci}\mathbf{D}(s)$, for $i = 1, 2, \ldots, p$; that is, the column degrees of $\mathbf{N}(s)$ are less than the corresponding column degrees of $\mathbf{D}(s)$. If $\hat{\mathbf{G}}(s)$ is proper, then $\delta_{ci}\mathbf{N}(s) \leq \delta_{ci}\mathbf{D}(s)$, for $i = 1, 2, \ldots, p$. The converse, however, is not necessarily true. For example, consider

$$\mathbf{N}(s)\mathbf{D}^{-1}(s) = [1 \ 2] \begin{bmatrix} s^2 & s-1 \\ s+1 & 1 \end{bmatrix}^{-1} = \begin{bmatrix} \dfrac{-2s-1}{1} & \dfrac{2s^2-s+1}{1} \end{bmatrix}$$

Although $\delta_{ci}\mathbf{N}(s) < \delta_{ci}\mathbf{D}(s)$ for $i = 1, 2$, $\mathbf{N}(s)\mathbf{D}^{-1}(s)$ is not strictly proper. The reason is that $\mathbf{D}(s)$ is not column reduced.

▶ **Theorem 7.8**

Let $\mathbf{N}(s)$ and $\mathbf{D}(s)$ be $q \times p$ and $p \times p$ polynomial matrices, and let $\mathbf{D}(s)$ be column reduced. Then the rational matrix $\mathbf{N}(s)\mathbf{D}^{-1}(s)$ is proper (strictly proper) if and only if

$$\delta_{ci}\mathbf{N}(s) \leq \delta_{ci}\mathbf{D}(s) \qquad [\delta_{ci}\mathbf{N}(s) < \delta_{ci}\mathbf{D}(s)]$$

for $i = 1, 2, \ldots, p$.

Proof: The necessity part of the theorem follows from the preceding example. We show the sufficiency. Following (7.80), we express

$$\mathbf{D}(s) = \mathbf{D}_{hc}\mathbf{H}_c(s) + \mathbf{D}_{lc}(s) = [\mathbf{D}_{hc} + \mathbf{D}_{lc}(s)\mathbf{H}_c^{-1}(s)]\mathbf{H}_c(s)$$

$$\mathbf{N}(s) = \mathbf{N}_{hc}\mathbf{H}_c(s) + \mathbf{N}_{lc}(s) = [\mathbf{N}_{hc} + \mathbf{N}_{lc}(s)\mathbf{H}_c^{-1}(s)]\mathbf{H}_c(s)$$

Then we have

$$\hat{\mathbf{G}}(s) := \mathbf{N}(s)\mathbf{D}^{-1}(s) = [\mathbf{N}_{hc} + \mathbf{N}_{lc}(s)\mathbf{H}_c^{-1}(s)][\mathbf{D}_{hc} + \mathbf{D}_{lc}(s)\mathbf{H}_c^{-1}(s)]^{-1}$$

Because $\mathbf{D}_{lc}(s)\mathbf{H}_c^{-1}(s)$ and $\mathbf{N}_{lc}(s)\mathbf{H}_c^{-1}(s)$ both approach zero as $s \to \infty$, we have

$$\lim_{s \to \infty} \hat{\mathbf{G}}(s) = \mathbf{N}_{hc}\mathbf{D}_{hc}^{-1}$$

where $\mathbf{D}_{hc}$ is nonsingular by assumption. Now if $\delta_{ci}\mathbf{N}(s) \le \delta_{ci}\mathbf{D}(s)$, then $\mathbf{N}_{hc}$ is a nonzero constant matrix. Thus $\hat{\mathbf{G}}(\infty)$ is a nonzero constant and $\hat{\mathbf{G}}(s)$ is proper. If $\delta_{ci}\mathbf{N}(s) < \delta_{ci}\mathbf{D}(s)$, then $\mathbf{N}_{hc}$ is a zero matrix. Thus $\hat{\mathbf{G}}(\infty) = \mathbf{0}$ and $\hat{\mathbf{G}}(s)$ is strictly proper. This establishes the theorem.   Q.E.D.

We state the dual of Theorem 7.8 without proof.

▶ **Corollary 7.8**

Let $\bar{\mathbf{N}}(s)$ and $\bar{\mathbf{D}}(s)$ be $q \times p$ and $q \times q$ polynomial matrices, and let $\bar{\mathbf{D}}(s)$ be row reduced. Then the rational matrix $\bar{\mathbf{D}}^{-1}(s)\bar{\mathbf{N}}(s)$ is proper (strictly proper) if and only if

$$\delta_{ri}\bar{\mathbf{N}}(s) \le \delta_{ri}\bar{\mathbf{D}}(s) \qquad [\delta_{ri}\bar{\mathbf{N}}(s) < \delta_{ri}\bar{\mathbf{D}}(s)]$$

for $i = 1, 2, \ldots, q$.

7.8.2 Computing Matrix Coprime Fractions

Given a right fraction $\mathbf{N}(s)\mathbf{D}^{-1}(s)$, one way to reduce it to a right coprime fraction is to compute its gcrd. This can be achieved by applying a sequence of elementary operations. Once a gcrd $\mathbf{R}(s)$ is computed, we compute $\hat{\mathbf{N}}(s) = \mathbf{N}(s)\mathbf{R}^{-1}(s)$ and $\hat{\mathbf{D}}(s) = \mathbf{D}(s)\mathbf{R}^{-1}(s)$. Then $\mathbf{N}(s)\mathbf{D}^{-1}(s) = \hat{\mathbf{N}}(s)\hat{\mathbf{D}}^{-1}(s)$ and $\hat{\mathbf{N}}(s)\hat{\mathbf{D}}^{-1}(s)$ is a right coprime fraction. If $\hat{\mathbf{D}}(s)$ is not column reduced, then additional manipulation is needed. This procedure will not be discussed here. The interested reader is referred to Reference [6, pp. 590–591].

We now extend the method of computing scalar coprime fractions in Section 7.3 to the matrix case. Consider a $q \times p$ proper rational matrix $\hat{\mathbf{G}}(s)$ expressed as

$$\hat{\mathbf{G}}(s) = \bar{\mathbf{D}}^{-1}(s)\bar{\mathbf{N}}(s) = \mathbf{N}(s)\mathbf{D}^{-1}(s) \tag{7.81}$$

In this section, we use variables with an overbar to denote left fractions, without an overbar to denote right fractions. Clearly (7.81) implies

$$\bar{\mathbf{N}}(s)\mathbf{D}(s) = \bar{\mathbf{D}}(s)\mathbf{N}(s)$$

and

$$\bar{\mathbf{D}}(s)(-\mathbf{N}(s)) + \bar{\mathbf{N}}(s)\mathbf{D}(s) = \mathbf{0} \tag{7.82}$$

We shall show that given a left fraction $\bar{\mathbf{D}}^{-1}(s)\bar{\mathbf{N}}(s)$, not necessarily left coprime, we can obtain a right coprime fraction $\mathbf{N}(s)\mathbf{D}^{-1}(s)$ by solving the polynomial matrix equation in (7.82). Instead of solving (7.82) directly, we shall change it into solving sets of linear algebraic equations. As in (7.27), we express the polynomial matrices as, assuming the highest degree to be 4 to simplify writing,

$$\bar{\mathbf{D}}(s) = \bar{\mathbf{D}}_0 + \bar{\mathbf{D}}_1 s + \bar{\mathbf{D}}_2 s^2 + \bar{\mathbf{D}}_3 s^3 + \bar{\mathbf{D}}_4 s^4$$

$$\bar{\mathbf{N}}(s) = \bar{\mathbf{N}}_0 + \bar{\mathbf{N}}_1 s + \bar{\mathbf{N}}_2 s^2 + \bar{\mathbf{N}}_3 s^3 + \bar{\mathbf{N}}_4 s^4$$

$$\mathbf{D}(s) = \mathbf{D}_0 + \mathbf{D}_1 s + \mathbf{D}_2 s^2 + \mathbf{D}_3 s^3$$

$$\mathbf{N}(s) = \mathbf{N}_0 + \mathbf{N}_1 s + \mathbf{N}_2 s^2 + \mathbf{N}_3 s^3$$

where $\bar{\mathbf{D}}_i$, $\bar{\mathbf{N}}_i$, $\mathbf{D}_i$, and $\mathbf{N}_i$ are, respectively, $q \times q$, $q \times p$, $p \times p$, and $q \times p$ constant matrices. The constant matrices $\bar{\mathbf{D}}_i$ and $\bar{\mathbf{N}}_i$ are known, and $\mathbf{D}_i$ and $\mathbf{N}_i$ are to be solved. Substituting these into (7.82) and equating to zero the constant matrices associated with s^k, for $k = 0, 1, \ldots,$ we obtain

$$\mathbf{SM} := \begin{bmatrix} \bar{\mathbf{D}}_0 & \bar{\mathbf{N}}_0 & 0 & 0 & 0 & 0 & 0 & 0 \\ \bar{\mathbf{D}}_1 & \bar{\mathbf{N}}_1 & \bar{\mathbf{D}}_0 & \bar{\mathbf{N}}_0 & 0 & 0 & 0 & 0 \\ \bar{\mathbf{D}}_2 & \bar{\mathbf{N}}_2 & \bar{\mathbf{D}}_1 & \bar{\mathbf{N}}_1 & \bar{\mathbf{D}}_0 & \bar{\mathbf{N}}_0 & 0 & 0 \\ \bar{\mathbf{D}}_3 & \bar{\mathbf{N}}_3 & \bar{\mathbf{D}}_2 & \bar{\mathbf{N}}_2 & \bar{\mathbf{D}}_1 & \bar{\mathbf{N}}_1 & \bar{\mathbf{D}}_0 & \bar{\mathbf{N}}_0 \\ \bar{\mathbf{D}}_4 & \bar{\mathbf{N}}_4 & \bar{\mathbf{D}}_3 & \bar{\mathbf{N}}_3 & \bar{\mathbf{D}}_2 & \bar{\mathbf{N}}_2 & \bar{\mathbf{D}}_1 & \bar{\mathbf{N}}_1 \\ 0 & 0 & \bar{\mathbf{D}}_4 & \bar{\mathbf{N}}_4 & \bar{\mathbf{D}}_3 & \bar{\mathbf{N}}_3 & \bar{\mathbf{D}}_2 & \bar{\mathbf{N}}_2 \\ 0 & 0 & 0 & 0 & \bar{\mathbf{D}}_4 & \bar{\mathbf{N}}_4 & \bar{\mathbf{D}}_3 & \bar{\mathbf{N}}_3 \\ 0 & 0 & 0 & 0 & 0 & 0 & \bar{\mathbf{D}}_4 & \bar{\mathbf{N}}_4 \end{bmatrix} \begin{bmatrix} -\mathbf{N}_0 \\ \mathbf{D}_0 \\ \ldots \\ -\mathbf{N}_1 \\ \mathbf{D}_1 \\ \ldots \\ -\mathbf{N}_2 \\ \mathbf{D}_2 \\ \ldots \\ -\mathbf{N}_3 \\ \mathbf{D}_3 \end{bmatrix} = 0 \quad (7.83)$$

This equation is the matrix version of (7.28) and the matrix $\mathbf{S}$ will be called a *generalized resultant*. Note that every $\bar{D}$-block column has q $\bar{D}$-columns and every $\bar{N}$-block column has p $\bar{N}$-columns. The generalized resultant $\mathbf{S}$ as shown has four pairs of $\bar{D}$- and $\bar{N}$-block columns; thus it has a total of $4(q + p)$ columns. It has eight block rows; each block rows has q rows. Thus the resultant has $8q$ rows.

We now discuss some general properties of $\mathbf{S}$ under the assumption that linearly independent columns of $\mathbf{S}$ from left to right have been found. It turns out that every $\bar{D}$-column in every $\bar{D}$-block column is linearly independent of its left-hand-side (LHS) columns. The situation for $\bar{N}$-columns, however, is different. Recall that there are p $\bar{N}$-columns in each $\bar{N}$-block column. We use $\bar{N}i$-column to denote the ith $\bar{N}$-column in each $\bar{N}$-block column. It turns out that if the $\bar{N}i$-column in some $\bar{N}$-block column is linearly dependent on its LHS columns, then all subsequent $\bar{N}i$-columns, because of the repetitive structure of $\mathbf{S}$, will be linearly dependent on its LHS columns. Let μ_i, $i = 1, 2, \ldots, p$, be the number of linearly independent $\bar{N}i$-columns in $\mathbf{S}$. They are called the *column indices* of $\hat{\mathbf{G}}(s)$. The first $\bar{N}i$-column that becomes linearly dependent on its LHS columns is called the *primary dependent $\bar{N}i$-column*. It is clear that the $(\mu_i + 1)$th $\bar{N}i$-column is the primary dependent column.

Corresponding to each primary dependent $\bar{N}i$-column, we compute the monic null vector (its last entry equals 1) of the submatrix that consists of the primary dependent $\bar{N}i$-column and all its LHS linearly independent columns. There are totally p such monic null vectors. From these monic null vectors, we can then obtain a right fraction. This fraction is right coprime because we use the least possible μ_i and the resulting $\mathbf{D}(s)$ has the smallest possible column degrees. In addition, $\mathbf{D}(s)$ automatically will be column reduced. The next example illustrates the procedure.

EXAMPLE 7.7 Find a right coprime fraction of the transfer matrix in (7.75) or

$$\hat{\mathbf{G}}(s) = \begin{bmatrix} \dfrac{4s - 10}{2s + 1} & \dfrac{3}{s + 2} \\ \dfrac{1}{(2s + 1)(s + 2)} & \dfrac{s + 1}{(s + 2)^2} \end{bmatrix} \tag{7.84}$$

First we must find a left fraction, not necessarily left coprime. Using the least common denominator of each row, we can readily obtain

$$\hat{\mathbf{G}}(s) = \begin{bmatrix} (2s + 1)(s + 2) & 0 \\ 0 & (2s + 1)(s + 2)^2 \end{bmatrix}^{-1}$$

$$\times \begin{bmatrix} (4s - 10)(s + 2) & 3(2s + 1) \\ s + 2 & (s + 1)(2s + 1) \end{bmatrix} =: \bar{\mathbf{D}}^{-1}(s)\bar{\mathbf{N}}(s)$$

Thus we have

$$\bar{\mathbf{D}}(s) = \begin{bmatrix} 2s^2 + 5s + 2 & 0 \\ 0 & 2s^3 + 9s^2 + 12s + 4 \end{bmatrix}$$

$$= \begin{bmatrix} 2 & 0 \\ 0 & 4 \end{bmatrix} + \begin{bmatrix} 5 & 0 \\ 0 & 12 \end{bmatrix} s + \begin{bmatrix} 2 & 0 \\ 0 & 9 \end{bmatrix} s^2 + \begin{bmatrix} 0 & 0 \\ 0 & 2 \end{bmatrix} s^3$$

and

$$\bar{\mathbf{N}}(s) = \begin{bmatrix} 4s^2 - 2s - 20 & 6s + 3 \\ s + 2 & 2s^2 + 3s + 1 \end{bmatrix}$$

$$= \begin{bmatrix} -20 & 3 \\ 2 & 1 \end{bmatrix} + \begin{bmatrix} -2 & 6 \\ 1 & 3 \end{bmatrix} s + \begin{bmatrix} 4 & 0 \\ 0 & 2 \end{bmatrix} s^2 + \begin{bmatrix} 0 & 0 \\ 0 & 0 \end{bmatrix} s^3$$

We form the generalized resultant and then use the QR decomposition discussed in Section 7.3.1 to search its linearly independent columns in order from left to right. Because it is simpler to key in the transpose of $\mathbf{S}$, we type

```
d1=[2 0 5 0 2 0 0 0];d2=[0 4 0 12 0 9 0 2];
n1=[-20 2 -2 1 4 0 0 0];n2=[2 1 1 3 0 2 0 0];
s=[d1 0 0 0 0;d2 0 0 0 0;n1 0 0 0 0;n2 0 0 0 0;...
   0 0 d1 0 0;0 0 d2 0 0;0 0 n1 0 0;0 0 n2 0 0;...
   0 0 0 0 d1;0 0 0 0 d2;0 0 0 0 n1;0 0 0 0 n2]';
[q,r]=qr(s)
```

We need only r; therefore the matrix q will not be shown. As discussed in Section 7.3.1, we need to know whether or not entries of r are zero in determining linear independence of columns; therefore all nonzero entries are represented by x, di, and ni. The result is

$$
r = \begin{bmatrix}
d1 & 0 & x & x & x & x & x & x & x & 0 & x & x \\
0 & d2 & x & x & x & x & x & x & 0 & x & x & x \\
0 & 0 & n1 & x & x & x & x & x & x & x & x & x \\
0 & 0 & 0 & n2 & x & x & x & x & x & x & x & x \\
0 & 0 & 0 & 0 & d1 & x & x & x & x & x & x & x \\
0 & 0 & 0 & 0 & 0 & d2 & x & x & x & x & x & x \\
0 & 0 & 0 & 0 & 0 & 0 & n1 & x & x & x & x & x \\
0 & 0 & 0 & 0 & 0 & 0 & 0 & 0 & x & x & x & 0 \\
0 & 0 & 0 & 0 & 0 & 0 & 0 & 0 & d1 & x & x & 0 \\
0 & 0 & 0 & 0 & 0 & 0 & 0 & 0 & 0 & d2 & 0 & 0 \\
0 & 0 & 0 & 0 & 0 & 0 & 0 & 0 & 0 & 0 & 0 & 0 \\
0 & 0 & 0 & 0 & 0 & 0 & 0 & 0 & 0 & 0 & 0 & 0
\end{bmatrix}
$$

We see that all D-columns are linearly independent of their LHS columns. There are two linearly independent $\bar{N}1$-columns and one linearly independent $\bar{N}2$-column. Thus we have $\mu_1 = 2$ and $\mu_2 = 1$. The eighth column of $\mathbf{S}$ is the primary dependent $\bar{N}2$-column. We compute a null vector of the submatrix that consists of the primary dependent $\bar{N}2$-column and all its LHS linearly independent columns as

```
z2=null([d1 0 0;d2 0 0;n1 0 0;n2 0 0;...
         0 0 d1;0 0 d2;0 0 n1;0 0 n2]');
```

and then normalize the last entry to 1 by typing

```
z2b=z2/z2(8)
```

which yields the first monic null vector as

```
z2b=[7 -1 1 2 -4 0 2 1]'
```

The eleventh column of $\mathbf{S}$ is the primary dependent $\bar{N}1$-column. We compute a null vector of the submatrix that consists of the primary dependent $\bar{N}1$-column and all its LHS linearly independent columns (that is, deleting the eighth column) as

```
z1=null([d1 0 0 0 0;d2 0 0 0 0;n1 0 0 0 0;n2 0 0 0 0;...
         0 0 d1 0 0;0 0 d2 0 0;0 0 n1 0 0;...
         0 0 0 0 d1;0 0 0 0 d2;0 0 0 0 n1]');
```

and then normalize the last entry to 1 by typing

```
z1b=z1/z1(10)
```

which yields the second monic null vector as

```
z1b=[10 -0.5 1 0 1 0 2.5 -2 0 1]'
```

Thus we have

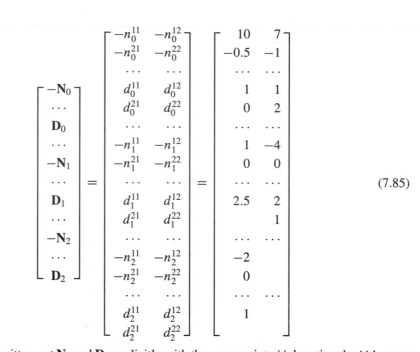

$$\begin{bmatrix} -\mathbf{N}_0 \\ \cdots \\ \mathbf{D}_0 \\ \cdots \\ -\mathbf{N}_1 \\ \cdots \\ \mathbf{D}_1 \\ \cdots \\ -\mathbf{N}_2 \\ \cdots \\ \mathbf{D}_2 \end{bmatrix} = \begin{bmatrix} -n_0^{11} & -n_0^{12} \\ -n_0^{21} & -n_0^{22} \\ \cdots & \cdots \\ d_0^{11} & d_0^{12} \\ d_0^{21} & d_0^{22} \\ \cdots & \cdots \\ -n_1^{11} & -n_1^{12} \\ -n_1^{21} & -n_1^{22} \\ \cdots & \cdots \\ d_1^{11} & d_1^{12} \\ d_1^{21} & d_1^{22} \\ \cdots & \cdots \\ -n_2^{11} & -n_2^{12} \\ -n_2^{21} & -n_2^{22} \\ \cdots & \cdots \\ d_2^{11} & d_2^{12} \\ d_2^{21} & d_2^{22} \end{bmatrix} = \begin{bmatrix} 10 & 7 \\ -0.5 & -1 \\ \cdots & \cdots \\ 1 & 1 \\ 0 & 2 \\ \cdots & \cdots \\ 1 & -4 \\ 0 & 0 \\ \cdots & \cdots \\ 2.5 & 2 \\ & 1 \\ \cdots & \cdots \\ -2 & \\ 0 & \\ \cdots & \cdots \\ 1 & \end{bmatrix} \tag{7.85}$$

where we have written out $\mathbf{N}_i$ and $\mathbf{D}_i$ explicitly with the superscripts ij denoting the ijth entry and the subscript denoting the degree. The two monic null vectors are arranged as shown. The order of the two null vectors can be interchanged, as we will discuss shortly. The empty entries are to be filled up with zeros. Note that the empty entry at the (8×1)th location is due to deleting the second $\bar{N}2$ linearly dependent column in computing the second null vector. By equating the corresponding entries in (7.85), we can readily obtain

$$\mathbf{D}(s) = \begin{bmatrix} 1 & 1 \\ 0 & 2 \end{bmatrix} + \begin{bmatrix} 2.5 & 2 \\ 0 & 1 \end{bmatrix} s + \begin{bmatrix} 1 & 0 \\ 0 & 0 \end{bmatrix} s^2$$
$$= \begin{bmatrix} s^2 + 2.5s + 1 & 2s + 1 \\ 0 & s + 2 \end{bmatrix}$$

and

$$\mathbf{N}(s) = \begin{bmatrix} -10 & -7 \\ 0.5 & 1 \end{bmatrix} + \begin{bmatrix} -1 & 4 \\ 0 & 0 \end{bmatrix} s + \begin{bmatrix} 2 & 0 \\ 0 & 0 \end{bmatrix} s^2$$
$$= \begin{bmatrix} 2s^2 - s - 10 & 4s - 7 \\ 0.5 & 1 \end{bmatrix}$$

Thus $\hat{\mathbf{G}}(s)$ in (7.84) has the following right coprime fraction

$$\hat{\mathbf{G}}(s) = \begin{bmatrix} (2s - 5)(s + 2) & 4s - 7 \\ 0.5 & 1 \end{bmatrix} \begin{bmatrix} (s + 2)(s + 0.5) & 2s + 1 \\ 0 & s + 2 \end{bmatrix}^{-1} \tag{7.86}$$

The $\mathbf{D}(s)$ in (7.86) is column reduced with column degrees $\mu_1 = 2$ and $\mu_2 = 1$. Thus we have deg det $\mathbf{D}(s) = 2 + 1 = 3$ and the degree of $\hat{\mathbf{G}}(s)$ in (7.84) is 3. The degree was computed in Example 7.6 as 3 by using Definition 7.1.

In general, if the generalized resultant has μ_i linearly independent $\bar{N}i$-columns, then $\mathbf{D}(s)$ computed using the preceding procedure is column reduced with column degrees μ_i. Thus we have

$$\deg \hat{\mathbf{G}}(s) = \deg \det \mathbf{D}(s) = \sum \mu_i$$

$$= \text{total number of linearly independent } \bar{N}\text{-columns in } \mathbf{S}$$

We next show that the order of column degrees is immaterial. In other words, the order of the columns of $\mathbf{N}(s)$ and $\mathbf{D}(s)$ can be changed. For example, consider the permutation matrix

$$\mathbf{P} = \begin{bmatrix} 0 & 0 & 1 \\ 1 & 0 & 0 \\ 0 & 1 & 0 \end{bmatrix}$$

and $\hat{\mathbf{D}}(s) = \mathbf{D}(s)\mathbf{P}$ and $\hat{\mathbf{N}}(s) = \mathbf{N}(s)\mathbf{P}$. Then the first, second, and third columns of $\mathbf{D}(s)$ and $\mathbf{N}(s)$ become the third, first, and second columns of $\hat{\mathbf{D}}(s)$ and $\hat{\mathbf{N}}(s)$. However, we have

$$\hat{\mathbf{G}}(s) = \hat{\mathbf{N}}(s)\hat{\mathbf{D}}^{-1}(s) = [\mathbf{N}(s)\mathbf{P}][\mathbf{D}(s)\mathbf{P}]^{-1} = \mathbf{N}(s)\mathbf{D}^{-1}(s)$$

This shows that the columns of $\mathbf{D}(s)$ and $\mathbf{N}(s)$ can arbitrarily be permutated. This is the same as permuting the order of the null vectors in (7.83). Thus the set of column degrees is an intrinsic property of a system just as the set of controllability indices is (Theorem 6.3). What has been discussed can be stated as a theorem. It is an extension of Theorem 7.4 to the matrix case.

▶ **Theorem 7.M4**

Let $\hat{\mathbf{G}}(s) = \bar{\mathbf{D}}^{-1}(s)\bar{\mathbf{N}}(s)$ be a left fraction, not necessarily left coprime. We use the coefficient matrices of $\bar{\mathbf{D}}(s)$ and $\bar{\mathbf{N}}(s)$ to form the generalized resultant $\mathbf{S}$ shown in (7.83) and search its linearly independent columns from left to right. Let $\mu_i, i = 1, 2, \ldots, p$, be the number of linearly independent $\bar{N}i$-columns. Then we have

$$\deg \hat{\mathbf{G}}(s) = \mu_1 + \mu_2 + \cdots + \mu_p \tag{7.87}$$

and a right coprime fraction $\mathbf{N}(s)\mathbf{D}^{-1}(s)$ can be obtained by computing p monic null vectors of the p matrices formed from each primary dependent $\bar{N}i$-column and all its LHS linearly independent columns.

The right coprime fraction obtained by solving the equation in (7.83) has one additional important property. After permutation, the column-degree coefficient matrix $\mathbf{D}_{hc}$ can always become a unit upper triangular matrix (an upper triangular matrix with 1 as its diagonal entries). Such a $\mathbf{D}(s)$ is said to be in *column echelon form*. See References [6, pp. 610–612; 13, pp. 483–487]. For the $\mathbf{D}(s)$ in (7.86), its column-degree coefficient matrix is

$$\mathbf{D}_{hc} = \begin{bmatrix} 1 & 2 \\ 0 & 1 \end{bmatrix}$$

It is unit upper triangular; thus the $\mathbf{D}(s)$ is in column echelon form. Although we need only column reducedness in subsequent discussions, if $\mathbf{D}(s)$ is in column echelon form, then the result in the next section will be nicer.

Dual to the preceding discussion, we can compute a left coprime fraction from a right fraction $\mathbf{N}(s)\mathbf{D}^{-1}(s)$, which is not necessarily right coprime. Then similar to (7.83), we form

$$[-\bar{\mathbf{N}}_0 \; \bar{\mathbf{D}}_0 \; \vdots \; -\bar{\mathbf{N}}_1 \; \bar{\mathbf{D}}_1 \; \vdots \; -\bar{\mathbf{N}}_2 \; \bar{\mathbf{D}}_2 \; \vdots \; -\bar{\mathbf{N}}_3 \; \bar{\mathbf{D}}_3]\mathbf{T} = 0 \tag{7.88}$$

with

$$\mathbf{T} := \begin{bmatrix} \mathbf{D}_0 & \mathbf{D}_1 & \mathbf{D}_2 & \mathbf{D}_3 & \mathbf{D}_4 & 0 & 0 & 0 \\ \mathbf{N}_0 & \mathbf{N}_1 & \mathbf{N}_2 & \mathbf{N}_3 & \mathbf{N}_4 & 0 & 0 & 0 \\ \cdots & \cdots & \cdots & \cdots & \cdots & \cdots & \cdots & \cdots \\ 0 & \mathbf{D}_0 & \mathbf{D}_1 & \mathbf{D}_2 & \mathbf{D}_3 & \mathbf{D}_4 & 0 & 0 \\ 0 & \mathbf{N}_0 & \mathbf{N}_1 & \mathbf{N}_2 & \mathbf{N}_3 & \mathbf{N}_4 & 0 & 0 \\ \cdots & \cdots & \cdots & \cdots & \cdots & \cdots & \cdots & \cdots \\ 0 & 0 & \mathbf{D}_0 & \mathbf{D}_1 & \mathbf{D}_2 & \mathbf{D}_3 & \mathbf{D}_4 & 0 \\ 0 & 0 & \mathbf{N}_0 & \mathbf{N}_1 & \mathbf{N}_2 & \mathbf{N}_3 & \mathbf{N}_4 & 0 \\ \cdots & \cdots & \cdots & \cdots & \cdots & \cdots & \cdots & \cdots \\ 0 & 0 & 0 & \mathbf{D}_0 & \mathbf{D}_1 & \mathbf{D}_2 & \mathbf{D}_3 & \mathbf{D}_4 \\ 0 & 0 & 0 & \mathbf{N}_0 & \mathbf{N}_1 & \mathbf{N}_2 & \mathbf{N}_3 & \mathbf{N}_4 \end{bmatrix} \tag{7.89}$$

We search linearly independent rows in order from top to bottom. Then all D-rows are linearly independent. Let the Ni-row denote the ith N-row in each N block-row. If an Ni-row becomes linearly dependent, then all Ni-rows in subsequent N block-rows are linearly dependent on their preceding rows. The first Ni-row that becomes linearly dependent is called a primary dependent Ni-row. Let ν_i, $i = 1, 2, \ldots, q$, be the number of linearly independent Ni-rows. They are called the *row indices* of $\hat{\mathbf{G}}(s)$. Then dual to Theorem 7.M4, we have the following corollary.

▶ **Corollary 7.M4**

Let $\hat{\mathbf{G}}(s) = \mathbf{N}(s)\mathbf{D}^{-1}(s)$ be a right fraction, not necessarily right coprime. We use the coefficient matrices of $\mathbf{D}(s)$ and $\mathbf{N}(s)$ to form the generlized resultant $\mathbf{T}$ shown in (7.89) and search its linearly independent rows in order from top to bottom. Let ν_i, for $i = 1, 2, \ldots, q$, be the number of linearly independent Ni-rows in $\mathbf{T}$. Then we have

$$\deg \hat{\mathbf{G}}(s) = \nu_1 + \nu_2 + \cdots + \nu_q$$

and a left coprime fraction $\bar{\mathbf{D}}^{-1}(s)\bar{\mathbf{N}}(s)$ can be obtained by computing q monic left null vectors of the q matrices formed from each primary dependent Ni-row and all its preceding linearly independent rows.

The polynomial matrix $\bar{\mathbf{D}}(s)$ obtained in Corollary 7.M4 is row reduced with row degrees $\{\nu_i, \ i = 1, 2, \ldots, q\}$. In fact, it is in *row echelon form*; that is, its row-degree coefficient matrix, after some row permutation, is a unit lower triangular matrix.

7.9 Realizations from Matrix Coprime Fractions

Not to be overwhelmed by notation, we discuss a realization of a 2×2 strictly proper rational matrix $\hat{\mathbf{G}}(s)$ expressed as

$$\hat{\mathbf{G}}(s) = \mathbf{N}(s)\mathbf{D}^{-1}(s) \tag{7.90}$$

where $\mathbf{N}(s)$ and $\mathbf{D}(s)$ are right coprime and $\mathbf{D}(s)$ is in column echelon form.[6] We further assume that the column degrees of $\mathbf{D}(s)$ are $\mu_1 = 4$ and $\mu_2 = 2$. First we define

$$\mathbf{H}(s) := \begin{bmatrix} s^{\mu_1} & 0 \\ 0 & s^{\mu_2} \end{bmatrix} = \begin{bmatrix} s^4 & 0 \\ 0 & s^2 \end{bmatrix} \tag{7.91}$$

and

$$\mathbf{L}(s) := \begin{bmatrix} s^{\mu_1-1} & 0 \\ \vdots & \vdots \\ 1 & 0 \\ 0 & s^{\mu_2-1} \\ \vdots & \vdots \\ 0 & 1 \end{bmatrix} = \begin{bmatrix} s^3 & 0 \\ s^2 & 0 \\ s & 0 \\ 1 & 0 \\ 0 & s \\ 0 & 1 \end{bmatrix} \tag{7.92}$$

The procedure for developing a realization for

$$\hat{\mathbf{y}}(s) = \hat{\mathbf{G}}(s)\hat{\mathbf{u}}(s) = \mathbf{N}(s)\mathbf{D}^{-1}\hat{\mathbf{u}}(s)$$

follows closely the scalar case from (7.3) through (7.9). First we introduce a new variable $\mathbf{v}(t)$ defined by $\hat{\mathbf{v}}(s) = \mathbf{D}^{-1}(s)\hat{\mathbf{u}}(s)$. Note that $\hat{\mathbf{v}}(s)$, called a pseudo state, is a 2×1 column vector. Then we have

$$\mathbf{D}(s)\hat{\mathbf{v}}(s) = \hat{\mathbf{u}}(s) \tag{7.93}$$

$$\hat{\mathbf{y}}(s) = \mathbf{N}(s)\hat{\mathbf{v}}(s) \tag{7.94}$$

Let us define state variables as

$$\hat{\mathbf{x}}(s) = \mathbf{L}(s)\hat{\mathbf{v}}(s) = \begin{bmatrix} s^{\mu_1-1} & 0 \\ \vdots & \vdots \\ 1 & 0 \\ 0 & s^{\mu_2-1} \\ \vdots & \vdots \\ 0 & 1 \end{bmatrix} \begin{bmatrix} \hat{v}_1(s) \\ \hat{v}_2(s) \end{bmatrix}$$

$$= \begin{bmatrix} s^3\hat{v}_1(s) \\ s^2\hat{v}_1(s) \\ s\hat{v}_1(s) \\ \hat{v}_1(s) \\ s\hat{v}_2(s) \\ \hat{v}_2(s) \end{bmatrix} =: \begin{bmatrix} x_1(s) \\ x_2(s) \\ x_3(s) \\ x_4(s) \\ x_5(s) \\ x_6(s) \end{bmatrix} \tag{7.95}$$

or, in the time domain,

$$x_1(t) = v_1^{(3)}(t) \qquad x_2(t) = \ddot{v}_1(t) \qquad x_3(t) = \dot{v}_1(t) \qquad x_4(t) = v_1(t)$$

$$x_5(t) = \dot{v}_2(t) \qquad x_6(t) = v_2(t)$$

---

6. All discussion is still applicable if $\mathbf{D}(s)$ is column reduced but not in echelon form.

This state vector has dimension $\mu_1 + \mu_2 = 6$. These definitions imply immediately

$$\dot{x}_2 = x_1 \qquad \dot{x}_3 = x_2 \qquad \dot{x}_4 = x_3 \qquad \dot{x}_6 = x_5 \qquad (7.96)$$

Next we use (7.93) to develop equations for $\dot{x}_1$ and $\dot{x}_5$. First we express $\mathbf{D}(s)$ as

$$\mathbf{D}(s) = \mathbf{D}_{hc}\mathbf{H}(s) + \mathbf{D}_{lc}\mathbf{L}(s) \qquad (7.97)$$

where $\mathbf{H}(s)$ and $\mathbf{L}(s)$ are defined in (7.91) and (7.92). Note that $\mathbf{D}_{hc}$ and $\mathbf{D}_{lc}$ are constant matrices and the column-degree coefficient matrix $\mathbf{D}_{hc}$ is a unit upper triangular matrix. Substituting (7.97) into (7.93) yields

$$[\mathbf{D}_{hc}\mathbf{H}(s) + \mathbf{D}_{lc}\mathbf{L}(s)]\hat{\mathbf{v}}(s) = \hat{\mathbf{u}}(s)$$

or

$$\mathbf{H}(s)\hat{\mathbf{v}}(s) + \mathbf{D}_{hc}^{-1}\mathbf{D}_{lc}\mathbf{L}(s)\hat{\mathbf{v}}(s) = \mathbf{D}_{hc}^{-1}\hat{\mathbf{u}}(s)$$

Thus we have, using (7.95),

$$\mathbf{H}(s)\hat{\mathbf{v}}(s) = -\mathbf{D}_{hc}^{-1}\mathbf{D}_{lc}\hat{\mathbf{x}}(s) + \mathbf{D}_{hc}^{-1}\hat{\mathbf{u}}(s) \qquad (7.98)$$

Let

$$\mathbf{D}_{hc}^{-1}\mathbf{D}_{lc} =: \begin{bmatrix} \alpha_{111} & \alpha_{112} & \alpha_{113} & \alpha_{114} & \alpha_{121} & \alpha_{122} \\ \alpha_{211} & \alpha_{212} & \alpha_{213} & \alpha_{214} & \alpha_{221} & \alpha_{222} \end{bmatrix} \qquad (7.99)$$

and

$$\mathbf{D}_{hc}^{-1} =: \begin{bmatrix} 1 & b_{12} \\ 0 & 1 \end{bmatrix} \qquad (7.100)$$

Note that the inverse of a unit upper triangular matrix is again a unit upper triangular matrix. Substituting (7.99) and (7.100) into (7.98), and using $s\hat{x}_1(s) = s^4\hat{v}_1(s)$, and $s\hat{x}_5(s) = s^2\hat{v}_2(s)$, we find

$$\begin{bmatrix} s\hat{x}_1(s) \\ s\hat{x}_5(s) \end{bmatrix} = -\begin{bmatrix} \alpha_{111} & \alpha_{112} & \alpha_{113} & \alpha_{114} & \alpha_{121} & \alpha_{122} \\ \alpha_{211} & \alpha_{212} & \alpha_{213} & \alpha_{214} & \alpha_{221} & \alpha_{222} \end{bmatrix} \hat{\mathbf{x}}(s)$$
$$+ \begin{bmatrix} 1 & b_{12} \\ 0 & 1 \end{bmatrix} \hat{\mathbf{u}}(s)$$

which becomes, in the time domain,

$$\begin{bmatrix} \dot{x}_1 \\ \dot{x}_5 \end{bmatrix} = -\begin{bmatrix} \alpha_{111} & \alpha_{112} & \alpha_{113} & \alpha_{114} & \alpha_{121} & \alpha_{122} \\ \alpha_{211} & \alpha_{212} & \alpha_{213} & \alpha_{214} & \alpha_{221} & \alpha_{222} \end{bmatrix} \mathbf{x} + \begin{bmatrix} 1 & b_{12} \\ 0 & 1 \end{bmatrix} \mathbf{u} \qquad (7.101)$$

If $\hat{\mathbf{G}}(s) = \mathbf{N}(s)\mathbf{D}^{-1}(s)$ is strictly proper, then the column degrees of $\mathbf{N}(s)$ are less than the corresponding column degrees of $\mathbf{D}(s)$. Thus we can express $\mathbf{N}(s)$ as

$$\mathbf{N}(s) = \begin{bmatrix} \beta_{111} & \beta_{112} & \beta_{113} & \beta_{114} & \beta_{121} & \beta_{122} \\ \beta_{211} & \beta_{212} & \beta_{213} & \beta_{214} & \beta_{221} & \beta_{222} \end{bmatrix} \mathbf{L}(s) \qquad (7.102)$$

Substituting this into (7.94) and using $\hat{\mathbf{x}}(s) = \mathbf{L}(s)\hat{\mathbf{v}}(s)$, we have

$$\hat{\mathbf{y}}(s) = \begin{bmatrix} \beta_{111} & \beta_{112} & \beta_{113} & \beta_{114} & \beta_{121} & \beta_{122} \\ \beta_{211} & \beta_{212} & \beta_{213} & \beta_{214} & \beta_{221} & \beta_{222} \end{bmatrix} \hat{\mathbf{x}}(s) \tag{7.103}$$

Combining (7.96), (7.101), and (7.103) yields the following realization for $\hat{\mathbf{G}}(s)$:

$$\dot{\mathbf{x}} = \begin{bmatrix} -\alpha_{111} & -\alpha_{112} & -\alpha_{113} & -\alpha_{114} & \vdots & -\alpha_{121} & -\alpha_{122} \\ 1 & 0 & 0 & 0 & \vdots & 0 & 0 \\ 0 & 1 & 0 & 0 & \vdots & 0 & 0 \\ 0 & 0 & 1 & 0 & \vdots & 0 & 0 \\ \cdots & \cdots & \cdots & \cdots & \cdots & \cdots & \cdots \\ -\alpha_{211} & -\alpha_{212} & -\alpha_{213} & -\alpha_{214} & \vdots & -\alpha_{221} & -\alpha_{222} \\ 0 & 0 & 0 & 0 & \vdots & 1 & 0 \end{bmatrix} \mathbf{x}$$

$$+ \begin{bmatrix} 1 & b_{12} \\ 0 & 0 \\ 0 & 0 \\ 0 & 0 \\ \cdots & \cdots \\ 0 & 1 \\ 0 & 0 \end{bmatrix} \mathbf{u} \tag{7.104}$$

$$\mathbf{y} = \begin{bmatrix} \beta_{111} & \beta_{112} & \beta_{113} & \beta_{114} & \vdots & \beta_{121} & \beta_{122} \\ \beta_{211} & \beta_{212} & \beta_{213} & \beta_{214} & \vdots & \beta_{221} & \beta_{222} \end{bmatrix} \mathbf{x}$$

This is a $(\mu_1 + \mu_2)$-dimensional state equation. The A-matrix has two companion-form diagonal blocks: one with order $\mu_1 = 4$ and the other with order $\mu_2 = 2$. The off-diagonal blocks are zeros except their first rows. This state equation is a generalization of the state equation in (7.9) to two-input two-output transfer matrices. We can easily show that (7.104) is always controllable and is called a *controllable-form* realization. Furthermore, the controllability indices are $\mu_1 = 4$ and $\mu_2 = 2$. As in (7.9), the state equation in (7.104) is observable if and only if $\mathbf{D}(s)$ and $\mathbf{N}(s)$ are right coprime. For its proof, see Reference [6, p. 282]. Because we start with a coprime fraction $\mathbf{N}(s)\mathbf{D}^{-1}$, the realization in (7.104) is observable as well. In conclusion, the realization in (7.104) is controllable and observable and its dimension equals $\mu_1 + \mu_2$ and, following Theorem 7.M4, the degree of $\hat{\mathbf{G}}(s)$. This establishes the second part of Theorem 7.M2, namely, a state equation is minimal or controllable and observable if and only if its dimension equals the degree of its transfer matrix.

EXAMPLE 7.8 Consider the transfer matrix in Example 7.6. We gave there a minimal realization that is obtained by reducing the nonminimal realization in (4.39). Now we will develop directly a minimal realization by using a coprime fraction. We first write the transfer matrix as

$$\hat{\mathbf{G}}(s) = \begin{bmatrix} \dfrac{4s - 10}{2s + 1} & \dfrac{3}{s + 2} \\ \dfrac{1}{(2s + 1)(s + 2)} & \dfrac{s + 1}{(s + 2)^2} \end{bmatrix} =: \hat{\mathbf{G}}(\infty) + \hat{\mathbf{G}}_{sp}(s)$$

$$= \begin{bmatrix} 2 & 0 \\ 0 & 0 \end{bmatrix} + \begin{bmatrix} -\dfrac{12}{2s + 1} & \dfrac{3}{s + 2} \\ \dfrac{1}{(2s + 1)(s + 2)} & \dfrac{s + 1}{(s + 2)^2} \end{bmatrix}$$

As in Example 7.7, we can find a right coprime fraction for the strictly proper part of $\hat{\mathbf{G}}(s)$ as

$$\hat{\mathbf{G}}_{sp}(s) = \begin{bmatrix} -6s - 12 & -9 \\ 0.5 & 1 \end{bmatrix} \begin{bmatrix} s^2 + 2.5s + 1 & 2s + 1 \\ 0 & s + 2 \end{bmatrix}^{-1}$$

Note that its denominator matrix is the same as the one in (7.86). Clearly we have $\mu_1 = 2$ and $\mu_2 = 1$. We define

$$\mathbf{H}(s) = \begin{bmatrix} s^2 & 0 \\ 0 & s \end{bmatrix} \qquad \mathbf{L}(s) = \begin{bmatrix} s & 0 \\ 1 & 0 \\ 0 & 1 \end{bmatrix}$$

Then we have

$$\mathbf{D}(s) = \begin{bmatrix} 1 & 2 \\ 0 & 1 \end{bmatrix} \mathbf{H}(s) + \begin{bmatrix} 2.5 & 1 & 1 \\ 0 & 0 & 2 \end{bmatrix} \mathbf{L}(s)$$

and

$$\mathbf{N}(s) = \begin{bmatrix} -6 & -12 & -9 \\ 0 & 0.5 & 1 \end{bmatrix} \mathbf{L}(s)$$

We compute

$$\mathbf{D}_{hc}^{-1} = \begin{bmatrix} 1 & 2 \\ 0 & 1 \end{bmatrix}^{-1} = \begin{bmatrix} 1 & -2 \\ 0 & 1 \end{bmatrix}$$

and

$$\mathbf{D}_{hc}^{-1}\mathbf{D}_{lc} = \begin{bmatrix} 1 & -2 \\ 0 & 1 \end{bmatrix} \begin{bmatrix} 2.5 & 1 & 1 \\ 0 & 0 & 2 \end{bmatrix} = \begin{bmatrix} 2.5 & 1 & -3 \\ 0 & 0 & 2 \end{bmatrix}$$

Thus a minimal realization of $\hat{\mathbf{G}}(s)$ is

$$\dot{\mathbf{x}} = \begin{bmatrix} -2.5 & -1 & \vdots & 3 \\ 1 & 0 & \vdots & 0 \\ \cdots & \cdots & \cdots & \cdots \\ 0 & 0 & \vdots & -2 \end{bmatrix} \mathbf{x} + \begin{bmatrix} 1 & -2 \\ 0 & 0 \\ \cdots & \cdots \\ 0 & 1 \end{bmatrix} \mathbf{u} \tag{7.105}$$

$$\mathbf{y} = \begin{bmatrix} -6 & -12 & \vdots & -9 \\ 0 & 0.5 & \vdots & 1 \end{bmatrix} \mathbf{x} + \begin{bmatrix} 2 & 0 \\ 0 & 0 \end{bmatrix} \mathbf{u}$$

This A-matrix has two companion-form diagonal blocks; one with order 2 and the other order 1. This three-dimensional realization is a minimal realization and is in controllable canonical form.

Dual to the preceding minimal realization, if we use $\hat{\mathbf{G}}(s) = \bar{\mathbf{D}}^{-1}(s)\bar{\mathbf{N}}(s)$, where $\bar{\mathbf{D}}(s)$ and $\bar{\mathbf{N}}(s)$ are left coprime and $\bar{\mathbf{D}}(s)$ is in row echelon form with row degrees $\{v_i, \ i = 1, 2, \ldots, q\}$, then we can obtain an observable-form realization with observability indices $\{v_i, \ i = 1, 2, \ldots, q\}$. This will not be repeated.

We summarize the main results in the following. As in the SISO case, an n-dimensional multivariable state equation is controllable and observable if its transfer matrix has degree n. If a proper transfer matrix is expressed as a right coprime fraction with column reducedness, then the realization obtained by using the preceding procedure will automatically be controllable and observable.

Let $(\mathbf{A}, \mathbf{B}, \mathbf{C}, \mathbf{D})$ be a minimal realization of $\hat{\mathbf{G}}(s)$ and let $\hat{\mathbf{G}}(s) = \bar{\mathbf{D}}^{-1}(s)\bar{\mathbf{N}}(s) = \mathbf{N}(s)\mathbf{D}^{-1}(s)$ be coprime fractions; $\bar{\mathbf{D}}(s)$ is row reduced, and $\mathbf{D}(s)$ is column reduced. Then we have

$$\mathbf{B}(s\mathbf{I} - \mathbf{A})^{-1}\mathbf{C} + \mathbf{D} = \mathbf{N}(s)\mathbf{D}^{-1}(s) = \bar{\mathbf{D}}^{-1}(s)\bar{\mathbf{N}}(s)$$

which implies

$$\frac{1}{\det(s\mathbf{I} - \mathbf{A})}\mathbf{B}[\text{Adj}(s\mathbf{I} - \mathbf{A})]\mathbf{C} + \mathbf{D} = \frac{1}{\det \mathbf{D}(s)}\mathbf{N}(s)[\text{Adj}(\mathbf{D}(s))]$$

$$= \frac{1}{\det \bar{\mathbf{D}}(s)}[\text{Adj}(\bar{\mathbf{D}}(s))]\bar{\mathbf{N}}(s)$$

Because the three polynomials $\det (s\mathbf{I} - \mathbf{A})$, $\det \mathbf{D}(s)$, and $\det \bar{\mathbf{D}}(s)$ have the same degree, they must denote the same polynomial except possibly different leading coefficients. Thus we conclude the following:

- deg $\hat{\mathbf{G}}(s) = $ deg det $\mathbf{D}(s) = $ deg det $\bar{\mathbf{D}}(s) = $ dim $\mathbf{A}$.
- The characteristic polynomial of $\hat{\mathbf{G}}(s) = k_1$ det $\mathbf{D}(s) = k_2$ det $\bar{\mathbf{D}}(s) = k_3$ det $(s\mathbf{I} - \mathbf{A})$ for some nonzero constant k_i.
- The set of column degrees of $\mathbf{D}(s)$ equals the set of controllability indices of $(\mathbf{A}, \mathbf{B})$.
- The set of row degrees of $\bar{\mathbf{D}}(s)$ equals the set of observability indices of $(\mathbf{A}, \mathbf{C})$.

We see that coprime fractions and controllable and observable state equations contain essentially the same information. Thus either description can be used in analysis and design.

7.10 Realizations from Matrix Markov Parameters

Consider a $q \times p$ strictly proper rational matrix $\hat{\mathbf{G}}(s)$. We expand it as

$$\hat{\mathbf{G}}(s) = \mathbf{H}(1)s^{-1} + \mathbf{H}(2)s^{-2} + \mathbf{H}(3)s^{-3} + \cdots \qquad (7.106)$$

where $\mathbf{H}(m)$ are $q \times p$ constant matrices. Let r be the degree of the least common denominator of all entries of $\hat{\mathbf{G}}(s)$. We form

$$
\mathbf{T} = \begin{bmatrix} \mathbf{H}(1) & \mathbf{H}(2) & \mathbf{H}(3) & \cdots & \mathbf{H}(r) \\ \mathbf{H}(2) & \mathbf{H}(3) & \mathbf{H}(4) & \cdots & \mathbf{H}(r+1) \\ \mathbf{H}(3) & \mathbf{H}(4) & \mathbf{H}(5) & \cdots & \mathbf{H}(r+2) \\ \vdots & \vdots & \vdots & \ddots & \vdots \\ \mathbf{H}(r) & \mathbf{H}(r+1) & \mathbf{H}(r+2) & \cdots & \mathbf{H}(2r-1) \end{bmatrix}
\tag{7.107}
$$

$$
\tilde{\mathbf{T}} = \begin{bmatrix} \mathbf{H}(2) & \mathbf{H}(3) & \mathbf{H}(4) & \cdots & \mathbf{H}(r+1) \\ \mathbf{H}(3) & \mathbf{H}(4) & \mathbf{H}(5) & \cdots & \mathbf{H}(r+2) \\ \mathbf{H}(4) & \mathbf{H}(5) & \mathbf{H}(6) & \cdots & \mathbf{H}(r+3) \\ \vdots & \vdots & \vdots & \ddots & \vdots \\ \mathbf{H}(r+1) & \mathbf{H}(r+2) & \mathbf{H}(r+3) & \cdots & \mathbf{H}(2r) \end{bmatrix}
\tag{7.108}
$$

Note that $\mathbf{T}$ and $\tilde{\mathbf{T}}$ have r block columns and r block rows and, consequently, have dimension $rq \times rp$. As in (7.53) and (7.55), we have

$$
\mathbf{T} = OC \quad \text{and} \quad \tilde{\mathbf{T}} = OAC
\tag{7.109}
$$

where O and C are some observability and controllability matrices of order $rq \times n$ and $n \times rp$, respectively. Note that n is not yet determined. Because r equals the degree of the minimal polynomial of any minimal realization of $\hat{\mathbf{G}}(s)$ and because of (6.16) and (6.34), the matrix $\mathbf{T}$ is sufficiently large to have rank n. This is stated as a theorem.

▶ **Theorem 7.M7**

A strictly proper rational matrix $\hat{\mathbf{G}}(s)$ has degree n if and only if the matrix $\mathbf{T}$ in (7.107) has rank n.

The singular-value decomposition method discussed in Section 7.5 can be applied to the matrix case with some modification. This is discussed in the following. First we use singular-value decomposition to express $\mathbf{T}$ as

$$
\mathbf{T} = \mathbf{K} \begin{bmatrix} \mathbf{\Lambda} & \mathbf{0} \\ \mathbf{0} & \mathbf{0} \end{bmatrix} \mathbf{L}'
\tag{7.110}
$$

where $\mathbf{K}$ and $\mathbf{L}$ are orthogonal matrices and $\mathbf{\Lambda} = \mathrm{diag}(\lambda_1, \lambda_2, \ldots, \lambda_n)$, where λ_i are the positive square roots of the positive eigenvalues of $\mathbf{T}'\mathbf{T}$. Clearly n is the rank of $\mathbf{T}$. Let $\bar{\mathbf{K}}$ denote the first n columns of $\mathbf{K}$ and $\bar{\mathbf{L}}'$ denote the first n rows of $\mathbf{L}'$. Then we can write $\mathbf{T}$ as

$$
\mathbf{T} = \bar{\mathbf{K}}\mathbf{\Lambda}\bar{\mathbf{L}}' = \bar{\mathbf{K}}\mathbf{\Lambda}^{1/2}\mathbf{\Lambda}^{1/2}\bar{\mathbf{L}}' =: OC
\tag{7.111}
$$

where

$$
O = \bar{\mathbf{K}}\mathbf{\Lambda}^{1/2} \quad \text{and} \quad C = \mathbf{\Lambda}^{1/2}\bar{\mathbf{L}}'
$$

Note that O is $nq \times n$ and C is $n \times np$. They are not square and their inverses are not defined. However, their pseudoinverses are defined. The pseudoinverse of O is, as defined in (7.69),

$$O^+ = [(\boldsymbol{\Lambda}^{1/2})'\bar{\mathbf{K}}'\bar{\mathbf{K}}\boldsymbol{\Lambda}^{1/2}]^{-1}(\boldsymbol{\Lambda}^{1/2})'\bar{\mathbf{K}}'$$

Because $\mathbf{K}$ is orthogonal, we have $\bar{\mathbf{K}}'\bar{\mathbf{K}} = \mathbf{I}$ and because $\boldsymbol{\Lambda}^{1/2}$ is symmetric, the pseudoinverse O^+ reduces to

$$O^+ = \boldsymbol{\Lambda}^{-1/2}\bar{\mathbf{K}}' \tag{7.112}$$

Similarly, we have

$$C^+ = \bar{\mathbf{L}}\boldsymbol{\Lambda}^{-1/2} \tag{7.113}$$

Then, as in (7.64) through (7.66), the triplet

$$\mathbf{A} = O^+\tilde{\mathbf{T}}C^+ \tag{7.114}$$

$$\mathbf{B} = \text{first } p \text{ columns of } C \tag{7.115}$$

$$\mathbf{C} = \text{first } q \text{ rows of } O \tag{7.116}$$

is a minimal realization of $\hat{\mathbf{G}}(s)$. This realization has the property

$$O'O = \boldsymbol{\Lambda}^{1/2}\bar{\mathbf{K}}'\bar{\mathbf{K}}\boldsymbol{\Lambda}^{1/2} = \boldsymbol{\Lambda}$$

and, using $\bar{\mathbf{L}}'\bar{\mathbf{L}} = \mathbf{I}$,

$$CC' = \boldsymbol{\Lambda}^{1/2}\bar{\mathbf{L}}'\bar{\mathbf{L}}\boldsymbol{\Lambda}^{1/2} = \boldsymbol{\Lambda}$$

Thus the realization is a balanced realization. The MATLAB procedure in Example 7.3 can be applied directly to the matrix case if the function inverse (`inv`) is replaced by the function pseudoinverse (`pinv`). We see once again that the procedure in the scalar case can be extended directly to the matrix case. We also mention that if we decompose $\mathbf{T} = OC$ in (7.111) differently, we will obtain a different realization. This will not be discussed.

7.11 Concluding Remarks

In addition to a number of minimal realizations, we introduced in this chapter coprime fractions (right fractions with column reducedness and left fractions with row reducedness). These fractions can readily be obtained by searching linearly independent vectors of generalized resultants and then solving monic null vectors of their submatrices. A fundamental result of this chapter is that controllable and observable state equations are essentially equivalent to coprime polynomial fractions, denoted as

controllable and observable state equations

⇕

coprime polynomial fractions

Thus either description can be used to describe a system. We use the former in the next chapter and the latter in Chapter 9 to carry out various designs.

A great deal more can be said regarding coprime polynomial fractions. For example, it is possible to show that all coprime fractions are related by unimodular matrices. Controllability and observability conditions can also be expressed in terms of coprimeness conditions. See References [4, 6, 13, 20]. The objectives of this chapter are to discuss a numerical method to compute coprime fractions and to introduce just enough background to carry out designs in Chapter 9.

In addition to polynomial fractions, it is possible to express transfer functions as stable rational function fractions. See References [9, 21]. Stable rational function fractions can be developed without discussing polynomial fractions; however, polynomial fractions can provide an efficient way of computing rational fractions. Thus this chapter is also useful in studying rational fractions.

PROBLEMS

7.1 Given

$$\hat{g}(s) = \frac{s - 1}{(s^2 - 1)(s + 2)}$$

find a three-dimensional controllable realization. Check its observability.

7.2 Find a three-dimensional observable realization for the transfer function in Problem 7.1. Check its controllability.

7.3 Find an uncontrollable and unobservable realization for the transfer function in Problem 7.1. Find also a minimal realization.

7.4 Use the Sylvester resultant to find the degree of the transfer function in Problem 7.1.

7.5 Use the Sylvester resultant to reduce $(2s - 1)/(4s^2 - 1)$ to a coprime fraction.

7.6 Form the Sylvester resultant of $\hat{g}(s) = (s + 2)/(s^2 + 2s)$ by arranging the coefficients of $N(s)$ and $D(s)$ in descending powers of s and then search linearly independent columns in order from left to right. Is it true that all D-columns are linearly independent of their LHS columns? Is it true that the degree of $\hat{g}(s)$ equals the number of linearly independent N-columns?

7.7 Consider

$$\hat{g}(s) = \frac{\beta_1 s + \beta_2}{s^2 + \alpha_1 s + \alpha_2} =: \frac{N(s)}{D(s)}$$

and its realization

$$\dot{\mathbf{x}} = \begin{bmatrix} -\alpha_1 & -\alpha_2 \\ 1 & 0 \end{bmatrix} \mathbf{x} + \begin{bmatrix} 1 \\ 0 \end{bmatrix} u \qquad y = [\beta_1 \ \beta_2] \mathbf{x}$$

Show that the state equation is observable if and only if the Sylvester resultant of $D(s)$ and $N(s)$ is nonsingular.

7.8 Repeat Problem 7.7 for a transfer function of degree 3 and its controllable-form realization.

7.9 Verify Theorem 7.7 for $\hat{g}(s) = 1/(s + 1)^2$.

7.10 Use the Markov parameters of $\hat{g}(s) = 1/(s+1)^2$ to find an irreducible companion-form realization.

7.11 Use the Markov parameters of $\hat{g}(s) = 1/(s+1)^2$ to find an irreducible balanced-form realization.

7.12 Show that the two state equations

$$\dot{\mathbf{x}} = \begin{bmatrix} 2 & 1 \\ 0 & 1 \end{bmatrix} \mathbf{x} + \begin{bmatrix} 1 \\ 0 \end{bmatrix} u \qquad y = [2 \ 2]\mathbf{x}$$

and

$$\dot{\mathbf{x}} = \begin{bmatrix} 2 & 0 \\ -1 & -1 \end{bmatrix} \mathbf{x} + \begin{bmatrix} 1 \\ 2 \end{bmatrix} u \qquad y = [2 \ 0]\mathbf{x}$$

are realizations of $(2s + 2)/(s^2 - s - 2)$. Are they minimal realizations? Are they algebraically equivalent?

7.13 Find the characteristic polynomials and degrees of the following proper rational matrices

$$\hat{\mathbf{G}}_1(s) = \begin{bmatrix} \dfrac{1}{s} & \dfrac{s+3}{s+1} \\ \dfrac{1}{s+3} & \dfrac{s}{s+1} \end{bmatrix} \qquad \hat{\mathbf{G}}_2(s) = \begin{bmatrix} \dfrac{1}{(s+1)^2} & \dfrac{1}{(s+1)(s+2)} \\ \dfrac{1}{s+2} & \dfrac{1}{(s+1)(s+2)} \end{bmatrix}$$

and

$$\hat{\mathbf{G}}_3(s) = \begin{bmatrix} \dfrac{1}{(s+1)^2} & \dfrac{s+3}{s+2} & \dfrac{1}{s+5} \\ \dfrac{1}{(s+3)^2} & \dfrac{s+1}{s+4} & \dfrac{1}{s} \end{bmatrix}$$

Note that each entry of $\hat{\mathbf{G}}_3(s)$ has different poles from other entries.

7.14 Use the left fraction

$$\hat{\mathbf{G}}(s) = \begin{bmatrix} s & 1 \\ -s & s \end{bmatrix}^{-1} \begin{bmatrix} 1 \\ -1 \end{bmatrix}$$

to form a generalized resultant as in (7.83), and then search its linearly independent columns in order from left to right. What is the number of linearly independent N-columns? What is the degree of $\hat{\mathbf{G}}(s)$? Find a right coprime fraction of $\hat{\mathbf{G}}(s)$. Is the given left fraction coprime?

7.15 Are all D-columns in the generalized resultant in Problem 7.14 linearly independent of their LHS columns? Now in forming the generalized resultant, the coefficient matrices of $D(s)$ and $N(s)$ are arranged in descending powers of s, instead of ascending powers of s as in Problem 7.14. Is it true that all D-columns are linearly independent of their LHS columns? Does the degree of $\hat{\mathbf{G}}(s)$ equal the number of linearly independent N-columns? Does Theorem 7.M4 hold?

7.16 Use the right coprime fraction of $\hat{\mathbf{G}}(s)$ obtained in Problem 7.14 to form a generalized resultant as in (7.89), search its linearly independent rows in order from top to bottom, and then find a left coprime fraction of $\hat{\mathbf{G}}(s)$.

7.17 Find a right coprime fraction of

$$\hat{\mathbf{G}}(s) = \begin{bmatrix} \dfrac{s^2 + 1}{s^3} & \dfrac{2s + 1}{s^2} \\ \dfrac{s + 2}{s^2} & \dfrac{2}{s} \end{bmatrix}$$

and then a minimal realization.

Chapter

8

State Feedback
and State Estimators

8.1 Introduction

The concepts of controllability and observability were used in the preceding two chapters to study the internal structure of systems and to establish the relationships between the internal and external descriptions. Now we discuss their implications in the design of feedback control systems.

Most control systems can be formulated as shown in Fig. 8.1, in which the *plant* and the reference signal $r(t)$ are given. The input $u(t)$ of the plant is called the *actuating signal* or *control signal*. The output $y(t)$ of the plant is called the *plant output* or *controlled signal*. The problem is to design an overall system so that the plant output will follow as closely as possible the reference signal $r(t)$. There are two types of control. If the actuating signal $u(t)$ depends only on the reference signal and is independent of the plant output, the control is called an *open-loop* control. If the actuating signal depends on both the reference signal and the plant output, the control is called a *closed-loop* or *feedback* control. The open-loop control is, in general, not satisfactory if there are plant parameter variations and/or there are noise and disturbance around the system. A properly designed feedback system, on the other hand,

Figure 8.1 Design of control systems.

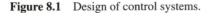
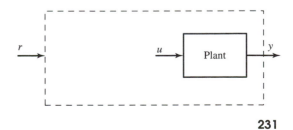

can reduce the effect of parameter variations and suppress noise and disturbance. Therefore feedback control is more widely used in practice.

This chapter studies designs using state-space equations. Designs using coprime fractions will be studied in the next chapter. We study first single-variable systems and then multivariable systems. We study only linear time-invariant systems.

8.2 State Feedback

Consider the n-dimensional single-variable state equation

$$\dot{\mathbf{x}} = \mathbf{A}\mathbf{x} + \mathbf{b}u$$

$$y = \mathbf{c}\mathbf{x}$$

(8.1)

where we have assumed $d = 0$ to simplify discussion. In state feedback, the input u is given by

$$u = r - \mathbf{k}\mathbf{x} = r - [k_1 \ k_2 \ \cdots \ k_n]\mathbf{x} = r - \sum_{i=1}^{n} k_i x_i$$

(8.2)

as shown in Fig. 8.2. Each feedback gain k_i is a real constant. This is called the *constant gain negative state feedback* or, simply, *state feedback*. Substituting (8.2) into (8.1) yields

$$\dot{\mathbf{x}} = (\mathbf{A} - \mathbf{b}\mathbf{k})\mathbf{x} + \mathbf{b}r$$

$$y = \mathbf{c}\mathbf{x}$$

(8.3)

▷ **Theorem 8.1**

The pair $(\mathbf{A} - \mathbf{b}\mathbf{k}, \mathbf{b})$, for any $1 \times n$ real constant vector $\mathbf{k}$, is controllable if and only if $(\mathbf{A}, \mathbf{b})$ is controllable.

⇒ **Proof:** We show the theorem for $n = 4$. Define

$$C = [\mathbf{b} \ \mathbf{A}\mathbf{b} \ \mathbf{A}^2\mathbf{b} \ \mathbf{A}^3\mathbf{b}]$$

and

$$C_f = [\mathbf{b} \ (\mathbf{A} - \mathbf{b}\mathbf{k})\mathbf{b} \ (\mathbf{A} - \mathbf{b}\mathbf{k})^2\mathbf{b} \ (\mathbf{A} - \mathbf{b}\mathbf{k})^3\mathbf{b}]$$

They are the controllability matrices of (8.1) and (8.3). It is straightforward to verify

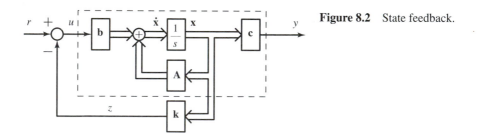

Figure 8.2   State feedback.

$$C_f = C \begin{bmatrix} 1 & -\mathbf{kb} & -\mathbf{k(A-bk)b} & -\mathbf{k(A-bk)^2b} \\ 0 & 1 & -\mathbf{kb} & -\mathbf{k(A-bk)b} \\ 0 & 0 & 1 & -\mathbf{kb} \\ 0 & 0 & 0 & 1 \end{bmatrix} \qquad (8.4)$$

Note that $\mathbf{k}$ is $1 \times n$ and $\mathbf{b}$ is $n \times 1$. Thus $\mathbf{kb}$ is scalar; so is every entry in the rightmost matrix in (8.4). Because the rightmost matrix is nonsingular for any $\mathbf{k}$, the rank of C_f equals the rank of C. Thus (8.3) is controllable if and only if (8.1) is controllable.

This theorem can also be established directly from the definition of controllability. Let $\mathbf{x}_0$ and $\mathbf{x}_1$ be two arbitrary states. If (8.1) is controllable, there exists an input u_1 that transfers $\mathbf{x}_0$ to $\mathbf{x}_1$ in a finite time. Now if we choose $r_1 = u_1 + \mathbf{kx}$, then the input r_1 of the state feedback system will transfer $\mathbf{x}_0$ to $\mathbf{x}_1$. Thus we conclude that if (8.1) is controllable, so is (8.3).

We see from Fig. 8.2 that the input r does not control the state $\mathbf{x}$ directly; it generates u to control $\mathbf{x}$. Therefore, if u cannnot control $\mathbf{x}$, neither can r. This establishes once again the theorem.   Q.E.D.

Although the controllability property is invariant under any state feedback, the observability property is not. This is demonstrated by the example that follows.

EXAMPLE 8.1 Consider the state equation

$$\dot{\mathbf{x}} = \begin{bmatrix} 1 & 2 \\ 3 & 1 \end{bmatrix} \mathbf{x} + \begin{bmatrix} 0 \\ 1 \end{bmatrix} u$$

$$y = [1 \ 2]\mathbf{x}$$

The state equation can readily be shown to be controllable and observable. Now we introduce the state feedback

$$u = r - [3 \ 1]\mathbf{x}$$

Then the state feedback equation becomes

$$\dot{\mathbf{x}} = \begin{bmatrix} 1 & 2 \\ 0 & 0 \end{bmatrix} \mathbf{x} + \begin{bmatrix} 0 \\ 1 \end{bmatrix} r$$

$$y = [1 \ 2]\mathbf{x}$$

Its controllability matrix is

$$C_f = \begin{bmatrix} 0 & 2 \\ 1 & 0 \end{bmatrix}$$

which is nonsingular. Thus the state feedback equation is controllable. Its observability matrix is

$$O_f = \begin{bmatrix} 1 & 2 \\ 1 & 2 \end{bmatrix}$$

which is singular. Thus the state feedback equation is not observable. The reason that the observability property may not be preserved in state feedback will be given later.

We use an example to discuss what can be achieved by state feedback.

EXAMPLE 8.2 Consider a plant described by

$$\dot{\mathbf{x}} = \begin{bmatrix} 1 & 3 \\ 3 & 1 \end{bmatrix} \mathbf{x} + \begin{bmatrix} 1 \\ 0 \end{bmatrix} u$$

The A-matrix has characteristic polynomial

$$\Delta(s) = (s-1)^2 - 9 = s^2 - 2s - 8 = (s-4)(s+2)$$

and, consequently, eigenvalues 4 and -2. It is unstable. Let us introduce state feedback $u = r - [k_1 \ k_2]\mathbf{x}$. Then the state feedback system is described by

$$\dot{\mathbf{x}} = \left(\begin{bmatrix} 1 & 3 \\ 3 & 1 \end{bmatrix} - \begin{bmatrix} k_1 & k_2 \\ 0 & 0 \end{bmatrix} \right) \mathbf{x} + \begin{bmatrix} 1 \\ 0 \end{bmatrix} r$$

$$= \begin{bmatrix} 1-k_1 & 3-k_2 \\ 3 & 1 \end{bmatrix} \mathbf{x} + \begin{bmatrix} 1 \\ 0 \end{bmatrix} r$$

This new A-matrix has characteristic polynomial

$$\Delta_f(s) = (s-1+k_1)(s-1) - 3(3-k_2)$$
$$= s^2 + (k_1-2)s + (3k_2 - k_1 - 8)$$

It is clear that the roots of $\Delta_f(s)$ or, equivalently, the eigenvalues of the state feedback system can be placed in any positions by selecting appropriate k_1 and k_2. For example, if the two eigenvalues are to be placed at $-1 \pm j2$, then the desired characteristic polynomial is $(s+1+j2)(s+1-j2) = s^2 + 2s + 5$. Equating $k_1 - 2 = 2$ and $3k_2 - k_1 - 8 = 5$ yields $k_1 = 4$ and $k_2 = 17/3$. Thus the state feedback gain [4 17/3] will shift the eigenvalues from $4, -2$ to $-1 \pm j2$.

This example shows that state feedback can be used to place eigenvalues in any positions. Moreover the feedback gain can be computed by direct substitution. This approach, however, will become very involved for three- or higher-dimensional state equations. More seriously, the approach will not reveal how the controllability condition comes into the design. Therefore a more systematic approach is desirable. Before proceeding, we need the following theorem. We state the theorem for $n = 4$; the theorem, however, holds for every positive integer n.

▶ **Theorem 8.2**

Consider the state equation in (8.1) with $n = 4$ and the characteristic polynomial

$$\Delta(s) = \det(s\mathbf{I} - \mathbf{A}) = s^4 + \alpha_1 s^3 + \alpha_2 s^2 + \alpha_3 s + \alpha_4 \tag{8.5}$$

If (8.1) is controllable, then it can be transformed by the transformation $\bar{\mathbf{x}} = \mathbf{P}\mathbf{x}$ with

$$\mathbf{Q} := \mathbf{P}^{-1} = [\mathbf{b} \ \mathbf{Ab} \ \mathbf{A}^2\mathbf{b} \ \mathbf{A}^3\mathbf{b}] \begin{bmatrix} 1 & \alpha_1 & \alpha_2 & \alpha_3 \\ 0 & 1 & \alpha_1 & \alpha_2 \\ 0 & 0 & 1 & \alpha_1 \\ 0 & 0 & 0 & 1 \end{bmatrix} \tag{8.6}$$

into the controllable canonical form

$$\dot{\bar{x}} = \bar{A}\bar{x} + \bar{b}u = \begin{bmatrix} -\alpha_1 & -\alpha_2 & -\alpha_3 & -\alpha_4 \\ 1 & 0 & 0 & 0 \\ 0 & 1 & 0 & 0 \\ 0 & 0 & 1 & 0 \end{bmatrix} \bar{x} + \begin{bmatrix} 1 \\ 0 \\ 0 \\ 0 \end{bmatrix} u \qquad (8.7)$$

$$y = \bar{c}\bar{x} = [\beta_1 \ \beta_2 \ \beta_3 \ \beta_4]\bar{x}$$

Furthermore, the transfer function of (8.1) with $n = 4$ equals

$$\hat{g}(s) = \frac{\beta_1 s^3 + \beta_2 s^2 + \beta_3 s + \beta_4}{s^4 + \alpha_1 s^3 + \alpha_2 s^2 + \alpha_3 s + \alpha_4} \qquad (8.8)$$

Proof: Let C and $\bar{C}$ be the controllability matrices of (8.1) and (8.7). In the SISO case, both C and $\bar{C}$ are square. If (8.1) is controllable or C is nonsingular, so is $\bar{C}$. And they are related by $\bar{C} = \mathbf{P}C$ (Theorem 6.2 and Equation (6.20)). Thus we have

$$\mathbf{P} = \bar{C}C^{-1} \quad \text{or} \quad \mathbf{Q} := \mathbf{P}^{-1} = C\bar{C}^{-1}$$

The controllability matrix $\bar{C}$ of (8.7) was computed in (7.10). Its inverse turns out to be

$$\bar{C}^{-1} = \begin{bmatrix} 1 & \alpha_1 & \alpha_2 & \alpha_3 \\ 0 & 1 & \alpha_1 & \alpha_2 \\ 0 & 0 & 1 & \alpha_1 \\ 0 & 0 & 0 & 1 \end{bmatrix} \qquad (8.9)$$

This can be verified by multiplying (8.9) with (7.10) to yield a unit matrix. Note that the constant term α_4 of (8.5) does not appear in (8.9). Substituting (8.9) into $\mathbf{Q} = C\bar{C}^{-1}$ yields (8.6). As shown in Section 7.2, the state equation in (8.7) is a realization of (8.8). Thus the transfer function of (8.7) and, consequently, of (8.1) equals (8.8). This establishes the theorem. Q.E.D.

With this theorem, we are ready to discuss eigenvalue assignment by state feedback.

▶ Theorem 8.3

If the n-dimensional state equation in (8.1) is controllable, then by state feedback $u = r - \mathbf{kx}$, where $\mathbf{k}$ is a $1 \times n$ real constant vector, the eigenvalues of $\mathbf{A} - \mathbf{bk}$ can arbitrarily be assigned provided that complex conjugate eigenvalues are assigned in pairs.

Proof: We again prove the theorem for $n = 4$. If (8.1) is controllable, it can be transformed into the controllable canonical form in (8.7). Let $\bar{A}$ and $\bar{b}$ denote the matrices in (8.7). Then we have $\bar{A} = \mathbf{P}A\mathbf{P}^{-1}$ and $\bar{b} = \mathbf{P}b$. Substituting $\bar{x} = \mathbf{P}x$ into the state feedback yields

$$u = r - \mathbf{kx} = r - \mathbf{k}\mathbf{P}^{-1}\bar{x} =: r - \bar{k}\bar{x}$$

where $\bar{k} := \mathbf{k}\mathbf{P}^{-1}$. Because $\bar{A} - \bar{b}\bar{k} = \mathbf{P}(A - bk)\mathbf{P}^{-1}$, $A - bk$ and $\bar{A} - \bar{b}\bar{k}$ have the same set of eigenvalues. From any set of desired eigenvalues, we can readily form

$$\Delta_f(s) = s^4 + \bar{\alpha}_1 s^3 + \bar{\alpha}_2 s^2 + \bar{\alpha}_3 s + \bar{\alpha}_4 \tag{8.10}$$

If $\bar{\mathbf{k}}$ is chosen as

$$\bar{\mathbf{k}} = [\bar{\alpha}_1 - \alpha_1 \quad \bar{\alpha}_2 -, \alpha_2 \quad \bar{\alpha}_3 - \alpha_3 \quad \bar{\alpha}_4 - \alpha_4] \tag{8.11}$$

the state feedback equation becomes

$$\dot{\bar{\mathbf{x}}} = (\bar{\mathbf{A}} - \bar{\mathbf{b}}\bar{\mathbf{k}})\bar{\mathbf{x}} + \bar{\mathbf{b}}r = \begin{bmatrix} -\bar{\alpha}_1 & -\bar{\alpha}_2 & -\bar{\alpha}_3 & -\bar{\alpha}_4 \\ 1 & 0 & 0 & 0 \\ 0 & 1 & 0 & 0 \\ 0 & 0 & 1 & 0 \end{bmatrix} \bar{\mathbf{x}} + \begin{bmatrix} 1 \\ 0 \\ 0 \\ 0 \end{bmatrix} r \tag{8.12}$$

$$y = [\beta_1 \quad \beta_2 \quad \beta_3 \quad \beta_4]\bar{\mathbf{x}}$$

Because of the companion form, the characteristic polynomial of $(\bar{\mathbf{A}} - \bar{\mathbf{b}}\bar{\mathbf{k}})$ and, consequently, of $(\mathbf{A} - \mathbf{b}\mathbf{k})$ equals (8.10). Thus the state feedback equation has the set of desired eigenvalues. The feedback gain $\mathbf{k}$ can be computed from

$$\mathbf{k} = \bar{\mathbf{k}}\mathbf{P} = \bar{\mathbf{k}}\bar{C}C^{-1} \tag{8.13}$$

with $\bar{\mathbf{k}}$ in (8.11), $\bar{C}^{-1}$ in (8.9), and $C = [\mathbf{b} \quad \mathbf{A}\mathbf{b} \quad \mathbf{A}^2\mathbf{b} \quad \mathbf{A}^3\mathbf{b}]$. Q.E.D.

We give an alternative derivation of the formula in (8.11). We compute

$$\Delta_f(s) = \det(s\mathbf{I} - \mathbf{A} + \mathbf{b}\mathbf{k}) = \det\left((s\mathbf{I} - \mathbf{A})[\mathbf{I} + (s\mathbf{I} - \mathbf{A})^{-1}\mathbf{b}\mathbf{k}]\right)$$

$$= \det(s\mathbf{I} - \mathbf{A})\det[\mathbf{I} + (s\mathbf{I} - \mathbf{A})^{-1}\mathbf{b}\mathbf{k}]$$

which becomes, using (8.5) and (3.64),

$$\Delta_f(s) = \Delta(s)[1 + \mathbf{k}(s\mathbf{I} - \mathbf{A})^{-1}\mathbf{b}]$$

Thus we have

$$\Delta_f(s) - \Delta(s) = \Delta(s)\mathbf{k}(s\mathbf{I} - \mathbf{A})^{-1}\mathbf{b} = \Delta(s)\bar{\mathbf{k}}(s\mathbf{I} - \bar{\mathbf{A}})^{-1}\bar{\mathbf{b}} \tag{8.14}$$

Let z be the output of the feedback gain shown in Fig. 8.2 and let $\bar{\mathbf{k}} = [\bar{k}_1 \quad \bar{k}_2 \quad \bar{k}_3 \quad \bar{k}_4]$. Because the transfer function from u to y in Fig. 8.2 equals

$$\bar{\mathbf{c}}(s\mathbf{I} - \bar{\mathbf{A}})^{-1}\bar{\mathbf{b}} = \frac{\beta_1 s^3 + \beta_2 s^2 + \beta_3 s + \beta_4}{\Delta(s)}$$

the transfer function from u to z should equal

$$\bar{\mathbf{k}}(s\mathbf{I} - \bar{\mathbf{A}})^{-1}\bar{\mathbf{b}} = \frac{\bar{k}_1 s^3 + \bar{k}_2 s^2 + \bar{k}_3 s + \bar{k}_4}{\Delta(s)} \tag{8.15}$$

Substituting (8.15), (8.5), and (8.10) into (8.14) yields

$$(\bar{\alpha}_1 - \alpha_1)s^3 + (\bar{\alpha}_2 - \alpha_2)s^2 + (\bar{\alpha}_3 - \alpha_3)s + (\bar{\alpha}_4 - \alpha_4) = \bar{k}_1 s^3 + \bar{k}_2 s^2 + \bar{k}_3 s + \bar{k}_4$$

This yields (8.11).

Feedback transfer function   Consider a plant described by $(\mathbf{A}, \mathbf{b}, \mathbf{c})$. If $(\mathbf{A}, \mathbf{b})$ is controllable, $(\mathbf{A}, \mathbf{b}, \mathbf{c})$ can be transformed into the controllable form in (8.7) and its transfer function can then be read out as, for $n = 4$,

$$\hat{g}(s) = \mathbf{c}(s\mathbf{I} - \mathbf{A})^{-1}\mathbf{b} = \frac{\beta_1 s^3 + \beta_2 s^2 + \beta_3 s + \beta_4}{s^4 + \alpha_1 s^3 + \alpha_2 s^2 + \alpha_3 s + \alpha_4} \tag{8.16}$$

After state feedback, the state equation becomes $(\mathbf{A} - \mathbf{bk}, \mathbf{b}, \mathbf{c})$ and is still of the controllable canonical form as shown in (8.12). Thus the feedback transfer function from r to y is

$$\hat{g}_f(s) = \mathbf{c}(s\mathbf{I} - \mathbf{A} + \mathbf{bk})^{-1}\mathbf{b} = \frac{\beta_1 s^3 + \beta_2 s^2 + \beta_3 s + \beta_4}{s^4 + \bar{\alpha}_1 s^3 + \bar{\alpha}_2 s^2 + \bar{\alpha}_3 s + \bar{\alpha}_4} \tag{8.17}$$

We see that the numerators of (8.16) and (8.17) are the same. In other words, state feedback does not affect the zeros of the plant transfer function. This is actually a general property of feedback: *feedback can shift the poles of a plant but has no effect on the zeros.* This can be used to explain hy a state feedback may alter the observability property of a state quation. If one or more poles are shifted to coincide with zeros of $\hat{g}(s)$, then the numerator and denominator of $\hat{g}_f(s)$ in (8.17) are not coprime. Thus the state equation in (8.12) and, equivalently, $(\mathbf{A} - \mathbf{bk}, \mathbf{c})$ are not observable (Theorem 7.1).

EXAMPLE 8.3 Consider the inverted pendulum studied in Example 6.2. Its state equation is, as derived in (6.11),

$$\dot{\mathbf{x}} = \begin{bmatrix} 0 & 1 & 0 & 0 \\ 0 & 0 & -1 & 0 \\ 0 & 0 & 0 & 1 \\ 0 & 0 & 5 & 0 \end{bmatrix} \mathbf{x} + \begin{bmatrix} 0 \\ 1 \\ 0 \\ -2 \end{bmatrix} u \tag{8.18}$$

$$y = [1 \ 0 \ 0 \ 0]\mathbf{x}$$

It is controllable; thus its eigenvalues can be assigned arbitrarily. Because the A-matrix is block triangular, its characteristic polynomial can be obtained by inspection as

$$\Delta(s) = s^2(s^2 - 5) = s^4 + 0 \cdot s^3 - 5s^2 + 0 \cdot s + 0$$

First we compute $\mathbf{P}$ that will transform (8.18) into the controllable canonical form. Using (8.6), we have

$$\mathbf{P}^{-1} = C\bar{C}^{-1} = \begin{bmatrix} 0 & 1 & 0 & 2 \\ 1 & 0 & 2 & 0 \\ 0 & -2 & 0 & -10 \\ -2 & 0 & -10 & 0 \end{bmatrix} \begin{bmatrix} 1 & 0 & -5 & 0 \\ 0 & 1 & 0 & -5 \\ 0 & 0 & 1 & 0 \\ 0 & 0 & 0 & 1 \end{bmatrix}$$

$$= \begin{bmatrix} 0 & 1 & 0 & -3 \\ 1 & 0 & -3 & 0 \\ 0 & -2 & 0 & 0 \\ -2 & 0 & 0 & 0 \end{bmatrix}$$

Its inverse is

$$\mathbf{P} = \begin{bmatrix} 0 & 0 & 0 & -\frac{1}{2} \\ 0 & 0 & -\frac{1}{2} & 0 \\ 0 & -\frac{1}{3} & 0 & -\frac{1}{6} \\ -\frac{1}{3} & 0 & -\frac{1}{6} & 0 \end{bmatrix}$$

Let the desired eigenvalues be $-1.5 \pm 0.5j$ and $-1 \pm j$. Then we have

$$\Delta_f(s) = (s + 1.5 - 0.5j)(s + 1.5 + 0.5j)(s + 1 - j)(s + 1 + j)$$
$$= s^4 + 5s^3 + 10.5s^2 + 11s + 5$$

Thus we have, using (8.11),

$$\bar{\mathbf{k}} = [5 - 0 \ \ 10.5 + 5 \ \ 11 - 0 \ \ 5 - 0] = [5 \ \ 15.5 \ \ 11 \ \ 5]$$

and

$$\mathbf{k} = \bar{\mathbf{k}}\mathbf{P} = [-\tfrac{5}{3} \ \ -\tfrac{11}{3} \ \ -\tfrac{103}{12} \ \ -\tfrac{13}{3}] \qquad (8.19)$$

This state feedback gain will shift the eigenvalues of the plant from $\{0, \ 0, \ \pm j\sqrt{5}\}$ to $\{-1.5 \pm 0.5j, \ -1 \pm j\}$.

The MATLAB function `place` computes state feedback gains for eigenvalue placement or assignment. For the example, we type

```
a=[0 1 0 0;0 0 -1 0;0 0 0 1;0 0 5 0];b=[0;1;0;-2];
p=[-1.5+0.5j -1.5-0.5j -1+j -1-j];
k=place(a,b,p)
```

which yields $[-1.6667 \ -3.6667 \ -8.5833 \ -4.3333]$. This is the gain in (8.19).

One may wonder at this point how to select a set of desired eigenvalues. This depends on the performance criteria, such as rise time, settling time, and overshoot, used in the design. Because the response of a system depends not only on poles but also on zeros, the zeros of the plant will also affect the selection. In addition, most physical systems will saturate or burn out if the magnitude of the actuating signal is very large. This will again affect the selection of desired poles. As a guide, we may place all eigenvalues inside the region denoted by C in Fig. 8.3(a). The region is bounded on the right by a vertical line. The larger the distance of the vertical line from the imaginary axis, the faster the response. The region is also bounded by two straight lines emanating from the origin with angle θ. The larger the angle, the larger the overshoot. See Reference [7]. If we place all eigenvalues at one point or group them in a very small region, then usually the response will be slow and the actuating signal will be large. Therefore it is better to place all eigenvalues evenly around a circle with radius r inside the sector as shown. The larger the radius, the faster the response; however, the actuating signal will also be larger. Furthermore, the bandwidth of the feedback system will be larger and the resulting system will be more susceptible to noise. Therefore a final selection may involve compromises among many conflicting requirements. One way to proceed is by computer simulation. Another way is to find the state feedback gain $\mathbf{k}$ to minimize the quadratic performance index

$$J = \int_0^\infty [\mathbf{x}'(t)\mathbf{Q}\mathbf{x}(t) + \mathbf{u}'(t)\mathbf{R}\mathbf{u}(t)] \, dt$$

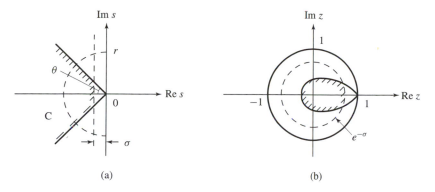

Figure 8.3   Desired eigenvalue location.

See Reference [1]. However, selecting **Q** and **R** requires trial and error. In conclusion, how to select a set of desired eigenvalues is not a simple problem.

We mention that Theorems 8.1 through 8.3—in fact, all theorems to be introduced later in this chapter—apply to the discrete-time case without any modification. The only difference is that the region in Fig. 8.3(a) must be replaced by the one in Fig. 8.3(b), which is obtained by the transformation $z = e^s$.

8.2.1 Solving the Lyapunov Equation

This subsection discusses a different method of computing state feedback gain for eigenvalue assignment. The method, however, has the restriction that the selected eigenvalues cannot contain any eigenvalues of **A**.

▶ **Procedure 8.1**

Consider controllable $(\mathbf{A}, \mathbf{b})$, where **A** is $n \times n$ and **b** is $n \times 1$. Find a $1 \times n$ real **k** such that $(\mathbf{A} - \mathbf{bk})$ has any set of desired eigenvalues that contains no eigenvalues of **A**.

1. Select an $n \times n$ matrix **F** that has the set of desired eigenvalues. The form of **F** can be chosen arbitrarily and will be discussed later.

2. Select an arbitrary $1 \times n$ vector $\bar{\mathbf{k}}$ such that $(\mathbf{F}, \bar{\mathbf{k}})$ is observable.

3. Solve the unique **T** in the Lyapunov equation $\mathbf{AT} - \mathbf{TF} = \mathbf{b}\bar{\mathbf{k}}$.

4. Compute the feedback gain $\mathbf{k} = \bar{\mathbf{k}}\mathbf{T}^{-1}$.

We justify first the procedure. If **T** is nonsingular, then $\bar{\mathbf{k}} = \mathbf{kT}$ and the Lyapunov equation $\mathbf{AT} - \mathbf{TF} = \mathbf{b}\bar{\mathbf{k}}$ implies

$$(\mathbf{A} - \mathbf{bk})\mathbf{T} = \mathbf{TF} \quad \text{or} \quad \mathbf{A} - \mathbf{bk} = \mathbf{TFT}^{-1}$$

Thus $(\mathbf{A} - \mathbf{bk})$ and **T** are similar and have the same set of eigenvalues. Thus the eigenvalues of $(\mathbf{A} - \mathbf{bk})$ can be assigned arbitrarily except those of **A**. As discussed in Section 3.7, if **A** and

F have no eigenvalues in common, then a solution **T** exists in $\mathbf{AT} - \mathbf{TF} = \mathbf{b}\bar{\mathbf{k}}$ for any $\bar{\mathbf{k}}$ and is unique. If **A** and **F** have common eigenvalues, a solution **T** may or may not exist depending on $\mathbf{b}\bar{\mathbf{k}}$. To remove this uncertainty, we require **A** and **F** to have no eigenvalues in common. What remains to be proved is the nonsingularity of **T**.

▶ **Theorem 8.4**

If **A** and **F** have no eigenvalues in common, then the unique solution **T** of $\mathbf{AT} - \mathbf{TF} = \mathbf{b}\bar{\mathbf{k}}$ is nonsingular if and only if $(\mathbf{A}, \mathbf{b})$ is controllable and $(\mathbf{F}, \bar{\mathbf{k}})$ is observable.

⟹ *Proof:* We prove the theorem for $n = 4$. Let the characteristic polynomial of **A** be

$$\Delta(s) = s^4 + \alpha_1 s^3 + \alpha_2 s^2 + \alpha_3 s + \alpha_4 \tag{8.20}$$

Then we have

$$\Delta(\mathbf{A}) = \mathbf{A}^4 + \alpha_1 \mathbf{A}^3 + \alpha_2 \mathbf{A}^2 + \alpha_3 \mathbf{A} + \alpha_4 \mathbf{I} = \mathbf{0}$$

(Cayley–Hamilton theorem). Let us consider

$$\Delta(\mathbf{F}) := \mathbf{F}^4 + \alpha_1 \mathbf{F}^3 + \alpha_2 \mathbf{F}^2 + \alpha_3 \mathbf{F} + \alpha_4 \mathbf{I} \tag{8.21}$$

If $\bar{\lambda}_i$ is an eigenvalue of **F**, then $\Delta(\bar{\lambda}_i)$ is an eigenvalue of $\Delta(\mathbf{F})$ (Problem 3.19). Because **A** and **F** have no eigenvalues in common, we have $\Delta(\bar{\lambda}_i) \neq 0$ for all eigenvalues of **F**. Because the determinant of a matrix equals the product of all its eigenvalues, we have

$$\det \Delta(\mathbf{F}) = \prod_i \Delta(\bar{\lambda}_i) \neq 0$$

Thus $\Delta(\mathbf{F})$ is nonsingular.

Substituting $\mathbf{AT} = \mathbf{TF} + \mathbf{b}\bar{\mathbf{k}}$ into $\mathbf{A}^2\mathbf{T} - \mathbf{AF}^2$ yields

$$\mathbf{A}^2\mathbf{T} - \mathbf{TF}^2 = \mathbf{A}(\mathbf{TF} + \mathbf{b}\bar{\mathbf{k}}) - \mathbf{TF}^2 = \mathbf{Ab}\bar{\mathbf{k}} + (\mathbf{AT} - \mathbf{TF})\mathbf{F}$$
$$= \mathbf{Ab}\bar{\mathbf{k}} + \mathbf{b}\bar{\mathbf{k}}\mathbf{F}$$

Proceeding forward, we can obtain the following set of equations:

$$\mathbf{IT} - \mathbf{TI} = \mathbf{0}$$
$$\mathbf{AT} - \mathbf{TF} = \mathbf{b}\bar{\mathbf{k}}$$
$$\mathbf{A}^2\mathbf{T} - \mathbf{TF}^2 = \mathbf{Ab}\bar{\mathbf{k}} + \mathbf{b}\bar{\mathbf{k}}\mathbf{F}$$
$$\mathbf{A}^3\mathbf{T} - \mathbf{TF}^3 = \mathbf{A}^2\mathbf{b}\bar{\mathbf{k}} + \mathbf{Ab}\bar{\mathbf{k}}\mathbf{F} + \mathbf{b}\bar{\mathbf{k}}\mathbf{F}^2$$
$$\mathbf{A}^4\mathbf{T} - \mathbf{TF}^4 = \mathbf{A}^3\mathbf{b}\bar{\mathbf{k}} + \mathbf{A}^2\mathbf{b}\bar{\mathbf{k}}\mathbf{F} + \mathbf{Ab}\bar{\mathbf{k}}\mathbf{F}^2 + \mathbf{b}\bar{\mathbf{k}}\mathbf{F}^3$$

We multiply the first equation by α_4, the second equation by α_3, the third equation by α_2, the fourth equation by α_1, and the last equation by 1, and then sum them up. After some manipulation, we finally obtain

$$\Delta(\mathbf{A})\mathbf{T} - \mathbf{T}\Delta(\mathbf{F}) = -\mathbf{T}\Delta(\mathbf{F})$$

$$= [\mathbf{b} \; \mathbf{Ab} \; \mathbf{A}^2\mathbf{b} \; \mathbf{A}^3\mathbf{b}] \begin{bmatrix} \alpha_3 & \alpha_2 & \alpha_1 & 1 \\ \alpha_2 & \alpha_1 & 1 & 0 \\ \alpha_1 & 1 & 0 & 0 \\ 1 & 0 & 0 & 0 \end{bmatrix} \begin{bmatrix} \bar{\mathbf{k}} \\ \bar{\mathbf{k}}\mathbf{F} \\ \bar{\mathbf{k}}\mathbf{F}^2 \\ \bar{\mathbf{k}}\mathbf{F}^3 \end{bmatrix} \tag{8.22}$$

where we have used $\Delta(\mathbf{A}) = \mathbf{0}$. If $(\mathbf{A}, \mathbf{b})$ is controllable and $(\mathbf{F}, \bar{\mathbf{k}})$ is observable, then all three matrices after the last equality are nonsingular. Thus (8.22) and the nonsingularity of $\Delta(\mathbf{F})$ imply that $\mathbf{T}$ is nonsingular. If $(\mathbf{A}, \mathbf{b})$ is uncontrollable and/or $(\mathbf{F}, \bar{\mathbf{k}})$ is unobservable, then the product of the three matrices is singular. Therefore $\mathbf{T}$ is singular. This establishes the theorem.   Q.E.D.

We now discuss the selection of $\mathbf{F}$ and $\bar{\mathbf{k}}$. Given a set of desired eigenvalues, there are infinitely many $\mathbf{F}$ that have the set of eigenvalues. If we form a polynomial from the set, we can use its coefficients to form a companion-form matrix $\mathbf{F}$ as shown in (7.14). For this $\mathbf{F}$, we can select $\bar{\mathbf{k}}$ as $[1 \; 0 \; \cdots \; 0]$ and $(\mathbf{F}, \bar{\mathbf{k}})$ is observable. If the desired eigenvalues are all distinct, we can also use the modal form discussed in Section 4.3.1. For example, if $n = 5$, and if the five distinct desired eigenvalues are selected as λ_1, $\alpha_1 \pm j\beta_1$, and $\alpha_2 \pm j\beta_2$, then we can select $\mathbf{F}$ as

$$\mathbf{F} = \begin{bmatrix} \lambda_1 & 0 & 0 & 0 & 0 \\ 0 & \alpha_1 & \beta_1 & 0 & 0 \\ 0 & -\beta_1 & \alpha_1 & 0 & 0 \\ 0 & 0 & 0 & \alpha_2 & \beta_2 \\ 0 & 0 & 0 & -\beta_2 & \alpha_2 \end{bmatrix} \tag{8.23}$$

It is a block-diagonal matrix. For this $\mathbf{F}$, if $\bar{\mathbf{k}}$ has at least one nonzero entry associated with each diagonal block such as $\bar{\mathbf{k}} = [1 \; 1 \; 0 \; 1 \; 0]$, $\bar{\mathbf{k}} = [1 \; 1 \; 0 \; 0 \; 1]$, or $\bar{\mathbf{k}} = [1 \; 1 \; 1 \; 1 \; 1]$, then $(\mathbf{F}, \bar{\mathbf{k}})$ is observable (Problem 6.16). Thus the first two steps of Procedure 8.1 are very simple. Once $\mathbf{F}$ and $\bar{\mathbf{k}}$ are selected, we may use the MATLAB function `lyap` to solve the Lyapunov equation in Step 3. Thus Procedure 8.1 is easy to carry out as the next example illustrates.

EXAMPLE 8.4 Consider the inverted pendulum studied in Example 8.3. The plant state equation is given in (8.18) and the desired eigenvalues were chosen as $-1 \pm j$ and $-1.5 \pm 0.5j$. We select $\mathbf{F}$ in modal form as

$$\mathbf{F} = \begin{bmatrix} -1 & 1 & 0 & 0 \\ -1 & -1 & 0 & 0 \\ 0 & 0 & -1.5 & 0.5 \\ 0 & 0 & -0.5 & -1.5 \end{bmatrix}$$

and $\bar{\mathbf{k}} = [1 \; 0 \; 1 \; 0]$. We type

```
a=[0 1 0 0;0 0 -1 0;0 0 0 1;0 0 5 0];b=[0;1;0;-2];
f=[-1 1 0 0;-1 -1 0 0;0 0 -1.5 0.5;0 0 -0.5 -1.5];
kb=[1 0 1 0];t=lyap(a,b,-b*kb);
k=kb*inv(t)
```

The answer is $[-1.6667 \ -3.6667 \ -8.5833 \ -4.3333]$, which is the same as the one obtained by using function `place`. If we use a different $\bar{\mathbf{k}} = [1 \ 1 \ 1 \ 1]$, we will obtain the same $\bar{\mathbf{k}}$. Note that the feedback gain is unique for the SISO case.

8.3 Regulation and Tracking

Consider the state feedback system shown in Fig. 8.2. Suppose the reference signal r is zero, and the response of the system is caused by some nonzero initial conditions. The problem is to find a state feedback gain so that the response will die out at a desired rate. This is called a *regulator problem*. This problem may arise when an aircraft is cruising at a fixed altitude H_0. Now, because of turbulance or other factors, the aircraft may deviate from the desired altitude. Bringing the deviation to zero is a regulator problem. This problem also arises in maintaining the liquid level in Fig. 2.14 at equilibrium.

A closely related problem is the tracking problem. Suppose the reference signal r is a constant or $r(t) = a$, for $t \geq 0$. The problem is to design an overall system so that $y(t)$ approaches $r(t) = a$ as t approaches infinity. This is called *asymptotic tracking* of a step reference input. It is clear that if $r(t) = a = 0$, then the tracking problem reduces to the regulator problem. Why do we then study these two problems separately? Indeed, if the same state equation is valid for all r, designing a system to track asymptotically a step reference input will automatically achieve regulation. However, a linear state equation is often obtained by shifting to an operating point and linearization, and the equation is valid only for r very small or zero; thus the study of the regulator problem is needed. We mention that a step reference input can be set by the position of a potentiometer and is therefore often referred to as set point. Maintaining a chamber at a desired temperature is often said to be regulating the temperature; it is actually tracking the desired temperature. Therefore no sharp distinction is made in practice between regulation and tracking a step reference input. Tracking a nonconstant reference signal is called a *servomechanism* problem and is a much more difficult problem.

Consider a plant described by $(\mathbf{A}, \mathbf{b}, \mathbf{c})$. If all eigenvalues of $\mathbf{A}$ lie inside the sector shown in Fig. 8.3, then the response caused by any initial conditions will decay rapidly to zero and no state feedback is needed. If $\mathbf{A}$ is stable but some eigenvalues are outside the sector, then the decay may be slow or too oscillatory. If $\mathbf{A}$ is unstable, then the response excited by any nonzero initial conditions will grow unbounded. In these situations, we may introduce state feedback to improve the behavior of the system. Let $u = r - \mathbf{kx}$. Then the state feedback equation becomes $(\mathbf{A} - \mathbf{bk}, \mathbf{b}, \mathbf{c})$ and the response caused by $\mathbf{x}(0)$ is

$$y(t) = \mathbf{c}e^{(\mathbf{A}-\mathbf{bk})t}\mathbf{x}(0)$$

If all eigenvalues of $(\mathbf{A} - \mathbf{bk})$ lie inside the sector in Fig. 8.3(a), then the output will decay rapidly to zero. Thus regulation can easily be achieved by introducing state feedback.

The tracking problem is slightly more complex. In general, in addition to state feedback, we need a feedforward gain p as

$$u(t) = pr(t) - \mathbf{kx}$$

Then the transfer function from r to y differs from the one in (8.17) only by the feedforward gain p. Thus we have

$$\hat{g}_f(s) = \frac{\hat{y}(s)}{\hat{r}(s)} = p\frac{\beta_1 s^3 + \beta_2 s^2 + \beta_3 s + \beta_4}{s^4 + \bar{\alpha}_1 s^3 + \bar{\alpha}_2 s^2 + \bar{\alpha}_3 s + \bar{\alpha}_4} \tag{8.24}$$

If $(\mathbf{A}, \mathbf{b})$ is controllable, all eigenvalues of $(\mathbf{A} - \mathbf{bk})$ or, equivalently, all poles of $\hat{g}_f(s)$ can be assigned arbitrarily, in particular, assigned to lie inside the sector in Fig. 8.3(a). Under this assumption, if the reference input is a step function with magnitude a, then the output $y(t)$ will approach the constant $\hat{g}_f(0) \cdot a$ as $t \to \infty$ (Theorem 5.2). Thus in order for $y(t)$ to track asymptotically any step reference input, we need

$$1 = \hat{g}_f(0) = p\frac{\beta_4}{\bar{\alpha}_4} \quad \text{or} \quad p = \frac{\bar{\alpha}_4}{\beta_4} \tag{8.25}$$

which requires $\beta_4 \neq 0$. From (8.16) and (8.17), we see that β_4 is the numerator constant term of the plant transfer function. Thus $\beta_4 \neq 0$ if and only if the plant transfer function $\hat{g}(s)$ has no zero at $s = 0$. In conclusion, if $\hat{g}(s)$ has one or more zeros at $s = 0$, tracking is not possible. If $\hat{g}(s)$ has no zero at $s = 0$, we introduce a feedforward gain as in (8.25). Then the resulting system will track asymptotically any step reference input.

We summarize the preceding discussion. Given $(\mathbf{A}, \mathbf{b}, \mathbf{c})$; if $(\mathbf{A}, \mathbf{b})$ is controllable, we may introduce state feedback to place the eigenvalues of $(\mathbf{A} - \mathbf{bk})$ in any desired positions and the resulting system will achieve regulation. If $(\mathbf{A}, \mathbf{b})$ is controllable and if $\mathbf{c}(s\mathbf{I} - \mathbf{A})^{-1}\mathbf{b}$ has no zero at $s = 0$, then after state feedback, we may introduce a feedforward gain as in (8.25). Then the resulting system can track asymptotically any step reference input.

8.3.1 Robust Tracking and Disturbance Rejection[1]

The state equation and transfer function developed to describe a plant may change due to change of load, environment, or aging. Thus plant parameter variations often occur in practice. The equation used in the design is often called the *nominal equation*. The feedforward gain p in (8.25), computed for the nominal plant transfer function, may not yield $\hat{g}_f(0) = 1$ for nonnominal plant transfer functions. Then the output will not track asymptotically any step reference input. Such a tracking is said to be *nonrobust*.

In this subsection we discuss a different design that can achieve robust tracking and disturbance rejection. Consider a plant described by (8.1). We now assume that a constant disturbance w with unknown magnitude enters at the plant input as shown in Fig. 8.4(a). Then the state equation must be modified as

$$\dot{\mathbf{x}} = \mathbf{Ax} + \mathbf{b}u + \mathbf{b}w$$
$$y = \mathbf{cx} \tag{8.26}$$

The problem is to design an overall system so that the output $y(t)$ will track asymptotically any step reference input even with the presence of a disturbance $w(t)$ and with plant parameter variations. This is called *robust tracking and disturbance rejection*. In order to achieve this design, in addition to introducing state feedback, we will introduce an integrator and a unity feedback from the output as shown in Fig. 8.4(a). Let the output of the integrator be denoted by

---

1. This section may be skipped without loss of continuity.

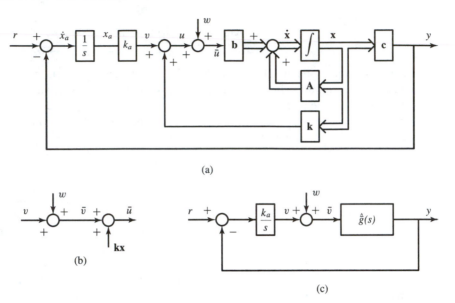

(a)

(b)

(c)

Figure 8.4 (a) State feedback with internal model. (b) Interchange of two summers. (c) Transfer-function block diagram.

$x_a(t)$, an augmented state variable. Then the system has the augmented state vector $[\mathbf{x}'\ x_a]'$. From Fig. 8.4(a), we have

$$\dot{x}_a = r - y = r - \mathbf{c}\mathbf{x} \tag{8.27}$$

$$u = [\mathbf{k}\ k_a] \begin{bmatrix} \mathbf{x} \\ x_a \end{bmatrix} \tag{8.28}$$

For convenience, the state is fed back positively to u as shown. Substituting these into (8.26) yields

$$\begin{bmatrix} \dot{\mathbf{x}} \\ \dot{x}_a \end{bmatrix} = \begin{bmatrix} \mathbf{A} + \mathbf{b}\mathbf{k} & \mathbf{b}k_a \\ -\mathbf{c} & 0 \end{bmatrix} \begin{bmatrix} \mathbf{x} \\ x_a \end{bmatrix} + \begin{bmatrix} \mathbf{0} \\ 1 \end{bmatrix} r + \begin{bmatrix} \mathbf{b} \\ 0 \end{bmatrix} w$$

$$y = [\mathbf{c}\ 0] \begin{bmatrix} \mathbf{x} \\ x_a \end{bmatrix} \tag{8.29}$$

This describes the system in Fig. 8.4(a).

▶ **Theorem 8.5**

If $(\mathbf{A}, \mathbf{b})$ is controllable and if $\hat{g}(s) = \mathbf{c}(s\mathbf{I} - \mathbf{A})^{-1}\mathbf{b}$ has no zero at $s = 0$, then all eigenvalues of the A-matrix in (8.29) can be assigned arbitrarily by selecting a feedback gain $[\mathbf{k}\ k_a]$.

Proof: We show the theorem for $n = 4$. We assume that $\mathbf{A}, \mathbf{b}$, and $\mathbf{c}$ have been transformed into the controllable canonical form in (8.7) and its transfer function equals (8.8). Then the plant transfer function has no zero at $s = 0$ if and only if $\beta_4 \neq 0$. We now show that the pair

$$\begin{bmatrix} \mathbf{A} & \mathbf{0} \\ -\mathbf{c} & 0 \end{bmatrix} \qquad \begin{bmatrix} \mathbf{b} \\ 0 \end{bmatrix} \tag{8.30}$$

is controllable if and only if $\beta_4 \neq 0$. Note that we have assumed $n = 4$; thus the dimension of (8.30) is five because of the additional augmented state variable x_a. The controllability matrix of (8.30) is

$$\begin{bmatrix} \mathbf{b} & \mathbf{Ab} & \mathbf{A}^2\mathbf{b} & \mathbf{A}^3\mathbf{b} & \mathbf{A}^4\mathbf{b} \\ 0 & -\mathbf{cb} & -\mathbf{cAb} & -\mathbf{cA}^2\mathbf{b} & -\mathbf{cA}^3\mathbf{b} \end{bmatrix}$$

$$= \begin{bmatrix} 1 & -\alpha_1 & \alpha_1^2 - \alpha_2 & -\alpha_1(\alpha_1^2 - \alpha_2) + \alpha_2\alpha_1 - \alpha_3 & a_{15} \\ 0 & 1 & -\alpha_1 & \alpha_1^2 - \alpha_2 & a_{25} \\ 0 & 0 & 1 & -\alpha_1 & a_{35} \\ 0 & 0 & 0 & 1 & a_{45} \\ 0 & -\beta_1 & \beta_1\alpha_1 - \beta_2 & -\beta_1(\alpha_1^2 - \alpha_2) + \beta_2\alpha_1 - \beta_3 & a_{55} \end{bmatrix}$$

where the last column is not written out to save space. The rank of a matrix will not change by elementary operations. Adding the second row multiplied by β_1 to the last row, and adding the third row multiplied by β_2 to the last row, and adding the fourth row multiplied by β_3 to the last row, we obtain

$$\begin{bmatrix} 1 & -\alpha_1 & \alpha_1^2 - \alpha_2 & -\alpha_1(\alpha_1^2 - \alpha_2) + \alpha_2\alpha_1 - \alpha_3 & a_{15} \\ 0 & 1 & -\alpha_1 & \alpha_1^2 - \alpha_2 & a_{25} \\ 0 & 0 & 1 & -\alpha_1 & a_{35} \\ 0 & 0 & 0 & 1 & a_{45} \\ 0 & 0 & 0 & 0 & -\beta_4 \end{bmatrix} \tag{8.31}$$

Its determinant is $-\beta_4$. Thus the matrix is nonsingular if and only if $\beta_4 \neq 0$. In conclusion, if $(\mathbf{A}, \mathbf{b})$ is controllable and if $\hat{g}(s)$ has no zero at $s = 0$, then the pair in (8.30) is controllable. It follows from Theorem 8.3 that all eigenvalues of the A-matrix in (8.29) can be assigned arbitrarily by selecting a feedback gain $[\mathbf{k} \quad k_a]$. Q.E.D.

We mention that the controllability of the pair in (8.30) can also be explained from pole–zero cancellations. If the plant transfer function has a zero at $s = 0$, then the tandem connection of the integrator, which has transfer function $1/s$, and the plant will involve the pole–zero cancellation of s and the state equation describing the connection will not be controllable. On the other hand, if the plant transfer function has no zero at $s = 0$, then there is no pole–zero cancellation and the connection will be controllable.

Consider again (8.29). We assume that a set of $n + 1$ desired stable eigenvalues or, equivalently, a desired polynomial $\Delta_f(s)$ of degree $n + 1$ has been selected and the feedback gain $[\mathbf{k} \quad k_a]$ has been found such that

$$\Delta_f(s) = \det \begin{bmatrix} s\mathbf{I} - \mathbf{A} - \mathbf{bk} & -\mathbf{b}k_a \\ \mathbf{c} & s \end{bmatrix} \tag{8.32}$$

Now we show that the output y will track asymptotically and robustly any step reference input $r(t) = a$ and reject any step disturbance with unknown magnitude. Instead of establishing the assertion directly from (8.29), we will develop an equivalent block diagram of Fig. 8.4(a) and then establish the assertion. First we interchange the two summers between v and $\bar{u}$ as shown in Fig. 8.4(b). This is permitted because we have $\bar{u} = v + \mathbf{k}x + w$ before and after the interchange. The transfer function from $\bar{v}$ to y is

$$\hat{\bar{g}}(s) := \frac{\bar{N}(s)}{\bar{D}(s)} := \mathbf{c}(s\mathbf{I} - \mathbf{A} - \mathbf{bk})^{-1}\mathbf{b} \tag{8.33}$$

with $\bar{D}(s) = \det(s\mathbf{I} - \mathbf{A} - \mathbf{bk})$. Thus Fig. 8.4(a) can be redrawn as shown in Fig. 8.4(c). We next establish the relationship between $\Delta_f(s)$ in (8.32) and $\hat{\bar{g}}(s)$ in (8.33). It is straightforward to verify the following equality:

$$\begin{bmatrix} \mathbf{I} & \mathbf{0} \\ -\mathbf{c}(s\mathbf{I} - \mathbf{A} - \mathbf{bk})^{-1} & 1 \end{bmatrix} \begin{bmatrix} s\mathbf{I} - \mathbf{A} - \mathbf{bk} & -\mathbf{b}k_a \\ \mathbf{c} & s \end{bmatrix}$$

$$= \begin{bmatrix} s\mathbf{I} - \mathbf{A} - \mathbf{bk} & -\mathbf{b}k_a \\ 0 & s + \mathbf{c}(s\mathbf{I} - \mathbf{A} - \mathbf{bk})^{-1}\mathbf{b}k_a \end{bmatrix}$$

Taking its determinants and using (8.32) and (8.33), we obtain

$$1 \cdot \Delta_f(s) = \bar{D}(s)\left(s + \frac{\bar{N}(s)}{\bar{D}(s)}k_a\right)$$

which implies

$$\Delta_f(s) = s\bar{D}(s) + k_a\bar{N}(s)$$

This is a key equation.

From Fig. 8.4(c), the transfer function from w to y can readily be computed as

$$\hat{g}_{yw} = \frac{\dfrac{\bar{N}(s)}{\bar{D}(s)}}{1 + \dfrac{k_a\bar{N}(s)}{s\bar{D}(s)}} = \frac{s\bar{N}(s)}{s\bar{D}(s) + k_a\bar{N}(s)} = \frac{s\bar{N}(s)}{\Delta_f(s)}$$

If the disturbance is $w(t) = \bar{w}$ for all $t \geq 0$, where $\bar{w}$ is an unknown constant, then $\hat{w}(s) = \bar{w}/s$ and the corresponding output is given by

$$\hat{y}_w(s) = \frac{s\bar{N}(s)}{\Delta_f(s)}\frac{\bar{w}}{s} = \frac{\bar{w}\bar{N}(s)}{\Delta_f(s)} \tag{8.34}$$

Because the pole s in (8.34) is canceled, all remaining poles of $\hat{y}_w(s)$ are stable poles. Therefore the corresponding time response, for any $\bar{w}$, will die out as $t \to \infty$. The only condition to achieve the disturbance rejection is that $\hat{y}_w(s)$ has only stable poles. Thus the rejection still holds, even if there are plant parameter variations and variations in the feedforward gain k_a and feedback gain $\mathbf{k}$, as long as the overall system remains stable. Thus the disturbance is suppressed at the output both asymptotically and robustly.

The transfer function from r to y is

$$\hat{g}_{yr}(s) = \frac{\dfrac{k_a}{s}\dfrac{\bar{N}(s)}{\bar{D}(s)}}{1 + \dfrac{k_a}{s}\dfrac{\bar{N}(s)}{\bar{D}(s)}} = \frac{k_a\bar{N}(s)}{s\bar{D}(s) + k_a\bar{N}(s)} = \frac{k_a\bar{N}(s)}{\Delta_f(s)}$$

We see that

$$\hat{g}_{yr}(0) = \frac{k_a \bar{N}(0)}{0 \cdot \bar{D}(0) + k_a \bar{N}(0)} = \frac{k_a \bar{N}(0)}{k_a \bar{N}(0)} = 1 \tag{8.35}$$

Equation (8.35) holds even when there are parameter perturbations in the plant transfer function and the gains. Thus asymptotic tracking of any step reference input is robust. Note that this robust tracking holds even for very large parameter perturbations as long as the overall system remains stable.

We see that the design is achieved by inserting an integrator as shown in Fig. 8.4. The integrator is in fact a model of the step reference input and constant disturbance. Thus it is called the *internal model principle*. This will be discussed further in the next chapter.

8.3.2 Stabilization

If a state equation is controllable, all eigenvalues can be assigned arbitrarily by introducing state feedback. We now discuss the case when the state equation is not controllable. Every uncontrollable state equation can be transformed into

$$\begin{bmatrix} \dot{\bar{x}}_c \\ \dot{\bar{x}}_{\bar{c}} \end{bmatrix} = \begin{bmatrix} \bar{A}_c & \bar{A}_{12} \\ 0 & \bar{A}_{\bar{c}} \end{bmatrix} \begin{bmatrix} \bar{x}_c \\ \bar{x}_{\bar{c}} \end{bmatrix} + \begin{bmatrix} \bar{b}_c \\ 0 \end{bmatrix} u \tag{8.36}$$

where $(\bar{A}_c, \bar{b}_c)$ is controllable (Theorem 6.6). Because the A-matrix is block triangular, the eigenvalues of the original A-matrix are the union of the eigenvalues of $\bar{A}_c$ and $\bar{A}_{\bar{c}}$. If we introduce the state feedback

$$u = r - \mathbf{kx} = r - \bar{\mathbf{k}}\bar{\mathbf{x}} = r - [\bar{\mathbf{k}}_1 \ \bar{\mathbf{k}}_2] \begin{bmatrix} \bar{x}_c \\ \bar{x}_{\bar{c}} \end{bmatrix}$$

where we have partitioned $\bar{\mathbf{k}}$ as in $\bar{\mathbf{x}}$, then (8.36) becomes

$$\begin{bmatrix} \dot{\bar{x}}_c \\ \dot{\bar{x}}_{\bar{c}} \end{bmatrix} = \begin{bmatrix} \bar{A}_c - \bar{b}_c\bar{k}_1 & \bar{A}_{12} - \bar{b}_c\bar{k}_2 \\ 0 & \bar{A}_{\bar{c}} \end{bmatrix} \begin{bmatrix} \bar{x}_c \\ \bar{x}_{\bar{c}} \end{bmatrix} + \begin{bmatrix} \bar{b}_c \\ 0 \end{bmatrix} r \tag{8.37}$$

We see that $\bar{A}_{\bar{c}}$ and, consequently, its eigenvalues are not affected by the state feedback. Thus we conclude that the controllability condition of $(\mathbf{A}, \mathbf{b})$ in Theorem 8.3 is not only sufficient but also necessary to assign *all* eigenvalues of $(\mathbf{A} - \mathbf{bk})$ to any desired positions.

Consider again the state equation in (8.36). If $\bar{A}_{\bar{c}}$ is stable, and if $(\bar{A}_c, \bar{b}_c)$ is controllable, then (8.36) is said to be *stabilizable*. We mention that the controllability condition for tracking and disturbance rejection can be replaced by the weaker condition of stabilizability. But in this case, we do not have complete control of the rate of tracking and rejection. If the uncontrollable stable eigenvalues have large imaginary parts or are close to the imaginary axis, then the tracking and rejection may not be satisfactory.

8.4 State Estimator

We introduced in the preceding sections state feedback under the implicit assumption that all state variables are available for feedback. This assumption may not hold in practice either because the state variables are not accessible for direct connection or because sensing devices or transducers are not available or very expensive. In this case, in order to apply state feedback, we must design a device, called a *state estimator* or *state observer*, so that the output of the device will generate an estimate of the state. In this section, we introduce full-dimensional state estimators which have the same dimension as the original state equation. We use the circumflex over a variable to denote an estimate of the variable. For example, $\hat{\mathbf{x}}$ is an estimate of $\mathbf{x}$ and $\hat{\bar{\mathbf{x}}}$ is an estimate of $\bar{\mathbf{x}}$.

Consider the n-dimensional state equation

$$\dot{\mathbf{x}} = \mathbf{Ax} + \mathbf{b}u$$
$$y = \mathbf{cx}$$

(8.38)

where $\mathbf{A}$, $\mathbf{b}$, and $\mathbf{c}$ are given and the input $u(t)$ and the output $y(t)$ are available to us. The state $\mathbf{x}$, however, is not available to us. The problem is to estimate $\mathbf{x}$ from u and y with the knowledge of $\mathbf{A}$, $\mathbf{b}$, and $\mathbf{c}$. If we know $\mathbf{A}$ and $\mathbf{b}$, we can duplicate the original system as

$$\dot{\hat{\mathbf{x}}} = \mathbf{A}\hat{\mathbf{x}} + \mathbf{b}u$$

(8.39)

and as shown in Fig. 8.5. Note that the original system could be an electromechanical system and the duplicated system could be an op-amp circuit. The duplication will be called an *open-loop* estimator. Now if (8.38) and (8.39) have the same initial state, then for any input, we have $\hat{\mathbf{x}}(t) = \mathbf{x}(t)$ for all $t \geq 0$. Therefore the remaining question is how to find the initial state of (8.38) and then set the initial state of (8.39) to that state. If (8.38) is observable, its initial state $\mathbf{x}(0)$ can be computed from u and y over any time interval, say, $[0, \ t_1]$. We can then compute the state at t_2 and set $\hat{\mathbf{x}}(t_2) = \mathbf{x}(t_2)$. Then we have $\hat{\mathbf{x}}(t) = \mathbf{x}(t)$ for all $t \geq t_2$. Thus if (8.38) is observable, an open-loop estimator can be used to generate the state vector.

There are, however, two disadvantages in using an open-loop estimator. First, the initial state must be computed and set each time we use the estimator. This is very inconvenient. Second, and more seriously, if the matrix $\mathbf{A}$ has eigenvalues with positive real parts, then even for a very small difference between $\mathbf{x}(t_0)$ and $\hat{\mathbf{x}}(t_0)$ for some t_0, which may be caused by disturbance or imperfect estimation of the initial state, the difference between $\mathbf{x}(t)$ and $\hat{\mathbf{x}}(t)$ will grow with time. Therefore the open-loop estimator is, in general, not satisfactory.

We see from Fig. 8.5 that even though the input and output of (8.38) are available, we

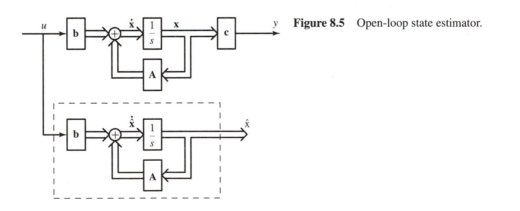

Figure 8.5 Open-loop state estimator.

Figure 8.6   Closed-loop state estimator.

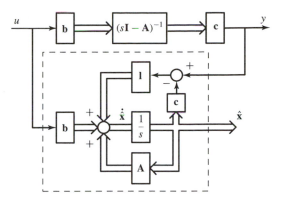

use only the input to drive the open-loop estimator. Now we shall modify the estimator in Fig. 8.5 to the one in Fig. 8.6, in which the output $y(t) = \mathbf{c}\mathbf{x}(t)$ of (8.38) is compared with $\mathbf{c}\hat{\mathbf{x}}(t)$. Their difference, passing through an $n \times 1$ constant gain vector $\mathbf{l}$, is used as a correcting term. If the difference is zero, no correction is needed. If the difference is nonzero and if the gain $\mathbf{l}$ is properly designed, the difference will drive the estimated state to the actual state. Such an estimator is called a *closed-loop* or an *asymptotic* estimator or, simply, an estimator.

The open-loop estimator in (8.39) is now modified as, following Fig. 8.6,

$$\dot{\hat{\mathbf{x}}} = \mathbf{A}\hat{\mathbf{x}} + \mathbf{b}u + \mathbf{l}(y - \mathbf{c}\hat{\mathbf{x}})$$

which can be written as

$$\dot{\hat{\mathbf{x}}} = (\mathbf{A} - \mathbf{lc})\hat{\mathbf{x}} + \mathbf{b}u + \mathbf{l}y \tag{8.40}$$

and is shown in Fig. 8.7. It has two inputs u and y and its output yields an estimated state $\hat{\mathbf{x}}$. Let us define

$$\mathbf{e}(t) := \mathbf{x}(t) - \hat{\mathbf{x}}(t)$$

It is the error between the actual state and the estimated state. Differentiating $\mathbf{e}$ and then substituting (8.38) and (8.40) into it, we obtain

Figure 8.7   Closed-loop state estimator.

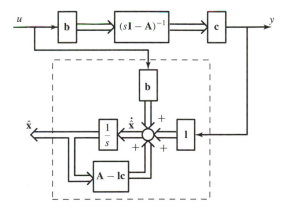

$$\dot{\mathbf{e}} = \dot{\mathbf{x}} - \dot{\hat{\mathbf{x}}} = \mathbf{A}\mathbf{x} + \mathbf{b}u - (\mathbf{A} - \mathbf{lc})\hat{\mathbf{x}} - \mathbf{b}u - \mathbf{l}(\mathbf{cx})$$
$$= (\mathbf{A} - \mathbf{lc})\mathbf{x} - (\mathbf{A} - \mathbf{lc})\hat{\mathbf{x}} = (\mathbf{A} - \mathbf{lc})(\mathbf{x} - \hat{\mathbf{x}})$$

or

$$\dot{\mathbf{e}} = (\mathbf{A} - \mathbf{lc})\mathbf{e} \tag{8.41}$$

This equation governs the estimation error. If all eigenvalues of $(\mathbf{A} - \mathbf{lc})$ can be assigned arbitrarily, then we can control the rate for $\mathbf{e}(t)$ to approach zero or, equivalently, for the estimated state to approach the actual state. For example, if all eigenvalues of $(\mathbf{A} - \mathbf{lc})$ have negative real parts smaller than $-\sigma$, then all entries of $\mathbf{e}$ will approach zero at rates faster than $e^{-\sigma t}$. Therefore, even if there is a large error between $\hat{\mathbf{x}}(t_0)$ and $\mathbf{x}(t_0)$ at initial time t_0, the estimated state will approach the actual state rapidly. Thus there is no need to compute the initial state of the original state equation. In conclusion, if all eigenvalues of $(\mathbf{A} - \mathbf{lc})$ are properly assigned, a closed-loop estimator is much more desirable than an open-loop estimator.

As in the state feedback, what constitutes the best eigenvalues is not a simple problem. Probably, they should be placed evenly along a circle inside the sector shown in Fig. 8.3(a). If an estimator is to be used in state feedback, then the estimator eigenvalues should be faster than the desired eigenvalues of the state feedback. Again, saturation and noise problems will impose constraints on the selection. One way to carry out the selection is by computer simulation.

▶ **Theorem 8.O3**

Consider the pair $(\mathbf{A}, \mathbf{c})$. All eigenvalues of $(\mathbf{A} - \mathbf{lc})$ can be assigned arbitrarily by selecting a real constant vector $\mathbf{l}$ if and only if $(\mathbf{A}, \mathbf{c})$ is observable.

This theorem can be established directly or indirectly by using the duality theorem. The pair $(\mathbf{A}, \mathbf{c})$ is observable if and only if $(\mathbf{A}', \mathbf{c}')$ is controllable. If $(\mathbf{A}', \mathbf{c}')$ is controllable, all eigenvalues of $(\mathbf{A}' - \mathbf{c}'\mathbf{k})$ can be assigned arbitrarily by selecting a constant gain vector $\mathbf{k}$. The transpose of $(\mathbf{A}' - \mathbf{c}'\mathbf{k})$ is $(\mathbf{A} - \mathbf{k}'\mathbf{c})$. Thus we have $\mathbf{l} = \mathbf{k}'$. In conclusion, the procedure for computing state feedback gains can be used to compute the gain $\mathbf{l}$ in state estimators.

Solving the Lyapunov equation   We discuss a different method of designing a state estimator for the n-dimensional state equation

$$\dot{\mathbf{x}} = \mathbf{A}\mathbf{x} + \mathbf{b}u$$
$$y = \mathbf{cx} \tag{8.42}$$

The method is dual to Procedure 8.1 in Section 8.2.1.

▶ **Procedure 8.O1**

1. Select an arbitrary $n \times n$ stable matrix $\mathbf{F}$ that has no eigenvalues in common with those of $\mathbf{A}$.

2. Select an arbitrary $n \times 1$ vector $\mathbf{l}$ such that $(\mathbf{F}, \mathbf{l})$ is controllable.

3. Solve the unique $\mathbf{T}$ in the Lyapunov equation $\mathbf{TA} - \mathbf{FT} = \mathbf{lc}$. This $\mathbf{T}$ is nonsingular following the dual of Theorem 8.4.

4. Then the state equation

$$\dot{\mathbf{z}} = \mathbf{Fz} + \mathbf{T}\mathbf{b}u + \mathbf{l}y \tag{8.43}$$

$$\hat{\mathbf{x}} = \mathbf{T}^{-1}\mathbf{z} \tag{8.44}$$

generates an estimate of **x**.

We first justify the procedure. Let us define

$$\mathbf{e} := \mathbf{z} - \mathbf{Tx}$$

Then we have, replacing **TA** by **FT** + **lc**,

$$\dot{\mathbf{e}} = \dot{\mathbf{z}} - \mathbf{T}\dot{\mathbf{x}} = \mathbf{Fz} + \mathbf{T}\mathbf{b}u + \mathbf{lcx} - \mathbf{TAx} - \mathbf{T}\mathbf{b}u$$

$$= \mathbf{Fz} + \mathbf{lcx} - (\mathbf{FT} + \mathbf{lc})\mathbf{x} = \mathbf{F}(\mathbf{z} - \mathbf{Tx}) = \mathbf{Fe}$$

If **F** is stable, for any **e**(0), the error vector **e**(t) approaches zero as $t \to \infty$. Thus **z** approaches **Tx** or, equivalently, $\mathbf{T}^{-1}\mathbf{z}$ is an estimate of **x**. All discussion in Section 8.2.1 applies here and will not be repeated.

8.4.1 Reduced-Dimensional State Estimator

Consider the state equation in (8.42). If it is observable, then it can be transformed, dual to Theorem 8.2, into the observable canonical form in (7.14). We see that y equals x_1, the first state variable. Therefore it is sufficient to construct an $(n-1)$-dimensional state estimator to estimate x_i for $i = 2, 3, \ldots, n$. This estimator with the output equation can then be used to estimate all n state variables. This estimator has a lesser dimension than (8.42) and is called a reduced-dimensional estimator.

Reduced-dimensional estimators can be designed by transformations or by solving Lyapunov equations. The latter approach is considerably simpler and will be discussed next. For the former approach, the interested reader is referred to Reference [6, pp. 361–363].

▶ **Procedure 8.R1**

1. Select an arbitrary $(n-1) \times (n-1)$ stable matrix **F** that has no eigenvalues in common with those of **A**.

2. Select an arbitrary $(n-1) \times 1$ vector **l** such that (**F**, **l**) is controllable.

3. Solve the unique **T** in the Lyapunov equation $\mathbf{TA} - \mathbf{FT} = \mathbf{lc}$. Note that **T** is an $(n-1) \times n$ matrix.

4. Then the $(n-1)$-dimensional state equation

$$\dot{\mathbf{z}} = \mathbf{Fz} + \mathbf{T}\mathbf{b}u + \mathbf{l}y \tag{8.45}$$

$$\hat{\mathbf{x}} = \begin{bmatrix} \mathbf{c} \\ \mathbf{T} \end{bmatrix}^{-1} \begin{bmatrix} y \\ \mathbf{z} \end{bmatrix} \tag{8.46}$$

is an estimate of **x**.

We first justify the procedure. We write (8.46) as

$$\begin{bmatrix} y \\ z \end{bmatrix} = \begin{bmatrix} \mathbf{c} \\ \mathbf{T} \end{bmatrix} \hat{\mathbf{x}} =: \mathbf{P}\hat{\mathbf{x}}$$

which implies $y = \mathbf{c}\hat{\mathbf{x}}$ and $\mathbf{z} = \mathbf{T}\hat{\mathbf{x}}$. Clearly y is an estimate of $\mathbf{cx}$. We now show that $\mathbf{z}$ is an estimate of $\mathbf{Tx}$. Define

$$\mathbf{e} = \mathbf{z} - \mathbf{Tx}$$

Then we have

$$\dot{\mathbf{e}} = \dot{\mathbf{z}} - \mathbf{T}\dot{\mathbf{x}} = \mathbf{Fz} + \mathbf{T}\mathbf{b}u + \mathbf{lcx} - \mathbf{TAx} - \mathbf{T}\mathbf{b}u = \mathbf{Fe}$$

Clearly if $\mathbf{F}$ is stable, then $\mathbf{e}(t) \to \mathbf{0}$ as $t \to \infty$. Thus $\mathbf{z}$ is an estimate of $\mathbf{Tx}$.

▶ **Theorem 8.6**

If $\mathbf{A}$ and $\mathbf{F}$ have no common eigenvalues, then the square matrix

$$\mathbf{P} = \begin{bmatrix} \mathbf{c} \\ \mathbf{T} \end{bmatrix}$$

where $\mathbf{T}$ is the unique solution of $\mathbf{TA} - \mathbf{FT} = \mathbf{lc}$, is nonsingular if and only if $(\mathbf{A}, \mathbf{c})$ is observable and $(\mathbf{F}, \mathbf{l})$ is controllable.

Proof: We prove the theorem for $n = 4$. The first part of the proof follows closely the proof of Theorem 8.4. Let

$$\Delta(s) = \det(s\mathbf{I} - \mathbf{A}) = s^4 + \alpha_1 s^3 + \alpha_2 s^2 + \alpha_3 s + \alpha_4$$

Then, dual to (8.22), we have

$$-\mathbf{T}\Delta(\mathbf{F}) = [\mathbf{l}\ \mathbf{Fl}\ \mathbf{F}^2\mathbf{l}\ \mathbf{F}^3\mathbf{l}] \begin{bmatrix} \alpha_3 & \alpha_2 & \alpha_1 & 1 \\ \alpha_2 & \alpha_1 & 1 & 0 \\ \alpha_1 & 1 & 0 & 0 \\ 1 & 0 & 0 & 0 \end{bmatrix} \begin{bmatrix} \mathbf{c} \\ \mathbf{cA} \\ \mathbf{cA}^2 \\ \mathbf{cA}^3 \end{bmatrix} \tag{8.47}$$

and $\Delta(\mathbf{F})$ is nonsingular if $\mathbf{A}$ and $\mathbf{F}$ have no common eigenvalues. Note that if $\mathbf{A}$ is 4×4, then $\mathbf{F}$ is 3×3. The rightmost matrix in (8.47) is the observability matrix of $(\mathbf{A}, \mathbf{c})$ and will be denoted by O. The first matrix after the equality is the controllability matrix of $(\mathbf{F}, \mathbf{l})$ with one extra column and will be denoted by C_4. The middle matrix will be denoted by Λ and is always nonsingular. Using these notations, we write $\mathbf{T}$ as $-\Delta^{-1}(\mathbf{F})C_4\Lambda O$ and $\mathbf{P}$ becomes

$$\mathbf{P} = \begin{bmatrix} \mathbf{c} \\ \mathbf{T} \end{bmatrix} = \begin{bmatrix} \mathbf{c} \\ -\Delta^{-1}(\mathbf{F})C_4\Lambda O \end{bmatrix}$$

$$= \begin{bmatrix} 1 & \mathbf{0} \\ \mathbf{0} & -\Delta^{-1}(\mathbf{F}) \end{bmatrix} \begin{bmatrix} \mathbf{c} \\ C_4\Lambda O \end{bmatrix} \tag{8.48}$$

Note that if $n = 4$, then $\mathbf{P}$, O, and Λ are 4×4; $\mathbf{T}$ and C_4 are 3×4 and $\Delta(\mathbf{F})$ is 3×3. If $(\mathbf{F}, \mathbf{l})$ is not controllable, C_4 has rank at most 2. Thus $\mathbf{T}$ has rank at most 2 and $\mathbf{P}$ is singular. If $(\mathbf{A}, \mathbf{c})$ is not observable, then there exists a nonzero 4×1 vector $\mathbf{r}$ such that

$O\mathbf{r} = \mathbf{0}$, which implies $\mathbf{cr} = 0$ and $\mathbf{Pr} = \mathbf{0}$. Thus $\mathbf{P}$ is singular. This shows the necessity of the theorem.

Next we show the sufficiency by contradiction. Suppose $\mathbf{P}$ is singular. Then there exists a nonzero vector $\mathbf{r}$ such that $\mathbf{Pr} = \mathbf{0}$, which implies

$$\begin{bmatrix} \mathbf{c} \\ C_4\Lambda O \end{bmatrix}\mathbf{r} = \begin{bmatrix} \mathbf{cr} \\ C_4\Lambda O\mathbf{r} \end{bmatrix} = \mathbf{0} \tag{8.49}$$

Define $\mathbf{a} := \Lambda O\mathbf{r} = [a_1\ a_2\ a_3\ a_4]' =: [\bar{\mathbf{a}}\ a_4]'$, where $\bar{\mathbf{a}}$ represents the first three entries of $\mathbf{a}$. Expressing it explicitly yields

$$\begin{bmatrix} a_1 \\ a_2 \\ a_3 \\ a_4 \end{bmatrix} = \begin{bmatrix} \alpha_3 & \alpha_2 & \alpha_1 & 1 \\ \alpha_2 & \alpha_1 & 1 & 0 \\ \alpha_1 & 1 & 0 & 0 \\ 1 & 0 & 0 & 0 \end{bmatrix}\begin{bmatrix} \mathbf{cr} \\ \mathbf{cAr} \\ \mathbf{cA}^2\mathbf{r} \\ \mathbf{cA}^3\mathbf{r} \end{bmatrix} = \begin{bmatrix} x \\ x \\ x \\ \mathbf{cr} \end{bmatrix}$$

where x denotes entries that are not needed in subsequent discussion. Thus we have $a_4 = \mathbf{cr}$. Clearly (8.49) implies $a_4 = \mathbf{cr} = 0$. Substituting $a_4 = 0$ into the lower part of (8.49) yields

$$C_4\Lambda O\mathbf{r} = C_4\mathbf{a} = C\bar{\mathbf{a}} = \mathbf{0} \tag{8.50}$$

where C is 3×3 and is the controllability matrix of $(\mathbf{F}, \mathbf{l})$ and $\bar{\mathbf{a}}$ is the first three entries of $\mathbf{a}$. If $(\mathbf{F}, \mathbf{l})$ is controllable, then $C\bar{\mathbf{a}} = \mathbf{0}$ implies $\bar{\mathbf{a}} = \mathbf{0}$. In conclusion, (8.49) and the controllability of $(\mathbf{F}, \mathbf{l})$ imply $\mathbf{a} = \mathbf{0}$.

Consider $\Lambda O\mathbf{r} = \mathbf{a} = \mathbf{0}$. The matrix Λ is always nonsingular. If $(\mathbf{A}, \mathbf{c})$ is observable, then O is nonsingular and $\Lambda O\mathbf{r} = \mathbf{0}$ implies $\mathbf{r} = \mathbf{0}$. This contradicts the hypothesis that $\mathbf{r}$ is nonzero. Thus if $(\mathbf{A}, \mathbf{c})$ is observable and $(\mathbf{F}, \mathbf{l})$ is controllable, then $\mathbf{P}$ is nonsingular. This establishes Theorem 8.6.   Q.E.D.

Designing state estimators by solving Lyapunov equations is convenient because the same procedure can be used to design full-dimensional and reduced-dimensional estimators. As we shall see in a later section, the same procedure can also be used to design estimators for multi-input multi-output systems.

8.5 Feedback from Estimated States

Consider a plant described by the n-dimensional state equation

$$\dot{\mathbf{x}} = \mathbf{Ax} + \mathbf{b}u$$
$$y = \mathbf{cx} \tag{8.51}$$

If $(\mathbf{A}, \mathbf{b})$ is controllable, state feedback $u = r - \mathbf{kx}$ can place the eigenvalues of $(\mathbf{A} - \mathbf{bk})$ in any desired positions. If the state variables are not available for feedback, we can design a state estimator. If $(\mathbf{A}, \mathbf{c})$ is observable, a full- or reduced-dimensional estimator with arbitrary eigenvalues can be constructed. We discuss here only full-dimensional estimators. Consider the n-dimensional state estimator

$$\dot{\hat{\mathbf{x}}} = (\mathbf{A} - \mathbf{lc})\hat{\mathbf{x}} + \mathbf{b}u + \mathbf{l}y \tag{8.52}$$

The estimated state in (8.52) can approach the actual state in (8.51) with any rate by selecting the vector $\mathbf{l}$.

The state feedback is designed for the state in (8.51). If $\mathbf{x}$ is not available, it is natural to apply the feedback gain to the estimated state as

$$u = r - \mathbf{k}\hat{\mathbf{x}} \tag{8.53}$$

as shown in Fig. 8.8. The connection is called the *controller-estimator* configuration. Three questions may be raised in this connection: (1) The eigenvalues of $(\mathbf{A} - \mathbf{bk})$ are obtained from $u = r - \mathbf{kx}$. Do we still have the same set of eigenvalues in using $u = r - \mathbf{k}\hat{\mathbf{x}}$? (2) Will the eigenvalues of the estimator be affected by the connection? (3) What is the effect of the estimator on the transfer function from r to y? To answer these questions, we must develop a state equation to describe the overall system in Fig. 8.8. Substituting (8.53) into (8.51) and (8.52) yields

$$\dot{\mathbf{x}} = \mathbf{A}\mathbf{x} - \mathbf{bk}\hat{\mathbf{x}} + \mathbf{b}r$$

$$\dot{\hat{\mathbf{x}}} = (\mathbf{A} - \mathbf{lc})\hat{\mathbf{x}} + \mathbf{b}(r - \mathbf{k}\hat{\mathbf{x}}) + \mathbf{lcx}$$

They can be combined as

$$\begin{bmatrix} \dot{\mathbf{x}} \\ \dot{\hat{\mathbf{x}}} \end{bmatrix} = \begin{bmatrix} \mathbf{A} & -\mathbf{bk} \\ \mathbf{lc} & \mathbf{A} - \mathbf{lc} - \mathbf{bk} \end{bmatrix} \begin{bmatrix} \mathbf{x} \\ \hat{\mathbf{x}} \end{bmatrix} + \begin{bmatrix} \mathbf{b} \\ \mathbf{b} \end{bmatrix} r$$

$$y = [\mathbf{c} \ \ 0] \begin{bmatrix} \mathbf{x} \\ \hat{\mathbf{x}} \end{bmatrix} \tag{8.54}$$

This $2n$-dimensional state equation describes the feedback system in Fig. 8.8. It is not easy to answer the posed questions from this equation. Let us introduce the following equivalence transformation:

$$\begin{bmatrix} \mathbf{x} \\ \mathbf{e} \end{bmatrix} = \begin{bmatrix} \mathbf{x} \\ \mathbf{x} - \hat{\mathbf{x}} \end{bmatrix} = \begin{bmatrix} \mathbf{I} & 0 \\ \mathbf{I} & -\mathbf{I} \end{bmatrix} \begin{bmatrix} \mathbf{x} \\ \hat{\mathbf{x}} \end{bmatrix} =: \mathbf{P} \begin{bmatrix} \mathbf{x} \\ \hat{\mathbf{x}} \end{bmatrix}$$

Computing $\mathbf{P}^{-1}$, which happens to equal $\mathbf{P}$, and then using (4.26), we can obtain the following equivalent state equation:

$$\begin{bmatrix} \dot{\mathbf{x}} \\ \dot{\mathbf{e}} \end{bmatrix} = \begin{bmatrix} \mathbf{A} - \mathbf{bk} & \mathbf{bk} \\ 0 & \mathbf{A} - \mathbf{lc} \end{bmatrix} \begin{bmatrix} \mathbf{x} \\ \mathbf{e} \end{bmatrix} + \begin{bmatrix} \mathbf{b} \\ 0 \end{bmatrix} r$$

$$y = [\mathbf{c} \ \ 0] \begin{bmatrix} \mathbf{x} \\ \mathbf{e} \end{bmatrix} \tag{8.55}$$

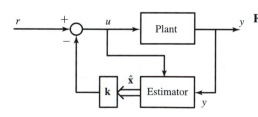

Figure 8.8   Controller-estimator configuration.

The A-matrix in (8.55) is block triangular; therefore its eigenvalues are the union of those of $(\mathbf{A} - \mathbf{bk})$ and $(\mathbf{A} - \mathbf{lc})$. Thus inserting the state estimator does not affect the eigenvalues of the original state feedback; nor are the eigenvalues of the state estimator affected by the connection. Thus the design of state feedback and the design of state estimator can be carried out independently. This is called the *separation property*.

The state equation in (8.55) is of the form shown in (6.40); thus (8.55) is not controllable and the transfer function of (8.55) equals the transfer function of the reduced equation

$$\dot{\mathbf{x}} = (\mathbf{A} - \mathbf{bk})\mathbf{x} + \mathbf{b}r \quad y = \mathbf{cx}$$

or

$$\hat{g}_f(s) = \mathbf{c}(s\mathbf{I} - \mathbf{A} + \mathbf{bk})^{-1}\mathbf{b}$$

(Theorem 6.6). This is the transfer function of the original state feedback system without using a state estimator. Therefore the estimator is completely canceled in the transfer function from r to y. This has a simple explanation. In computing transfer functions, all initial states are assumed to be zero. Consequently, we have $\mathbf{x}(0) = \hat{\mathbf{x}}(0) = \mathbf{0}$, which implies $\mathbf{x}(t) = \hat{\mathbf{x}}(t)$ for all t. Thus, as far as the transfer function from r to y is concerned, there is no difference whether a state estimator is employed or not.

8.6 State Feedback—Multivariable Case

This section extends state feedback to multivariable systems. Consider a plant described by the n-dimensional p-input state equation

$$\dot{\mathbf{x}} = \mathbf{Ax} + \mathbf{Bu}$$
$$y = \mathbf{Cx}$$

$$(8.56)$$

In state feedback, the input $\mathbf{u}$ is given by

$$\mathbf{u} = \mathbf{r} - \mathbf{Kx} \tag{8.57}$$

where $\mathbf{K}$ is a $p \times n$ real constant matrix and $\mathbf{r}$ is a reference signal. Substituting (8.57) into (8.56) yields

$$\dot{\mathbf{x}} = (\mathbf{A} - \mathbf{BK})\mathbf{x} + \mathbf{Br}$$
$$y = \mathbf{Cx}$$

$$(8.58)$$

▶ **Theorem 8.M1**

The pair $(\mathbf{A} - \mathbf{BK}, \mathbf{B})$, for any $p \times n$ real constant matrix $\mathbf{K}$, is controllable if and only if $(\mathbf{A}, \mathbf{B})$ is controllable.

The proof of this theorem follows closely the proof of Theorem 8.1. The only difference is that we must modify (8.4) as

$$C_f = C \begin{bmatrix} \mathbf{I}_p & -\mathbf{KB} & -\mathbf{K(A-BK)B} & -\mathbf{K(A-BK)^2 B} \\ 0 & \mathbf{I}_p & -\mathbf{KB} & -\mathbf{K(A-BK)B} \\ 0 & 0 & \mathbf{I}_p & -\mathbf{KB} \\ 0 & 0 & 0 & \mathbf{I}_p \end{bmatrix}$$

where C_f and C are $n \times np$ controllability matrices with $n = 4$ and $\mathbf{I}_p$ is the unit matrix of order p. Because the rightmost $4p \times 4p$ matrix is nonsingular, C_f has rank n if and only if C has rank n. Thus the controllability property is preserved in any state feedback. As in the SISO case, the observability property, however, may not be preserved. Next we extend Theorem 8.3 to the matrix case

▶ **Theorem 8.M3**

All eigenvalues of $(\mathbf{A} - \mathbf{BK})$ can be assigned arbitrarily (provided complex conjugate eigenvalues are assigned in pairs) by selecting a real constant $\mathbf{K}$ if and only if $(\mathbf{A}, \mathbf{B})$ is controllable.

If $(\mathbf{A}, \mathbf{B})$ is not controllable, then $(\mathbf{A}, \mathbf{B})$ can be transformed into the form shown in (8.36) and the eigenvalues of $\bar{\mathbf{A}}_{\bar{c}}$ will not be affected by any state feedback. This shows the necessity of the theorem. The sufficiency will be established constructively in the next three subsections.

8.6.1 Cyclic Design

In this design, we change the multi-input problem into a single-input problem and then apply Theorem 8.3. A matrix $\mathbf{A}$ is called *cyclic* if its characteristic polynomial equals its minimal polynomial. From the discussion in Section 3.6, we can conclude that $\mathbf{A}$ is cyclic if and only if the Jordan form of $\mathbf{A}$ has one and only one Jordan block associated with each distinct eigenvalue.

▶ **Theorem 8.7**

If the n-dimensional p-input pair $(\mathbf{A}, \mathbf{B})$ is controllable and if $\mathbf{A}$ is cyclic, then for almost any $p \times 1$ vector $\mathbf{v}$, the single-input pair $(\mathbf{A}, \mathbf{Bv})$ is controllable.

We argue intuitively the validity of this theorem. Controllability is invariant under any equivalence transformation; thus we may assume $\mathbf{A}$ to be in Jordan form. To see the basic idea, we use the following example:

$$\mathbf{A} = \begin{bmatrix} 2 & 1 & 0 & 0 & 0 \\ 0 & 2 & 1 & 0 & 0 \\ 0 & 0 & 2 & 0 & 0 \\ 0 & 0 & 0 & -1 & 1 \\ 0 & 0 & 0 & 0 & -1 \end{bmatrix} \quad \mathbf{B} = \begin{bmatrix} 0 & 1 \\ 0 & 0 \\ 1 & 2 \\ 4 & 3 \\ 1 & 0 \end{bmatrix} \quad \mathbf{Bv} = \mathbf{B} \begin{bmatrix} v_1 \\ v_2 \end{bmatrix} = \begin{bmatrix} x \\ x \\ \alpha \\ x \\ \beta \end{bmatrix} \quad (8.59)$$

There is only one Jordan block associated with each distinct eigenvalue; thus $\mathbf{A}$ is cyclic. The condition for $(\mathbf{A}, \mathbf{B})$ to be controllable is that the third and the last rows of $\mathbf{B}$ are nonzero (Theorem 6.8).

The necessary and sufficient conditions for $(\mathbf{A}, \mathbf{Bv})$ to be controllable are $\alpha \neq 0$ and $\beta \neq 0$ in (8.59). Because $\alpha = v_1 + 2v_2$ and $\beta = v_1$, either α or β is zero if and only if $v_1 = 0$ or $v_1/v_2 = -2/1$. Thus any $\mathbf{v}$ other than $v_1 = 0$ and $v_1 = -2v_2$ will make $(\mathbf{A}, \mathbf{Bv})$ controllable. The vector $\mathbf{v}$ can assume any value in the two-dimensional real space shown in Fig. 8.9. The conditions $v_1 = 0$ and $v_1 = -2v_2$ constitute two straight lines as shown. The probability for an arbitrarily selected $\mathbf{v}$ to lie on either straight line is zero. This establishes Theorem 8.6. The cyclicity assumption in this theorem is essential. For example, the pair

$$\mathbf{A} = \begin{bmatrix} 2 & 1 & 0 \\ 0 & 2 & 0 \\ 0 & 0 & 2 \end{bmatrix} \qquad \mathbf{B} = \begin{bmatrix} 2 & 1 \\ 0 & 2 \\ 1 & 0 \end{bmatrix}$$

is controllable (Theorem 6.8). However, there is no $\mathbf{v}$ such that $(\mathbf{A}, \mathbf{Bv})$ is controllable (Corollary 6.8).

If all eigenvalues of $\mathbf{A}$ are distinct, then there is only one Jordan block associated with each eigenvalue. Thus a sufficient condition for $\mathbf{A}$ to be cyclic is that all eigenvalues of $\mathbf{A}$ are distinct.

▶ **Theorem 8.8**

If $(\mathbf{A}, \mathbf{B})$ is controllable, then for almost any $p \times n$ real constant matrix $\mathbf{K}$, the matrix $(\mathbf{A} - \mathbf{BK})$ has only distinct eigenvalues and is, consequently, cyclic.

We show intuitively the theorem for $n = 4$. Let the characteristic polynomial of $\mathbf{A} - \mathbf{BK}$ be

$$\Delta_f(s) = s^4 + a_1 s^3 + a_2 s^2 + a_3 s + a_4$$

where the a_i are functions of the entries of $\mathbf{K}$. The differentiation of $\Delta_f(s)$ with respect to s yields

$$\Delta_f'(s) = 4s^3 + 3a_1 s^2 + 2a_2 s + a_3$$

If $\Delta_f(s)$ has repeated roots, then $\Delta_f(s)$ and $\Delta_f'(s)$ are not coprime. The necessary and sufficient condition for them to be not coprime is that their Sylvester resultant is singular or

Figure 8.9 Two-dimensional real space.

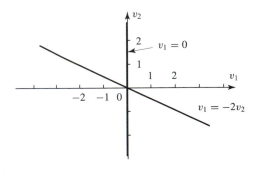

$$\det \begin{bmatrix} a_4 & a_3 & 0 & 0 & 0 & 0 & 0 & 0 \\ a_3 & 2a_2 & a_4 & a_3 & 0 & 0 & 0 & 0 \\ a_2 & 3a_1 & a_3 & 2a_2 & a_4 & a_3 & 0 & 0 \\ a_1 & 4 & a_2 & 3a_1 & a_3 & 2a_2 & a_4 & a_3 \\ 1 & 0 & a_1 & 4 & a_2 & 3a_1 & a_3 & 2a_2 \\ 0 & 0 & 1 & 0 & a_1 & 4 & a_2 & 3a_1 \\ 0 & 0 & 0 & 0 & 1 & 0 & a_1 & 4 \\ 0 & 0 & 0 & 0 & 0 & 0 & 1 & 0 \end{bmatrix} = b(k_{ij}) = 0$$

See (7.28). It is clear that all possible solutions of $b(k_{ij}) = 0$ constitute a very small subset of all real k_{ij}. Thus if we select an arbitrary $\mathbf{K}$, the probability for its entries to meet $b(k_{ij}) = 0$ is 0. Thus all eigenvalues of $(\mathbf{A} - \mathbf{BK})$ will be distinct. This establishes the theorem.

With these two theorems, we can now find a $\mathbf{K}$ to place all eigenvalues of $(\mathbf{A} - \mathbf{BK})$ in any desired positions. If $\mathbf{A}$ is not cyclic, we introduce $\mathbf{u} = \mathbf{w} - \mathbf{K}_1\mathbf{x}$, as shown in Fig. 8.10, such that $\bar{\mathbf{A}} := \mathbf{A} - \mathbf{BK}_1$ in

$$\dot{\mathbf{x}} = (\mathbf{A} - \mathbf{BK}_1)\mathbf{x} + \mathbf{Bw} =: \bar{\mathbf{A}}\mathbf{x} + \mathbf{Bw} \tag{8.60}$$

is cyclic. Because $(\mathbf{A}, \mathbf{B})$ is controllable, so is $(\bar{\mathbf{A}}, \mathbf{B})$. Thus there exists a $p \times 1$ real vector $\mathbf{v}$ such that $(\bar{\mathbf{A}}, \mathbf{Bv})$ is controllable.[2] Next we introduce another state feedback $\mathbf{w} = \mathbf{r} - \mathbf{K}_2\mathbf{x}$ with $\mathbf{K}_2 = \mathbf{vk}$, where $\mathbf{k}$ is a $1 \times n$ real vector. Then (8.60) becomes

$$\dot{\mathbf{x}} = (\bar{\mathbf{A}} - \mathbf{BK}_2)\mathbf{x} + \mathbf{Br} = (\bar{\mathbf{A}} - \mathbf{Bvk})\mathbf{x} + \mathbf{Br}$$

Because the single-input pair $(\bar{\mathbf{A}}, \mathbf{Bv})$ is controllable, the eigenvalues of $(\bar{\mathbf{A}} - \mathbf{Bvk})$ can

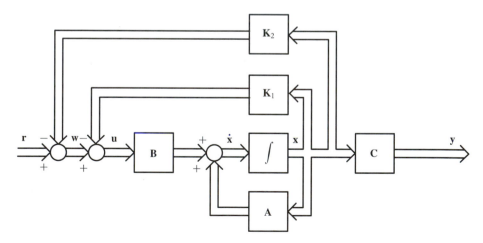

Figure 8.10 State feedback by cyclic design.

---

2. The choices of $\mathbf{K}_1$ and $\mathbf{v}$ are not unique. They can be chosen arbitrarily and the probability is 1 that they will meet the requirements. In Theorem 7.5 of Reference [5], a procedure is given to choose $\mathbf{K}_1$ and $\mathbf{V}$ with no uncertainty. The computation, however, is complicated.

be assigned arbitrarily by selecting a $\mathbf{k}$ (Theorem 8.3). Combining the two state feedback $\mathbf{u} = \mathbf{w} - \mathbf{K}_1\mathbf{x}$ and $\mathbf{w} = \mathbf{r} - \mathbf{K}_2\mathbf{x}$ as

$$\mathbf{u} = \mathbf{r} - (\mathbf{K}_1 + \mathbf{K}_2)\mathbf{x} =: \mathbf{r} - \mathbf{Kx}$$

we obtain a $\mathbf{K} := \mathbf{K}_1 + \mathbf{K}_2$ that achieves arbitrary eigenvalue assignment. This establishes Theorem 8.M3.

8.6.2 Lyapunov-Equation Method

This section will extend the procedure of computing feedback gain in Section 8.2.1 to the multivariable case. Consider an n-dimensional p-input pair $(\mathbf{A}, \mathbf{B})$. Find a $p \times n$ real constant matrix $\mathbf{K}$ so that $(\mathbf{A} - \mathbf{BK})$ has any set of desired eigenvalues as long as the set does not contain any eigenvalue of $\mathbf{A}$.

▶ **Procedure 8.M1**

1. Select an $n \times n$ matrix $\mathbf{F}$ with a set of desired eigenvalues that contains no eigenvalues of $\mathbf{A}$.

2. Select an arbitrary $p \times n$ matrix $\bar{\mathbf{K}}$ such that $(\mathbf{F}, \bar{\mathbf{K}})$ is observable.

3. Solve the unique $\mathbf{T}$ in the Lyapunov equation $\mathbf{AT} - \mathbf{TF} = \mathbf{B\bar{K}}$.

4. If $\mathbf{T}$ is singular, select a different $\bar{\mathbf{K}}$ and repeat the process. If $\mathbf{T}$ is nonsingular, we compute $\mathbf{K} = \bar{\mathbf{K}}\mathbf{T}^{-1}$, and $(\mathbf{A} - \mathbf{BK})$ has the set of desired eigenvalues.

   If $\mathbf{T}$ is nonsingular, the Lyapunov equation and $\mathbf{KT} = \bar{\mathbf{K}}$ imply

$$(\mathbf{A} - \mathbf{BK})\mathbf{T} = \mathbf{TF} \quad \text{or} \quad \mathbf{A} - \mathbf{BK} = \mathbf{TFT}^{-1}$$

Thus $(\mathbf{A} - \mathbf{BK})$ and $\mathbf{F}$ are similar and have the same set of eigenvalues. Unlike the SISO case where $\mathbf{T}$ is always nonsingular, the $\mathbf{T}$ here may not be nonsingular even if $(\mathbf{A}, \mathbf{B})$ is controllable and $(\mathbf{F}, \bar{\mathbf{K}})$ is observable. In other words, the two conditions are necessary but not sufficient for $\mathbf{T}$ to be nonsingular.

▶ **Theorem 8.M4**

If $\mathbf{A}$ and $\mathbf{F}$ have no eigenvalues in common, then the unique solution $\mathbf{T}$ of $\mathbf{AT} - \mathbf{TF} = \mathbf{B\bar{K}}$ is nonsingular only if $(\mathbf{A}, \mathbf{B})$ is controllable and $(\mathbf{F}, \bar{\mathbf{K}})$ is observable.

Proof: The proof of Theorem 8.4 applies here except that (8.22) must be modified as, for $n = 4$,

$$-\mathbf{T}\Delta(\mathbf{F}) = [\mathbf{B} \ \mathbf{AB} \ \mathbf{A}^2\mathbf{B} \ \mathbf{A}^3\mathbf{B}] \begin{bmatrix} \alpha_3\mathbf{I} & \alpha_2\mathbf{I} & \alpha_1\mathbf{I} & \mathbf{I} \\ \alpha_2\mathbf{I} & \alpha_1\mathbf{I} & \mathbf{I} & \mathbf{0} \\ \alpha_1\mathbf{I} & \mathbf{I} & \mathbf{0} & \mathbf{0} \\ \mathbf{I} & \mathbf{0} & \mathbf{0} & \mathbf{0} \end{bmatrix} \begin{bmatrix} \bar{\mathbf{K}} \\ \bar{\mathbf{K}}\mathbf{F} \\ \bar{\mathbf{K}}\mathbf{F}^2 \\ \bar{\mathbf{K}}\mathbf{F}^3 \end{bmatrix}$$

or

$$-\mathbf{T}\Delta(\mathbf{F}) = C\Sigma O \tag{8.61}$$

where $\Delta(\mathbf{F})$ is nonsingular and C, Σ, and O are, respectively, $n \times np$, $np \times np$, and $np \times n$. If C or O has rank less than n, then $\mathbf{T}$ is singular following (3.61). However, the conditions that C and O have rank n do not imply the nonsingularity of $\mathbf{T}$. Thus the controllability of $(\mathbf{A}, \mathbf{B})$ and observability of $(\mathbf{F}, \bar{\mathbf{K}})$ are only necessary conditions for $\mathbf{T}$ to be nonsingular. This establishes Theorem 8.M4.   Q.E.D.

Given a controllable $(\mathbf{A}, \mathbf{B})$, it is possible to construct an observable $(\mathbf{F}, \bar{\mathbf{K}})$ so that the $\mathbf{T}$ in Theorem 8.M4 is singular. However, after selecting $\mathbf{F}$, if $\bar{\mathbf{K}}$ is selected randomly and if $(\mathbf{F}, \bar{\mathbf{K}})$ is observable, it is believed that the probability for $\mathbf{T}$ to be nonsingular is 1. Therefore solving the Lyapunov equation is a viable method of computing a feedback gain matrix to achieve arbitrary eigenvalue assignment. As in the SISO case, we may choose $\mathbf{F}$ in companion form or in modal form as shown in (8.23). If $\mathbf{F}$ is chosen as in (8.23), then we can select $\bar{\mathbf{K}}$ as

$$\bar{\mathbf{K}} = \begin{bmatrix} 1 & 1 & 0 & 0 & 0 \\ 0 & 0 & 0 & 1 & 0 \end{bmatrix} \quad \text{or} \quad \bar{\mathbf{K}} = \begin{bmatrix} 0 & 0 & 1 & 0 & 0 \\ 1 & 0 & 0 & 1 & 0 \end{bmatrix}$$

(see Problem 6.16). Once $\mathbf{F}$ and $\bar{\mathbf{K}}$ are chosen, we can then use the MATLAB function `lyap` to solve the Lyapunov equation. Thus the procedure can easily be carried out.

8.6.3 Canonical-Form Method

We introduced in the preceding subsections two methods of computing a feedback gain matrix to achieve arbitrary eigenvalue assignment. The methods are relatively simple; however, they will not reveal the structure of the resulting feedback system. In this subsection, we discuss a different design that will reveal the effect of state feedback on the transfer matrix. We also give a transfer matrix interpretation of state feedback.

In this design, we must transform $(\mathbf{A}, \mathbf{B})$ into a controllable canonical form. It is an extension of Theorem 8.2 to the multivariable case. Although the basic idea is the same, the procedure can become very involved. Therefore we will skip the details and present the final result. To simplify the discussion, we assume that (8.56) has dimension 6, two inputs, and two outputs. We first search linearly independent columns of $C = [\mathbf{B} \ \mathbf{AB} \ \cdots \ \mathbf{A}^5\mathbf{B}]$ in order from left to right. It is assumed that its controllability indices are $\mu_1 = 4$ and $\mu_2 = 2$. Then there exists a nonsingular matrix $\mathbf{P}$ and $\bar{\mathbf{x}} = \mathbf{Px}$ will transform (8.56) into the controllable canonical form

$$\dot{\bar{\mathbf{x}}} = \begin{bmatrix} -\alpha_{111} & -\alpha_{112} & -\alpha_{113} & -\alpha_{114} & \vdots & -\alpha_{121} & -\alpha_{122} \\ 1 & 0 & 0 & 0 & \vdots & 0 & 0 \\ 0 & 1 & 0 & 0 & \vdots & 0 & 0 \\ 0 & 0 & 1 & 0 & \vdots & 0 & 0 \\ \cdots & \cdots & \cdots & \cdots & \cdots & \cdots & \cdots \\ -\alpha_{211} & -\alpha_{212} & -\alpha_{213} & -\alpha_{214} & \vdots & -\alpha_{221} & -\alpha_{222} \\ 0 & 0 & 0 & 0 & \vdots & 1 & 0 \end{bmatrix} \bar{\mathbf{x}}$$

$$+ \begin{bmatrix} 1 & b_{12} \\ 0 & 0 \\ 0 & 0 \\ 0 & 0 \\ \cdots & \cdots \\ 0 & 1 \\ 0 & 0 \end{bmatrix} \mathbf{u} \tag{8.62}$$

$$\mathbf{y} = \begin{bmatrix} \beta_{111} & \beta_{112} & \beta_{113} & \beta_{114} & \beta_{121} & \beta_{122} \\ \beta_{211} & \beta_{212} & \beta_{213} & \beta_{214} & \beta_{221} & \beta_{222} \end{bmatrix} \bar{\mathbf{x}}$$

Note that this form is identical to the one in (7.104).

We now discuss how to find a feedback gain matrix to achieve arbitrary eigenvalue assignment. From a given set of six desired eigenvalues, we can form

$$\Delta_f(s) = (s^4 + \bar{\alpha}_{111}s^3 + \bar{\alpha}_{112}s^2 + \bar{\alpha}_{113}s + \bar{\alpha}_{114})(s^2 + \bar{\alpha}_{221}s + \bar{\alpha}_{222}) \tag{8.63}$$

Let us select $\bar{\mathbf{K}}$ as

$$\bar{\mathbf{K}} = \begin{bmatrix} 1 & b_{12} \\ 0 & 1 \end{bmatrix}^{-1} \begin{bmatrix} \bar{\alpha}_{111} - \alpha_{111} & \bar{\alpha}_{112} - \alpha_{112} & \bar{\alpha}_{113} - \alpha_{113} \\ \bar{\alpha}_{211} - \alpha_{211} & \bar{\alpha}_{212} - \alpha_{212} & \bar{\alpha}_{213} - \alpha_{213} \end{bmatrix}$$

$$\begin{bmatrix} \bar{\alpha}_{114} - \alpha_{114} & -\alpha_{121} & -\alpha_{122} \\ \bar{\alpha}_{214} - \alpha_{214} & \bar{\alpha}_{221} - \alpha_{221} & \bar{\alpha}_{222} - \alpha_{222} \end{bmatrix} \tag{8.64}$$

Then it is straightforward to verify the following

$$\bar{\mathbf{A}} - \bar{\mathbf{B}}\bar{\mathbf{K}} = \begin{bmatrix} -\bar{\alpha}_{111} & -\bar{\alpha}_{112} & -\bar{\alpha}_{113} & -\bar{\alpha}_{114} & \vdots & 0 & 0 \\ 1 & 0 & 0 & 0 & \vdots & 0 & 0 \\ 0 & 1 & 0 & 0 & \vdots & 0 & 0 \\ 0 & 0 & 1 & 0 & \vdots & 0 & 0 \\ \cdots & \cdots & \cdots & \cdots & \cdots & \cdots & \cdots \\ -\bar{\alpha}_{211} & -\bar{\alpha}_{212} & -\bar{\alpha}_{213} & -\bar{\alpha}_{214} & \vdots & -\bar{\alpha}_{221} & -\bar{\alpha}_{222} \\ 0 & 0 & 0 & 0 & \vdots & 1 & 0 \end{bmatrix} \tag{8.65}$$

Because $(\bar{\mathbf{A}} - \bar{\mathbf{B}}\bar{\mathbf{K}})$ is block triangular, for any $\bar{\alpha}_{21i}$, $i = 1, 2, 3, 4$, its characteristic polynomial equals the product of the characteristic polynomials of the two diagonal blocks of orders 4 and 2. Because the diagonal blocks are of companion form, we conclude that the characteristic polynomial of $(\bar{\mathbf{A}} - \bar{\mathbf{B}}\bar{\mathbf{K}})$ equals the one in (8.63). If $\mathbf{K} = \bar{\mathbf{K}}\mathbf{P}$, then $(\bar{\mathbf{A}} - \bar{\mathbf{B}}\bar{\mathbf{K}}) = \mathbf{P}(\mathbf{A} - \mathbf{B}\mathbf{K})\mathbf{P}^{-1}$. Thus the feedback gain $\mathbf{K} = \bar{\mathbf{K}}\mathbf{P}$ will place the eigenvalues of $(\mathbf{A} - \mathbf{B}\mathbf{K})$ in the desired locations. This establishes once again Theorem 8.M3.

Unlike the single-input case, where the feedback gain is unique, the feedback gain matrix in the multi-input case is not unique. For example, the $\bar{\mathbf{K}}$ in (8.64) yields a lower block-triangular matrix in $(\bar{\mathbf{A}} - \bar{\mathbf{B}}\bar{\mathbf{K}})$. It is possible to select a different $\bar{\mathbf{K}}$ to yield an upper block-triangular matrix or a block-diagonal matrix. Furthermore, a different grouping of (8.63) will again yield a different $\bar{\mathbf{K}}$.

8.6.4 Effect on Transfer Matrices[3]

In the single-variable case, state feedback can shift the poles of a plant transfer function $\hat{g}(s)$ to any positions and yet has no effect on the zeros. Or, equivalently, state feedback can change the denominator coefficients, except the leading coefficient 1, to any values but has no effect on the numerator coefficients. Although we can establish a similar result for the multivariable case from (8.62) and (8.65), it is instructive to do so by using the result in Section 7.9. Following the notation in Section 7.9, we express $\hat{\mathbf{G}}(s) = \mathbf{C}(s\mathbf{I} - \mathbf{A})^{-1}\mathbf{B}$ as

$$\hat{\mathbf{G}}(s) = \mathbf{N}(s)\mathbf{D}^{-1}(s) \tag{8.66}$$

or

$$\hat{\mathbf{y}}(s) = \mathbf{N}(s)\mathbf{D}^{-1}(s)\hat{\mathbf{u}}(s) \tag{8.67}$$

where $\mathbf{N}(s)$ and $\mathbf{D}(s)$ are right coprime and $\mathbf{D}(s)$ is column reduced. Define

$$\mathbf{D}(s)\hat{\mathbf{v}}(s) = \hat{\mathbf{u}}(s) \tag{8.68}$$

as in (7.93). Then we have

$$\hat{\mathbf{y}}(s) = \mathbf{N}(s)\hat{\mathbf{v}}(s) \tag{8.69}$$

Let $\mathbf{H}(s)$ and $\mathbf{L}(s)$ be defined as in (7.91) and (7.92). Then the state vector in (8.62) is

$$\hat{\mathbf{x}}(s) = \mathbf{L}(s)\hat{\mathbf{v}}(s)$$

Thus the state feedback becomes, in the Laplace-transform domain,

$$\hat{\mathbf{u}}(s) = \hat{\mathbf{r}}(s) - \mathbf{K}\hat{\mathbf{x}}(s) = \hat{\mathbf{r}}(s) - \mathbf{K}\mathbf{L}(s)\hat{\mathbf{v}}(s) \tag{8.70}$$

and can be represented as shown in Fig. 8.11.

Let us express $\mathbf{D}(s)$ as

$$\mathbf{D}(s) = \mathbf{D}_{hc}\mathbf{H}(s) + \mathbf{D}_{lc}\mathbf{L}(s) \tag{8.71}$$

Substituting (8.71) and (8.70) into (8.68) yields

$$[\mathbf{D}_{hc}\mathbf{H}(s) + \mathbf{D}_{lc}\mathbf{L}(s)]\,\hat{\mathbf{v}}(s) = \hat{\mathbf{r}}(s) - \mathbf{K}\mathbf{L}(s)\hat{\mathbf{v}}(s)$$

which implies

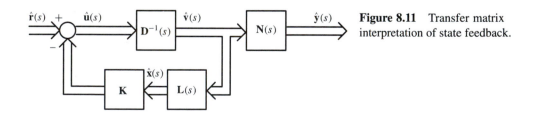

Figure 8.11 Transfer matrix interpretation of state feedback.

---

3. This subsection may be skipped without loss of continuity. The material in Section 7.9 is needed to study this subsection.

$$[\mathbf{D}_{hc}\mathbf{H}(s) + (\mathbf{D}_{lc} + \mathbf{K})\mathbf{L}(s)]\,\hat{\mathbf{v}}(s) = \hat{\mathbf{r}}(s)$$

Substituting this into (8.69) yields

$$\hat{\mathbf{y}}(s) = \mathbf{N}(s)\,[\mathbf{D}_{hc}\mathbf{H}(s) + (\mathbf{D}_{lc} + \mathbf{K})\mathbf{L}(s)]^{-1}\,\hat{\mathbf{r}}(s)$$

Thus the transfer matrix from $\mathbf{r}$ to $\mathbf{y}$ is

$$\hat{\mathbf{G}}_f(s) = \mathbf{N}(s)\,[\mathbf{D}_{hc}\mathbf{H}(s) + (\mathbf{D}_{lc} + \mathbf{K})\mathbf{L}(s)]^{-1} \tag{8.72}$$

The state feedback changes the plant transfer matrix $\mathbf{N}(s)\mathbf{D}^{-1}(s)$ to the one in (8.72). We see that the numerator matrix $\mathbf{N}(s)$ is not affected by the state feedback. Neither are the column degree $\mathbf{H}(s)$ and the column-degree coefficient matrix $\mathbf{D}_{hc}$ affected by the state feedback. However, all coefficients associated with $\mathbf{L}(s)$ can be assigned arbitrarily by selecting a $\mathbf{K}$. This is similar to the SISO case.

It is possible to extend the robust tracking and disturbance rejection discussed in Section 8.3 to the multivariable case. It is simpler, however, to do so by using coprime fractions; therefore it will not be discussed here.

8.7 State Estimators—Multivariable Case

All discussion for state estimators in the single-variable case applies to the multivariable case; therefore the discussion will be brief. Consider the n-dimensional p-input q-output state equation

$$\dot{\mathbf{x}} = \mathbf{Ax} + \mathbf{Bu}$$
$$\mathbf{y} = \mathbf{Cx} \tag{8.73}$$

The problem is to use available input $\mathbf{u}$ and output $\mathbf{y}$ to drive a system whose output gives an estimate of the state $\mathbf{x}$. We extend (8.40) to the multivariable case as

$$\dot{\hat{\mathbf{x}}} = (\mathbf{A} - \mathbf{LC})\hat{\mathbf{x}} + \mathbf{Bu} + \mathbf{Ly} \tag{8.74}$$

This is a full-dimensional state estimator. Let us define the error vector as

$$\mathbf{e}(t) := \mathbf{x}(t) - \hat{\mathbf{x}}(t) \tag{8.75}$$

Then we have, as in (8.41),

$$\dot{\mathbf{e}} = (\mathbf{A} - \mathbf{LC})\mathbf{e} \tag{8.76}$$

If $(\mathbf{A}, \mathbf{C})$ is observable, then all eigenvalues of $(\mathbf{A} - \mathbf{LC})$ can be assigned arbitrarily by choosing an $\mathbf{L}$. Thus the convergence rate for the estimated state $\hat{\mathbf{x}}$ to approach the actual state $\mathbf{x}$ can be as fast as desired. As in the SISO case, the three methods of computing state feedback gain $\mathbf{K}$ in Sections 8.6.1 through 8.6.3 can be applied here to compute $\mathbf{L}$.

Next we discuss reduced-dimensional state estimators. The next procedure is an extension of Procedure 8.R1 to the multivariable case.

▶ **Procedure 8.MR1**

Consider the n-dimensional q-output observable pair $(\mathbf{A}, \mathbf{C})$. It is assumed that $\mathbf{C}$ has rank q.

1. Select an arbitrary $(n - q) \times (n - q)$ stable matrix $\mathbf{F}$ that has no eigenvalues in common with those of $\mathbf{A}$.
2. Select an arbitrary $(n - q) \times q$ matrix $\mathbf{L}$ such that $(\mathbf{F}, \mathbf{L})$ is controllable.
3. Solve the unique $(n - q) \times n$ matrix $\mathbf{T}$ in the Lyapunov equation $\mathbf{TA} - \mathbf{FT} = \mathbf{LC}$.
4. If the square matrix of order n

$$\mathbf{P} = \begin{bmatrix} \mathbf{C} \\ \mathbf{T} \end{bmatrix} \tag{8.77}$$

is singular, go back to Step 2 and repeat the process. If $\mathbf{P}$ is nonsingular, then the $(n - q)$-dimensional state equation

$$\dot{\mathbf{z}} = \mathbf{Fz} + \mathbf{TBu} + \mathbf{Ly} \tag{8.78}$$

$$\hat{\mathbf{x}} = \begin{bmatrix} \mathbf{C} \\ \mathbf{T} \end{bmatrix}^{-1} \begin{bmatrix} \mathbf{y} \\ \mathbf{z} \end{bmatrix} \tag{8.79}$$

generates an estimate of $\mathbf{x}$.

We first justify the procedure. We write (8.79) as

$$\begin{bmatrix} \mathbf{y} \\ \mathbf{z} \end{bmatrix} = \begin{bmatrix} \mathbf{C} \\ \mathbf{T} \end{bmatrix} \hat{\mathbf{x}}$$

which implies $\mathbf{y} = \mathbf{C}\hat{\mathbf{x}}$ and $\mathbf{z} = \mathbf{T}\hat{\mathbf{x}}$. Clearly $\mathbf{y}$ is an estimate of $\mathbf{Cx}$. We now show that $\mathbf{z}$ is an estimate of $\mathbf{Tx}$. Let us define

$$\mathbf{e} := \mathbf{z} - \mathbf{Tx}$$

Then we have

$$\dot{\mathbf{e}} = \dot{\mathbf{z}} - \mathbf{T}\dot{\mathbf{x}} = \mathbf{Fz} + \mathbf{TBu} + \mathbf{LCx} - \mathbf{TAx} - \mathbf{TBu}$$

$$= \mathbf{Fz} + (\mathbf{LC} - \mathbf{TA})\mathbf{x} = \mathbf{F}(\mathbf{z} - \mathbf{Tx}) = \mathbf{Fe}$$

If $\mathbf{F}$ is stable, then $\mathbf{e}(t) \to \mathbf{0}$ as $t \to \infty$. Thus $\mathbf{z}$ is an estimate of $\mathbf{Tx}$.

▶ **Theorem 8.M6**

If $\mathbf{A}$ and $\mathbf{F}$ have no common eigenvalues, then the square matrix

$$\mathbf{P} := \begin{bmatrix} \mathbf{C} \\ \mathbf{T} \end{bmatrix}$$

where $\mathbf{T}$ is the unique solution of $\mathbf{TA} - \mathbf{FT} = \mathbf{LC}$, is nonsingular only if $(\mathbf{A}, \mathbf{C})$ is observable and $(\mathbf{F}, \mathbf{L})$ is controllable.

This theorem can be proved by combining the proofs of Theorems 8.M4 and 8.6. Unlike Theorem 8.6, where the conditions are necessary and sufficient for $\mathbf{P}$ to be nonsingular, the conditions here are only necessary. Given $(\mathbf{A}, \mathbf{C})$, it is possible to construct a controllable pair $(\mathbf{F}, \mathbf{L})$ so that $\mathbf{P}$ is singular. However, after selecting $\mathbf{F}$, if $\mathbf{L}$ is selected randomly and if $(\mathbf{F}, \mathbf{L})$ is controllable, it is believed that the probability for $\mathbf{P}$ to be nonsingular is 1.

8.8 Feedback from Estimated States–Multivariable Case

This section will extend the separation property discussed in Section 8.5 to the multivariable case. We use the reduced-dimensional state estimator; therefore the development is more complex.

Consider the n-dimensional state equation

$$\dot{\mathbf{x}} = \mathbf{A}\mathbf{x} + \mathbf{B}\mathbf{u}$$
$$\mathbf{y} = \mathbf{C}\mathbf{x} \tag{8.80}$$

and the $(n-q)$-dimensional state estimator in (8.78) and (8.79). First we compute the inverse of $\mathbf{P}$ in (8.77) and then partition it as $[\mathbf{Q}_1 \quad \mathbf{Q}_2]$, where $\mathbf{Q}_1$ is $n \times q$ and $\mathbf{Q}_2$ is $n \times (n-q)$; that is,

$$[\mathbf{Q}_1 \quad \mathbf{Q}_2]\begin{bmatrix}\mathbf{C}\\\mathbf{T}\end{bmatrix} = \mathbf{Q}_1\mathbf{C} + \mathbf{Q}_2\mathbf{T} = \mathbf{I} \tag{8.81}$$

Then the $(n-q)$-dimensional state estimator in (8.78) and (8.79) can be written as

$$\dot{\mathbf{z}} = \mathbf{F}\mathbf{z} + \mathbf{T}\mathbf{B}\mathbf{u} + \mathbf{L}\mathbf{y} \tag{8.82}$$
$$\hat{\mathbf{x}} = \mathbf{Q}_1\mathbf{y} + \mathbf{Q}_2\mathbf{z} \tag{8.83}$$

If the original state is not available for state feedback, we apply the feedback gain matrix to $\hat{\mathbf{x}}$ to yield

$$\mathbf{u} = \mathbf{r} - \mathbf{K}\hat{\mathbf{x}} = \mathbf{r} - \mathbf{K}\mathbf{Q}_1\mathbf{y} - \mathbf{K}\mathbf{Q}_2\mathbf{z} \tag{8.84}$$

Substituting this into (8.80) and (8.82) yields

$$\dot{\mathbf{x}} = \mathbf{A}\mathbf{x} + \mathbf{B}(\mathbf{r} - \mathbf{K}\mathbf{Q}_1\mathbf{C}\mathbf{x} - \mathbf{K}\mathbf{Q}_2\mathbf{z})$$
$$= (\mathbf{A} - \mathbf{B}\mathbf{K}\mathbf{Q}_1\mathbf{C})\mathbf{x} - \mathbf{B}\mathbf{K}\mathbf{Q}_2\mathbf{z} + \mathbf{B}\mathbf{r} \tag{8.85}$$
$$\dot{\mathbf{z}} = \mathbf{F}\mathbf{z} + \mathbf{T}\mathbf{B}(\mathbf{r} - \mathbf{K}\mathbf{Q}_1\mathbf{C}\mathbf{x} - \mathbf{K}\mathbf{Q}_2\mathbf{z}) + \mathbf{L}\mathbf{C}\mathbf{x}$$
$$= (\mathbf{L}\mathbf{C} - \mathbf{T}\mathbf{B}\mathbf{K}\mathbf{Q}_1\mathbf{C})\mathbf{x} + (\mathbf{F} - \mathbf{T}\mathbf{B}\mathbf{K}\mathbf{Q}_2)\mathbf{z} + \mathbf{T}\mathbf{B}\mathbf{r} \tag{8.86}$$

They can be combined as

$$\begin{bmatrix}\dot{\mathbf{x}}\\\dot{\mathbf{z}}\end{bmatrix} = \begin{bmatrix}\mathbf{A} - \mathbf{B}\mathbf{K}\mathbf{Q}_1\mathbf{C} & -\mathbf{B}\mathbf{K}\mathbf{Q}_2\\\mathbf{L}\mathbf{C} - \mathbf{T}\mathbf{B}\mathbf{K}\mathbf{Q}_1\mathbf{C} & \mathbf{F} - \mathbf{T}\mathbf{B}\mathbf{K}\mathbf{Q}_2\end{bmatrix}\begin{bmatrix}\mathbf{x}\\\mathbf{z}\end{bmatrix} + \begin{bmatrix}\mathbf{B}\\\mathbf{T}\mathbf{B}\end{bmatrix}\mathbf{r}$$
$$\mathbf{y} = [\mathbf{C} \quad \mathbf{0}]\begin{bmatrix}\mathbf{x}\\\mathbf{z}\end{bmatrix} \tag{8.87}$$

This $(2n-q)$-dimensional state equation describes the feedback system in Fig. 8.8. As in the SISO case, let us carry out the following equivalence transformation

$$\begin{bmatrix} \mathbf{x} \\ \mathbf{e} \end{bmatrix} = \begin{bmatrix} \mathbf{x} \\ \mathbf{z} - \mathbf{Tx} \end{bmatrix} = \begin{bmatrix} \mathbf{I}_n & \mathbf{0} \\ -\mathbf{T} & \mathbf{I}_{n-q} \end{bmatrix} \begin{bmatrix} \mathbf{x} \\ \mathbf{z} \end{bmatrix}$$

After some manipulation and using $\mathbf{TA} - \mathbf{FT} = \mathbf{LC}$ and (8.81), we can finally obtain the following equivalent state equation

$$\begin{bmatrix} \dot{\mathbf{x}} \\ \dot{\mathbf{e}} \end{bmatrix} = \begin{bmatrix} \mathbf{A} - \mathbf{BK} & -\mathbf{BKQ}_2 \\ \mathbf{0} & \mathbf{F} \end{bmatrix} \begin{bmatrix} \mathbf{x} \\ \mathbf{e} \end{bmatrix} + \begin{bmatrix} \mathbf{B} \\ \mathbf{0} \end{bmatrix} \mathbf{r}$$

$$y = [\mathbf{C} \ \ \mathbf{0}] \begin{bmatrix} \mathbf{x} \\ \mathbf{e} \end{bmatrix}$$

(8.88)

This equation is similar to (8.55) for the single-variable case. Therefore all discussion there applies, without any modification, to the multivariable case. In other words, the design of a state feedback and the design of a state estimator can be carried out independently. This is the *separation property*. Furthermore, all eigenvalues of $\mathbf{F}$ are not controllable from $\mathbf{r}$ and the transfer matrix from $\mathbf{r}$ to $\mathbf{y}$ equals

$$\hat{\mathbf{G}}_f(s) = \mathbf{C}(s\mathbf{I} - \mathbf{A} + \mathbf{BK})^{-1}\mathbf{B}$$

PROBLEMS

8.1 Given

$$\dot{\mathbf{x}} = \begin{bmatrix} 2 & 1 \\ -1 & 1 \end{bmatrix} \mathbf{x} + \begin{bmatrix} 1 \\ 2 \end{bmatrix} u \qquad y = [1 \ \ 1]\mathbf{x}$$

find the state feedback gain $\mathbf{k}$ so that the state feedback system has -1 and -2 as its eigenvalues. Compute $\mathbf{k}$ directly without using any equivalence transformation.

8.2 Repeat Problem 8.1 by using (8.13).

8.3 Repeat Problem 8.1 by solving a Lyapunov equation.

8.4 Find the state feedback gain for the state equation

$$\dot{\mathbf{x}} = \begin{bmatrix} 1 & 1 & -2 \\ 0 & 1 & 1 \\ 0 & 0 & 1 \end{bmatrix} \mathbf{x} + \begin{bmatrix} 1 \\ 0 \\ 1 \end{bmatrix} u$$

so that the resulting system has eigenvalues -2 and $-1 \pm j1$. Use the method you think is the simplest by hand to carry out the design.

8.5 Consider a system with transfer function

$$\hat{g}(s) = \frac{(s-1)(s+2)}{(s+1)(s-2)(s+3)}$$

Is it possible to change the transfer function to

$$\hat{g}_f(s) = \frac{s-1}{(s+2)(s+3)}$$

by state feedback? Is the resulting system BIBO stable? Asymptotically stable?

8.6 Consider a system with transfer function

$$\hat{g}(s) = \frac{(s-1)(s+2)}{(s+1)(s-2)(s+3)}$$

Is it possible to change the transfer function to

$$\hat{g}_f(s) = \frac{1}{s+3}$$

by state feedback? Is the resulting system BIBO stable? Asymptotically stable?

8.7 Consider the continuous-time state equation

$$\dot{\mathbf{x}} = \begin{bmatrix} 1 & 1 & -2 \\ 0 & 1 & 1 \\ 0 & 0 & 1 \end{bmatrix} \mathbf{x} + \begin{bmatrix} 1 \\ 0 \\ 1 \end{bmatrix} u$$

$$y = [2 \ 0 \ 0]\mathbf{x}$$

Let $u = pr - \mathbf{kx}$. Find the feedforward gain p and state feedback gain $\mathbf{k}$ so that the resulting system has eigenvalues -2 and $-1 \pm j1$ and will track asymptotically any step reference input.

8.8 Consider the discrete-time state equation

$$\mathbf{x}[k+1] = \begin{bmatrix} 1 & 1 & -2 \\ 0 & 1 & 1 \\ 0 & 0 & 1 \end{bmatrix} \mathbf{x}[k] + \begin{bmatrix} 1 \\ 0 \\ 1 \end{bmatrix} u[k]$$

$$y[k] = [2 \ 0 \ 0]\mathbf{x}[k]$$

Find the state feedback gain so that the resulting system has all eigenvalues at $z = 0$. Show that for any initial state, the zero-input response of the feedback system becomes identically zero for $k \geq 3$.

8.9 Consider the discrete-time state equation in Problem 8.8. Let $u[k] = pr[k] - \mathbf{kx}[k]$, where p is a feedforward gain. For the $\mathbf{k}$ in Problem 8.8, find a gain p so that the output will track any step reference input. Show also that $y[k] = r[k]$ for $k \geq 3$. Thus exact tracking is achieved in a finite number of sampling periods instead of asymptotically. This is possible if all poles of the resulting system are placed at $z = 0$. This is called the *dead-beat* design.

8.10 Consider the uncontrollable state equation

$$\dot{\mathbf{x}} = \begin{bmatrix} 2 & 1 & 0 & 0 \\ 0 & 2 & 0 & 0 \\ 0 & 0 & -1 & 0 \\ 0 & 0 & 0 & -1 \end{bmatrix} \mathbf{x} + \begin{bmatrix} 0 \\ 1 \\ 1 \\ 1 \end{bmatrix} u$$

Is it possible to find a gain $\mathbf{k}$ so that the equation with state feedback $u = r - \mathbf{kx}$ has eigenvalues $-2, -2, -1, -1$? Is it possible to have eigenvalues $-2, -2, -2, -1$? How about $-2, -2, -2, -2$? Is the equation stabilizable?

8.11 Design a full-dimensional and a reduced-dimensional state estimator for the state equation in Problem 8.1. Select the eigenvalues of the estimators from $\{-3, -2 \pm j2\}$.

8.12 Consider the state equation in Problem 8.1. Compute the transfer function from r to y of the state feedback system. Compute the transfer function from r to y if the feedback gain is applied to the estimated state of the full-dimensional estimator designed in Problem 8.11. Compute the transfer function from r to y if the feedback gain is applied to the estimated state of the reduced-dimensional state estimator also designed in Problem 8.11. Are the three overall transfer functions the same?

8.13 Let

$$\mathbf{A} = \begin{bmatrix} 0 & 1 & 0 & 0 \\ 0 & 0 & 1 & 0 \\ -3 & 1 & 2 & 3 \\ 2 & 1 & 0 & 0 \end{bmatrix} \qquad \mathbf{B} = \begin{bmatrix} 0 & 0 \\ 0 & 0 \\ 1 & 2 \\ 0 & 2 \end{bmatrix}$$

Find two different constant matrices $\mathbf{K}$ such that $(\mathbf{A} - \mathbf{BK})$ has eigenvalues $-4 \pm 3j$ and $-5 \pm 4j$.

Chapter

9

Pole Placement and Model Matching

9.1 Introduction

We first give reasons for introducing this chapter. Chapter 6 discusses state-space analysis (controllability and observability) and Chapter 8 introduces state-space design (state feedback and state estimators). In Chapter 7 coprime fractions were discussed. Therefore it is logical to discuss in this chapter their applications in design.

One way to introduce coprime fraction design is to develop the Bezout identity and to parameterize all stabilization compensators. See References [3, 6, 9, 13, 20]. This approach is important in some optimization problems but is not necessarily convenient for all designs. See Reference [8]. We study in this chapter only designs of minimum-degree compensators to achieve pole placement and model matching. We will change the problems into solving linear algebraic equations. Using only Theorem 3.2 and its corollary, we can establish all needed results. Therefore we can bypass the Bezout identity and some polynomial theorems and simplify the discussion.

Most control systems can be formulated as shown in Fig. 8.1. That is, given a plant with input u and output y and a reference signal r, design an overall system so that the output y will follow the reference signal r as closely as possible. The plant input u is also called the actuating signal and the plant output y, the controlled signal. If the actuating signal u depends only on the reference signal r as shown in Fig. 9.1(a), it is called an open-loop control. If u depends on r and y, then it is called a closed-loop or feedback control. The open-loop control is, in general, not satisfactory if there are plant parameter variations due to changes of load, environment, or aging. It is also very sensitive to noise and disturbance, which often exist in the real world. Therefore open-loop control is used less often in practice.

269

There are many possible feedback configurations. The simplest is the unity-feedback configuration shown in Fig. 9.1(b) in which the constant gain p and the compensator with transfer function $C(s)$ are to be designed. Clearly we have

$$\hat{u}(s) = C(s)[p\hat{r}(s) - \hat{y}(s)] \tag{9.1}$$

Because p is a constant, the reference signal r and the plant output y drive essentially the same compensator to generate an actuating signal. Thus the configuration is said to have one degree of freedom. Clearly the open-loop configuration also has one degree of freedom.

The connection of state feedback and state estimator in Fig. 8.8 can be redrawn as shown in Fig. 9.1(c). Simple manipulation yields

$$\hat{u}(s) = \frac{1}{1 + C_1(s)}\hat{r}(s) - \frac{C_2(s)}{1 + C_1(s)}\hat{y}(s) \tag{9.2}$$

We see that r and y drive two independent compensators to generate a u. Thus the configuration is said to have two degrees of freedom.

A more natural two-degree-of-freedom configuration can be obtained by modifying (9.1) as

$$\hat{u}(s) = C_1(s)\hat{r}(s) - C_2(s)\hat{y}(s) \tag{9.3}$$

and is plotted in Fig. 9.1(d). This is the most general control signal because each of r and y drives a compensator, which we have freedom in designing. Thus no configuration has three degrees of freedom. There are many possible two-degree-of-freedom configurations; see, for example, Reference [12]. We call the one in Fig. 9.1(d) the *two-parameter configuration*; the one in Fig. 9.1(c) the *controller-estimator* or *plant-input–output-feedback* configuration. Because the two-parameter configuration seems to be more natural and more suitable for practical application, we study only this configuration in this chapter. For designs using the plant-input–output-feedback configuration, see Reference [6].

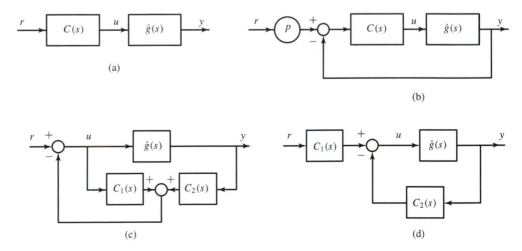

Figure 9.1 Control configurations.

The plants studied in this chapter will be limited to those describable by strictly proper rational functions or matrices. We also assume that every transfer matrix has full rank in the sense that if $\hat{\mathbf{G}}(s)$ is $q \times p$, then it has a $q \times q$ or $p \times p$ submatrix with a nonzero determinant. If $\hat{\mathbf{G}}(s)$ is square, then its determinant is nonzero or its inverse exists. This is equivalent to the assumption that if $(\mathbf{A}, \mathbf{B}, \mathbf{C})$ is a minimal realization of the transfer matrix, then $\mathbf{B}$ has full column rank and $\mathbf{C}$ has full row rank.

The design to be introduced in this chapter is based on coprime polynomial fractions of rational matrices. Thus the concept of coprimeness and the method of computing coprime fractions introduced in Sections 7.1 through 7.3 and 7.6 through 7.8 are needed for studying this chapter. The rest of Chapter 7 and the entire Chapter 8, however, are not needed here. In this chapter, we will change the design problem into solving sets of linear algebraic equations. Thus the method is called the *linear algebraic* method in Reference [7].

For convenience, we first introduce some terminology. Every transfer function $\hat{g}(s) = N(s)/D(s)$ is assumed to be a coprime fraction. Then every root of $D(s)$ is a pole and every root of $N(s)$ is a zero. A pole is called a stable pole if it has a negative real part; an unstable pole if it has a zero or positive real part. We also define

- Minimum-phase zeros: zeros with negative real parts
- Nonminimum-phase zeros: zeros with zero or positive real parts

Although some texts call them stable and unstable zeros, they have nothing to do with stability. A transfer function with only minimum-phase zeros has the smallest phase among all transfer functions with the same amplitude characteristics. See Reference [7, pp. 284–285]. Thus we use the aforementioned terminology. A polynomial is called a *Hurwitz polynomial* if all its roots have negative real parts.

9.1.1 Compensator Equations—Classical Method

Consider the equation

$$A(s)D(s) + B(s)N(s) = F(s) \tag{9.4}$$

where $D(s)$, $N(s)$, and $F(s)$ are given polynomials and $A(s)$ and $B(s)$ are unknown polynomials to be solved. Mathematically speaking, this problem is equivalent to the problem of solving integer solutions A and B in $AD + BN = F$, where D, N, and F are given integers. This is a very old mathematical problem and has been associated with mathematicians such as Diophantine, Bezout, and Aryabhatta.[1] To avoid controversy, we follow Reference [3] and call it a *compensator equation*. All design problems in this chapter can be reduced to solving compensator equations. Thus the equation is of paramount importance.

We first discuss the existence condition and general solutions of the equation. What will be discussed, however, is not needed in subsequent sections and the reader may glance through this subsection.

---

1. See Reference [21, last page of Preface].

▶ **Theorem 9.1**

Given polynomials $D(s)$ and $N(s)$, polynomial solutions $A(s)$ and $B(s)$ exist in (9.4) for any polynomial $F(s)$ if and only if $D(s)$ and $N(s)$ are coprime.

Suppose $D(s)$ and $N(s)$ are not coprime and contain the same factor $s + a$. Then the factor $s + a$ will appear in $F(s)$. Thus if $F(s)$ does not contain the factor, no solutions exist in (9.4). This shows the necessity of the theorem.

If $D(s)$ and $N(s)$ are coprime, there exist polynomials $\bar{A}(s)$ and $\bar{B}(s)$ such that

$$\bar{A}(s)D(s) + \bar{B}(s)N(s) = 1 \tag{9.5}$$

Its matrix version is called the *Bezout identity* in Reference [13]. The polynomials $\bar{A}(s)$ and $\bar{B}(s)$ can be obtained by the Euclidean algorithm and will not be discussed here. See Reference [6, pp. 578–580]. For example, if $D(s) = s^2 - 1$ and $N(s) = s - 2$, then $\bar{A}(s) = 1/3$ and $\bar{B}(s) = -(s + 2)/3$ meet (9.5). For any polynomial $F(s)$, (9.5) implies

$$F(s)\bar{A}(s)D(s) + F(s)\bar{B}(s)N(s) = F(s) \tag{9.6}$$

Thus $A(s) = F(s)\bar{A}(s)$ and $B(s) = F(s)\bar{B}(s)$ are solutions. This shows the sufficiency of the theorem.

Next we discuss general solutions. For any $D(s)$ and $N(s)$, there exist two polynomials $\hat{A}(s)$ and $\hat{B}(s)$ such that

$$\hat{A}(s)D(s) + \hat{B}(s)N(s) = 0 \tag{9.7}$$

Obviously $\hat{A}(s) = -N(s)$ and $\hat{B}(s) = D(s)$ are such solutions. Then for any polynomial $Q(s)$,

$$A(s) = \bar{A}(s)F(s) + Q(s)\hat{A}(s) \qquad B(s) = \bar{B}(s)F(s) + Q(s)\hat{B}(s) \tag{9.8}$$

are general solutions of (9.4). This can easily be verified by substituting (9.8) into (9.4) and using (9.5) and (9.7).

EXAMPLE 9.1 Given $D(s) = s^2 - 1$, $N(s) = s - 2$, and $F(s) = s^3 + 4s^2 + 6s + 4$, then

$$A(s) = \tfrac{1}{3}(s^3 + 4s^2 + 6s + 4) + Q(s)(-s + 2)$$
$$B(s) = -\tfrac{1}{3}(s + 2)(s^3 + 4s^2 + 6s + 4) + Q(s)(s^2 - 1) \tag{9.9}$$

for any polynomial $Q(s)$, are solutions of (9.4).

Although the classical method can yield general solutions, the solutions are not necessarily convenient to use in design. For example, we may be interested in solving $A(s)$ and $B(s)$ with least degrees to meet (9.4). For the polynomials in Example 9.1, after some manipulation, we find that if $Q(s) = (s^2 + 6s + 15)/3$, then (9.9) reduces to

$$A(s) = s + 34/3 \qquad B(s) = (-22s - 23)/3 \tag{9.10}$$

They are the least-degree solutions of Example 9.1. In this chapter, instead of solving the compensator equation directly as shown, we will change it into solving a set of linear algebraic equations as in Section 7.3. By so doing, we can bypass some polynomial theorems.

9.2 Unity-Feedback Configuration—Pole Placement

Consider the unity-feedback system shown in Fig. 9.1(b). The plant transfer function $\hat{g}(s)$ is assumed to be strictly proper and of degree n. The problem is to design a proper compensator $C(s)$ of least possible degree m so that the resulting overall system has any set of $n+m$ desired poles. Because all transfer functions are required to have real coefficients, complex conjugate poles must be assigned in pairs. This will be a standing assumption throughout this chapter.

Let $\hat{g}(s) = N(s)/D(s)$ and $C(s) = B(s)/A(s)$. Then the overall transfer function from r to y in Fig. 9.1(b) is

$$\hat{g}_o(s) = \frac{pC(s)\hat{g}(s)}{1 + C(s)\hat{g}(s)} = \frac{p\dfrac{B(s)}{A(s)}\dfrac{N(s)}{D(s)}}{1 + \dfrac{B(s)}{A(s)}\dfrac{N(s)}{D(s)}}$$

$$= \frac{pB(s)N(s)}{A(s)D(s) + B(s)N(s)} \tag{9.11}$$

In pole assignment, we are interested in assigning all poles of $\hat{g}_o(s)$ or, equivalently, all roots of $A(s)D(s) + B(s)N(s)$. In this design, nothing is said about the zeros of $\hat{g}_o(s)$. As we can see from (9.11), the design not only has no effect on the plant zeros (roots of $N(s)$) but also introduces new zeros (roots of $B(s)$) into the overall transfer function. On the other hand, the poles of the plant and compensator are shifted from $D(s)$ and $A(s)$ to the roots of $A(s)D(s) + B(s)N(s)$. Thus *feedback can shift poles but has no effect on zeros.*

Given a set of desired poles, we can readily form a polynomial $F(s)$ that has the desired poles as its roots. Then the pole-placement problem becomes one of solving the polynomial equation

$$A(s)D(s) + B(s)N(s) = F(s) \tag{9.12}$$

Instead of solving (9.12) directly, we will transform it into solving a set of linear algebraic equations. Let $\deg N(s) < \deg D(s) = n$ and $\deg B(s) \le \deg A(s) = m$. Then $F(s)$ in (9.12) has degree at most $n+m$. Let us write

$$D(s) = D_0 + D_1 s + D_2 s^2 + \cdots + D_n s^n \qquad D_n \ne 0$$
$$N(s) = N_0 + N_1 s + N_2 s^2 + \cdots + N_n s^n$$
$$A(s) = A_0 + A_1 s + A_2 s^2 + \cdots + A_m s^m$$
$$B(s) = B_0 + B_1 s + B_2 s^2 + \cdots + B_m s^m$$
$$F(s) = F_0 + F_1 s + F_2 s^2 + \cdots + F_{n+m} s^{n+m}$$

where all coefficients are real constants, not necessarily nonzero. Substituting these into (9.12) and matching the coefficients of like powers of s, we obtain

$$A_0 D_0 + B_0 N_0 = F_0$$

$$A_0 D_1 + B_0 N_1 + A_1 D_0 + B_1 N_0 = F_1$$

$$\vdots$$

$$A_m D_n + B_m N_n = F_{n+m}$$

There are a total of $(n + m + 1)$ equations. They can be arranged in matrix form as

$$[A_0 \ B_0 \ A_1 \ B_1 \ \cdots \ A_m \ B_m]\mathbf{S}_m = [F_0 \ F_1 \ F_2 \ \cdots \ F_{n+m}] \tag{9.13}$$

with

$$\mathbf{S}_m := \begin{bmatrix} D_0 & D_1 & \cdots & D_n & 0 & \cdots & 0 \\ N_0 & N_1 & \cdots & N_n & 0 & \cdots & 0 \\ \cdots & \cdots & \cdots & \cdots & \cdots & \cdots & \cdots \\ 0 & D_0 & \cdots & D_{n-1} & D_n & \cdots & 0 \\ 0 & N_0 & \cdots & N_{n-1} & N_n & \cdots & 0 \\ \cdots & \cdots & \cdots & \cdots & \cdots & \cdots & \cdots \\ \vdots & \vdots & & \vdots & \vdots & & \vdots \\ \cdots & \cdots & \cdots & \cdots & \cdots & \cdots & \cdots \\ 0 & 0 & \cdots & 0 & D_0 & \cdots & D_n \\ 0 & 0 & \cdots & 0 & N_0 & \cdots & N_n \end{bmatrix} \tag{9.14}$$

If we take the transpose of (9.13), then it becomes the standard form studied in Theorems 3.1 and 3.2. We use the form in (9.13) because it can be extended directly to the matrix case. The matrix $\mathbf{S}_m$ has $2(m + 1)$ rows and $(n + m + 1)$ columns and is formed from the coefficients of $D(s)$ and $N(s)$. The first two rows are simply the coefficients of $D(s)$ and $N(s)$ arranged in ascending powers of s. The next two rows are the first two rows shifted to the right by one position. We repeat the process until we have $(m + 1)$ sets of coefficients. The left-hand-side row vector of (9.13) consists of the coefficients of the compensator $C(s)$ to be solved. If $C(s)$ has degree m, then the row vector has $2(m + 1)$ entries. The right-hand-side row vector of (9.13) consists of the coefficients of $F(s)$. Now solving the compensator equation in (9.12) becomes solving the linear algebraic equation in (9.13).

Applying Corollary 3.2, we conclude that (9.13) has a solution for any $F(s)$ if and only if $\mathbf{S}_m$ has full column rank. A necessary condition for $\mathbf{S}_m$ to have full column rank is that $\mathbf{S}_m$ is square or has more rows than columns, that is,

$$2(m + 1) \geq n + m + 1 \quad \text{or} \quad m \geq n - 1$$

If $m < n - 1$, then $\mathbf{S}_m$ does not have full column rank and solutions may exist for some $F(s)$, but not for every $F(s)$. Thus if the degree of the compensator is less than $n - 1$, it is not possible to achieve arbitrary pole placement.

If $m = n - 1$, $\mathbf{S}_{n-1}$ becomes a square matrix of order $2n$. It is the transpose of the Sylvester resultant in (7.28) with $n = 4$. As discussed in Section 7.3, $\mathbf{S}_{n-1}$ is nonsingular if and only if $D(s)$ and $N(s)$ are coprime. Thus if $D(s)$ and $N(s)$ are coprime, then $\mathbf{S}_{n-1}$ has rank $2n$ (full column rank). Now if m increases by 1, the number of columns increases by 1 but the the number of rows increases by 2. Because $D_n \neq 0$, the new D row is linearly independent of its preceding rows. Thus the $2(n + 1) \times (2n + 1)$ matrix $\mathbf{S}_n$ has rank $(2n + 1)$ (full column rank). Repeating the argument, we conclude that if $D(s)$ and $N(s)$ are coprime and if $m \geq n - 1$, then the matrix $\mathbf{S}_m$ in (9.14) has full column rank.

▶ **Theorem 9.2**

Consider the unity-feedback system shown in Fig. 9.1(b). The plant is described by a strictly proper transfer function $\hat{g}(s) = N(s)/D(s)$ with $N(s)$ and $D(s)$ coprime and deg $N(s) <$ deg $D(s) = n$. Let $m \geq n - 1$. Then for any polynomial $F(s)$ of degree $(n + m)$, there exists a proper compensator $C(s) = B(s)/A(s)$ of degree m such that the overall transfer function equals

$$\hat{g}_o(s) = \frac{pN(s)B(s)}{A(s)D(s) + B(s)N(s)} = \frac{pN(s)B(s)}{F(s)}$$

Furthermore, the compensator can be obtained by solving the linear algebraic equation in (9.13).

As discussed earlier, the matrix $\mathbf{S}_m$ has full column rank for $m \geq n - 1$; therefore, for any $(n + m)$ desired poles or, equivalently, for any $F(s)$ of degree $(n + m)$, solutions exist in (9.13). Next we show that $B(s)/A(s)$ is proper or $A_m \neq 0$. If $N(s)/D(s)$ is strictly proper, then $N_n = 0$ and the last equation of (9.13) reduces to

$$A_m D_n + B_m N_n = D_n A_m = F_{n+m}$$

Because $F(s)$ has degree $(n + m)$, we have $F_{n+m} \neq 0$ and, consequently, $A_m \neq 0$. This establishes the theorem. If $m = n - 1$, the compensator is unique; if $m > n - 1$, compensators are not unique and free parameters can be used to achieve other design objectives, as we will discuss later.

9.2.1 Regulation and Tracking

Pole placement can be used to achieve the regulation and tracking discussed in Section 8.3. In the regulator problem, we have $r = 0$ and the problem is to design a compensator $C(s)$ so that the response excited by any nonzero initial state will die out at a desired rate. For this problem, if all poles of $\hat{g}_o(s)$ are selected to have negative real parts, then for any gain p, in particular $p = 1$ (no feedforward gain is needed), the overall system will achieve regulation.

We discuss next the tracking problem. Let the reference signal be a step function with magnitude a. Then $\hat{r}(s) = a/s$ and the output $\hat{y}(s)$ equals

$$\hat{y}(s) = \hat{g}_o(s)\hat{r}(s) = \hat{g}_o(s)\frac{a}{s}$$

If $\hat{g}_o(s)$ is BIBO stable, the output will approach the constant $\hat{g}_o(0)a$ (Theorem 5.2). This can also be obtained by employing the final-value theorem of the Laplace transform as

$$\lim_{t \to \infty} y(t) = \lim_{s \to 0} s\hat{y}(s) = \hat{g}_o(0)a$$

Thus in order to track asymptotically any step reference input, $\hat{g}_o(s)$ must be BIBO stable and $\hat{g}_o(0) = 1$. The transfer function from r to y in Fig. 9.1(b) is $\hat{g}_o(s) = pN(s)B(s)/F(s)$. Thus we have

$$\hat{g}_o(0) = p\frac{N(0)B(0)}{F(0)} = p\frac{B_0 N_0}{F_0}$$

which implies

$$p = \frac{F_0}{B_0 N_0} \tag{9.15}$$

Thus in order to track any step reference input, we require $B_0 \neq 0$ and $N_0 \neq 0$. The constant B_0 is a coefficient of the compensator and can be designed to be nonzero. The coefficient N_0 is the constant term of the plant numerator. Thus if the plant transfer function has one or more zeros at $s = 0$, then $N_0 = 0$ and the plant cannot be designed to track any step reference input. This is consistent with the discussion in Section 8.3.

If the reference signal is a ramp function or $r(t) = at$, for $t \geq 0$, then using a similar argument, we can show that the overall transfer function $\hat{g}_o(s)$ must be BIBO stable and has the properties $\hat{g}_o(0) = 1$ and $\hat{g}'_o(0) = 0$ (Problems 9.13 and 9.14). This is summarized in the following.

- Regulation $\Leftrightarrow \hat{g}_o(s)$ BIBO stable.
- Tracking step reference input $\Leftrightarrow \hat{g}_o(s)$ BIBO stable and $\hat{g}_o(0) = 1$.
- Tracking ramp reference input $\Leftrightarrow \hat{g}_o(s)$ BIBO stable, $\hat{g}_o(0) = 1$, and $\hat{g}'_o(0) = 0$.

EXAMPLE 9.2 Given a plant with transfer function $\hat{g}(s) = (s - 2)/(s^2 - 1)$, find a proper compensator $C(s)$ and a gain p in the unity-feedback configuration in Fig. 9.1(b) so that the output y will track asymptotically any step reference input.

The plant transfer function has degree $n = 2$. Thus if we choose $m = 1$, all three poles of the overall system can be assigned arbitrarily. Let the three poles be selected as $-2, -1 \pm j1$; they spread evenly in the sector shown in Fig. 8.3(a). Then we have

$$F(s) = (s + 2)(s + 1 + j1)(s + 1 - j1) = (s + 2)(s^2 + 2s + 2) = s^3 + 4s^2 + 6s + 4$$

We use the coefficients of $D(s) = -1 + 0 \cdot s + 1 \cdot s^2$ and $N(s) = -2 + 1 \cdot s + 0 \cdot s^2$ to form (9.13) as

$$[A_0 \ B_0 \ A_1 \ B_1] \begin{bmatrix} -1 & 0 & 1 & 0 \\ -2 & 1 & 0 & 0 \\ \cdots & \cdots & \cdots & \cdots \\ 0 & -1 & 0 & 1 \\ 0 & -2 & 1 & 0 \end{bmatrix} = [4 \ \ 6 \ \ 4 \ \ 1]$$

Its solution is

$$A_1 = 1 \qquad A_0 = 34/3 \qquad B_1 = -22/3 \qquad B_0 = -23/3$$

This solution can easily be obtained using the MATLAB function / (slash), which denotes matrix right division. Thus we have[2]

---

2. This is the solution obtained in (9.10). This process of solving the polynomial equation in (9.13) is considerably simpler than the procedure discussed in Section 9.1.1.

$$A(s) = s + 34/3 \qquad B(s) = (-22/3)s - 23/3 = (-22s - 23)/3$$

and the compensator

$$C(s) = \frac{B(s)}{A(s)} = \frac{-(23 + 22s)/3}{34/3 + s} = \frac{-22s - 23}{3s + 34} \tag{9.16}$$

will place the three poles of the overall system at -2 and $-1 \pm j1$. If the system is designed to achieve regulation, we set $p = 1$ (no feedforward gain is needed) and the design is completed. To design tracking, we check whether or not $N_0 \neq 0$. This is the case; thus we can find a p so that the overall system will track asymptotically any step reference input. We use (9.15) to compute p:

$$p = \frac{F_0}{B_0 N_0} = \frac{4}{(-23/3)(-2)} = \frac{6}{23} \tag{9.17}$$

Thus the overall transfer function from r to y is

$$\hat{g}_o(s) = \frac{6}{23} \frac{[-(22s + 23)/3](s - 2)}{(s^3 + 4s^3 + 6s + 4)} = \frac{-2(22s + 23)(s - 2)}{23(s^3 + 4s^2 + 6s + 4)} \tag{9.18}$$

Because $\hat{g}_o(s)$ is BIBO stable and $\hat{g}_o(0) = 1$, the overall system will track any step reference input.

9.2.2 Robust Tracking and Disturbance Rejection

Consider the design problem in Example 9.2. Suppose after the design is completed, the plant transfer function $\hat{g}(s)$ changes, due to load variations, to $\hat{\bar{g}}(s) = (s - 2.1)/(s^2 - 0.95)$. Then the overall transfer function becomes

$$\hat{\bar{g}}_o(s) = \frac{pC(s)\hat{\bar{g}}(s)}{1 + C(s)\hat{\bar{g}}(s)} = \frac{6}{23} \frac{\dfrac{-22s - 23}{3s + 34} \dfrac{s - 2.1}{s - 0.95}}{1 + \dfrac{-22s - 23}{3s + 34} \dfrac{s - 2.1}{s - 0.95}}$$

$$= \frac{-6(22s + 23)(s - 2.1)}{23(3s^3 + 12s^2 + 20.35s + 16)} \tag{9.19}$$

This $\hat{\bar{g}}_o(s)$ is still BIBO stable, but $\hat{\bar{g}}_o(0) = (6 \cdot 23 \cdot 2.1)/(23 \cdot 16) = 0.7875 \neq 1$. If the reference input is a unit step function, the output will approach 0.7875 as $t \to \infty$. There is a tracking error of over 20%. Thus the overall system will no longer track any step reference input after the plant parameter variations, and the design is said to be nonrobust.

In this subsection, we discuss a design that can achieve robust tracking and disturbance rejection. Consider the system shown in Fig. 9.2, in which a disturbance enters at the plant input as shown. The problem is to design an overall system so that the plant output y will track asymptotically a class of reference signal r even with the presence of the disturbance and with plant parameter variations. This is called *robust tracking and disturbance rejection*.

Before proceeding, we discuss the nature of the reference signal $r(t)$ and the disturbance $w(t)$. If both $r(t)$ and $w(t)$ approach zero as $t \to \infty$, then the design is automatically

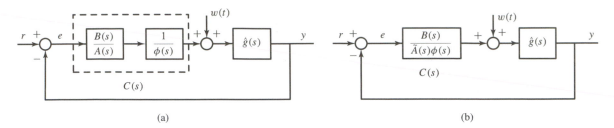

Figure 9.2   Robust tracking and disturbance rejection.

achieved if the overall system in Fig. 9.2 is designed to be BIBO stable. To exclude this trivial case, we assume that $r(t)$ and $w(t)$ do not approach zero as $t \to \infty$. If we have no knowledge whatsoever about the nature of $r(t)$ and $w(t)$, it is not possible to achieve asymptotic tracking and disturbance rejection. Therefore we need some information of $r(t)$ and $w(t)$ before carrying out the design. We assume that the Laplace transforms of $r(t)$ and $w(t)$ are given by

$$\hat{r}(s) = \mathcal{L}[r(t)] = \frac{N_r(s)}{D_r(s)} \qquad \hat{w}(s) = \mathcal{L}[w(t)] = \frac{N_w(s)}{D_w(s)} \tag{9.20}$$

where $D_r(s)$ and $D_w(s)$ are known polynomials; however, $N_r(s)$ and $N_w(s)$ are unknown to us. For example, if $r(t)$ is a step function with unknown magnitude a, then $\hat{r}(s) = a/s$. Suppose the disturbance is $w(t) = b + c\sin(\omega_o t + d)$; it consists of a constant biasing with unknown magnitude b and a sinusoid with known frequency ω_o but unknown amplitude c and phase d. Then we have $\hat{w}(s) = N_w(s)/s(s^2 + \omega_o^2)$. Let $\phi(s)$ be the least common denominator of the unstable poles of $\hat{r}(s)$ and $\hat{w}(s)$. The stable poles are excluded because they have no effect on y as $t \to \infty$. Thus all roots of $\phi(s)$ have zero or positive real parts. For the examples just discussed, we have $\phi(s) = s(s^2 + \omega_o^2)$.

▷ **Theorem 9.3**

Consider the unity-feedback system shown in Fig. 9.2(a) with a strictly proper plant transfer function $\hat{g}(s) = N(s)/D(s)$. It is assumed that $D(s)$ and $N(s)$ are coprime. The reference signal $r(t)$ and disturbance $w(t)$ are modeled as $\hat{r}(s) = N_r(s)/D_r(s)$ and $\hat{w}(s) = N_w(s)/D_w(s)$. Let $\phi(s)$ be the least common denominator of the unstable poles of $\hat{r}(s)$ and $\hat{w}(s)$. If no root of $\phi(s)$ is a zero of $\hat{g}(s)$, then there exists a proper compensator such that the overall system will track $r(t)$ and reject $w(t)$, both asymptotically and robustly.

▷ ***Proof:*** If no root of $\phi(s)$ is a zero of $\hat{g}(s) = N(s)/D(s)$, then $D(s)\phi(s)$ and $N(s)$ are coprime. Thus there exists a proper compensator $B(s)/A(s)$ such that the polynomial $F(s)$ in

$$A(s)D(s)\phi(s) + B(s)N(s) = F(s)$$

has any desired roots, in particular, has all roots lying inside the sector shown in Fig. 8.3(a). We claim that the compensator

$$C(s) = \frac{B(s)}{A(s)\phi(s)}$$

as shown in Fig. 9.2(a) will achieve the design. Let us compute the transfer function from w to y:

$$\hat{g}_{yw}(s) = \frac{N(s)/D(s)}{1 + (B(s)/A(s)\phi(s))(N(s)/D(s))}$$

$$= \frac{N(s)A(s)\phi(s)}{A(s)D(s)\phi(s) + B(s)N(s)} = \frac{N(s)A(s)\phi(s)}{F(s)}$$

Thus the output excited by $w(t)$ equals

$$\hat{y}_w(s) = \hat{g}_{yw}(s)\hat{w}(s) = \frac{N(s)A(s)\phi(s)}{F(s)} \frac{N_w(s)}{D_w(s)} \qquad (9.21)$$

Because all unstable roots of $D_w(s)$ are canceled by $\phi(s)$, all poles of $\hat{y}_w(s)$ have negative real parts. Thus we have $y_w(t) \to 0$ as $t \to \infty$. In other words, the response excited by $w(t)$ is asymptotically suppressed at the output.

Next we compute the output $\hat{y}_r(s)$ excited by $\hat{r}(s)$:

$$\hat{y}_r(s) = \hat{g}_{yr}(s)\hat{r}(s) = \frac{B(s)N(s)}{A(s)D(s)\phi(s) + B(s)N(s)}\hat{r}(s)$$

Thus we have

$$\hat{e}(s) := \hat{r}(s) - \hat{y}_r(s) = (1 - \hat{g}_{yr}(s))\hat{r}(s)$$

$$= \frac{A(s)D(s)\phi(s)}{F(s)} \frac{N_r(s)}{D_r(s)} \qquad (9.22)$$

Again all unstable roots of $D_r(s)$ are canceled by $\phi(s)$ in (9.22). Thus we conclude $r(t) - y_r(t) \to 0$ as $t \to \infty$. Because of linearity, we have $y(t) = y_w(t) + y_r(t)$ and $r(t) - y(t) \to 0$ as $t \to \infty$. This shows asymptotic tracking and disturbance rejection. From (9.21) and (9.22), we see that even if the parameters of $D(s)$, $N(s)$, $A(s)$, and $B(s)$ change, as long as the overall system remains BIBO stable and the unstable roots of $D_r(s)$ and $D_w(s)$ are canceled by $\phi(s)$, the system still achieve tracking and rejection. Thus the design is robust.   Q.E.D.

This robust design consists of two steps. First find a model $1/\phi(s)$ of the reference signal and disturbance and then carry out pole-placement design. Inserting the model inside the loop is referred to as the *internal model principle*. If the model $1/\phi(s)$ is not located in the forward path from w to y and from r to e, then $\phi(s)$ will appear in the numerators of $\hat{g}_{yw}(s)$ and $\hat{g}_{er}(s)$ (see Problem 9.7) and cancel the unstable poles of $\hat{w}(s)$ and $\hat{r}(s)$, as shown in (9.21) and (9.22). Thus the design is achieved by unstable pole–zero cancellations of $\phi(s)$. It is important to mention that there are no unstable pole–zero cancellations in the pole-placement design and the resulting unity-feedback system is totally stable, which will be defined in Section 9.3. Thus the internal model principle can be used in practical design.

In classical control system design, if a plant transfer function or a compensator transfer function is of type 1 (has one pole at $s = 0$), and the unity-feedback system is designed to be

BIBO stable, then the overall system will track asymptotically and robustly any step reference input. This is a special case of the internal model principle.

EXAMPLE 9.3 Consider the plant in Example 9.2 or $\hat{g}(s) = (s - 2)/(s^2 - 1)$. Design a unity-feedback system with a set of desired poles to track robustly any step reference input.

First we introduce the internal model $\phi(s) = 1/s$. Then $B(s)/A(s)$ in Fig. 9.2(a) can be solved from

$$A(s)D(s)\phi(s) + B(s)N(s) = F(s)$$

Because $\tilde{D}(s) := D(s)\phi(s)$ has degree 3, we may select $A(s)$ and $B(s)$ to have degree 2. Then $F(s)$ has degree 5. If we select five desired poles as -2, $-2 \pm j1$, and $-1 \pm j2$, then we have

$$F(s) = (s + 2)(s^2 + 4s + 5)(s^2 + 2s + 5)$$
$$= s^5 + 8s^4 + 30s^3 + 66s^2 + 85s + 50$$

Using the coefficients of $\tilde{D}(s) = (s^2 - 1)s = 0 - s + 0 \cdot s^2 + s^3$ and $N(s) = -2 + s + 0 \cdot s^2 + 0 \cdot s^3$, we form

$$[A_0 \ B_0 \ A_1 \ B_1 \ A_2 \ B_2]
\begin{bmatrix}
0 & -1 & 0 & 1 & 0 & 0 \\
-2 & 1 & 0 & 0 & 0 & 0 \\
\cdots & \cdots & \cdots & \cdots & \cdots & \cdots \\
0 & 0 & -1 & 0 & 1 & 0 \\
0 & -2 & 1 & 0 & 0 & 0 \\
\cdots & \cdots & \cdots & \cdots & \cdots & \cdots \\
0 & 0 & 0 & -1 & 0 & 1 \\
0 & 0 & -2 & 1 & 0 & 0
\end{bmatrix}
= [50 \ 85 \ 66 \ 30 \ 8 \ 1]$$

Its solution is $[127.3 \ -25 \ \ 0 \ -118.7 \ \ 1 \ -96.3]$. Thus we have

$$\frac{B(s)}{A(s)} = \frac{-96.3s^2 - 118.7s - 25}{s^2 + 127.3}$$

and the compensator is

$$C(s) = \frac{B(s)}{A(s)\phi(s)} = \frac{-96.3s^2 - 118.7s - 25}{(s^2 + 127.3)s}$$

Using this compensator of degree 3, the unity-feedback system in Fig. 9.2(a) will track robustly any step reference input and has the set of desired poles.

9.2.3 Embedding Internal Models

The design in the preceding subsection was achieved by first introducing an internal model $1/\phi(s)$ and then designing a proper $B(s)/A(s)$. Thus the compensator $B(s)/A(s)\phi(s)$ is always strictly proper. In this subsection, we discuss a method of designing a biproper compensator whose denominator will include the internal model as a factor as shown in Fig. 9.2(b). By so doing, the degree of compensators can be reduced.

Consider

$$A(s)D(s) + B(s)N(s) = F(s)$$

If deg $D(s) = n$ and if deg $A(s) = n - 1$, then the solution $A(s)$ and $B(s)$ is unique. If we increase the degree of $A(s)$ by one, then solutions are not unique, and there is one free parameter we can select. Using the free parameter, we may be able to include an internal model in the compensator, as the next example illustrates.

EXAMPLE 9.4 Consider again the design problem in Example 9.2. The degree of $D(s)$ is 2. If $A(s)$ has degree 1, then the solution is unique. Let us select $A(s)$ to have degree 2. Then $F(s)$ must have degree 4 and can be selected as

$$F(s) = (s^2 + 4s + 5)(s^2 + 2s + 5) = s^4 + 6s^3 + 18s^2 + 30s + 25$$

We form

$$
[A_0 \ B_0 \ A_1 \ B_1 \ A_2 \ B_2]
\begin{bmatrix}
-1 & 0 & 1 & 0 & 0 \\
-2 & 1 & 0 & 0 & 0 \\
\cdots & \cdots & \cdots & \cdots & \cdots \\
0 & -1 & 0 & 1 & 0 \\
0 & -2 & 1 & 0 & 0 \\
\cdots & \cdots & \cdots & \cdots & \cdots \\
0 & 0 & -1 & 0 & 1 \\
0 & 0 & -2 & 1 & 0
\end{bmatrix}
= [25 \ 30 \ 18 \ 6 \ 1] \qquad (9.23)
$$

In order for the proper compensator

$$C(s) = \frac{B_0 + B_1 s + B_2 s^2}{A_0 + A_1 s + A_2 s^2}$$

to have $1/s$ as a factor, we require $A_0 = 0$. There are five equations and six unknowns in (9.23). Thus one of the unknowns can be arbitrarily assigned. Let us select $A_0 = 0$. This is equivalent to deleting the first row of the 6×5 matrix in (9.23). The remaining 5×5 matrix is nonsingular, and the remaining five unknowns can be solved uniquely. The solution is

$$[A_0 \ B_0 \ A_1 \ B_1 \ A_2 \ B_2] = [0 \ -12.5 \ 34.8 \ -38.7 \ 1 \ -28.8]$$

Thus the compensator is

$$C(s) = \frac{B(s)}{A(s)} = \frac{-28.8s^2 - 38.7s - 12.5}{s^2 + 34.8s}$$

This biproper compensator can achieve robust tracking. This compensator has degree 2, one less than the one obtained in Example 9.3. Thus this is a better design.

In the preceding example, we mentioned that one of the unknowns in (9.23) can be arbitrarily assigned. This does not mean that any one of them can be arbitrarily assigned. For example, if we assign $A_2 = 0$ or, equivalently, delete the fifth row of the 6×5 matrix in (9.23), then the remaining square matrix is singular and no solution may exist. In Example 9.4, if we select $A_0 = 0$ and if the remaining equation in (9.23) does not have a solution, then we must increase the degree of the compensator and repeat the design. Another way to carry out the design is to find the general solution of (9.23). Using Corollary 3.2, we can express the general solution as

$$[A_0 \ B_0 \ A_1 \ B_1 \ A_2 \ B_2] = [1 \ -13 \ 34.3 \ -38.7 \ 1 \ -28.3] + \alpha[2 \ -1 \ -1 \ 0 \ 0 \ 1]$$

with one free parameter α. If we select $\alpha = -0.5$, then $A_0 = 0$ and we will obtain the same compensator.

We give one more example and discuss a different method of embedding $\phi(s)$ in the compensator.

EXAMPLE 9.5 Consider the unity-feedback system in Fig. 9.2(b) with $\hat{g}(s) = 1/s$. Design a proper compensator $C(s) = B(s)/A(s)$ so that the system will track asymptotically any step reference input and reject disturbance $w(t) = a \sin(2t + \theta)$ with unknown a and θ.

In order to achieve the design, the polynomial $A(s)$ must contain the disturbance model $(s^2 + 4)$. Note that the reference model s is not needed because the plant already contains the factor. Consider

$$A(s)D(s) + B(s)N(s) = F(s)$$

For this equation, we have deg $D(s) = n = 1$. Thus if $m = n - 1 = 0$, then the solution is unique and we have no freedom in assigning $A(s)$. If $m = 2$, then we have two free parameters that can be used to assign $A(s)$. Let

$$A(s) = \tilde{A}_0(s^2 + 4) \qquad B(s) = B_0 + B_1 s + B_2 s^2$$

Define

$$\tilde{D}(s) = D(s)(s^2 + 4) = \tilde{D}_0 + \tilde{D}_1 s + \tilde{D}_2 s^2 + \tilde{D}_3 s^3 = 0 + 4s + 0 \cdot s^2 + s^3$$

We write $A(s)D(s) + B(s)N(s) = F(s)$ as

$$\tilde{A}_0 \tilde{D}(s) + B(s)N(s) = F(s)$$

Equating its coefficients, we obtain

$$[\tilde{A}_0 \ B_0 \ B_1 \ B_2] \begin{bmatrix} \tilde{D}_0 & \tilde{D}_1 & \tilde{D}_2 & \tilde{D}_3 \\ N_0 & N_1 & 0 & 0 \\ 0 & N_0 & N_1 & 0 \\ 0 & 0 & N_0 & N_1 \end{bmatrix} = [F_0 \ F_1 \ F_2 \ F_3]$$

For this example, if we select

$$F(s) = (s + 2)(s^2 + 2s + 2) = s^3 + 4s^2 + 6s + 4$$

then the equation becomes

$$[\tilde{A}_0 \ B_0 \ B_1 \ B_2] \begin{bmatrix} 0 & 4 & 0 & 1 \\ 1 & 0 & 0 & 0 \\ 0 & 1 & 0 & 0 \\ 0 & 0 & 1 & 1 \end{bmatrix} = [4 \ 6 \ 4 \ 1]$$

Its solution is [1 4 2 4]. Thus the compensator is

$$C(s) = \frac{B(s)}{A(s)} = \frac{4s^2 + 2s + 4}{1 \times (s^2 + 4)} = \frac{4s^2 + 2s + 4}{s^2 + 4}$$

This biproper compensator will place the poles of the unity-feedback system in the assigned positions, track any step reference input, and reject the disturbance $a \sin(2t + \theta)$, both asymptotically and robustly.

9.3 Implementable Transfer Functions

Consider again the design problem posed in Fig. 8.1 with a given plant transfer function $\hat{g}(s)$. Now the problem is the following: given a desired overall transfer function $\hat{g}_o(s)$, find a feedback configuration and compensators so that the transfer function from r to y equals $\hat{g}_o(s)$. This is called the *model matching* problem. This problem is clearly different from the pole-placement problem. In pole placement, we specify only poles; its design will introduce some zeros over which we have no control. In model matching, we specify not only poles but also zeros. Thus model matching can be considered as pole-and-zero placement and should yield a better design.

Given a proper plant transfer function $\hat{g}(s)$, we claim that $\hat{g}_o(s) = 1$ is the best possible overall system we can design. Indeed, if $\hat{g}_o(s) = 1$, then $y(t) = r(t)$ for $t \geq 0$ and for any $r(t)$. Thus the overall system can track immediately (not asymptotically) any reference input no matter how erratic $r(t)$ is. Note that although $y(t) = r(t)$, the power levels at the reference input and plant output may be different. The reference signal may be provided by turning a knob by hand; the plant output may be the angular position of an antenna with weight over several tons.

Although $\hat{g}_o(s) = 1$ is the best overall system, we may not be able to match it for a given plant. The reason is that in matching or implementation, there are some physical constraints that every overall system should meet. These constraints are listed in the following:

1. All compensators used have proper rational transfer functions.
2. The configuration selected has *no plant leakage* in the sense that all forward paths from r to y pass through the plant.
3. The closed-loop transfer function of every possible input–output pair is proper and BIBO stable.

Every compensator with a proper rational transfer function can be implemented using the op-amp circuit elements shown in Fig. 2.6. If a compensator has an improper transfer function, then its implementation requires the use of pure differentiators, which are not standard op-amp circuit elements. Thus compensators used in practice are often required to have proper transfer functions. The second constraint requires that all power passes through the plant and no compensator be introduced in parallel with the plant. All configurations in Fig. 9.1 meet this constraint. In practice, noise and disturbance may exist in every component. For example, noise may be generated in using potentiometers because of brush jumps and wire irregularity. The load of an antenna may change because of gusting or air turbulence. These will be modeled as exogenous inputs entering the input and output terminals of every block as shown in Fig. 9.3.

Clearly we cannot disregard the effects of these exogenous inputs on the system. Although the plant output is the signal we want to control, we should be concerned with all variables inside the system. For example, suppose the closed-loop transfer function from r to u is not BIBO stable; then any r will excite an unbounded u and the system will either saturate or burn out. If the closed-loop transfer function from n_1 to u is improper, and if n_1 contains high-frequency noise, then the noise will be greatly amplified at u and the amplified noise will drive the system crazy. Thus the closed-loop transfer function of every possible input–output pair of the overall system should be proper and BIBO stable. An overall system is said to be *well posed* if the closed-loop transfer function of every possible input–output pair is proper; it is *totally stable* if the closed-loop transfer function of every possible input–output pair is BIBO stable.

Total stability can readily be met in design. If the overall transfer function from r to y is BIBO stable and if there is no unstable pole–zero cancellation in the system, then the overall system is totally stable. For example, consider the system shown in Fig. 9.3(a). The overall transfer function from r to y is

$$\hat{g}_{yr}(s) = \frac{1}{s+1}$$

which is BIBO stable. However, the system is not totally stable because it involves an unstable pole–zero cancellation of $(s-2)$. The closed-loop transfer function from n_2 to y is $s/(s-2)(s+1)$, which is not BIBO stable. Thus the output will grow unbounded if noise n_2, even very small, enters the system. Thus we require BIBO stability not only of $\hat{g}_o(s)$ but also of every possible closed-loop transfer function. Note that whether or not $\hat{g}(s)$ and $C(s)$ are BIBO stable is immaterial.

The condition for the unity-feedback configuration in Fig. 9.3 to be well posed is $C(\infty)\hat{g}(\infty) \neq -1$ (Problem 9.9). This can readily be established by using Mason's formula. See Reference [7, pp. 200–201]. For example, for the unity-feedback system in Fig. 9.3(b), we have $C(\infty)\hat{g}(\infty) = (-1/2) \times 2 = -1$. Thus the system is not well posed. Indeed, the closed-loop transfer function from r to y is

$$\hat{g}_o(s) = \frac{(-s+2)(2s+2)}{s+3}$$

which is improper. The condition for the two-parameter configuration in Fig. 9.1(d) to be well posed is $\hat{g}(\infty)C_2(\infty) \neq -1$. In the unity-feedback and two-parameter configurations, if $\hat{g}(s)$ is strictly proper or $\hat{g}(\infty) = 0$, then $\hat{g}(\infty)C(\infty) = 0 \neq -1$ for any proper $C(s)$ and the overall systems will automatically be well posed. In conclusion, total stability and well-posedness can easily be met in design. Nevertheless, they do impose some restrictions on $\hat{g}_o(s)$.

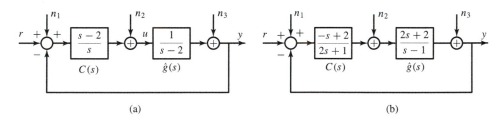

(a)   (b)

Figure 9.3   Feedback systems.

> **Definition 9.1** *Given a plant with proper transfer function* $\hat{g}(s)$, *an overall transfer function* $\hat{g}_o(s)$ *is said to be* implementable *if there exists a no-plant-leakage configuration and proper compensators so that the transfer function from* r *to* y *in Fig. 8.1 equals* $\hat{g}_o(s)$ *and the overall system is well posed and totally stable.*

If an overall transfer function $\hat{g}_o(s)$ is not implementable, then no matter what configuration is used to implement it, the design will violate at least one of the aforementioned constraints. Therefore, in model matching, the selected $\hat{g}_o(s)$ must be implementable; otherwise, it is not possible to implement it in practice.

▶ **Theorem 9.4**

Consider a plant with proper transfer function $\hat{g}(s)$. Then $\hat{g}_o(s)$ is implementable if and only if $\hat{g}_o(s)$ and

$$\hat{t}(s) := \frac{\hat{g}_o(s)}{\hat{g}(s)} \tag{9.24}$$

are proper and BIBO stable.

▶ **Corollary 9.4**

Consider a plant with proper transfer function $\hat{g}(s) = N(s)/D(s)$. Then $\hat{g}_o(s) = E(s)/F(s)$ is implementable if and only if

1. All roots of $F(s)$ have negative real parts ($F(s)$ is Hurwitz).

2. Deg $F(s)$ − deg $E(s) \geq$ deg $D(s)$ − deg $N(s)$ (pole–zero excess inequality).

3. All zeros of $N(s)$ with zero or positive real parts are retained in $E(s)$ (retainment of nonminimum-phase zeros).

We first develop Corollary 9.4 from Theorem 9.4. If $\hat{g}_o(s) = E(s)/F(s)$ is BIBO stable, then all roots of $F(s)$ have negative real parts. This is condition (1). We write (9.24) as

$$\hat{t}(s) = \frac{\hat{g}_o(s)}{\hat{g}(s)} = \frac{E(s)D(s)}{F(s)N(s)}$$

The condition for $\hat{t}(s)$ to be proper is

$$\deg F(s) + \deg \ N(s) \geq \deg \ E(s) + \deg \ D(s)$$

which implies (2). In order for $\hat{t}(s)$ to be BIBO stable, all roots of $N(s)$ with zero or positive real parts must be canceled by the roots of $E(s)$. Thus $E(s)$ must contain the nonminimum-phase zeros of $N(s)$. This is condition (3). Thus Corollary 9.4 follows directly Theorem 9.4.

Now we show the necessity of Theorem 9.4. For any configuration that has no plant leakage, if the closed-loop transfer function from r to y is $\hat{g}_o(s)$, then we have

$$\hat{y}(s) = \hat{g}_o(s)\hat{r}(s) = \hat{g}(s)\hat{u}(s)$$

which implies

$$\hat{u}(s) = \frac{\hat{g}_o(s)}{\hat{g}(s)}\hat{r}(s) = \hat{t}(s)\hat{r}(s)$$

Thus the closed-loop transfer function from r to u is $\hat{t}(s)$. Total stability requires every closed-loop transfer function to be BIBO stable. Thus $\hat{g}_o(s)$ and $\hat{t}(s)$ must be BIBO stable. Well-posedness requires every closed-loop transfer function to be proper. Thus $\hat{g}_o(s)$ and $\hat{t}(s)$ must be proper. This establishes the necessity of the theorem. The sufficiency of the theorem will be established constructively in the next subsection. Note that if $\hat{g}(s)$ and $\hat{t}(s)$ are proper, then $\hat{g}_o(s) = \hat{g}(s)\hat{t}(s)$ is proper. Thus the condition for $\hat{g}_o(s)$ to be proper can be dropped from Theorem 9.4.

In pole placement, the design will always introduce some zeros over which we have no control. In model matching, other than retaining nonminimum-phase zeros and meeting the pole–zero excess inequality, we have complete freedom in selecting poles and zeros: any pole inside the open left-half s-plane and any zero in the entire s-plane. Thus model matching can be considered as pole-and-zero placement and should yield a better overall system than pole-placement design.

Given a plant transfer function $\hat{g}(s)$, how to select an implementable model $\hat{g}_o(s)$ is not a simple problem. For a discussion of this problem, see Reference [7, Chapter 9].

9.3.1 Model Matching—Two-Parameter Configuration

This section discusses the implementation of $\hat{g}_o(s) = \hat{g}(s)\hat{t}(s)$. Clearly, if $C(s) = \hat{t}(s)$ in Fig. 9.1(a), then the open-loop configuration has $\hat{g}_o(s)$ as its transfer function. This implementation may involve unstable pole–zero cancellations and, consequently, may not be totally stable. Even if it is totally stable, the configuration can be very sensitive to plant parameter variations. Therefore the open-loop configuration should not be used. The unity-feedback configuration in Fig. 9.1(b) can be used to achieve every pole placement; however it cannot be used to achieve every model matching, as the next example shows.

EXAMPLE 9.6 Consider a plant with transfer function $\hat{g}(s) = (s-2)/(s^2-1)$. We can readily show that

$$\hat{g}_o(s) = \frac{-(s-2)}{s^2 + 2s + 2} \tag{9.25}$$

is implementable. Because $\hat{g}_o(0) = 1$, the plant output will track asymptotically any step reference input. Suppose we use the unity-feedback configuration with $p = 1$ to implement $\hat{g}_o(s)$. Then from

$$\hat{g}_o(s) = \frac{C(s)\hat{g}(s)}{1 + C(s)\hat{g}(s)}$$

we can compute the compensator as

$$C(s) = \frac{\hat{g}_o(s)}{\hat{g}(s)[1 - \hat{g}_o(s)]} = \frac{-(s^2-1)}{s(s+3)}$$

This compensator is proper. However, the tandem connection of $C(s)$ and $\hat{g}(s)$ involves the pole–zero cancellation of $(s^2 - 1) = (s + 1)(s - 1)$. The cancellation of the stable pole $s + 1$ will not cause any serious problem in the overall system. However, the cancellation of the unstable pole $s - 1$ will make the overall system not totally stable. Thus the implementation is not acceptable.

Model matching in general involves some pole–zero cancellations. The same situation arises in state-feedback state-estimator design; all eigenvalues of the estimator are not controllable from the reference input and are canceled in the overall transfer function. However, because we have complete freedom in selecting the eigenvalues of the estimator, if we select them properly, the cancellation will not cause any problem in design. In using the unity-feedback configuration in model matching, as we saw in the preceding example, the canceled poles are dictated by the plant transfer function. Thus, if a plant transfer function has poles with positive real parts, the cancellation will involve unstable poles. Therefore the unity-feedback configuration, in general, cannot be used in model matching.

The open-loop and the unity-feedback configurations in Figs. 9.1(a) and 9.1(b) have one degree of freedom and cannot be used to achieve every model matching. The configurations in Figs. 9.1(c) and 9.1(d) both have two degrees of freedom. In using either configuration, we have complete freedom in assigning canceled poles; therefore both can be used to achieve every model matching. Because the two-parameter configuration in Fig. 9.1(d) seems to be more natural and more suitable for practical implementation, we discuss only that configuration here. For model matching using the configuration in Fig. 9.1(c), see Reference [6].

Consider the two-parameter configuration in Fig. 9.1(d). Let

$$C_1(s) = \frac{L(s)}{A_1(s)} \qquad C_2(s) = \frac{M(s)}{A_2(s)}$$

where $L(s)$, $M(s)$, $A_1(s)$, and $A_2(s)$ are polynomials. We call $C_1(s)$ the *feedforward compensator* and $C_2(s)$ the *feedback compensator*. In general, $A_1(s)$ and $A_2(s)$ need not be the same. It turns out that even if they are chosen to be the same, the configuration can still be used to achieve any model matching. Furthermore, a simple design procedure can be developed. Therefore we assume $A_1(s) = A_2(s) = A(s)$ and the compensators become

$$C_1(s) = \frac{L(s)}{A(s)} \qquad C_2(s) = \frac{M(s)}{A(s)} \tag{9.26}$$

The transfer function from r to y in Fig. 9.1(d) then becomes

$$\hat{g}_o(s) = C_1(s)\frac{\hat{g}(s)}{1 + \hat{g}(s)C_2(s)} = \frac{L(s)}{A(s)}\frac{\dfrac{N(s)}{D(s)}}{1 + \dfrac{N(s)}{D(s)}\dfrac{M(s)}{A(s)}}$$

$$= \frac{L(s)N(s)}{A(s)D(s) + M(s)N(s)} \tag{9.27}$$

Thus in model matching, we search for proper $L(s)/A(s)$ and $M(s)/A(s)$ to meet

$$\hat{g}_o(s) = \frac{E(s)}{F(s)} = \frac{L(s)N(s)}{A(s)D(s) + M(s)N(s)} \tag{9.28}$$

Note that the two-parameter configuration has no plant leakage. If the plant transfer function $\hat{g}(s)$ is strictly proper as assumed and if $C_2(s) = M(s)/A(s)$ is proper, then the overall system is automatically well posed. The question of total stability will be discussed in the next subsection.

Problem Given $\hat{g}(s) = N(s)/D(s)$, where $N(s)$ and $D(s)$ are coprime and deg $N(s) < \deg D(s) = n$, and given an implementable $\hat{g}_o(s) = E(s)/F(s)$, find proper $L(s)/A(s)$ and $M(s)/A(s)$ to meet (9.28).

▶ **Procedure 9.1**

1. Compute

$$\frac{\hat{g}_o(s)}{N(s)} = \frac{E(s)}{F(s)N(s)} =: \frac{\bar{E}(s)}{\bar{F}(s)} \tag{9.29}$$

where $\bar{E}(s)$ and $\bar{F}(s)$ are coprime. Since $E(s)$ and $F(s)$ are implicitly assumed to be coprime, common factors may exist only between $E(s)$ and $N(s)$. Cancel all common factors between them and denote the rest as $\bar{E}(s)$ and $\bar{F}(s)$. Note that if $E(s) = N(s)$, then $\bar{F}(s) = F(s)$ and $\bar{E}(s) = 1$. Using (9.29), we rewrite (9.28) as

$$\hat{g}_o(s) = \frac{\bar{E}(s)N(s)}{\bar{F}(s)} = \frac{L(s)N(s)}{A(s)D(s) + M(s)N(s)} \tag{9.30}$$

From this equation, we may be tempted to set $L(s) = \bar{E}(s)$ and solve for $A(s)$ and $M(s)$ from $\bar{F}(s) = A(s)D(s) + M(s)N(s)$. However, no proper $C_2(s) = M(s)/A(s)$ may exist in the equation. See Problem 9.1. Thus we need some additional manipulation.

2. Introduce an arbitrary Hurwitz polynomial $\hat{F}(s)$ such that the degree of $\bar{F}(s)\hat{F}(s)$ is $2n - 1$ or higher. In other words, if deg $\bar{F}(s) = p$, then deg $\hat{F}(s) \geq 2n - 1 - p$. Because the polynomial $\hat{F}(s)$ will be canceled in the design, its roots should be chosen to lie inside the sector shown in Fig. 8.3(a).

3. Rewrite (9.30) as

$$\hat{g}_o(s) = \frac{\bar{E}(s)\hat{F}(s)N(s)}{\bar{F}(s)\hat{F}(s)} = \frac{L(s)N(s)}{A(s)D(s) + M(s)N(s)} \tag{9.31}$$

Now we set

$$L(s) = \bar{E}(s)\hat{F}(s) \tag{9.32}$$

and solve $A(s)$ and $M(s)$ from

$$A(s)D(s) + M(s)N(s) = \bar{F}(s)\hat{F}(s) \tag{9.33}$$

If we write

$$A(s) = A_0 + A_1 s + A_2 s^2 + \cdots + A_m s^m$$
$$M(s) = M_0 + B_1 s + M_2 s^2 + \cdots + M_m s^m$$
$$\bar{F}(s)\hat{F}(s) = F_0 + F_1 s + F_2 s^2 + \cdots + F_{n+m} s^{n+m}$$

with $m \geq n - 1$, then $A(s)$ and $M(s)$ can be obtained by solving

$$[A_0 \ M_0 \ A_1 \ M_1 \ \cdots \ A_m \ M_m]\mathbf{S}_m = [F_0 \ F_1 \ F_2 \ \cdots \ F_{n+m}] \qquad (9.34)$$

with

$$\mathbf{S}_m := \begin{bmatrix} D_0 & D_1 & \cdots & D_n & 0 & \cdots & 0 \\ N_0 & N_1 & \cdots & N_n & 0 & \cdots & 0 \\ \cdots & \cdots & \cdots & \cdots & \cdots & \cdots & \cdots \\ 0 & D_0 & \cdots & D_{n-1} & D_n & \cdots & 0 \\ 0 & N_0 & \cdots & N_{n-1} & N_n & \cdots & 0 \\ \cdots & \cdots & \cdots & \cdots & \cdots & \cdots & \cdots \\ \vdots & \vdots & \ddots & \vdots & \vdots & \ddots & \vdots \\ \cdots & \cdots & \cdots & \cdots & \cdots & \cdots & \cdots \\ 0 & 0 & \cdots & 0 & D_0 & \cdots & D_n \\ 0 & 0 & \cdots & 0 & N_0 & \cdots & N_n \end{bmatrix}$$

The computed compensators $L(s)/A(s)$ and $M(s)/A(s)$ are proper.

We justify the procedure. By introducing $\hat{F}(s)$, the degree of $\bar{F}(s)\hat{F}(s)$ is $2n - 1$ or higher and, following Theorem 9.2, solutions $A(s)$ and $M(s)$ with deg $M(s) \leq$ deg $A(s) = m$ and $m \geq n - 1$ exist in (9.34) for any $\bar{F}(s)\hat{F}(s)$. Thus the compensator $M(s)/A(s)$ is proper. Note that if we do not introduce $\hat{F}(s)$, proper compensator $M(s)/A(s)$ may not exist in (9.34).

Next we show deg $L(s) \leq$ deg $A(s)$. Applying the pole–zero excess inequality to (9.31) and using (9.32), we have

$$\deg(\bar{F}(s)\hat{F}(s)) - \deg N(s) - \deg L(s) \geq \deg D(s) - \deg N(s)$$

which implies

$$\deg L(s) \leq \deg(\bar{F}(s)\hat{F}(s)) - \deg D(s) = \deg A(s)$$

Thus the compensator $L(s)/A(s)$ is proper.

EXAMPLE 9.7 Consider the model matching problem studied in Example 9.6. That is, given $\hat{g}(s) = (s - 2)/(s^2 - 1)$, match $\hat{g}_o(s) = -(s - 2)/(s^2 + 2s + 2)$. We implement it in the two-parameter configuration shown in Fig. 9.1(d). First we compute

$$\frac{\hat{g}_o(s)}{N(s)} = \frac{-(s - 2)}{(s^2 + 2s + 2)(s - 2)} = \frac{-1}{s^2 + 2s + 2} =: \frac{\bar{E}(s)}{\bar{F}(s)}$$

Because the degree of $\bar{F}(s)$ is 2, we select arbitrarily $\hat{F}(s) = s + 4$ so that the degree of $\bar{F}(s)\hat{F}(s)$ is $3 = 2n - 1$. Thus we have

$$L(s) = \bar{E}(s)\hat{F}(s) = -(s + 4) \qquad (9.35)$$

and $A(s)$ and $M(s)$ can be solved from

$$A(s)D(s) + M(s)N(s) = \bar{F}(s)\hat{F}(s) = (s^2 + 2s + 2)(s + 4)$$
$$= s^3 + 6s^2 + 10s + 8$$

or

$$[A_0 \ M_0 \ A_1 \ M_1] \begin{bmatrix} -1 & 0 & 1 & 0 \\ -2 & 1 & 0 & 0 \\ \cdots & \cdots & \cdots & \cdots \\ 0 & -1 & 0 & 1 \\ 0 & -2 & 1 & 0 \end{bmatrix} = [8 \ 10 \ 6 \ 1]$$

The solution is $A_0 = 18$, $A_1 = 1$, $M_0 = -13$, and $M_1 = -12$. Thus we have $A(s) = 18 + s$ and $M(s) = -13 - 12s$ and the compensators are

$$C_1(s) = \frac{L(s)}{A(s)} = \frac{-(s+4)}{s+18} \qquad C_2(s) = \frac{M(s)}{A(s)} = \frac{-(12s+13)}{s+18}$$

This completes the design. Note that, because $\hat{g}_o(0) = 1$, the output of the feedback system will track any step reference input.

EXAMPLE 9.8 Given $\hat{g}(s) = (s-2)/(s^2-1)$, match

$$\hat{g}_o(s) = \frac{-(s-2)(4s+2)}{(s^2+2s+2)(s+2)} = \frac{-4s^2+6s+4}{s^3+4s^2+6s+4}$$

This $\hat{g}_o(s)$ is BIBO stable and has the property $\hat{g}_o(0) = 1$ and $\hat{g}'_o(s) = 0$; thus the overall system will track asymptotically not only any step reference input but also any ramp input. See Problems 9.13 and 9.14. This $\hat{g}_o(s)$ meets all three conditions in Corollary 9.4; thus it is implementable. We use the two-parameter configuration. First we compute

$$\frac{\hat{g}_o(s)}{N(s)} = \frac{-(s-2)(4s+2)}{(s^2+2s+2)(s+2)(s-2)} = \frac{-(4s+2)}{s^3+4s^2+6s+4} =: \frac{\bar{E}(s)}{\bar{F}(s)}$$

Because the degree of $\bar{F}(s)$ is 3, which equals $2n - 1 = 3$, there is no need to introduce $\hat{F}(s)$ and we set $\hat{F}(s) = 1$. Thus we have

$$L(s) = \hat{F}(s)\bar{E}(s) = -(4s+2)$$

and $A(s)$ and $M(s)$ can be solved from

$$[A_0 \ M_0 \ A_1 \ M_1] \begin{bmatrix} -1 & 0 & 1 & 0 \\ -2 & 1 & 0 & 0 \\ \cdots & \cdots & \cdots & \cdots \\ 0 & -1 & 0 & 1 \\ 0 & -2 & 1 & 0 \end{bmatrix} = [4 \ 6 \ 4 \ 1]$$

as $A_0 = 1$, $A_1 = 34/3$, $M_0 = -23/3$, and $M_1 = -22/3$. Thus the compensators are

$$C_1(s) = \frac{-(4s+2)}{s+34/3} \qquad C_2(s) = \frac{-(22s+23)}{3s+34}$$

This completes the design. Note that this design does not involve any pole–zero cancellation because $\hat{F}(s) = 1$.

9.3.2 Implementation of Two-Parameter Compensators

Given a plant with transfer function $\hat{g}(s)$ and an implementable model $\hat{g}_o(s)$, we can implement the model in the two-parameter configuration shown in Fig. 9.1(d) and redrawn in Fig. 9.4(a). The compensators $C_1(s) = L(s)/A(s)$ and $C_2(s) = M(s)/A(s)$ can be obtained by using Procedure 9.1. To complete the design, the compensators must be built or implemented. This is discussed in this subsection.

Consider the configuration in Fig. 9.4(a). The denominator $A(s)$ of $C_1(s)$ is obtained by solving the compensator equation in (9.33) and may or may not be a Hurwitz polynomial. See Problem 9.12. If it is not a Hurwitz polynomial and if we implement $C_1(s)$ as shown in Fig. 9.4(a), then the output of $C_1(s)$ will grow without bound and the overall system is not totally stable. Therefore, in general, we should not implement the two compensators as shown in Fig. 9.4(a). If we move $C_2(s)$ outside the loop as shown in Fig. 9.4(b), then the design will involve the cancellation of $M(s)$. Because $M(s)$ is also obtained by solving (9.33), we have no direct control of $M(s)$. Thus the design is in general not acceptable. If we move $C_1(s)$ inside the loop, then the configuration becomes the one shown in Fig. 9.4(c). We see that the connection involves the pole–zero cancellation of $L(s) = \hat{F}(s)\bar{E}(s)$. We have freedom in selecting $\hat{F}(s)$. The polynomial $\bar{E}(s)$ is part of $E(s)$, which, other than the nonminimum-phase zeros of $N(s)$, we can also select. The nonminimum-phase zeros, however, are completely canceled in $\bar{E}(s)$. Thus $L(s)$ can be Hurwitz[3] and the implementation in Fig. 9.4(c) can be totally stable and is

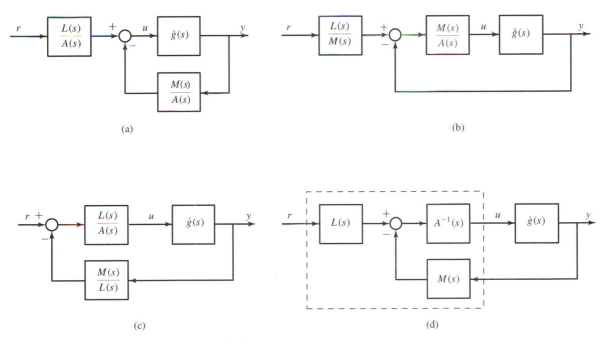

(a)

(b)

(c)

(d)

Figure 9.4 Two-degrees-of-freedom configurations.

3. This may not be true in the multivariable case.

acceptable. However, because the two compensators $L(s)/A(s)$ and $M(s)/L(s)$ have different denominators, their implementations require a total of $2m$ integrators. We discuss next a better implementation that requires only m integrators and involves only the cancellation of $\hat{F}(s)$.

Consider

$$\hat{u}(s) = C_1(s)\hat{r}(s) - C_2(s)\hat{y}(s) = \frac{L(s)}{A(s)}\hat{r}(s) - \frac{M(s)}{A(s)}\hat{y}(s)$$

$$= A^{-1}(s)[L(s) \quad - M(s)]\begin{bmatrix} \hat{r}(s) \\ \hat{y}(s) \end{bmatrix}$$

This can be plotted as shown in Fig. 9.4(d). Thus we can consider the two compensators as a single compensator with two inputs and one output with transfer matrix

$$\mathbf{C}(s) = [C_1(s) \quad - C_2(s)] = A^{-1}(s)[L(s) \quad - M(s)] \tag{9.36}$$

If we find a minimal realization of (9.36), then its dimension is m and the two compensators can be implemented using only m integrators. As we can see from (9.31), the design involves only the cancellation of $\hat{F}(s)$. Thus the implementation in Fig. 9.4(d) is superior to the one in Fig. 9.4(c). We see that the four configurations in Fig. 9.4 all have two degrees of freedom and are mathematically equivalent. However, they can be different in actual implementation.

EXAMPLE 9.9 Implement the compensators in Example 9.8 using an op-amp circuit. We write

$$\hat{u}(s) = C_1(s)\hat{r}(s) - C_2(s)\hat{y}(s) = \begin{bmatrix} \dfrac{-(4s+2)}{s+11.33} & \dfrac{7.33s+7.67}{s+11.33} \end{bmatrix}\begin{bmatrix} \hat{r}(s) \\ \hat{y}(s) \end{bmatrix}$$

$$= \left([-4 \quad 7.33] + \frac{1}{s+11.33}[43.33 \quad -75.38] \right)\begin{bmatrix} \hat{r}(s) \\ \hat{y}(s) \end{bmatrix}$$

Its state-space realization is, using the formula in Problem 4.10,

$$\dot{x} = -11.33x + [43.33 \quad -75.38]\begin{bmatrix} r \\ y \end{bmatrix}$$

$$u = x + [-4 \quad 7.33]\begin{bmatrix} r \\ y \end{bmatrix}$$

(See Problem 4.14.) This one-dimensional state equation can be realized as shown in Fig. 9.5. This completes the implementation of the compensators.

9.4 Multivariable Unity-Feedback Systems

This section extends the pole placement discussed in Section 9.2 to the multivariable case. Consider the unity-feedback system shown in Fig. 9.6. The plant has p inputs and q outputs and is described by a $q \times p$ strictly proper rational matrix $\hat{\mathbf{G}}(s)$. The compensator $\mathbf{C}(s)$ to be designed must have q inputs and p outputs in order for the connection to be possible. Thus $\mathbf{C}(s)$ is required to be a $p \times q$ proper rational matrix. The matrix $\mathbf{P}$ is a $q \times q$ constant gain matrix. For the time being, we assume $\mathbf{P} = \mathbf{I}_q$. Let the transfer matrix from $\mathbf{r}$ to $\mathbf{y}$ be denoted by $\hat{\mathbf{G}}_o(s)$, a $q \times q$ matrix. Then we have

Figure 9.5 Op-amp circuit
implementation.

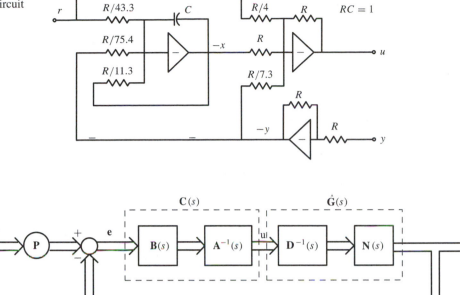

Figure 9.6 Multivariable unity feedback system with $\mathbf{P} = \mathbf{I}_q$.

$$\hat{\mathbf{G}}_o(s) = [\mathbf{I}_q + \hat{\mathbf{G}}(s)\mathbf{C}(s)]^{-1}\hat{\mathbf{G}}(s)\mathbf{C}(s)$$
$$= \hat{\mathbf{G}}(s)\mathbf{C}(s)[\mathbf{I}_q + \hat{\mathbf{G}}(s)\mathbf{C}(s)]^{-1}$$
$$= \hat{\mathbf{G}}(s)[\mathbf{I}_p + \mathbf{C}(s)\hat{\mathbf{G}}(s)]^{-1}\mathbf{C}(s) \qquad (9.37)$$

The first equality is obtained from $\hat{\mathbf{y}}(s) = \hat{\mathbf{G}}(s)\mathbf{C}(s)[\hat{\mathbf{r}}(s) - \hat{\mathbf{y}}(s)]$; the second one from $\hat{\mathbf{e}}(s) = \hat{\mathbf{r}}(s) - \hat{\mathbf{G}}(s)\mathbf{C}(s)\hat{\mathbf{e}}(s)$; and the third one from $\hat{\mathbf{u}}(s) = \mathbf{C}(s)[\hat{\mathbf{r}}(s) - \hat{\mathbf{G}}(s)\hat{\mathbf{u}}(s)]$. They can also be verified directly. For example, pre- and postmultiplying by $[\mathbf{I}_q + \hat{\mathbf{G}}(s)\mathbf{C}(s)]$ in the first two equations yield

$$\hat{\mathbf{G}}(s)\mathbf{C}(s)[\mathbf{I}_q + \hat{\mathbf{G}}(s)\mathbf{C}(s)] = [\mathbf{I}_q + \hat{\mathbf{G}}(s)\mathbf{C}(s)]\hat{\mathbf{G}}(s)\mathbf{C}(s)$$

which is an identity. This establishes the second equality. The third equality can similarly be established.

Let $\hat{\mathbf{G}}(s) = \mathbf{N}(s)\mathbf{D}^{-1}(s)$ be a right coprime fraction and let $\mathbf{C}(s) = \mathbf{A}^{-1}(s)\mathbf{B}(s)$ be a left fraction to be designed. Then (9.37) implies

$$\hat{\mathbf{G}}_o(s) = \mathbf{N}(s)\mathbf{D}^{-1}(s)[\mathbf{I} + \mathbf{A}^{-1}(s)\mathbf{B}(s)\mathbf{N}(s)\mathbf{D}^{-1}(s)]^{-1}\mathbf{A}^{-1}(s)\mathbf{B}(s)$$
$$= \mathbf{N}(s)\mathbf{D}^{-1}(s)\left\{\mathbf{A}^{-1}(s)[\mathbf{A}(s)\mathbf{D}(s) + \mathbf{B}(s)\mathbf{N}(s)]\mathbf{D}^{-1}(s)\right\}^{-1}\mathbf{A}^{-1}(s)\mathbf{B}(s)$$
$$= \mathbf{N}(s))[\mathbf{A}(s)\mathbf{D}(s) + \mathbf{B}(s)\mathbf{N}(s)]^{-1}\mathbf{B}(s)$$
$$= \mathbf{N}(s)\mathbf{F}^{-1}(s)\mathbf{B}(s) \qquad (9.38)$$

where

$$\mathbf{A}(s)\mathbf{D}(s) + \mathbf{B}(s)\mathbf{N}(s) = \mathbf{F}(s) \tag{9.39}$$

It is a polynomial matrix equation. Thus the design problem can be stated as follows: given $p \times p$ $\mathbf{D}(s)$ and $q \times p$ $\mathbf{N}(s)$ and an arbitrary $p \times p$ $\mathbf{F}(s)$, find $p \times p$ $\mathbf{A}(s)$ and $p \times q$ $\mathbf{B}(s)$ to meet (9.39). This is the matrix version of the polynomial compensator equation in (9.12).

▶ **Theorem 9.M1**

Given polynomial matrices $\mathbf{D}(s)$ and $\mathbf{N}(s)$, polynomial matrix solutions $\mathbf{A}(s)$ and $\mathbf{B}(s)$ exist in (9.39) for any polynomial matrix $\mathbf{F}(s)$ if and only if $\mathbf{D}(s)$ and $\mathbf{N}(s)$ are right coprime.

Suppose $\mathbf{D}(s)$ and $\mathbf{N}(s)$ are not right coprime, then there exists a nonunimodular polynomial matrix $\mathbf{R}(s)$ such that $\mathbf{D}(s) = \hat{\mathbf{D}}(s)\mathbf{R}(s)$ and $\mathbf{N}(s) = \hat{\mathbf{N}}(s)\mathbf{R}(s)$. Then $\mathbf{F}(s)$ in (9.39) must be of the form $\hat{\mathbf{F}}(s)\mathbf{R}(s)$, for some polynomial matrix $\hat{\mathbf{F}}(s)$. Thus if $\mathbf{F}(s)$ cannot be expressed in such a form, no solutions exist in (9.39). This shows the necessity of the theorem. If $\mathbf{D}(s)$ and $\mathbf{N}(s)$ are right coprime, there exist polynomial matrices $\bar{\mathbf{A}}(s)$ and $\bar{\mathbf{B}}(s)$ such that

$$\bar{\mathbf{A}}(s)\mathbf{D}(s) + \bar{\mathbf{B}}(s)\mathbf{N}(s) = \mathbf{I}$$

The polynomial matrices $\bar{\mathbf{A}}(s)$ and $\bar{\mathbf{B}}(s)$ can be obtained by a sequence of elementary operations. See Reference [6, pp. 587–595]. Thus $\mathbf{A}(s) = \mathbf{F}(s)\bar{\mathbf{A}}(s)$ and $\mathbf{B}(s) = \mathbf{F}(s)\bar{\mathbf{B}}(s)$ are solutions of (9.39) for any $\mathbf{F}(s)$. This establishes Theorem 9.M1. As in the scalar case, it is possible to develop general solutions for (9.39). However, the general solutions are not convenient to use in our design. Thus they will not be discussed.

Next we will change solving (9.39) into solving a set of linear algebraic equations. Consider $\hat{\mathbf{G}}(s) = \mathbf{N}(s)\mathbf{D}^{-1}(s)$, where $\mathbf{D}(s)$ and $\mathbf{N}(s)$ are right coprime and $\mathbf{D}(s)$ is column reduced. Let μ_i be the degree of the ith column of $\mathbf{D}(s)$. Then we have, as discussed in Section 7.8.2,

$$\deg \hat{\mathbf{G}}(s) = \deg \det \mathbf{D}(s) = \mu_1 + \mu_2 + \cdots + \mu_p =: n \tag{9.40}$$

Let $\mu := \max(\mu_1, \mu_2, \ldots, \mu_p)$. Then we can express $\mathbf{D}(s)$ and $\mathbf{N}(s)$ as

$$\mathbf{D}(s) = \mathbf{D}_0 + \mathbf{D}_1 s + \mathbf{D}_2 s^2 + \cdots + \mathbf{D}_\mu s^\mu \qquad \mathbf{D}_\mu \neq 0$$
$$\mathbf{N}(s) = \mathbf{N}_0 + \mathbf{N}_1 s + \mathbf{N}_2 s^2 + \cdots + \mathbf{N}_\mu s^\mu$$

Note that $\mathbf{D}_\mu$ is singular unless $\mu_1 = \mu_2 = \cdots = \mu_p$. Note also that $\mathbf{N}_\mu = 0$, following the strict properness assumption of $\hat{\mathbf{G}}(s)$. We also express $\mathbf{A}(s)$, $\mathbf{B}(s)$, and $\mathbf{F}(s)$ as

$$\mathbf{A}(s) = \mathbf{A}_0 + \mathbf{A}_1 s + \mathbf{A}_2 s^2 + \cdots + \mathbf{A}_m s^m$$
$$\mathbf{B}(s) = \mathbf{B}_0 + \mathbf{B}_1 s + \mathbf{B}_2 s^2 + \cdots + \mathbf{B}_m s^m$$
$$\mathbf{F}(s) = \mathbf{F}_0 + \mathbf{F}_1 s + \mathbf{F}_2 s^2 + \cdots + \mathbf{F}_{\mu+m} s^{\mu+m}$$

Substituting these into (9.39) and matching the coefficients of like powers of s, we obtain

$$[\mathbf{A}_0 \ \mathbf{B}_0 \ \mathbf{A}_1 \ \mathbf{B}_1 \ \cdots \ \mathbf{A}_m \ \mathbf{B}_m]\mathbf{S}_m = [\mathbf{F}_0 \ \mathbf{F}_1 \ \cdots \ \mathbf{F}_{\mu+m}] =: \bar{\mathbf{F}} \tag{9.41}$$

where

$$
\mathbf{S}_m := \begin{bmatrix}
\mathbf{D}_0 & \mathbf{D}_1 & \cdots & \mathbf{D}_\mu & \mathbf{0} & \mathbf{0} & \cdots & \mathbf{0} \\
\cdots & \cdots & \cdots & \cdots & \cdots & \cdots & \cdots & \cdots \\
\mathbf{N}_0 & \mathbf{N}_1 & \cdots & \mathbf{N}_\mu & \mathbf{0} & \mathbf{0} & \cdots & \mathbf{0} \\
\hline
\mathbf{0} & \mathbf{D}_0 & \cdots & \mathbf{D}_{\mu-1} & \mathbf{D}_\mu & \mathbf{0} & \cdots & \mathbf{0} \\
\cdots & \cdots & \cdots & \cdots & \cdots & \cdots & \cdots & \cdots \\
\mathbf{0} & \mathbf{N}_0 & \cdots & \mathbf{N}_{\mu-1} & \mathbf{N}_\mu & \mathbf{0} & \cdots & \mathbf{0} \\
\hline
\vdots & \vdots & & \vdots & \vdots & \vdots & & \vdots \\
\hline
\mathbf{0} & \mathbf{0} & \cdots & \mathbf{0} & \mathbf{D}_0 & \mathbf{D}_1 & \cdots & \mathbf{D}_\mu \\
\cdots & \cdots & \cdots & \cdots & \cdots & \cdots & \cdots & \cdots \\
\mathbf{0} & \mathbf{0} & \cdots & \mathbf{0} & \mathbf{N}_0 & \mathbf{N}_1 & \cdots & \mathbf{N}_\mu
\end{bmatrix}
\tag{9.42}
$$

The matrix $\mathbf{S}_m$ has $m+1$ block rows; each block row consists of p D-rows and q N-rows. Thus $\mathbf{S}_m$ has $(m+1)(p+q)$ number of rows. Let us search linearly independent rows of $\mathbf{S}_m$ in order from top to bottom. It turns out that if $\mathbf{N}(s)\mathbf{D}^{-1}(s)$ is proper, then all D-rows are linearly independent of their previous rows. An N-row can be linearly independent of its previous rows. However, if an N-row becomes linearly dependent, then, because of the structure of $\mathbf{S}_m$, the same N-rows in subsequent N-block rows will be linearly dependent. Let ν_i be the number of linearly independent ith N-rows and let

$$
\nu := \max\{\nu_1, \nu_2, \ldots, \nu_q\}
$$

It is called the *row index* of $\hat{\mathbf{G}}(s)$. Then all q N-rows in the last N-block row of $\mathbf{S}_\nu$ are linearly dependent on their previous rows. Thus $\mathbf{S}_{\nu-1}$ contains all linearly independent N-rows and its total number equals, as discussed in Section 7.8.2, the degree of $\hat{\mathbf{G}}(s)$, that is,

$$
\nu_1 + \nu_2 + \cdots + \nu_q = n
\tag{9.43}
$$

Because all D-rows are linearly independent and there are a total of $p\nu$ D-rows in $\mathbf{S}_{\nu-1}$, we conclude that $\mathbf{S}_{\nu-1}$ has $n + p\nu$ independent rows or rank $n + p\nu$.

Let us consider

$$
\mathbf{S}_0 = \begin{bmatrix}
\mathbf{D}_0 & \mathbf{D}_1 & \cdots & \mathbf{D}_{\mu-1} & \mathbf{D}_\mu \\
\mathbf{N}_0 & \mathbf{N}_1 & \cdots & \mathbf{N}_{\mu-1} & \mathbf{N}_\mu
\end{bmatrix}
$$

It has $p(\mu+1)$ number of columns; however, it has at least a total of $\sum_{i=1}^{p}(\mu - \mu_i)$ zero columns. In the matrix $\mathbf{S}_1$, some new zero columns will appear in the rightmost block column. However, some zero columns in $\mathbf{S}_0$ will not be zero columns in $\mathbf{S}_1$. Thus the number of zero columns in $\mathbf{S}_1$ remains as

$$
\alpha := \sum_{i=1}^{p}(\mu - \mu_i) = p\mu - (\mu_1 + \mu_2 + \cdots + \mu_p) = p\mu - n
\tag{9.44}
$$

In fact, this is the number of zero columns in $\mathbf{S}_m$, $m = 2, 3, \ldots$. Let $\tilde{\mathbf{S}}_{\mu-1}$ be the matrix $\mathbf{S}_{\mu-1}$ after deleting these zero columns. Because the number of columns in $\mathbf{S}_m$ is $p(\mu + 1 + m)$, the number of columns in $\tilde{\mathbf{S}}_{\nu-1}$ is

$$\beta := p(\mu + 1 + \nu - 1) - (p\mu - n) = p\nu + n \tag{9.45}$$

The rank of $\tilde{\mathbf{S}}_{\mu-1}$ clearly equals the rank of $\mathbf{S}_{\mu-1}$ or $p\nu + n$. Thus $\tilde{\mathbf{S}}_{\mu-1}$ has full column rank. Now if m increases by 1, the rank and the number of the columns of $\tilde{\mathbf{S}}_\mu$ both increase by p (because the p new D-rows are all linearly independent of their previous rows); thus $\tilde{\mathbf{S}}_\mu$ still has full column rank. Proceeding forward, we conclude that $\tilde{\mathbf{S}}_m$, for $m \geq \mu - 1$, has full column rank.

Let us define

$$\mathbf{H}_c(s) := \mathrm{diag}(s^{\mu_1}, s^{\mu_2}, \ldots, s^{\mu_p}) \tag{9.46}$$

and

$$\mathbf{H}_r(s) = \mathrm{diag}(s^{m_1}, s^{m_2}, \ldots, s^{m_p}) \tag{9.47}$$

Then we have the following matrix version of Theorem 9.2.

▶ **Theorem 9.M2**

Consider the unity-feedback system shown in Fig. 9.6 with $\mathbf{P} = \mathbf{I}_q$. The plant is described by a $q \times p$ strictly proper rational matrix $\hat{\mathbf{G}}(s)$. Let $\hat{\mathbf{G}}(s)$ be factored as $\hat{\mathbf{G}}(s) = \mathbf{N}(s)\mathbf{D}^{-1}(s)$, where $\mathbf{D}(s)$ and $\mathbf{N}(s)$ are right coprime and $\mathbf{D}(s)$ is column reduced with column degrees μ_i, $i = 1, 2, \ldots, p$. Let ν be the row index of $\hat{\mathbf{G}}(s)$ and let $m_i \geq \nu - 1$ for $i = 1, 2, \ldots, p$. Then for any $p \times p$ polynominal matrix $\mathbf{F}(s)$, such that

$$\lim_{s \to \infty} \mathbf{H}_r^{-1}(s)\mathbf{F}(s)\mathbf{H}_c^{-1}(s) = \mathbf{F}_h \tag{9.48}$$

is a nonsingular constant matrix, there exists a $p \times q$ proper compensator $\mathbf{A}^{-1}(s)\mathbf{B}(s)$, where $\mathbf{A}(s)$ is row reduced with row degrees m_i, such that the transfer matrix from $\mathbf{r}$ to $\mathbf{y}$ equals

$$\hat{\mathbf{G}}_o(s) = \mathbf{N}(s)\mathbf{F}^{-1}(s)\mathbf{B}(s)$$

Furthermore, the compensator can be obtained by solving sets of linear algebraic equations in (9.41).

Proof: Let $m = \max(m_1, m_2, \ldots, m_p)$. Consider the constant matrix

$$\bar{\mathbf{F}} := [\mathbf{F}_0 \ \mathbf{F}_1 \ \mathbf{F}_2 \ \cdots \ \mathbf{F}_{m+\mu}]$$

It is formed from the coefficient matrices of $\mathbf{F}(s)$ and has order $p \times (m + \mu + 1)$. Clearly $\mathbf{F}(s)$ has column degrees at most $m + \mu_i$. Thus $\bar{\mathbf{F}}$ has at least α number of zero columns, where α is given in (9.44). Furthermore, the positions of these zero columns coincide with those of $\mathbf{S}_m$. Let $\tilde{\mathbf{F}}$ be the constant matrix $\bar{\mathbf{F}}$ after deleting these zero columns. Now consider

$$[\mathbf{A}_0 \ \mathbf{B}_0 \ \mathbf{A}_1 \ \mathbf{B}_1 \ \cdots \ \mathbf{A}_m \ \mathbf{B}_m]\tilde{\mathbf{S}}_m = \tilde{\mathbf{F}} \tag{9.49}$$

It is obtained from (9.41) by deleting α number of zero columns in $\mathbf{S}_m$ and the corresponding zero columns in $\bar{\mathbf{F}}$. Now because $\tilde{\mathbf{S}}_m$ has full column rank if $m \geq \nu - 1$, we conclude that for any $\mathbf{F}(s)$ of column degrees at most $m + \mu_i$, solutions $\mathbf{A}_i$ and $\mathbf{B}_i$ exist in (9.49). Or, equivalently, polynomial matrices $\mathbf{A}(s)$ and $\mathbf{B}(s)$ of row degree m or less exist in (9.49). Note that generally $\tilde{\mathbf{S}}_m$ has more rows than columns; therefore solutions of (9.49) are not unique.

Next we show that $\mathbf{A}^{-1}(s)\mathbf{B}(s)$ is proper. Note that $\mathbf{D}_\mu$ is, in general, singular and the method of proving Theorem 9.2 cannot be used here. Using $\mathbf{H}_r(s)$ and $\mathbf{H}_c(s)$, we write, as in (7.80),

$$\mathbf{D}(s) = [\mathbf{D}_{hc} + \mathbf{D}_{lc}(s)\mathbf{H}_c^{-1}(s)]\mathbf{H}_c(s)$$

$$\mathbf{N}(s) = [\mathbf{N}_{hc} + \mathbf{N}_{lc}(s)\mathbf{H}_c^{-1}(s)]\mathbf{H}_c(s)$$

$$\mathbf{A}(s) = \mathbf{H}_r(s)[\mathbf{A}_{hr} + \mathbf{H}_r^{-1}(s)\mathbf{A}_{lr}(s)]$$

$$\mathbf{B}(s) = \mathbf{H}_r(s)[\mathbf{B}_{hr} + \mathbf{H}_r^{-1}(s)\mathbf{B}_{lr}(s)]$$

$$\mathbf{F}(s) = \mathbf{H}_r(s)[\mathbf{F}_h + \mathbf{H}_r^{-1}(s)\mathbf{F}_l(s)\mathbf{H}_c^{-1}(s)]\mathbf{H}_c(s)$$

where $\mathbf{D}_{lc}(s)\mathbf{H}_c^{-1}(s), \mathbf{N}_{lc}(s)\mathbf{H}_c^{-1}(s), \mathbf{H}_r^{-1}(s)\mathbf{A}_{lr}(s), \mathbf{H}_r^{-1}(s)\mathbf{B}_{lr}(s)$, and $\mathbf{H}_r^{-1}(s)\mathbf{F}_l(s)\mathbf{H}_c^{-1}(s)$ are all strictly proper rational functions. Substituting the above into (9.39) yields, at $s = \infty$,

$$\mathbf{A}_{hr}\mathbf{D}_{hc} + \mathbf{B}_{hr}\mathbf{N}_{hc} = \mathbf{F}_h$$

which reduces to, because $\mathbf{N}_{hc} = \mathbf{0}$ following strict properness of $\hat{\mathbf{G}}(s)$,

$$\mathbf{A}_{hr}\mathbf{D}_{hc} = \mathbf{F}_h$$

Because $\mathbf{D}(s)$ is column reduced, $\mathbf{D}_{hc}$ is nonsingular. The constant matrix $\mathbf{F}_h$ is nonsingular by assumption; thus $\mathbf{A}_{hr} = \mathbf{F}_h\mathbf{D}_{hc}^{-1}$ is nonsingular and $\mathbf{A}(s)$ is row reduced. Therefore $\mathbf{A}^{-1}(s)\mathbf{B}(s)$ is proper (Corollary 7.8). This establishes the theorem.   Q.E.D.

A polynomial matrix $\mathbf{F}(s)$ meeting (9.48) is said to be *row–column reduced* with row degrees m_i and column degrees μ_i. If $m_1 = m_2 = \cdots = m_p = m$, then the row–column reducedness is the same as column reducedness with column degrees $m + \mu_i$. In application, we can select $\mathbf{F}(s)$ to be diagonal or triangular with polynomials with desired roots as its diagonal entries. Then $\mathbf{F}^{-1}(s)$ and, consequently, $\hat{\mathbf{G}}_o(s)$ have the desired roots as their poles.

Consider again $\mathbf{S}_{\nu-1}$. It is of order $(p + q)\nu \times (\mu + \nu)p$. It has $\alpha = p\mu - n$ number of zero columns. Thus the matrix $\tilde{\mathbf{S}}_{\nu-1}$ is of order $(p + q)\nu \times [(\mu + \nu)p - (p\mu - n)]$ or $(p + q)\nu \times (\nu p + n)$. The matrix $\tilde{\mathbf{S}}_{\nu-1}$ contains $p\nu$ linearly independent D-rows but contains only $\nu_1 + \cdots + \nu_q = n$ linearly independent N-rows. Thus $\tilde{\mathbf{S}}_{\nu-1}$ contains

$$\gamma := (p + q)\nu - p\nu - n = q\nu - n$$

linearly dependent N-rows. Let $\check{\mathbf{S}}_{\nu-1}$ be the matrix $\tilde{\mathbf{S}}_{\nu-1}$ after deleting these linearly dependent N-rows. Then the matrix $\check{\mathbf{S}}_{\nu-1}$ is of order

$$[(p + q)\nu - (q\nu - n)] \times (\nu p + n) = (\nu p + n) \times (\nu p + n)$$

Thus $\check{\mathbf{S}}_{\nu-1}$ is square and nonsingular.

Consider (9.49) with $m = \nu - 1$:

$$\mathbf{K}\tilde{\mathbf{S}}_{\nu-1} := [\mathbf{A}_0 \ \mathbf{B}_0 \ \mathbf{A}_1 \ \mathbf{B}_1 \ \cdots \ \mathbf{A}_{\nu-1} \ \mathbf{B}_{\nu-1}]\tilde{\mathbf{S}}_{\nu-1} = \tilde{\mathbf{F}}$$

It actually consists of the following p sets of linear algebraic equations

$$\mathbf{k}_i\tilde{\mathbf{S}}_{\nu-1} = \tilde{\mathbf{f}}_i \qquad i = 1, 2, \ldots, p \qquad (9.50)$$

where $\mathbf{k}_i$ and $\tilde{\mathbf{f}}_i$ are the ith row of $\mathbf{K}$ and $\tilde{\mathbf{F}}$, respectively. Because $\tilde{\mathbf{S}}_{\nu-1}$ has full column rank, for any $\tilde{\mathbf{f}}_i$, solutions $\mathbf{k}_i$ exist in (9.50). Because $\tilde{\mathbf{S}}_{\nu-1}$ has more γ rows than columns, the general solution of (9.50) contains γ free parameters (Corollary 3.2). If m in $\mathbf{S}_m$ increases by 1 from $\nu - 1$ to ν, then the number of rows of $\tilde{\mathbf{S}}_\nu$ increases by $(p + q)$ but the rank of $\tilde{\mathbf{S}}_\nu$ increases only by p. In this case, the number of free parameters will increase from γ to $\gamma + q$. Thus in the MIMO case, we have a great deal of freedom in carrying out the design.

We discuss a special case of (9.50). The matrix $\tilde{\mathbf{S}}_{\nu-1}$ has γ linearly dependent N-rows. If we delete these linearly dependent N-rows from $\tilde{\mathbf{S}}_{\nu-1}$ and assign the corresponding columns in $\mathbf{B}_i$ as zero, then (9.50) becomes

$$[\mathbf{A}_0 \ \bar{\mathbf{B}}_0 \ \cdots \ \mathbf{A}_{\nu-1} \ \bar{\mathbf{B}}_{\nu-1}]\check{\mathbf{S}}_{\nu-1} = \tilde{\mathbf{F}}$$

where $\check{\mathbf{S}}_{\nu-1}$ is, as discussed earlier, square and nonsingular. Thus the solution is unique. This is illustrated in the next example.

EXAMPLE 9.10 Consider a plant with the strictly proper transfer matrix

$$\hat{\mathbf{G}}(s) = \begin{bmatrix} 1/s^2 & 1/s \\ 0 & 1/s \end{bmatrix} = \begin{bmatrix} 1 & 1 \\ 0 & 1 \end{bmatrix} \begin{bmatrix} s^2 & 0 \\ 0 & s \end{bmatrix}^{-1} =: \mathbf{N}(s)\mathbf{D}^{-1}(s) \qquad (9.51)$$

The fraction is right coprime and $\mathbf{D}(s)$ is column reduced with column degrees $\mu_1 = 2$ and $\mu_2 = 1$. We write

$$\mathbf{D}(s) = \begin{bmatrix} 0 & 0 \\ 0 & 0 \end{bmatrix} + \begin{bmatrix} 0 & 0 \\ 0 & 1 \end{bmatrix}s + \begin{bmatrix} 1 & 0 \\ 0 & 0 \end{bmatrix}s^2$$

and

$$\mathbf{N}(s) = \begin{bmatrix} 1 & 1 \\ 0 & 1 \end{bmatrix} + \begin{bmatrix} 0 & 0 \\ 0 & 0 \end{bmatrix}s + \begin{bmatrix} 0 & 0 \\ 0 & 0 \end{bmatrix}s^2$$

We use MATLAB to compute the row index. The QR decomposition discussed in Section 7.3.1 can reveal linearly independent columns, from left to right, of a matrix. Here we need linearly independent rows, from top to bottom, of $\mathbf{S}_m$; therefore we will apply QR decomposition to the transpose of $\mathbf{S}_m$. We type

```
d1=[0 0 0 0 1 0];d2=[0 0 0 1 0 0];
n1=[1 1 0 0 0 0];n2=[0 1 0 0 0 0];
s1=[d1 0 0;d2 0 0;n1 0 0;n2 0 0;...
    0 0 d1;0 0 d2;0 0 n1;0 0 n2];
[q,r]=qr(s1')
```

which yields, as in Example 7.7,

$$
r = \begin{bmatrix}
d1 & 0 & 0 & 0 & 0 & 0 & 0 & 0 \\
0 & d2 & 0 & 0 & 0 & 0 & x & x \\
0 & 0 & n1 & x & 0 & 0 & 0 & 0 \\
0 & 0 & 0 & n2 & 0 & 0 & 0 & 0 \\
0 & 0 & 0 & 0 & d1 & 0 & 0 & 0 \\
0 & 0 & 0 & 0 & 0 & d2 & 0 & 0 \\
0 & 0 & 0 & 0 & 0 & 0 & n1 & 0 \\
0 & 0 & 0 & 0 & 0 & 0 & 0 & 0
\end{bmatrix}
$$

The matrix q is not needed and is not typed. In the matrix r, we use x, di, and ni to denote nonzero entries. The nonzero diagonal entries of r yield the linearly independent columns of $\mathbf{S}_1'$ or, equivalently, linearly independent rows of $\mathbf{S}_1$. We see that there are two linearly independent $N1$-rows and one linearly independent $N2$-row. The degree of $\hat{\mathbf{G}}(s)$ is 3 and we have found three linearly independent N-rows. Therefore there is no need to search further and we have $\nu_1 = 2$ and $\nu_2 = 1$. Thus the row index is $\nu = 2$. We select $m_1 = m_2 = m = \nu - 1 = 1$. Thus for any column-reduced $\mathbf{F}(s)$ of column degrees $m + \mu_1 = 3$ and $m + \mu_2 = 2$, we can find a proper compensator such that the resulting unity-feedback system has $\mathbf{F}(s)$ as its denominator matrix. Let us choose arbitrarily

$$
\begin{aligned}
\mathbf{F}(s) &= \begin{bmatrix} (s^2 + 4s + 5)(s + 3) & 0 \\ 0 & s^2 + 2s + 5 \end{bmatrix} \\
&= \begin{bmatrix} 15 + 17s + 7s^2 + s^3 & 0 \\ 0 & 5 + 2s + s^2 \end{bmatrix}
\end{aligned}
\tag{9.52}
$$

and form (9.41) with $m = \nu - 1 = 1$:

$$
[\mathbf{A}_0 \ \mathbf{B}_0 \ \mathbf{A}_1 \ \mathbf{B}_1]
\begin{bmatrix}
0 & 0 & 0 & 0 & 1 & 0 & 0 & 0 \\
0 & 0 & 0 & 1 & 0 & 0 & 0 & 0 \\
\cdots & \cdots & \cdots & \cdots & \cdots & \cdots & \cdots & \cdots \\
1 & 1 & 0 & 0 & 0 & 0 & 0 & 0 \\
0 & 1 & 0 & 0 & 0 & 0 & 0 & 0 \\
\hline
0 & 0 & 0 & 0 & 0 & 0 & 1 & 0 \\
0 & 0 & 0 & 0 & 0 & 1 & 0 & 0 \\
\cdots & \cdots & \cdots & \cdots & \cdots & \cdots & \cdots & \cdots \\
0 & 0 & 1 & 1 & 0 & 0 & 0 & 0 \\
0 & 0 & 0 & 1 & 0 & 0 & 1 & 0
\end{bmatrix}
$$

$$
= \begin{bmatrix} 15 & 0 & 17 & 0 & 7 & 0 & 1 & 0 \\ 0 & 5 & 0 & 2 & 0 & 1 & 0 & 0 \end{bmatrix} = \bar{\mathbf{F}}
\tag{9.53}
$$

The α in (9.44) is 1 for this problem. Thus $\mathbf{S}_1$ and $\bar{\mathbf{F}}$ both have one zero column as we can see in (9.53). After deleting the zero column, the remaining $\tilde{\mathbf{S}}_1$ has full column rank and, for any $\tilde{\mathbf{F}}$, solutions exist in (9.53). The matrix $\tilde{\mathbf{S}}_1$ has order 8×7 and solutions are not unique.

In searching the row index, we knew that the last row of $\tilde{\mathbf{S}}_1$ is a linearly dependent row. If we delete the row, then we must assign the second column of $\mathbf{B}_1$ as $\mathbf{0}$ and the solution will be unique. We type

```
d1=[0 0 0 0 0 1];d2=[0 0 0 1 0 0];
n1=[1 1 0 0 0 0];n2=[0 1 0 0 0 0];
d1t=[0 0 0 0 1];d2t=[0 0 0 1 0];n1t=[0 1 0 0 0];
s1t=[d1 0;d2 0;n1 0;n2 0;0 0 d1t;0 0 d2t;0 0 n1t];
f1t=[15 0 17 0 7 1];
f1t/s1t
```

which yields $[7\ -17\ 15\ -15\ 1\ 0\ 17]$. Computing once again for the second row of $\bar{\mathbf{F}}$, we can finally obtain

$$[\mathbf{A}_0\ \mathbf{B}_0\ \mathbf{A}_1\ \mathbf{B}_1] = \begin{bmatrix} 7 & -17 & 15 & -15 & 1 & 0 & 17 & 0 \\ 0 & 2 & 0 & 5 & 0 & 1 & 0 & 0 \end{bmatrix}$$

Note that MATLAB yields the first 7 columns; the last column $\mathbf{0}$ is assigned by us (due to deleting the last row of $\tilde{\mathbf{S}}_1$). Thus we have

$$\mathbf{A}(s) = \begin{bmatrix} 7+s & -17 \\ 0 & 2+s \end{bmatrix} \qquad \mathbf{B}(s) = \begin{bmatrix} 15+17s & -15 \\ 0 & 5 \end{bmatrix}$$

and the proper compensator

$$\mathbf{C}(s) = \begin{bmatrix} s+7 & -17 \\ 0 & s+2 \end{bmatrix}^{-1} \begin{bmatrix} 17s+15 & -15 \\ 0 & 5 \end{bmatrix}$$

will yield the overall transfer matrix

$$\hat{\mathbf{G}}_o(s) = \begin{bmatrix} 1 & 1 \\ 0 & 1 \end{bmatrix} \begin{bmatrix} (s^2+4s+5)(s+3) & 0 \\ 0 & s^2+2s+5 \end{bmatrix}^{-1}$$
$$\times \begin{bmatrix} 17s+15 & -15 \\ 0 & 5 \end{bmatrix} \qquad (9.54)$$

This transfer matrix has the desired poles. This completes the design.

The design in Theorem 9.M2 is carried out by using a right coprime fraction of $\hat{\mathbf{G}}(s)$. We state next the result by using a left coprime fraction of $\hat{\mathbf{G}}(s)$.

▶ **Corollary 9.M2**

Consider the unity-feedback system shown in Fig. 9.7. The plant is described by a $q \times p$ strictly proper rational matrix $\hat{\mathbf{G}}(s)$. Let $\hat{\mathbf{G}}(s)$ be factored as $\hat{\mathbf{G}}(s) = \bar{\mathbf{D}}^{-1}(s)\bar{\mathbf{N}}(s)$, where $\bar{\mathbf{D}}(s)$ and $\bar{\mathbf{N}}(s)$ are left coprime and $\bar{\mathbf{D}}(s)$ is row reduced with row degrees ν_i, $i = 1, 2, \ldots, q$. Let μ be the column index of $\hat{\mathbf{G}}(s)$ and let $m_i \geq \mu - 1$. Then for any $q \times q$ row-column reduced polynominal matrix $\bar{\mathbf{F}}(s)$ such that

$$\lim_{s \to \infty} \text{diag}(s^{-\nu_1}, s^{-\nu_2}, \ldots, s^{-\nu_q})\bar{\mathbf{F}}(s)\text{diag}(s^{-m_1}, s^{-m_2}, \ldots, s^{-m_q}) = \bar{\mathbf{F}}_h$$

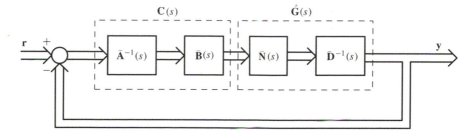

Figure 9.7 Unity feedback with $\hat{\mathbf{G}}(s) = \bar{\mathbf{D}}^{-1}(s)\bar{\mathbf{N}}(s)$.

is a nonsingular constant matrix, there exists a $p \times q$ proper compensator $\mathbf{C}(s) = \bar{\mathbf{B}}(s)\bar{\mathbf{A}}^{-1}(s)$, where $\bar{\mathbf{A}}(s)$ is column reduced with column degrees m_i, to meet

$$\bar{\mathbf{D}}(s)\bar{\mathbf{A}}(s) + \bar{\mathbf{N}}(s)\bar{\mathbf{B}}(s) = \bar{\mathbf{F}}(s) \tag{9.55}$$

and the transfer matrix from $\mathbf{r}$ to $\mathbf{y}$ equals

$$\hat{\mathbf{G}}_o(s) = \mathbf{I} - \bar{\mathbf{A}}(s)\bar{\mathbf{F}}^{-1}(s)\bar{\mathbf{D}}(s) \tag{9.56}$$

Substituting $\hat{\mathbf{G}}(s) = \bar{\mathbf{D}}^{-1}\bar{\mathbf{N}}(s)$ and $\mathbf{C}(s) = \bar{\mathbf{B}}(s)\bar{\mathbf{A}}^{-1}(s)$ into the first equation in (9.37) yields

$$
\begin{aligned}
\hat{\mathbf{G}}_o(s) &= [\mathbf{I} + \bar{\mathbf{D}}^{-1}(s)\bar{\mathbf{N}}(s)\bar{\mathbf{B}}(s)\bar{\mathbf{A}}^{-1}(s)]^{-1}\bar{\mathbf{D}}^{-1}(s)\bar{\mathbf{N}}(s)\bar{\mathbf{B}}(s)\bar{\mathbf{A}}^{-1}(s) \\
&= \bar{\mathbf{A}}(s)[\bar{\mathbf{D}}(s)\bar{\mathbf{A}}(s) + \bar{\mathbf{N}}(s)\bar{\mathbf{B}}(s)]^{-1}\bar{\mathbf{N}}(s)\bar{\mathbf{B}}(s)\bar{\mathbf{A}}^{-1}(s)
\end{aligned}
$$

which becomes, after substituting (9.55),

$$
\begin{aligned}
\hat{\mathbf{G}}_o(s) &= \bar{\mathbf{A}}(s)\bar{\mathbf{F}}^{-1}(s)[\bar{\mathbf{F}}(s) - \bar{\mathbf{D}}(s)\bar{\mathbf{A}}(s)]\bar{\mathbf{A}}^{-1}(s) \\
&= \mathbf{I} - \bar{\mathbf{A}}(s)\bar{\mathbf{F}}^{-1}(s)\bar{\mathbf{D}}(s)
\end{aligned}
$$

This establishes the transfer matrix from $\mathbf{r}$ to $\mathbf{y}$ in the theorem. The design in Corollary 9.M2 hinges on solving (9.55). Note that the transpose of (9.55) becomes (9.39); left coprime and row reducedness become right coprime and column reducedness. Thus the linear algebraic equation in (9.41) can be used to solve the transpose of (9.55). We can also solve (9.55) directly by forming

$$
\mathbf{T}_m
\begin{bmatrix}
\bar{\mathbf{B}}_0 \\
\bar{\mathbf{A}}_0 \\
\bar{\mathbf{B}}_1 \\
\bar{\mathbf{A}}_1 \\
\vdots \\
\bar{\mathbf{B}}_m \\
\bar{\mathbf{A}}_m
\end{bmatrix}
=
\begin{bmatrix}
\bar{\mathbf{D}}_0 & \bar{\mathbf{N}}_0 & \vdots & \mathbf{0} & \mathbf{0} & \vdots & \cdots & \vdots & \mathbf{0} & \mathbf{0} \\
\bar{\mathbf{D}}_1 & \bar{\mathbf{N}}_1 & \vdots & \bar{\mathbf{D}}_0 & \bar{\mathbf{N}}_0 & \vdots & \cdots & \vdots & \mathbf{0} & \mathbf{0} \\
\vdots & \vdots & \vdots & \vdots & \vdots & \vdots & \vdots & \vdots & \vdots & \vdots \\
\bar{\mathbf{D}}_n & \bar{\mathbf{N}}_n & \vdots & \bar{\mathbf{D}}_{n-1} & \bar{\mathbf{N}}_{n-1} & \vdots & \cdots & \vdots & \mathbf{0} & \mathbf{0} \\
\mathbf{0} & \mathbf{0} & \vdots & \bar{\mathbf{D}}_n & \bar{\mathbf{N}}_n & \vdots & \cdots & \vdots & \bar{\mathbf{D}}_0 & \bar{\mathbf{N}}_0 \\
\vdots & \vdots & \vdots & \vdots & \vdots & \vdots & \vdots & \vdots & \vdots & \vdots \\
\mathbf{0} & \mathbf{0} & \vdots & \mathbf{0} & \mathbf{0} & \vdots & \cdots & \vdots & \bar{\mathbf{D}}_n & \bar{\mathbf{N}}_n
\end{bmatrix}
$$

$$
\times \begin{bmatrix} \bar{\mathbf{B}}_0 \\ \bar{\mathbf{A}}_0 \\ \bar{\mathbf{B}}_1 \\ \bar{\mathbf{A}}_1 \\ \vdots \\ \bar{\mathbf{B}}_m \\ \bar{\mathbf{A}}_m \end{bmatrix} = \begin{bmatrix} \bar{\mathbf{F}}_0 \\ \bar{\mathbf{F}}_1 \\ \bar{\mathbf{F}}_2 \\ \vdots \\ \bar{\mathbf{F}}_{n+m} \end{bmatrix} \tag{9.57}
$$

We search linearly independent columns of $\mathbf{T}_m$ in order from left to right. Let μ be the column index of $\hat{\mathbf{G}}(s)$ or the least integer such that $\mathbf{T}_{\mu-1}$ contains n linearly independent $\bar{N}$-columns. Then the compensator can be solved from (9.57) with $m = \mu - 1$.

9.4.1 Regulation and Tracking

As in the SISO case, pole placement can be used to achieve regulation and tracking in multivariable systems. In the regulator problem, we have $\mathbf{r} \equiv \mathbf{0}$ and if all poles of the overall system are assigned to have negative real parts, then the responses caused by any nonzero initial state will decay to zero. Furthermore, the rate of decaying can be controlled by the locations of the poles; the larger the negative real parts, the faster the decay.

Next we discuss tracking of any step reference input. In this design, we generally require a feedforward constant gain matrix $\mathbf{P}$. Suppose the compensator in Fig. 9.6 has been designed by using Theorem 9.M2. Then the $q \times q$ transfer matrix from $\mathbf{r}$ to $\mathbf{y}$ is given by

$$
\hat{\mathbf{G}}_o(s) = \mathbf{N}(s)\mathbf{F}^{-1}(s)\mathbf{B}(s)\mathbf{P} \tag{9.58}
$$

If $\hat{\mathbf{G}}_o(s)$ is BIBO stable, then the steady-state response excited by $\mathbf{r}(t) = \mathbf{d}$, for $t \geq 0$, or $\hat{\mathbf{r}}(s) = \mathbf{d}s^{-1}$, where $\mathbf{d}$ is an arbitrary $q \times 1$ constant vector, can be computed as, using the final-value theorem of the Laplace transform,

$$
\lim_{t \to \infty} \mathbf{y}(t) = \lim_{s \to 0} s\hat{\mathbf{G}}_o(s)\mathbf{d}s^{-1} = \hat{\mathbf{G}}_o(0)\mathbf{d}
$$

Thus we conclude that in order for $\mathbf{y}(t)$ to track asymptotically any step reference input, we need, in addition to BIBO stability,

$$
\hat{\mathbf{G}}_o(0) = \mathbf{N}(0)\mathbf{F}^{-1}(0)\mathbf{B}(0)\mathbf{P} = \mathbf{I}_q \tag{9.59}
$$

Before discussing the conditions for meeting (9.59), we need the concept of transmission zeros.

Transmission zeros   Consider a $q \times p$ proper rational matrix $\hat{\mathbf{G}}(s) = \mathbf{N}(s)\mathbf{D}^{-1}(s)$, where $\mathbf{N}(s)$ and $\mathbf{D}(s)$ are right coprime. A number λ, real or complex, is called a *transmission zero* of $\hat{\mathbf{G}}(s)$ if the rank of $\mathbf{N}(\lambda)$ is smaller than $\min(p, q)$.

EXAMPLE 9.11 Consider the 3×2 proper rational matrix

$$
\hat{\mathbf{G}}_1(s) = \begin{bmatrix} \dfrac{s}{s+2} & 0 \\ 0 & \dfrac{s+1}{s^2} \\ \dfrac{s+1}{s+2} & \dfrac{1}{s} \end{bmatrix} = \begin{bmatrix} s & 0 \\ 0 & s+1 \\ s+1 & s \end{bmatrix} \begin{bmatrix} s+2 & 0 \\ 0 & s^2 \end{bmatrix}^{-1}
$$

This $\mathbf{N}(s)$ has rank 2 at every s; thus $\hat{\mathbf{G}}_1(s)$ has no transmission zero. Consider the 2×2 proper rational matrix

$$\hat{\mathbf{G}}_2(s) = \begin{bmatrix} \dfrac{s}{s+2} & 0 \\ 0 & \dfrac{s+2}{s} \end{bmatrix} = \begin{bmatrix} s & 0 \\ 0 & s+2 \end{bmatrix} \begin{bmatrix} s+2 & 0 \\ 0 & s \end{bmatrix}^{-1}$$

This $\mathbf{N}(s)$ has rank 1 at $s = 0$ and $s = -2$. Thus $\hat{\mathbf{G}}_2(s)$ has two transmission zeros at 0 and -2. Note that $\hat{\mathbf{G}}(s)$ has poles also at 0 and -2.

From this example, we see that a transmission zero cannot be defined directly from $\hat{\mathbf{G}}(s)$; it must be defined from its coprime fraction. Either a right coprime or left coprime fraction can be used and each yields the same set of transmission zeros. Note that if $\hat{\mathbf{G}}(s)$ is square and if $\hat{\mathbf{G}}(s) = \mathbf{N}(s)\mathbf{D}^{-1}(s) = \bar{\mathbf{D}}^{-1}(s)\bar{\mathbf{N}}(s)$, where $\mathbf{N}(s)$ and $\mathbf{D}(s)$ are right coprime and $\bar{\mathbf{D}}(s)$ and $\bar{\mathbf{N}}(s)$ are left coprime, then the transmission zeros of $\hat{\mathbf{G}}(s)$ are the roots of $\det \mathbf{N}(s)$ or the roots of $\det \bar{\mathbf{N}}(s)$. Transmission zeros can also be defined from a minimal realization of $\hat{\mathbf{G}}(s)$. Let $(\mathbf{A}, \mathbf{B}, \mathbf{C}, \mathbf{D})$ be any n-dimensional minimal realization of a $q \times p$ proper rational matrix $\hat{\mathbf{G}}(s)$. Then the transmission zeros are those λ such that

$$\text{rank} \begin{bmatrix} \lambda\mathbf{I} - \mathbf{A} & \mathbf{B} \\ -\mathbf{C} & \mathbf{D} \end{bmatrix} < n + \min(p, q)$$

This is used in the MATLAB function `tzero` to compute transmission zeros. For a more detailed discussion of transmission zeros, see Reference [6, pp. 623–635].

Now we are ready to discuss the conditions for achieving tracking or for meeting (9.59). Note that $\mathbf{N}(s)$, $\mathbf{F}(s)$, and $\mathbf{B}(s)$ are $q \times p$, $p \times p$, and $p \times q$. Because $\mathbf{I}_q$ has rank q, a necessary condition for (9.59) to hold is that the $q \times p$ matrix $\mathbf{N}(0)$ has rank q. Necessary conditions for $\rho(\mathbf{N}(0)) = q$ are $p \geq q$ and $s = 0$ is not a transmission zero of $\hat{\mathbf{G}}(s)$. Thus we conclude that in order for the unity-feedback configuration in Fig. 9.6 to achieve asymptotic tracking, the plant must have the following two properties:

- The plant has the same or a greater number of inputs than outputs.
- The plant transfer function has no transmission zero at $s = 0$.

Under these conditions, $\mathbf{N}(0)$ has rank q. Because we have freedom in selecting $\mathbf{F}(s)$, we can select it such that $\mathbf{B}(0)$ has rank q and the $q \times q$ constant matrix $\mathbf{N}(0)\mathbf{F}^{-1}(0)\mathbf{B}(0)$ is nonsingular. Under these conditions, the constant gain matrix $\mathbf{P}$ can be computed as

$$\mathbf{P} = [\mathbf{N}(0)\mathbf{F}^{-1}(0)\mathbf{B}(0)]^{-1} \tag{9.60}$$

Then we have $\hat{\mathbf{G}}_o(0) = \mathbf{I}_q$, and the unity-feedback system in Fig. 9.6 with $\mathbf{P}$ in (9.60) will track asymptotically any step reference input.

9.4.2 Robust Tracking and Disturbance Rejection

As in the SISO case, the asymptotic tracking design in the preceding section is not robust. In this section, we discuss a different design. To simplify the discussion, we study only plants with

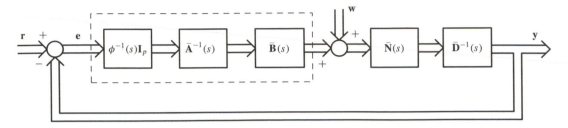

Figure 9.8 Robust tracking and disturbance rejection.

an equal number of input terminals and output terminals or $p = q$. Consider the unity-feedback system shown in Fig. 9.8. The plant is described by a $p \times p$ strictly proper transfer matrix factored in left coprime fraction as $\hat{\mathbf{G}}(s) = \bar{\mathbf{D}}^{-1}(s)\bar{\mathbf{N}}(s)$. It is assumed that $p \times 1$ disturbance $\mathbf{w}(t)$ enters the plant input as shown. The problem is to design a compensator so that the output $\mathbf{y}(t)$ will track asymptotically a class of $p \times 1$ reference signals $\mathbf{r}(t)$ even with the presence of the disturbance $\mathbf{w}(t)$ and with plant parameter variations. This is called *robust tracking and disturbance rejection.*

As in the SISO case, we need some information on $\mathbf{r}(t)$ and $\mathbf{w}(t)$ before carrying out the design. We assume that the Laplace transforms of $\mathbf{r}(t)$ and $\mathbf{w}(t)$ are given by

$$\hat{\mathbf{r}}(s) = \mathcal{L}[\mathbf{r}(t)] = \mathbf{N}_r(s)D_r^{-1}(s) \qquad \hat{\mathbf{w}}(s) = \mathcal{L}[\mathbf{w}(t)] = \mathbf{N}_w(s)D_w^{-1}(s) \qquad (9.61)$$

where $D_r(s)$ and $D_w(s)$ are known polynomials; however, $\mathbf{N}_r(s)$ and $\mathbf{N}_w(s)$ are unknown to us. Let $\phi(s)$ be the least common denominator of the unstable poles of $\hat{\mathbf{r}}(s)$ and $\hat{\mathbf{w}}(s)$. The stable poles are excluded because they have no effect on $\mathbf{y}(t)$ as $t \to \infty$. We introduce the internal model $\phi^{-1}(s)\mathbf{I}_p$ as shown in Fig. 9.8. If $\bar{\mathbf{D}}(s)$ and $\bar{\mathbf{N}}(s)$ are left coprime and if no root of $\phi(s)$ is a transmission zero of $\hat{\mathbf{G}}(s)$ or, equivalently, $\phi(s)$ and det $\bar{\mathbf{N}}(s)$ are coprime, then it can be shown that $\bar{\mathbf{D}}(s)\phi(s)$ and $\bar{\mathbf{N}}(s)$ are left coprime. See Reference [6, p. 443]. Then Corollary 9.M2 implies that there exists a proper compensator $\mathbf{C}(s) = \bar{\mathbf{B}}(s)\bar{\mathbf{A}}^{-1}(s)$ such that

$$\phi(s)\bar{\mathbf{D}}(s)\bar{\mathbf{A}}(s) + \bar{\mathbf{N}}(s)\bar{\mathbf{B}}(s) = \bar{\mathbf{F}}(s) \qquad (9.62)$$

for any $\bar{\mathbf{F}}(s)$ meeting the condition in Corollary 9.M2. Clearly $\bar{\mathbf{F}}(s)$ can be chosen to be diagaonl with the roots of its diagonal entries lying inside the sector shown in Fig. 8.3(a). The unity-feedback system in Fig. 9.8 so designed will track asymptotically and robustly the reference signal $\mathbf{r}(t)$ and reject the disturbance $\mathbf{w}(t)$. This is stated as a theorem.

▶ **Theorem 9.M3**

Consider the unity-feedback system shown in Fig. 9.8 where the plant has a $p \times p$ strictly proper transfer matrix $\hat{\mathbf{G}}(s) = \bar{\mathbf{D}}^{-1}(s)\bar{\mathbf{N}}(s)$. It is assumed that $\bar{\mathbf{D}}(s)$ and $\bar{\mathbf{N}}(s)$ are left coprime and $\bar{\mathbf{D}}(s)$ is row reduced with row degrees ν_i, $i = 1, 2, \ldots, p$. The reference signal $\mathbf{r}(t)$ and disturbance $\mathbf{w}(t)$ are modeled as $\hat{\mathbf{r}}(s) = \mathbf{N}_r(s)D_r^{-1}(s)$ and $\hat{\mathbf{w}}(s) = \mathbf{N}_w(s)D_w^{-1}(s)$. Let $\phi(s)$ be the least common denominator of the unstable poles of $\hat{\mathbf{r}}(s)$ and $\hat{\mathbf{w}}(s)$. If no root of $\phi(s)$ is a transmission zero of $\hat{\mathbf{G}}(s)$, then there exists a proper compensator $\mathbf{C}(s) = \bar{\mathbf{B}}(s)(\bar{\mathbf{A}}(s)\phi(s))^{-1}$ such that the overall system will robustly and asymptotically track the reference signal $\mathbf{r}(t)$ and reject the disturbance $\mathbf{w}(t)$.

Proof: First we show that the system will reject the disturbance at the output. Let us assume $\mathbf{r} = 0$ and compute the output $\hat{\mathbf{y}}_w(s)$ excited by $\hat{\mathbf{w}}(s)$. Clearly we have

$$\hat{\mathbf{y}}_w(s) = \bar{\mathbf{D}}^{-1}(s)\bar{\mathbf{N}}(s)[\hat{\mathbf{w}}(s) - \bar{\mathbf{B}}(s)\bar{\mathbf{A}}^{-1}(s)\phi^{-1}(s)\hat{\mathbf{y}}_w(s)]$$

which implies

$$\begin{aligned}
\hat{\mathbf{y}}_w(s) &= [\mathbf{I} + \bar{\mathbf{D}}^{-1}(s)\bar{\mathbf{N}}(s)\bar{\mathbf{B}}(s)\bar{\mathbf{A}}^{-1}(s)\phi^{-1}(s)]^{-1}\bar{\mathbf{D}}^{-1}(s)\bar{\mathbf{N}}(s)\hat{\mathbf{w}}(s) \\
&= \left[\bar{\mathbf{D}}^{-1}(s)[\bar{\mathbf{D}}(s)\phi(s)\bar{\mathbf{A}}(s) + \bar{\mathbf{N}}(s)\bar{\mathbf{B}}(s)]\bar{\mathbf{A}}^{-1}(s)\phi^{-1}(s)\right]^{-1} \\
&\quad \times \bar{\mathbf{D}}^{-1}(s)\bar{\mathbf{N}}(s)\hat{\mathbf{w}}(s) \\
&= \phi(s)\bar{\mathbf{A}}(s)[\bar{\mathbf{D}}(s)\phi(s)\bar{\mathbf{A}}(s) + \bar{\mathbf{N}}(s)\bar{\mathbf{B}}(s)]^{-1}\bar{\mathbf{N}}(s)\hat{\mathbf{w}}(s)
\end{aligned}$$

Thus we have, using (9.61) and (9.62),

$$\hat{\mathbf{y}}_w(s) = \bar{\mathbf{A}}(s)\bar{\mathbf{F}}^{-1}(s)\bar{\mathbf{N}}(s)\phi(s)\mathbf{N}_w(s)D_w^{-1}(s) \tag{9.63}$$

Because all unstable roots of $D_w(s)$ are canceled by $\phi(s)$, all poles of $\hat{\mathbf{y}}_w(s)$ have negative real parts. Thus we have $\mathbf{y}_w(t) \rightarrow 0$ as $t \rightarrow \infty$ and the response excited by $\mathbf{w}(t)$ is suppressed asymptotically at the output.

Next we compute the error $\hat{\mathbf{e}}_r(s)$ excited by the reference signal $\hat{\mathbf{r}}(s)$:

$$\hat{\mathbf{e}}_r(s) = \hat{\mathbf{r}}(s) - \bar{\mathbf{D}}^{-1}(s)\bar{\mathbf{N}}(s)\bar{\mathbf{B}}(s)\bar{\mathbf{A}}^{-1}(s)\phi^{-1}(s)\hat{\mathbf{e}}_r(s)$$

which implies

$$\begin{aligned}
\hat{\mathbf{e}}_r(s) &= [\mathbf{I} + \bar{\mathbf{D}}^{-1}(s)\bar{\mathbf{N}}(s)\bar{\mathbf{B}}(s)\bar{\mathbf{A}}^{-1}(s)\phi^{-1}(s)]^{-1}\hat{\mathbf{r}}(s) \\
&= \phi(s)\bar{\mathbf{A}}(s)[\bar{\mathbf{D}}(s)\phi(s)\bar{\mathbf{A}}(s) + \bar{\mathbf{N}}(s)\bar{\mathbf{B}}(s)]^{-1}\bar{\mathbf{D}}(s)\hat{\mathbf{r}}(s) \\
&= \bar{\mathbf{A}}(s)\bar{\mathbf{F}}^{-1}(s)\bar{\mathbf{D}}(s)\phi(s)\mathbf{N}_r(s)D_r^{-1}(s) \tag{9.64}
\end{aligned}$$

Because all unstable roots of $D_r(s)$ are canceled by $\phi(s)$, the error vector $\hat{\mathbf{e}}_r(s)$ has only stable poles. Thus its time response approaches zero as $t \rightarrow \infty$. Consequently, the output $\mathbf{y}(t)$ will track asymptotically the reference signal $\mathbf{r}(t)$. The tracking and disturbance rejection are accomplished by inserting the internal model $\phi^{-1}(s)\mathbf{I}_p$. If there is no perturbation in the internal model, the tracking property holds for any plant and compensator parameter perturbations, even large ones, as long as the unity-feedback system remains BIBO stable. Thus the design is robust. This establishes the theorem.   Q.E.D.

In the robust design, because of the internal model, $\phi(s)$ becomes zeros of every nonzero entry of the transfer matrices from $\mathbf{w}$ to $\mathbf{y}$ and from $\mathbf{r}$ to $\mathbf{e}$. Such zeros are called *blocking zeros*. These blocking zeros cancel all unstable poles of $\hat{\mathbf{w}}(s)$ and $\hat{\mathbf{r}}(s)$; thus the responses due to these unstable poles are completely blocked at the output. It is clear that every blocking zero is a transmission zero. The converse, however, is not true. To conclude this section, we mention that if we use a right coprime fraction for $\hat{\mathbf{G}}(s)$, insert an internal model and stabilize it, we can show only disturbance rejection. Because of the noncommutative property of matrices, we are not able to establish robust tracking. However, it is believed that the system still achieves robust tracking. The design discussed in Section 9.2.3 can also be extended to the multivariable case; the design, however, will be more complex and will not be discussed.

9.5 Multivariable Model Matching—Two-Parameter Configuration

In this section, we extend the SISO model matching to the multivariable case. We study only plants with square and nonsingular strictly proper transfer matrices. As in the SISO case, given a plant transfer matrix $\hat{\mathbf{G}}(s)$, a model $\hat{\mathbf{G}}_o(s)$ is said to be *implementable* if there exists a no-plant-leakage configuration and proper compensators so that the resulting system has the overall transfer matrix $\hat{\mathbf{G}}_o(s)$ and is totally stable and well posed. The next theorem extends Theorem 9.4 to the matrix case.

▶ **Theorem 9.M4**

Consider a plant with $p \times p$ strictly proper transfer matrix $\hat{\mathbf{G}}(s)$. Then a $p \times p$ transfer matrix $\hat{\mathbf{G}}_o(s)$ is implementable if and only if $\hat{\mathbf{G}}_o(s)$ and

$$\hat{\mathbf{T}}(s) := \hat{\mathbf{G}}^{-1}(s)\hat{\mathbf{G}}_o(s) \tag{9.65}$$

are proper and BIBO stable.[4]

For any no-plant-leakage configuration, the closed-loop transfer matrix from $\mathbf{r}$ to $\mathbf{u}$ is $\hat{\mathbf{T}}(s)$. Thus well posedness and total stability require $\hat{\mathbf{G}}_o(s)$ and $\hat{\mathbf{T}}(s)$ to be proper and BIBO stable. This shows the necessity of Theorem 9.M4. Let us write (9.65) as

$$\hat{\mathbf{G}}_o(s) = \hat{\mathbf{G}}(s)\hat{\mathbf{T}}(s) \tag{9.66}$$

Then (9.66) can be implemented in the open-loop configuration in Fig. 9.1(a) with $\mathbf{C}(s) = \hat{\mathbf{T}}(s)$. This design, however, is not acceptable either because it is not totally stable or because it is very sensitive to plant parameter variations. If we implement it in the unity-feedback configuration, we have no freedom in assigning canceled poles. Thus the configuration may not be acceptable. In the unity-feedback configuration, we have

$$\hat{\mathbf{u}}(s) = \mathbf{C}(s)[\hat{\mathbf{r}}(s) - \hat{\mathbf{y}}(s)]$$

Now we extend it to

$$\hat{\mathbf{u}}(s) = \mathbf{C}_1(s)\hat{\mathbf{r}}(s) - \mathbf{C}_2(s)\hat{\mathbf{y}}(s) \tag{9.67}$$

This is a two-degrees-of-freedom configuration. As in the SISO case, we may select $\mathbf{C}_1(s)$ and $\mathbf{C}_2(s)$ to have the same denominator matrix as

$$\mathbf{C}_1(s) = \mathbf{A}^{-1}(s)\mathbf{L}(s) \qquad \mathbf{C}_2(s) = \mathbf{A}^{-1}(s)\mathbf{M}(s) \tag{9.68}$$

Then the two-parameter configuration can be plotted as shown in Fig. 9.9. From the figure, we have

$$\hat{\mathbf{u}}(s) = \mathbf{A}^{-1}(s)[\mathbf{L}(s)\hat{\mathbf{r}}(s) - \mathbf{M}(s)\mathbf{N}(s)\mathbf{D}^{-1}(s)\hat{\mathbf{u}}(s)]$$

which implies

---

4. If $\hat{\mathbf{G}}(s)$ is not square, then $\hat{\mathbf{G}}_o(s)$ is implementable if and only if $\hat{\mathbf{G}}_o(s)$ is proper, is BIBO stable, and can be expressed as $\hat{\mathbf{G}}_o(s) = \hat{\mathbf{G}}(s)\hat{\mathbf{T}}(s)$, where $\hat{\mathbf{T}}(s)$ is proper and BIBO stable. See Reference [6, pp. 517–523].

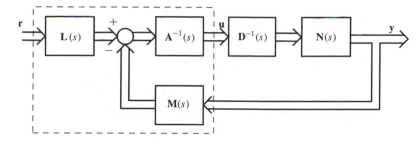

Figure 9.9 Two-parameter configuration.

$$\hat{\mathbf{u}}(s) = [\mathbf{I} + \mathbf{A}^{-1}(s)\mathbf{M}(s)\mathbf{N}(s)\mathbf{D}^{-1}(s)]^{-1}\mathbf{A}^{-1}(s)\mathbf{L}(s)\hat{\mathbf{r}}(s)$$

$$= \mathbf{D}(s)[\mathbf{A}(s)\mathbf{D}(s) + \mathbf{M}(s)\mathbf{N}(s)]^{-1}\mathbf{L}(s)\hat{\mathbf{r}}(s)$$

Thus we have

$$\hat{\mathbf{y}}(s) = \mathbf{N}(s)\mathbf{D}^{-1}(s)\hat{\mathbf{u}}(s) = \mathbf{N}(s)[\mathbf{A}(s)\mathbf{D}(s) + \mathbf{M}(s)\mathbf{N}(s)]^{-1}\mathbf{L}(s)\hat{\mathbf{r}}(s)$$

and the transfer matrix from **r** to **y** is

$$\hat{\mathbf{G}}_o(s) = \mathbf{N}(s)[\mathbf{A}(s)\mathbf{D}(s) + \mathbf{M}(s)\mathbf{N}(s)]^{-1}\mathbf{L}(s) \tag{9.69}$$

Thus model matching becomes the problem of solving $\mathbf{A}(s)$, $\mathbf{M}(s)$ and $\mathbf{L}(s)$ in (9.69).

Problem *Given a $p \times p$ strictly proper rational matrix $\hat{\mathbf{G}}(s) = \mathbf{N}(s)\mathbf{D}^{-1}(s)$, where $\mathbf{N}(s)$ and $\mathbf{D}(s)$ are right coprime and $\mathbf{D}(s)$ is column reduced with column degrees μ_i, $i = 1, 2, \ldots, p$, and given any implementable $\hat{\mathbf{G}}_o(s)$, find proper compensators $\mathbf{A}^{-1}(s)\mathbf{L}(s)$ and $\mathbf{A}^{-1}(s)\mathbf{M}(s)$ in Fig. 9.9 to implement $\hat{\mathbf{G}}_o(s)$.*

▶ **Procedure 9.M1**

1. Compute

$$\mathbf{N}^{-1}(s)\hat{\mathbf{G}}_o(s) = \bar{\mathbf{F}}^{-1}(s)\bar{\mathbf{E}}(s) \tag{9.70}$$

where $\bar{\mathbf{F}}(s)$ and $\bar{\mathbf{E}}(s)$ are left coprime and $\bar{\mathbf{F}}(s)$ is row reduced.

2. Compute the row index ν of $\hat{\mathbf{G}}(s) = \mathbf{N}(s)\mathbf{D}^{-1}(s)$. This can be achieved by using QR decomposition.
3. Select

$$\hat{\mathbf{F}}(s) = \text{diag}(\alpha_1(s), \alpha_2(s), \ldots, \alpha_p(s)) \tag{9.71}$$

where $\alpha_i(s)$ are arbitrary Hurwitz polynomials, such that $\hat{\mathbf{F}}(s)\bar{\mathbf{F}}(s)$ is row–column reduced with column degrees μ_i and row degrees m_i with

$$m_i \geq \nu - 1 \tag{9.72}$$

for $i = 1, 2, \ldots, p$.

4. Set

$$\mathbf{L}(s) = \hat{\mathbf{F}}(s)\bar{\mathbf{E}}(s) \tag{9.73}$$

and solve $\mathbf{A}(s)$ and $\mathbf{M}(s)$ from

$$\mathbf{A}(s)\mathbf{D}(s) + \mathbf{M}(s)\mathbf{N}(s) = \hat{\mathbf{F}}(s)\bar{\mathbf{F}}(s) =: \mathbf{F}(s) \tag{9.74}$$

Then proper compensators $\mathbf{A}^{-1}(s)\mathbf{L}(s)$ and $\mathbf{A}^{-1}(s)\mathbf{M}(s)$ can be obtained to achieve the model matching.

This procedure reduces to Procedure 9.1 if $\hat{\mathbf{G}}(s)$ is scalar. We first justify the procedure. Substituting (9.73) and (9.74) into (9.69) yields

$$\hat{\mathbf{G}}_o(s) = \mathbf{N}(s)[\hat{\mathbf{F}}(s)\bar{\mathbf{F}}(s)]^{-1}\hat{\mathbf{F}}(s)\bar{\mathbf{E}}(s) = \mathbf{N}(s)\bar{\mathbf{F}}^{-1}(s)\bar{\mathbf{E}}(s)$$

This is (9.70). Thus the compensators implement $\hat{\mathbf{G}}_o(s)$. Define

$$\mathbf{H}_c(s) = \mathrm{diag}(s^{\mu_1}, s^{\mu_2}, \ldots, s^{\mu_p}) \qquad \mathbf{H}_r(s) = \mathrm{diag}(s^{m_1}, s^{m_2}, \ldots, s^{m_p})$$

By assumption, the matrix

$$\lim_{s \to \infty} \mathbf{H}_r^{-1}(s)\mathbf{F}(s)\mathbf{H}_c^{-1}(s) =: \mathbf{F}_h$$

is a nonsingular constant matrix. Thus solutions $\mathbf{A}(s)$, having row degrees m_i and being row reduced, and $\mathbf{M}(s)$, having row degrees m_i or less, exist in (9.74) (Theorem 9.M2). Thus $\mathbf{A}^{-1}(s)\mathbf{M}(s)$ is proper. To show that $\mathbf{A}^{-1}(s)\mathbf{L}(s)$ is proper, we consider

$$\begin{aligned}
\hat{\mathbf{T}}(s) &= \hat{\mathbf{G}}^{-1}(s)\hat{\mathbf{G}}_o(s) = \mathbf{D}(s)\mathbf{N}^{-1}(s)\hat{\mathbf{G}}_o(s) \\
&= \mathbf{D}(s)\bar{\mathbf{F}}^{-1}(s)\bar{\mathbf{E}}(s) = \mathbf{D}(s)[\hat{\mathbf{F}}(s)\bar{\mathbf{F}}(s)]^{-1}\hat{\mathbf{F}}(s)\bar{\mathbf{E}}(s) \\
&= \mathbf{D}(s)[\mathbf{A}(s)\mathbf{D}(s) + \mathbf{M}(s)\mathbf{N}(s)]^{-1}\mathbf{L}(s) \\
&= \mathbf{D}(s)\left\{\mathbf{A}(s)[\mathbf{I} + \mathbf{A}^{-1}(s)\mathbf{M}(s)\mathbf{N}(s)\mathbf{D}^{-1}(s)]\mathbf{D}(s)\right\}^{-1}\mathbf{L}(s) \\
&= [\mathbf{I} + \mathbf{A}^{-1}(s)\mathbf{M}(s)\hat{\mathbf{G}}(s)]^{-1}\mathbf{A}^{-1}(s)\mathbf{L}(s)
\end{aligned}$$

which implies, because $\hat{\mathbf{G}}(s) = \mathbf{N}(s)\mathbf{D}^{-1}(s)$ is strictly proper and $\mathbf{A}^{-1}(s)\mathbf{M}(s)$ is proper,

$$\lim_{s \to \infty} \hat{\mathbf{T}}(s) = \lim_{s \to \infty} \mathbf{A}^{-1}(s)\mathbf{L}(s)$$

Because $\mathbf{T}(\infty)$ is finite by assumption, we conclude that $\mathbf{A}^{-1}(s)\mathbf{L}(s)$ is proper. If $\hat{\mathbf{G}}(s)$ is strictly proper and if all compensators are proper, then the two-parameter configuration is automatically well posed. The design involves the pole–zero cancellation of $\hat{\mathbf{F}}(s)$, which we can select. If $\hat{\mathbf{F}}(s)$ is selected as diagonal with Hurwitz polynomials as its entries, then pole–zero cancellations involve only stable poles and the system is totally stable. This completes the justification of the design procedure.

EXAMPLE 9.12 Consider a plant with transfer matrix

$$\hat{\mathbf{G}}(s) = \begin{bmatrix} 1/s^2 & 1/s \\ 0 & 1/s \end{bmatrix} = \begin{bmatrix} 1 & 1 \\ 0 & 1 \end{bmatrix} \begin{bmatrix} s^2 & 0 \\ 0 & s \end{bmatrix}^{-1} \tag{9.75}$$

It has column degrees $\mu_1 = 2$ and $\mu_2 = 1$. Let us select a model as

$$\hat{\mathbf{G}}_o(s) = \begin{bmatrix} \dfrac{2}{s^2 + 2s + 2} & 0 \\ 0 & \dfrac{2}{s^2 + 2s + 2} \end{bmatrix} \tag{9.76}$$

which is proper and BIBO stable. To check whether or not $\hat{\mathbf{G}}_o(s)$ is implementable, we compute

$$\hat{\mathbf{T}}(s) := \hat{\mathbf{G}}^{-1}(s)\hat{\mathbf{G}}_o(s) = \begin{bmatrix} s^2 & -s^2 \\ 0 & s \end{bmatrix} \hat{\mathbf{G}}_o(s) = \begin{bmatrix} \dfrac{2s^2}{s^2 + 2s + 2} & \dfrac{-2s^2}{s^2 + 2s + 2} \\ 0 & \dfrac{2s}{s^2 + 2s + 2} \end{bmatrix}$$

which is proper and BIBO stable. Thus $\hat{\mathbf{G}}_o(s)$ is implementable. We compute

$$\mathbf{N}^{-1}(s)\hat{\mathbf{G}}_o(s) = \begin{bmatrix} 1 & -1 \\ 0 & 1 \end{bmatrix} \begin{bmatrix} \dfrac{2}{s^2 + 2s + 2} & 0 \\ 0 & \dfrac{2}{s^2 + 2s + 2} \end{bmatrix}$$

$$= \begin{bmatrix} \dfrac{2}{s^2 + 2s + 2} & \dfrac{-2}{s^2 + 2s + 2} \\ 0 & \dfrac{2}{s^2 + 2s + 2} \end{bmatrix}$$

$$= \begin{bmatrix} s^2 + 2s + 2 & 0 \\ 0 & s^2 + 2s + 2 \end{bmatrix}^{-1} \begin{bmatrix} 2 & -2 \\ 0 & 2 \end{bmatrix}$$

$$=: \bar{\mathbf{F}}^{-1}(s)\bar{\mathbf{E}}(s)$$

For this example, the degree of $\mathbf{N}^{-1}(s)\hat{\mathbf{G}}_o(s)$ can easily be computed as 4. The determinant of $\bar{\mathbf{F}}(s)$ has degree 4; thus the pair $\bar{\mathbf{F}}(s)$ and $\bar{\mathbf{E}}(s)$ are left coprime. It is clear that $\bar{\mathbf{F}}(s)$ is row reduced with row degrees $r_1 = r_2 = 2$. The row index of $\hat{\mathbf{G}}(s)$ was computed in Example 9.10 as $\nu = 2$. Let us select

$$\hat{\mathbf{F}}(s) = \mathrm{diag}((s + 2), 1)$$

Then we have

$$\hat{\mathbf{F}}(s)\bar{\mathbf{F}}(s) = \begin{bmatrix} (s^2 + 2s + 2)(s + 2) & 0 \\ 0 & s^2 + 2s + 2 \end{bmatrix}$$

$$= \begin{bmatrix} 4 + 6s + 4s^2 + s^3 & 0 \\ 0 & 2 + 2s + s^2 \end{bmatrix} \tag{9.77}$$

It is row–column reduced with row degrees $\{m_1 = m_2 = 1 = \nu - 1\}$ and column degrees $\{\mu_1 = 2, \mu_2 = 1\}$. Note that without introducing $\hat{\mathbf{F}}(s)$, proper compensators may not exist. We set

$$\mathbf{L}(s) = \hat{\mathbf{F}}(s)\bar{\mathbf{E}}(s) = \begin{bmatrix} s + 2 & 0 \\ 0 & 1 \end{bmatrix} \begin{bmatrix} 2 & -2 \\ 0 & 2 \end{bmatrix}$$

$$= \begin{bmatrix} 2(s + 2) & -2(s + 2) \\ 0 & 2 \end{bmatrix} \tag{9.78}$$

and solve $\mathbf{A}(s)$ and $\mathbf{M}(s)$ from

$$\mathbf{A}(s)\mathbf{D}(s) + \mathbf{M}(s)\mathbf{N}(s) = \hat{\mathbf{F}}(s)\bar{\mathbf{F}}(s) =: \mathbf{F}(s) \tag{9.79}$$

From the coefficient matrices of $\mathbf{D}(s)$ and $\mathbf{N}(s)$ and the coefficient matrices of (9.77), we can readily form

$$[\mathbf{A}_0 \ \mathbf{M}_0 \ \mathbf{A}_1 \ \mathbf{M}_1]
\begin{bmatrix}
0 & 0 & 0 & 0 & 1 & 0 & 0 & 0 \\
0 & 0 & 0 & 1 & 0 & 0 & 0 & 0 \\
\cdots & \cdots & \cdots & \cdots & \cdots & \cdots & \cdots & \cdots \\
1 & 1 & 0 & 0 & 0 & 0 & 0 & 0 \\
0 & 1 & 0 & 0 & 0 & 0 & 0 & 0 \\
\hline
0 & 0 & 0 & 0 & 0 & 0 & 1 & 0 \\
0 & 0 & 0 & 0 & 0 & 1 & 0 & 0 \\
\cdots & \cdots & \cdots & \cdots & \cdots & \cdots & \cdots & \cdots \\
0 & 0 & 1 & 1 & 0 & 0 & 0 & 0 \\
0 & 0 & 0 & 1 & 0 & 0 & 0 & 0
\end{bmatrix}$$

$$= \begin{bmatrix} 4 & 0 & 6 & 0 & 4 & 0 & 1 & 0 \\ 0 & 2 & 0 & 2 & 0 & 1 & 0 & 0 \end{bmatrix} \tag{9.80}$$

As discussed in Example 9.10, if we delete the last column of $\mathbf{S}_1$, then the remaining $\tilde{\mathbf{S}}_1$ has full column rank and for any $\mathbf{F}(s)$, after deleting the last zero column, solutions exist in (9.80). Now if we delete the last N-row of $\tilde{\mathbf{S}}_1$, which is linearly dependent on its previous row, the set of solutions is unique and can be obtained, using MATLAB, as

$$[\mathbf{A}_0 \ \mathbf{M}_0 \ \mathbf{A}_1 \ \mathbf{M}_1] = \begin{bmatrix} 4 & -6 & 4 & -4 & 1 & 0 & 6 & 0 \\ 0 & 2 & 0 & 2 & 0 & 1 & 0 & 0 \end{bmatrix} \tag{9.81}$$

Note that the last zero column is by assignment because of the deletion of the last N-row in $\tilde{\mathbf{S}}_1$. Thus we have

$$\mathbf{A}(s) = \begin{bmatrix} 4+s & -6 \\ 0 & 2+s \end{bmatrix} \qquad \mathbf{M}(s) = \begin{bmatrix} 4+6s & -4 \\ 0 & 2 \end{bmatrix} \tag{9.82}$$

The two compensators $\mathbf{A}^{-1}(s)\mathbf{M}(s)$ and $\mathbf{A}^{-1}(s)\mathbf{L}(s)$ are clearly proper. This completes the design. As a check, we compute

$$\hat{\mathbf{G}}_o(s) = \mathbf{N}(s)\mathbf{F}^{-1}(s)\mathbf{L}(s) = \begin{bmatrix} \dfrac{2(s+2)}{s^3 + 4s^2 + 6s + 4} & 0 \\ 0 & \dfrac{2}{s^2 + 2s + 2} \end{bmatrix}$$

$$= \begin{bmatrix} \dfrac{2}{s^2 + 2s + 2} & 0 \\ 0 & \dfrac{2}{s^2 + 2s + 2} \end{bmatrix}$$

This is the desired model. Note that the design involves the cancellation of $(s + 2)$, which we can select. Thus the design is satisfactory.

Let us discuss a special case of model matching. Given $\hat{\mathbf{G}}(s) = \mathbf{N}(s)\mathbf{D}^{-1}(s)$, let us select $\hat{\mathbf{T}}(s) = \mathbf{D}(s)\mathbf{D}_f^{-1}(s)$, where $\mathbf{D}_f(s)$ has the same column degrees and the same column-degree coefficient matrix as $\mathbf{D}(s)$. Then $\hat{\mathbf{T}}(s)$ is proper and $\hat{\mathbf{G}}_o(s) = \hat{\mathbf{G}}(s)\hat{\mathbf{T}}(s) = \mathbf{N}(s)\mathbf{D}_f^{-1}(s)$. This is the feedback transfer matrix discussed in (8.72). Thus the state feedback design discussed in Chapter 8 can also be carried out by using Procedure 9.M1.

9.5.1 Decoupling

Consider a $p \times p$ strictly proper rational matrix $\hat{\mathbf{G}}(s) = \mathbf{N}(s)\mathbf{D}^{-1}(s)$. We have assumed that $\hat{\mathbf{G}}(s)$ is nonsingular. Thus $\hat{\mathbf{G}}^{-1}(s) = \mathbf{D}(s)\mathbf{N}^{-1}(s)$ is well defined; however, it is in general improper. Let us select $\hat{\mathbf{T}}(s)$ as

$$\hat{\mathbf{T}}(s) = \hat{\mathbf{G}}^{-1}(s)\,\mathrm{diag}(d_1^{-1}(s), d_2^{-1}(s), \ldots, d_p^{-1}(s)) \tag{9.83}$$

where $d_i(s)$ are Hurwitz polynomials of least degrees to make $\hat{\mathbf{T}}(s)$ proper. Define

$$\boldsymbol{\Sigma}(s) = \mathrm{diag}(d_1(s), d_2(s), \ldots, d_p(s)) \tag{9.84}$$

Then we can write $\hat{\mathbf{T}}(s)$ as

$$\hat{\mathbf{T}}(s) = \mathbf{D}(s)\mathbf{N}^{-1}(s)\boldsymbol{\Sigma}^{-1}(s) = \mathbf{D}(s)[\boldsymbol{\Sigma}(s)\mathbf{N}(s)]^{-1} \tag{9.85}$$

If all transmission zeros of $\hat{\mathbf{G}}(s)$ or, equivalently, all roots of $\det \mathbf{N}(s)$ have negative real parts, then $\hat{\mathbf{T}}(s)$ is proper and BIBO stable. Thus the overall transfer matrix

$$\begin{aligned} \hat{\mathbf{G}}_o(s) = \hat{\mathbf{G}}(s)\hat{\mathbf{T}}(s) &= \mathbf{N}(s)\mathbf{D}^{-1}(s)\mathbf{D}(s)[\boldsymbol{\Sigma}(s)\mathbf{N}(s)]^{-1} \\ &= \mathbf{N}(s)[\boldsymbol{\Sigma}(s)\mathbf{N}(s)]^{-1} = \boldsymbol{\Sigma}^{-1}(s) \end{aligned} \tag{9.86}$$

is implementable. This overall transfer matrix is a diagonal matrix and is said to be *decoupled*. This is the design in Example 9.12.

If $\hat{\mathbf{G}}(s)$ has nonminimum-phase transmission zeros or transmission zeros with zero or positive real parts, then the preceding design cannot be employed. However, with some modification, it is still possible to design a decoupled overall system. Consider again $\hat{\mathbf{G}}(s) = \mathbf{N}(s)\mathbf{D}^{-1}(s)$. We factor $\mathbf{N}(s)$ as

$$\mathbf{N}(s) = \mathbf{N}_1(s)\mathbf{N}_2(s)$$

with

$$\mathbf{N}_1(s) = \mathrm{diag}(\beta_{11}(s), \beta_{12}(s), \ldots, \beta_{1p}(s))$$

where $\beta_{1i}(s)$ is the greatest common divisor of the ith row of $\mathbf{N}(s)$. Let us compute $\mathbf{N}_2^{-1}(s)$, and let $\beta_{2i}(s)$ be the least common denominator of the *unstable* poles of the ith column of $\mathbf{N}_2^{-1}(s)$. Define

$$\mathbf{N}_{2d} := \mathrm{diag}(\beta_{21}(s), \beta_{22}(s), \ldots, \beta_{2p}(s))$$

Then the matrix

$$\bar{\mathbf{N}}_2(s) := \mathbf{N}_2^{-1}(s)\mathbf{N}_{2d}(s)$$

has no unstable poles. Now we select $\hat{\mathbf{T}}(s)$ as

$$\hat{\mathbf{T}}(s) = \mathbf{D}(s)\bar{\mathbf{N}}_2(s)\boldsymbol{\Sigma}^{-1}(s) \tag{9.87}$$

with

$$\boldsymbol{\Sigma}(s) = \text{diag}(d_1(s), d_2(s), \ldots, d_p(s))$$

where $d_i(s)$ are Hurwitz polynomials of least degrees to make $\hat{\mathbf{T}}(s)$ proper. Because $\bar{\mathbf{N}}_2(s)$ has only stable poles, and $d_i(s)$ are Hurwitz, $\hat{\mathbf{T}}(s)$ is BIBO stable. Consider

$$\begin{aligned}\hat{\mathbf{G}}_o(s) = \hat{\mathbf{G}}(s)\hat{\mathbf{T}}(s) &= \mathbf{N}_1(s)\mathbf{N}_2(s)\mathbf{D}^{-1}(s)\mathbf{D}(s)\bar{\mathbf{N}}_2(s)\boldsymbol{\Sigma}^{-1}(s) \\ &= \mathbf{N}_1(s)\mathbf{N}_{2d}(s)\boldsymbol{\Sigma}^{-1}(s) \\ &= \text{diag}\left(\frac{\beta_1(s)}{d_1(s)}, \frac{\beta_2(s)}{d_2(s)}, \ldots, \frac{\beta_p(s)}{d_p(s)}\right)\end{aligned} \tag{9.88}$$

where $\beta_i(s) = \beta_{1i}(s)\beta_{2i}(s)$. It is proper because both $\mathbf{T}(s)$ and $\hat{\mathbf{G}}(s)$ are proper. It is clearly BIBO stable. Thus $\hat{\mathbf{G}}_o(s)$ is implementable and is a decoupled system.

EXAMPLE 9.13 Consider

$$\hat{\mathbf{G}}(s) = \mathbf{N}(s)\mathbf{D}^{-1}(s) = \begin{bmatrix} s & 1 \\ s-1 & s-1 \end{bmatrix}\begin{bmatrix} s^3+1 & 1 \\ 0 & s^2 \end{bmatrix}^{-1}$$

We compute $\det \mathbf{N}(s) = (s-1)(s-1) = (s-1)^2$. The plant has two nonminimum-phase transmission zeros. We factor $\mathbf{N}(s)$ as

$$\mathbf{N}(s) = \mathbf{N}_1(s)\mathbf{N}_2(s) = \begin{bmatrix} 1 & 0 \\ 0 & s-1 \end{bmatrix}\begin{bmatrix} s & 1 \\ 1 & 1 \end{bmatrix}$$

with $\mathbf{N}_1(s) = \text{diag}(1, (s-1))$, and compute

$$\mathbf{N}_2^{-1}(s) = \frac{1}{(s-1)}\begin{bmatrix} 1 & -1 \\ -1 & s \end{bmatrix}$$

If we select

$$\mathbf{N}_{2d} = \text{diag}((s-1), (s-1)) \tag{9.89}$$

then the rational matrix

$$\bar{\mathbf{N}}_2(s) = \mathbf{N}_2^{-1}(s)\mathbf{N}_{2d}(s) = \begin{bmatrix} 1 & -1 \\ -1 & s \end{bmatrix}$$

has no unstable poles. We compute

$$\begin{aligned}\mathbf{D}(s)\bar{\mathbf{N}}_2(s) &= \begin{bmatrix} s^3+1 & 1 \\ 0 & s^2 \end{bmatrix}\begin{bmatrix} 1 & -1 \\ -1 & s \end{bmatrix} \\ &= \begin{bmatrix} s^3 & -s^3+s-1 \\ -s^2 & s^3+1 \end{bmatrix}\end{aligned}$$

If we choose

$$\mathbf{\Sigma}(s) = \text{diag}((s^2 + 2s + 2)(s + 2), \ (s^2 + 2s + 2)(s + 2))$$

then

$$\hat{\mathbf{T}}(s) = \mathbf{D}(s)\bar{\mathbf{N}}_2(s)\mathbf{\Sigma}^{-1}(s)$$

is proper. Thus the overall transfer matrix

$$\hat{\mathbf{G}}_o(s) = \hat{\mathbf{G}}(s)\hat{\mathbf{T}}(s) = \text{diag}\left(\frac{s-1}{(s^2 + 2s + 2)(s + 2)}, \ \frac{(s-1)^2}{(s^2 + 2s + 2)(s + 2)}\right)$$

is implementable. This is a decoupled system. This decoupled system will not track any step reference input. Thus we modify it as

$$\hat{\mathbf{G}}_o(s) = \text{diag}\left(\frac{-4(s-1)}{(s^2 + 2s + 2)(s + 2)}, \ \frac{4(s-1)^2}{(s^2 + 2s + 2)(s + 2)}\right) \tag{9.90}$$

which has $\hat{\mathbf{G}}_o(0) = \mathbf{I}$ and will track any step reference input.

Next we implement (9.90) in the two-parameter configuration. We follow Procedure 9.M1. To save space, we define $d(s) := (s^2 + 2s + 2)(s + 2)$. First we compute

$$\mathbf{N}^{-1}(s)\hat{\mathbf{G}}_o(s) = \begin{bmatrix} s & 1 \\ s-1 & s-1 \end{bmatrix}^{-1} \begin{bmatrix} \dfrac{-4(s-1)}{d(s)} & 0 \\ 0 & \dfrac{4(s+1)^2}{d(s)} \end{bmatrix}$$

$$= \frac{1}{(s-1)^2} \begin{bmatrix} s-1 & -1 \\ 1-s & s \end{bmatrix} \begin{bmatrix} \dfrac{-4(s-1)}{d(s)} & 0 \\ 0 & \dfrac{4(s-1)^2}{d(s)} \end{bmatrix}$$

$$= \begin{bmatrix} \dfrac{-4}{d(s)} & \dfrac{-4}{d(s)} \\ \dfrac{4}{d(s)} & \dfrac{4s}{d(s)} \end{bmatrix} = \begin{bmatrix} d(s) & 0 \\ 0 & d(s) \end{bmatrix}^{-1} \begin{bmatrix} -4 & -4 \\ 4 & 4s \end{bmatrix}$$

$$=: \bar{\mathbf{F}}^{-1}(s)\bar{\mathbf{E}}(s)$$

It is a left coprime fraction.

From the plant transfer matrix, we have

$$\mathbf{D}(s) = \begin{bmatrix} 1 & 1 \\ 0 & 0 \end{bmatrix} + \begin{bmatrix} 0 & 0 \\ 0 & 0 \end{bmatrix} s + \begin{bmatrix} 0 & 0 \\ 0 & 1 \end{bmatrix} s^2 + \begin{bmatrix} 1 & 0 \\ 0 & 0 \end{bmatrix} s^3$$

and

$$\mathbf{N}(s) = \begin{bmatrix} 0 & 1 \\ -1 & -1 \end{bmatrix} + \begin{bmatrix} 1 & 0 \\ 1 & 1 \end{bmatrix} s + \begin{bmatrix} 0 & 0 \\ 0 & 0 \end{bmatrix} s^2 + \begin{bmatrix} 0 & 0 \\ 0 & 0 \end{bmatrix} s^3$$

We use QR decomposition to compute the row index of $\hat{\mathbf{G}}(s)$. We type

```
d1=[1 1 0 0 0 0 1 0];d2=[0 0 0 0 0 1 0 0];
n1=[0 1 1 0 0 0 0 0];n2=[-1 -1 1 1 0 0 0 0];
```

```
s2=[d1 0 0 0 0;d2 0 0 0 0;n1 0 0 0 0;n2 0 0 0 0;...
    0 0 d1 0 0;0 0 d2 0 0;0 0 n1 0 0;0 0 n2 0 0;...
    0 0 0 0 d1;0 0 0 0 d2;0 0 0 0 n1;0 0 0 0 n2];
[q,r]=qr(s2')
```

From the matrix r (not shown), we can see that there are three linearly independent $N1$-rows and two linearly independent $N2$-rows. Thus we have $\nu_1 = 3$ and $\nu_2 = 2$ and the row index equals $\nu = 3$. If we select

$$\hat{\mathbf{F}}(s) = \text{diag}\left((s+3)^2, \ (s+3)\right)$$

then $\hat{\mathbf{F}}(s)\bar{\mathbf{F}}(s)$ is row–column reduced with row degrees $\{2, 2\}$ and column degrees $\{3, 2\}$. We set

$$\mathbf{L}(s) = \hat{\mathbf{F}}(s)\bar{\mathbf{E}}(s) = \begin{bmatrix} -4(s+3)^2 & -4(s+3)^2 \\ 4(s+3) & 4s(s+3) \end{bmatrix} \tag{9.91}$$

and solve $\mathbf{A}(s)$ and $\mathbf{M}(s)$ from

$$\mathbf{A}(s)\mathbf{D}(s) + \mathbf{M}(s)\mathbf{N}(s) = \hat{\mathbf{F}}(s)\bar{\mathbf{F}}(s) := \mathbf{F}(s)$$

Using MATLAB, they can be solved as

$$\mathbf{A}(s) = \begin{bmatrix} s^2 + 10s + 329 & 100 \\ -46 & s^2 + 7s + 6 \end{bmatrix} \tag{9.92}$$

and

$$\mathbf{M}(s) = \begin{bmatrix} -290s^2 - 114s - 36 & 189s + 293 \\ 46s^2 + 34s + 12 & -34s - 46 \end{bmatrix} \tag{9.93}$$

The compensators $\mathbf{A}^{-1}(s)\mathbf{M}(s)$ and $\mathbf{A}^{-1}(s)\mathbf{L}(s)$ are clearly proper. This completes the design.

The model matching discussed can be modified in several ways. For example, if stable roots of $\det \mathbf{N}_2(s)$ are not inside the sector shown in Fig. 8.3(a), they can be included in β_{2i}. Then they will be retained in $\hat{\mathbf{G}}_o(s)$ and will not be canceled in the design. Instead of decoupling the plant for each pair of input and output, we may decouple it for a group of inputs and a group of outputs. In this case, the resulting overall transfer matrix is a block-diagonal matrix. These modifications are straightforward and will not be discussed.

9.6 Concluding Remarks

In this chapter, we used coprime fractions to carry out designs to achieve pole placement or model matching. For pole placement, the unity-feedback configuration shown in Fig. 9.1(a), a one-degree-of-freedom configuration, can be used. If a plant has degree n, then any pole placement can be achieved by using a compensator of degree $n - 1$ or larger. If the degree of a compensator is larger than the minimum required, the extra degrees can be used to achieve robust tracking, disturbance rejection, or other design objectives.

Model matching generally involves pole–zero cancellations. One-degree-of-freedom configurations cannot be used here because we have no freedom in selecting canceled poles. Any two-degree-of-freedom configuration can be used because we have freedom in selecting canceled poles. This text discusses only the two-parameter configuration shown in Fig. 9.4.

All designs in this chapter are achieved by solving sets of linear algebraic equations. The basic idea and procedure are the same for both the SISO and MIMO cases. All discussion in this chapter can be applied, without any modification, to discrete-time systems; the only difference is that desired poles must be chosen to lie inside the region shown in Fig. 8.3(b) instead of in Fig. 8.3(a).

This chapter studies only strictly proper $\hat{\mathbf{G}}(s)$. If $\hat{\mathbf{G}}(s)$ is proper, the basic idea and procedure are still applicable, but the degree of compensators must be increased to ensure properness of compensators and well posedness of overall systems. See Reference [6]. The model matching in Section 9.5 can also be extended to nonsquare plant transfer matrices. See also Reference [6].

The controller-estimator design in Chapter 8 can be carried out using polynomial fractions. See References [6, pp. 506–514; 7, pp. 461–465]. Conversely, because of the equivalence of controllable and observable state equations and coprime fractions, we should be able to use state equations to carry out all designs in this chapter. The state-space design, however, will be more complex and less intuitively transparent, as we may conclude from comparing the designs in Sections 8.3.1 and 9.2.2.

The state-space approach first appeared in the 1960s, and by the 1980s the concepts of controllability and observability and controller-estimator design were integrated into most undergraduate control texts. The polynomial fraction approach was developed in the 1970s; its underlying concept of coprimeness, however, is an ancient one. Even though the concepts and design procedures of the coprime fraction approach are as simple as, if not simpler than, the state-space approach, the approach appears to be less widely known. It is hoped that this chapter has demonstrated its simplicity and usefulness and will help in its dissemination.

PROBLEMS

9.1   Consider

$$A(s)D(s) + B(s)N(s) = s^2 + 2s + 2$$

where $D(s)$ and $N(s)$ are given in Example 9.1. Do solutions $A(s)$ and $B(s)$ exist in the equation? Can you find solutions with deg $B(s) \leq$ deg $A(s)$ in the equation?

9.2   Given a plant with transfer function $\hat{g}(s) = (s - 1)/(s^2 - 4)$, find a compensator in the unity-feedback configuration so that the overall system has desired poles at -2 and $-1 \pm j1$. Also find a feedforward gain so that the resulting system will track any step reference input.

9.3   Suppose the plant transfer function in Problem 9.2 changes to $\hat{\bar{g}}(s) = (s - 0.9)/(s^2 - 4.1)$ after the design is completed. Can the overall system still track asymptotically any step reference input? If not, give two different designs, one with a compensator of degree 3 and another with degree 2, that will track asymptotically and robustly any step reference input. Do you need additional desired poles? If yes, place them at -3.

9.4 Repeat Problem 9.2 for a plant with transfer function $\hat{g}(s) = (s-1)/s(s-2)$. Do you need a feedforward gain to achieve tracking of any step reference input? Give your reason.

9.5 Suppose the plant transfer function in Problem 9.4 changes to $\hat{\bar{g}}(s) = (s-0.9)/s(s-2.1)$ after the design is completed. Can the overall system still track any step reference input? Is the design robust?

9.6 Consider a plant with transfer function $\hat{g}(s) = 1/(s-1)$. Suppose a disturbance of form $w(t) = a\sin(2t+\theta)$, with unknown amplitude a and phase θ, enters the plant as shown in Fig. 9.2. Design a biproper compensator of degree 3 in the feedback system so that the output will track asymptotically any step reference input and reject the disturbance. Place the desired poles at $-1 \pm j2$ and $-2 \pm j1$.

9.7 Consider the unity feedback system shown in Fig. 9.10. The plant transfer function is $\hat{g}(s) = 2/s(s+1)$. Can the output track robustly any step reference input? Can the output reject any step disturbance $w(t) = a$? Why?

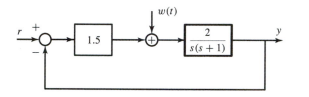

Figure 9.10

9.8 Consider the unity-feedback system shown in Fig. 9.11(a). Is the transfer function from r to y BIBO stable? Is the system totally stable? If not, find an input–output pair whose closed-loop transfer function is not BIBO stable.

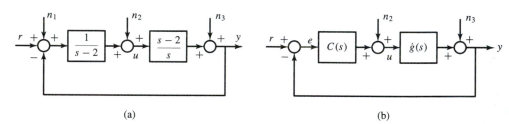

(a)   (b)

Figure 9.11

9.9 Consider the unity-feedback system shown in Fig. 9.11(b). (1) Show that the closed-loop transfer function of every possible input–output pair contains the factor $(1+C(s)\hat{g}(s))^{-1}$. (2) Show that $(1+C(s)\hat{g}(s))^{-1}$ is proper if and only if

$$1 + C(\infty)\hat{g}(\infty) \neq 0$$

(3) Show that if $C(s)$ and $\hat{g}(s)$ are proper, and if $C(\infty)\hat{g}(\infty) \neq -1$, then the system is well posed.

9.10 Given $\hat{g}(s) = (s^2 - 1)/(s^3 + a_1s^2 + a_2s + a_3)$, which of the following $\hat{g}_o(s)$ are implementable for any a_i and b_i.

$$\frac{s-1}{(s+1)^2} \qquad \frac{s+1}{(s+2)(s+3)} \qquad \frac{s^2-1}{(s-2)^3}$$

$$\frac{(s^2-1)}{(s+2)^2} \qquad \frac{(s-1)(b_0s+b_1)}{(s+2)^2(s^2+2s+2)} \qquad \frac{1}{1}$$

9.11 Given $\hat{g}(s) = (s-1)/s(s-2)$, implement the model $\hat{g}_o(s) = -2(s-1)/(s^2+2s+2)$ in the open-loop and the unity-feedback configurations. Are they totally stable? Can the implementations be used in practice?

9.12 Implement Problem 9.11 in the two-parameter configuration. Select the poles to be canceled at $s = -3$. Is $A(s)$ a Hurwitz polynomial? Can you implement the two compensators as shown in Fig. 9.4(a)? Implement the two compensators in Fig. 9.4(d) and draw its op-amp circuit.

9.13 Given a BIBO stable $\hat{g}_o(s)$, show that the steady-state response $y_{ss}(t) := \lim_{t \to \infty} y(t)$ excited by the ramp reference input $r(t) = at$ for $t \geq 0$ is given by

$$y_{ss}(t) = \hat{g}_o(0)at + \hat{g}_o'(0)a$$

Thus if the output is to track asymptotically the ramp reference input we require $\hat{g}_o(0) = 1$ and $\hat{g}_o'(0) = 0$.

9.14 Given a BIBO stable

$$\hat{g}_o(s) = \frac{b_0 + b_1 s + \cdots + b_m s^m}{a_0 + a_1 s + \cdots + a_n s^n}$$

with $n \geq m$, show that $\hat{g}_o(0) = 1$ and $\hat{g}_o'(0) = 0$ if and only if $a_0 = b_0$ and $a_1 = b_1$.

9.15 Given a plant with transfer function $\hat{g}(s) = (s+3)(s-2)/(s^3+2s-1)$, (1) find conditions on b_1, b_0, and a so that

$$\hat{g}_o(s) = \frac{b_1 s + b_0}{s^2 + 2s + a}$$

is implementable; and (2) determine if

$$\hat{g}_o(s) = \frac{(s-2)(b_1 s + b_0)}{(s+2)(s^2+2s+2)}$$

is implementable. Find conditions on b_1 and b_0 so that the overall transfer function will track any ramp reference input.

9.16 Consider a plant with transfer matrix

$$\hat{\mathbf{G}}(s) = \begin{bmatrix} \dfrac{s+1}{s(s-1)} \\[2ex] \dfrac{1}{s^2-1} \end{bmatrix}$$

Find a compensator in the unity-feedback configuration so that the poles of the overall system are located at -2, $-1 \pm j$ and the rest at $s = -3$. Can you find a feedforward gain so that the overall system will track asymptotically any step reference input?

9.17 Repeat Problem 9.16 for a plant with transfer matrix

$$\hat{\mathbf{G}}(s) = \left[\frac{s+1}{s(s-1)} \quad \frac{1}{s^2-1} \right]$$

9.18 Repeat Problem 9.16 for a plant with transfer matrix

$$\hat{\mathbf{G}}(s) = \begin{bmatrix} \dfrac{s-2}{s^2-1} & \dfrac{1}{s-1} \\ \dfrac{1}{s} & \dfrac{2}{s-1} \end{bmatrix}$$

9.19 Given a plant with the transfer matrix in Problem 9.18, is

$$\hat{\mathbf{G}}_o(s) = \begin{bmatrix} \dfrac{4(s^2-4s+1)}{(s^2+2s+2)(s+2)} & 0 \\ 0 & \dfrac{4(s^2-4s+1)}{(s^2+2s+2)(s+2)} \end{bmatrix}$$

implementable? If yes, implement it.

9.20 Diagonalize a plant with transfer matrix

$$\hat{\mathbf{G}}(s) = \begin{bmatrix} 1 & s \\ 1 & 1 \end{bmatrix} \begin{bmatrix} 1 & s^2+1 \\ s & 0 \end{bmatrix}^{-1}$$

Set the denominator of each diagonal entry of the resulting overall system as $s^2 + 2s + 2$.

References

We list only the references used in preparing this edition. The list does not indicate original sources of the results presented. For more extensive lists of references, see References [2, 6, 13, 15].

1. Anderson, B. D. O., and Moore, J. B., *Optimal Control—Linear Quadratic Methods*, Englewood Cliffs, NJ: Prentice Hall, 1990.
2. Antsaklis, A. J., and Michel, A. N., *Linear Systems*, New York: McGraw-Hill, 1997.
3. Callier, F. M., and Desoer, C. A., *Multivariable Feedback Systems*, New York: Springer-Verlag, 1982.
4. Callier, F. M., and Desoer, C. A., *Linear System Theory*, New York: Springer-Verlag, 1991.
5. Chen, C. T., *Introduction to Linear System Theory*, New York: Holt, Rinehart & Winston, 1970.
6. Chen, C. T., *Linear System Theory and Design*, New York: Oxford University Press, 1984.
7. Chen, C. T., *Analog and Digital Control System Design: Transfer Function, State-Space, and Algebraic Methods*, New York: Oxford University Press, 1993.
8. Chen, C. T., and Liu, C. S., "Design of Control Systems: A Comparative Study," *IEEE Control System Magazine*, vol. 14, pp. 47–51, 1994.
9. Doyle, J. C., Francis, B. A., and Tannenbaum, A. R., *Feedback Control Theory*, New York: Macmillan, 1992.
10. Gantmacher, F. R., *The Theory of Matrices*, vols. 1 and 2, New York: Chelsea, 1959.
11. Golub, G. H., and Van Loan, C. F., *Matrix Computations* 3rd ed., Baltimore: The Johns Hopkins University Press, 1996.
12. Howze, J. W., and Bhattacharyya, S. P., "Robust tracking, error feedback, and two-degree-of-freedom controllers," *IEEE Trans. Automatic Control*, vol. 42, pp. 980–983, 1997.
13. Kailath, T., *Linear Systems*, Englewood Cliffs, NJ: Prentice Hall, 1990.
14. Kurcera, V., *Analysis and Design of Discrete Linear Control Systems*, London: Prentice Hall, 1991.
15. Rugh, W., *Linear System Theory*, 2nd ed., Upper Saddle River, NJ: Prentice Hall, 1996.
16. Strang, G., *Linear Algebra and Its Application*, 3rd ed., San Diego: Harcourt, Brace, Javanovich, 1988.
17. *The Student Edition of MATLAB, Version 4*, Englewood Cliffs, NJ: Prentice Hall, 1995.
18. Tongue, B. H., *Principles of Vibration*, New York: Oxford University Press, 1996.
19. Tsui, C. C., *Robust Control System Design*, New York: Marcel Dekker, 1996.
20. Vardulakis, A. I. G., *Linear Multivariable Control*, Chichester: John Wiley, 1991.
21. Vidyasagar, M., *Control System Synthesis: A Factorization Approach*, Cambridge, MA: MIT Press, 1985.
22. Wolovich, W. A., *Linear Multivariable Systems*, New York: Springer-Verlag, 1974.
23. Zhou, K., Doyle, J. C., and Glover, K., *Robust and Optimal Control*, Upper Saddle River, NJ: Prentice Hall, 1995.

Answers to Selected Problems

CHAPTER 2

2.1 (a) Linear. (b) and (c) Nonlinear. In (b), shift the operating point to $(0, y_o)$; then $(u, \bar{y})$ is linear, where $\bar{y} = y - y_o$.

2.3 Linear, time varying, causal.

2.5 No, yes, no for $\mathbf{x}(0) \neq \mathbf{0}$. Yes, yes, yes for $\mathbf{x}(0) = \mathbf{0}$. The reason is that the superposition property must also apply to the initial states.

2.9 $y(t) = 0$ for $t < 0$ and $t > 4$.

$$y(t) = \begin{cases} 0.5t^2 & \text{for } 0 \le t < 1 \\ -1.5t^2 + 4t - 2 & \text{for } 1 \le t \le 2 \\ -y(4-t) & \text{for } 2 < t \le 4 \end{cases}$$

2.10 $\hat{g}(s) = 1/(s+3)$, $g(t) = e^{-3t}$ for $t \ge 0$.

2.12

$$\hat{\mathbf{G}}(s) = \begin{bmatrix} D_{11}(s) & D_{12}(s) \\ D_{21}(s) & D_{22}(s) \end{bmatrix}^{-1} \begin{bmatrix} N_{11}(s) & N_{12}(s) \\ N_{21}(s) & N_{22}(s) \end{bmatrix}$$

2.15 (a) Define $x_1 = \theta$ and $x_2 = \dot{\theta}$. Then $\dot{x}_1 = x_2$, $\dot{x}_2 = -(g/l)\sin x_1 - (u/ml)\cos x_1$. If θ is small, then

$$\dot{\mathbf{x}} = \begin{bmatrix} 0 & 1 \\ -g/l & 0 \end{bmatrix} \mathbf{x} + \begin{bmatrix} 0 \\ -1/mg \end{bmatrix} u$$

It is a linear state equation.
(b) Define $x_1 = \theta_1$, $x_2 = \dot{\theta}_1$, $x_3 = \theta_2$, and $x_4 = \dot{\theta}_2$. Then

$$\dot{x}_1 = x_2$$
$$\dot{x}_2 = -(g/l_1)\sin x_1 + (m_2g/m_1l)\cos x_3 \sin(x_3 - x_1)$$
$$\qquad + (1/m_1l)\sin x_3 \sin(x_3 - x_1) \cdot u$$
$$\dot{x}_3 = x_4$$
$$\dot{x}_4 = -(g/l_2)\sin x_3 + (1/m_2l_2)(\cos x_3)u$$

This is a set of nonlinear equations. If $\theta_i \approx 0$ and $\theta_i\theta_j \approx 0$, then it can be linearized as

$$\dot{\mathbf{x}} = \begin{bmatrix} 0 & 1 & 0 & 0 \\ -g(m_1 + m_2)/m_1 l_1 & 0 & m_2 g/m_1 l_1 & 0 \\ 0 & 0 & 0 & 1 \\ 0 & 0 & -g/l_2 & 0 \end{bmatrix} \mathbf{x} + \begin{bmatrix} 0 \\ 0 \\ 0 \\ 1/m_2 l_2 \end{bmatrix} u$$

2.16 From

$$m\ddot{h} = f_1 - f_2 = k_1 \theta - k_2 u$$
$$I\ddot{\theta} + b\dot{\theta} = (l_1 + l_2) f_2 - l_1 f_1$$

we can readily obtain a state equation. Assuming $I = 0$ and taking their Laplace transforms can yield the transfer function.

2.18 $\hat{g}_1(s) = \hat{y}_1(s)/\hat{u}(s) = 1/(A_1 R_1 s + 1)$, $\hat{g}_2(s) = \hat{y}(s)/\hat{y}_1(s) = 1/(A_2 R_2 s + 1)$. Yes, $\hat{y}(s)/\hat{u}(s) = \hat{g}_1(s)\hat{g}_2(s)$.

2.20

$$\dot{\mathbf{x}} = \begin{bmatrix} 0 & 0 & 1/C_1 \\ 0 & 0 & 1/C_2 \\ -1/L_1 & -1/L_1 & -(R_1 + R_2)/L_1 \end{bmatrix} \mathbf{x}$$
$$+ \begin{bmatrix} 0 & -1/C_1 \\ 0 & 0 \\ 1/L_1 & R_1/L_1 \end{bmatrix} \begin{bmatrix} u_1 \\ u_2 \end{bmatrix}$$
$$y = [-1 \ -1 \ -R_1]\mathbf{x} + [1 \ R_1]\mathbf{u}$$

$$\hat{g}_1(s) = \frac{\hat{y}(s)}{\hat{u}_1(s)} = \frac{s^2 + (R_2/L_1)s}{s^2 + \left(\dfrac{R_1 + R_2}{L_1}\right)s + \left(\dfrac{1}{C_1} + \dfrac{1}{C_2}\right)\dfrac{1}{L_1}}$$

$$\hat{g}_2(s) = \frac{\hat{y}(s)}{\hat{u}_2(s)} = \frac{(R_1 s + (1/C_1))(s + (R_2/L_1))}{s^2 + \left(\dfrac{R_1 + R_2}{L_1}\right)s + \left(\dfrac{1}{C_1} + \dfrac{1}{C_2}\right)\dfrac{1}{L_1}}$$

$$\hat{y}(s) = \hat{g}_1(s)\hat{u}_1(s) + \hat{g}_2(s)\hat{u}_2(s)$$

CHAPTER 3

3.1 $\left[\frac{1}{3} \ \frac{8}{3}\right]'$, $[-2 \ 1.5]'$.

3.3

$$\mathbf{q}_1 = \frac{1}{3.74}\begin{bmatrix} 2 \\ -3 \\ 1 \end{bmatrix} \qquad \mathbf{q}_2 = \frac{1}{1.732}\begin{bmatrix} 1 \\ 1 \\ 1 \end{bmatrix}$$

3.5 $\rho(\mathbf{A}_1) = 2$, nullity$(\mathbf{A}_1)$=1; 3, 0; 3, 1.

3.7 $\mathbf{x} = [1 \; 1]$ is a solution. Unique. No solution if $y = [1 \; 1 \; 1]'$.

3.9 $\alpha_1 = 4/11, \alpha_2 = 16/11$. The solution

$$\mathbf{x} = \left[\frac{4}{11} \quad \frac{-8}{11} \quad \frac{-4}{11} \quad \frac{-16}{11} \right]'$$

has the smallest 2-norm.

3.12

$$\bar{\mathbf{A}} = \begin{bmatrix} 0 & 0 & 0 & -8 \\ 1 & 0 & 0 & 20 \\ 0 & 1 & 0 & -18 \\ 0 & 0 & 1 & 7 \end{bmatrix}$$

3.13

$$\mathbf{Q}_3 = \begin{bmatrix} 1 & 0 & -1 \\ 0 & 1 & 0 \\ 0 & 0 & 1 \end{bmatrix} \quad \hat{\mathbf{A}}_3 = \begin{bmatrix} 1 & 0 & 0 \\ 0 & 1 & 0 \\ 0 & 0 & 2 \end{bmatrix}$$

$$\mathbf{Q}_4 = \begin{bmatrix} 5 & 4 & 0 \\ 0 & 20 & 1 \\ 0 & -25 & 0 \end{bmatrix} \quad \hat{\mathbf{A}}_4 = \begin{bmatrix} 0 & 1 & 0 \\ 0 & 0 & 1 \\ 0 & 0 & 0 \end{bmatrix}$$

3.18

$$\Delta_1(\lambda) = (\lambda - \lambda_1)^3(\lambda - \lambda_2) \quad \psi_1(\lambda) = \Delta_1(\lambda)$$
$$\Delta_2(\lambda) = (\lambda - \lambda_1)^4 \quad \psi_2(\lambda) = (\lambda - \lambda_1)^3$$
$$\Delta_3(\lambda) = (\lambda - \lambda_1)^4 \quad \psi_3(\lambda) = (\lambda - \lambda_1)^2$$
$$\Delta_4(\lambda) = (\lambda - \lambda_1)^4 \quad \psi_4(\lambda) = (\lambda - \lambda_1)$$

3.21

$$\mathbf{A}^{10} = \begin{bmatrix} 1 & 1 & 9 \\ 0 & 0 & 1 \\ 0 & 0 & 1 \end{bmatrix} \quad \mathbf{A}^{103} = \begin{bmatrix} 1 & 1 & 102 \\ 0 & 0 & 1 \\ 0 & 0 & 1 \end{bmatrix}$$

$$e^{\mathbf{A}t} = \begin{bmatrix} e^t & e^t - 1 & te^t - e^t + 1 \\ 0 & 1 & e^t - 1 \\ 0 & 0 & e^t \end{bmatrix}$$

3.22

$$e^{\mathbf{A}_4 t} = \begin{bmatrix} 1 & 4t + 2.5t^2 & 3t + 2t^2 \\ 0 & 1 + 20t & 16t \\ 0 & -25t & 1 - 20t \end{bmatrix}$$

3.24

$$\mathbf{B} = \begin{bmatrix} \ln \lambda_1 & 0 & 0 \\ 0 & \ln \lambda_2 & 0 \\ 0 & 0 & \ln \lambda_3 \end{bmatrix}$$

$$\mathbf{B} = \begin{bmatrix} \ln \lambda & 1/\lambda & 0 \\ 0 & \ln \lambda & 0 \\ 0 & 0 & \ln \lambda \end{bmatrix}$$

3.32 Eigenvalues: 0, 0. No solution for $\mathbf{C}_1$. For any m_1, $[m_1 \ \ 3 - m_1]'$ is a solution for $\mathbf{C}_2$.

3.34 $\sqrt{6}$, 1. 4.7, 1.7.

CHAPTER 4

4.2 $y(t) = 5e^{-t} \sin t$ for $t \geq 0$.

4.3 For $T = \pi$,

$$\mathbf{x}[k+1] = \begin{bmatrix} -0.0432 & 0 \\ 0 & -0.0432 \end{bmatrix} \mathbf{x}[k] + \begin{bmatrix} 1.5648 \\ -1.0432 \end{bmatrix} u[k]$$

$$y[k] = [2 \ \ 3]\mathbf{x}[k]$$

4.5 MATLAB yields $|y|_{max} = 0.55$, $|x_1|_{max} = 0.5$, $|x_2|_{max} = 1.05$, and $|x_3|_{max} = 0.52$ for unit step input. Define $\bar{x}_1 = x_1, \bar{x}_2 = 0.5x_2$, and $\bar{x}_3 = x_3$. Then

$$\dot{\bar{\mathbf{x}}} = \begin{bmatrix} -2 & 0 & 0 \\ 0.5 & 0 & 0.5 \\ 0 & -4 & -2 \end{bmatrix} \bar{\mathbf{x}} + \begin{bmatrix} 1 \\ 0 \\ 1 \end{bmatrix} u \qquad y = [1 \ \ -2 \ \ 0]\bar{\mathbf{x}}$$

The largest permissible a is $10/0.55 = 18.2$.

4.8 They are not equivalent but are zero-state equivalent.

4.11 Using (4.34), we have

$$\dot{\mathbf{x}} = \begin{bmatrix} -3 & 0 & -2 & 0 \\ 0 & -3 & 0 & -2 \\ 1 & 0 & 0 & 0 \\ 0 & 1 & 0 & 0 \end{bmatrix} \mathbf{x} + \begin{bmatrix} 1 & 0 \\ 0 & 1 \\ 0 & 0 \\ 0 & 0 \end{bmatrix} u$$

$$y = \begin{bmatrix} 2 & 2 & 4 & -3 \\ -3 & -2 & -6 & -2 \end{bmatrix} \mathbf{x} + \begin{bmatrix} 0 & 0 \\ 1 & 1 \end{bmatrix} u$$

4.13

$$\dot{\mathbf{x}} = \begin{bmatrix} -3 & 1 & 0 & 0 \\ -2 & 0 & 0 & 0 \\ 0 & 0 & -3 & 1 \\ 0 & 0 & -2 & 0 \end{bmatrix} \mathbf{x} + \begin{bmatrix} 2 & 2 \\ 4 & -3 \\ -3 & -2 \\ -6 & -2 \end{bmatrix} u$$

$$y = \begin{bmatrix} 1 & 0 & 0 & 0 \\ 0 & 0 & 1 & 0 \end{bmatrix} x + \begin{bmatrix} 0 & 0 \\ 1 & 1 \end{bmatrix} u$$

Both have dimension 4.

4.16

$$X(t) = \begin{bmatrix} 1 & \int_0^t e^{0.5\tau^2} d\tau \\ 0 & e^{0.5t^2} \end{bmatrix}$$

$$\Phi(t, t_0) = \begin{bmatrix} 1 & -e^{0.5t^2} \int_{t_0}^t e^{0.5\tau^2} d\tau \\ 0 & e^{0.5(t^2-t_0^2)} \end{bmatrix}$$

$$X(t) = \begin{bmatrix} e^{-t} & e^t \\ 0 & 2e^{-t} \end{bmatrix}$$

$$\Phi(t, t_0) = \begin{bmatrix} e^{-(t-t_0)} & 0.5(e^t e^{t_0} - e^{-t} e^{t_0}) \\ 0 & e^{-(t-t_0)} \end{bmatrix}$$

4.20

$$\Phi(t, t_0) = \begin{bmatrix} e^{\cos t - \cos t_0} & 0 \\ 0 & e^{-\sin t + \sin t_0} \end{bmatrix}$$

4.23 Let $\bar{x}(t) = P(t)x(t)$ with

$$P(t) = \begin{bmatrix} e^{-\cos t} & 0 \\ 0 & e^{\sin t} \end{bmatrix}$$

Then $\dot{\bar{x}}(t) = 0 \cdot \bar{x} = 0$.

4.25

$$\dot{x} = 0 \cdot x + \begin{bmatrix} t^2 e^{-\lambda t} \\ -2t e^{-\lambda t} \\ e^{-\lambda t} \end{bmatrix} u \quad y = [e^{\lambda t} \quad t e^{\lambda t} \quad t^2 e^{\lambda t}]x$$

$$\dot{x} = \begin{bmatrix} 3\lambda & -3\lambda^2 & \lambda^3 \\ 1 & 0 & 0 \\ 0 & 1 & 0 \end{bmatrix} x + \begin{bmatrix} 1 \\ 0 \\ 0 \end{bmatrix} u \quad y = [0 \ 0 \ 2]x$$

CHAPTER 5

5.1 The bounded input $u(t) = \sin t$ excites $y(t) = 0.5t \sin t$, which is not bounded. Thus the system is not BIBO stable.

5.3 No. Yes.

5.6 If $u(t) = 3$, then $y(t) \to -6$. If $u(t) = \sin 2t$, then $y(t) \to 1.26 \sin(2t + 1.25)$.

5.8 Yes.

5.10 Not asymptotically stable. Its minimal polynomial can be found from its Jordan form as $\psi(\lambda) = \lambda(\lambda + 1)$. $\lambda = 0$ is a simple eigenvalue, thus the equation is marginally stable.

5.13 Not asymptotically stable. Its minimal polynomial can be found from its Jordan form as $\psi(\lambda) = (\lambda - 1)^2(\lambda + 1)$. $\lambda = 1$ is a repeated eigenvalue, thus the equation is not marginally stable.

5.15 If $\mathbf{N} = \mathbf{I}$, then

$$\mathbf{M} = \begin{bmatrix} 2.2 & 1.6 \\ 1.6 & 4.8 \end{bmatrix}$$

It is positive definite; thus all eigenvalues have magnitudes less than 1.

5.17 Because $\mathbf{x}'\mathbf{Mx} = \mathbf{x}'[0.5(\mathbf{M}+\mathbf{M}')]\mathbf{x}$, the only way to check positive definiteness of nonsymmetric $\mathbf{M}$ is to check positive definiteness of symmetric $0.5(\mathbf{M} + \mathbf{M}')$.

5.20 Both are BIBO stable.

5.22 BIBO stable, marginally stable, not asymptotically stable. $P(t)$ is not a Lyapunov transformation.

CHAPTER 6

6.2 Controllable, observable.

6.5

$$\dot{\mathbf{x}} = \begin{bmatrix} -1 & 0 \\ 0 & -1 \end{bmatrix} \mathbf{x} + \begin{bmatrix} 1 \\ 0 \end{bmatrix} u \qquad y = [0 \quad -1]\mathbf{x} + 2u$$

Not controllable, not observable.

6.7 $\mu_i = 1$ for all i and $\mu = 1$.

6.9 $y = 2u$.

6.14 Controllable, not observable.

6.17 Using x_1 and x_2 yields

$$\dot{\mathbf{x}} = \begin{bmatrix} -2/11 & 0 \\ 3/22 & 0 \end{bmatrix} \mathbf{x} + \begin{bmatrix} -2/11 \\ 3/22 \end{bmatrix} u \qquad y = [-1 \quad -1]\mathbf{x}$$

This two-dimensional equation is not controllable but is observable.
Using x_1, x_2, and x_3 yields

$$\dot{\mathbf{x}} = \begin{bmatrix} -2/11 & 0 & 0 \\ 3/22 & 0 & 0 \\ 1/22 & 0 & 0 \end{bmatrix} \mathbf{x} + \begin{bmatrix} -2/11 \\ 3/22 \\ 1/22 \end{bmatrix} u \qquad y = [0 \ 0 \ 1]\mathbf{x}$$

This three-dimensional equation is not controllable and not observable.

6.19 For $T = \pi$, not controllable and not observable.

6.21 Not controllable at any t. Observable at every t.

CHAPTER 7

7.1

$$\dot{\mathbf{x}} = \begin{bmatrix} -2 & 1 & 2 \\ 1 & 0 & 0 \\ 0 & 1 & 0 \end{bmatrix} \mathbf{x} + \begin{bmatrix} 1 \\ 0 \\ 0 \end{bmatrix} u \qquad y = [0 \ 1 \ -1] \mathbf{x}$$

Not observable.

7.3

$$\dot{\mathbf{x}} = \begin{bmatrix} -2 & 1 & 2 & a_1 \\ 1 & 0 & 0 & a_2 \\ 0 & 1 & 0 & a_3 \\ 0 & 0 & 0 & a_4 \end{bmatrix} \mathbf{x} + \begin{bmatrix} 1 \\ 0 \\ 0 \\ 0 \end{bmatrix} u \qquad y = [0 \ 1 \ -1 \ c_4] \mathbf{x}$$

For any a_i and c_4, it is an uncontrollable and unobservable realization.

$$\dot{\mathbf{x}} = \begin{bmatrix} -3 & -2 \\ 1 & 0 \end{bmatrix} \mathbf{x} + \begin{bmatrix} 1 \\ 0 \end{bmatrix} u \qquad y = [0 \ 1] \mathbf{x}$$

A controllable and observable realization.

7.5

Solve the monic null vector of

$$\begin{bmatrix} -1 & -1 & 0 & 0 \\ 0 & 2 & -1 & -1 \\ 4 & 0 & 0 & 2 \\ 0 & 0 & 4 & 0 \end{bmatrix} \begin{bmatrix} -N_0 \\ D_0 \\ -N_1 \\ D_1 \end{bmatrix} = \mathbf{0}$$

as $[-0.5 \ 0.5 \ 0 \ 1]'$. Thus $0.5/(0.5 + s) = 1/(2s + 1)$.

7.10

$$\dot{\mathbf{x}} = \begin{bmatrix} 0 & 1 \\ -2 & -1 \end{bmatrix} \mathbf{x} + \begin{bmatrix} 0 \\ 1 \end{bmatrix} u \qquad y = [1 \ 0] \mathbf{x}$$

7.13

$$\Delta_1(s) = s(s + 1)(s + 3) \qquad \deg = 3$$
$$\Delta_2(s) = (s + 1)^3 (s + 2)^2 \qquad \deg = 5$$

7.17 Its right coprime fraction can be computed from any left fraction as

$$\hat{\mathbf{G}}(s) = \begin{bmatrix} 2.5 & s + 0.5 \\ 2.5 & s + 2.5 \end{bmatrix} \begin{bmatrix} 0.5s & s^2 + 0.5s \\ s - 0.5 & -0.5 \end{bmatrix}^{-1}$$

or, by interchanging their two columns,

$$\hat{G}(s) = \begin{bmatrix} s+0.5 & 2.5 \\ s+2.5 & 2.5 \end{bmatrix} \begin{bmatrix} s^2+0.5s & 0.5s \\ -0.5 & s-0.5 \end{bmatrix}^{-1}$$

Using the latter, we can obtain

$$\dot{x} = \begin{bmatrix} -0.5 & -0.25 & -0.25 \\ 1 & 0 & 0 \\ 0 & 0.5 & 0.5 \end{bmatrix} x + \begin{bmatrix} 1 & -0.5 \\ 0 & 0 \\ 0 & 1 \end{bmatrix} u$$

$$y = \begin{bmatrix} 1 & 0.5 & 2.5 \\ 1 & 2.5 & 2.5 \end{bmatrix} x$$

CHAPTER 8

8.1 $k = [4\ 1]$.

8.3   For

$$F = \begin{bmatrix} -1 & 0 \\ 0 & -2 \end{bmatrix} \qquad \bar{k} = [1\ 1]$$

we have

$$T = \begin{bmatrix} 0 & 1/13 \\ 1 & 9/13 \end{bmatrix} \qquad k = \bar{k}T^{-1} = [4\ 1]$$

8.5   Yes, yes, yes.

8.7 $u = pr - kx$ with $k = [15\ 47\ -8]$ and $p = 0.5$.

8.9 $u[k] = pr[k] - kx[k]$ with $k = [1\ 5\ 2]$ and $p = 0.5$. The overall transfer function is

$$\hat{g}_f(z) = \frac{0.5(2z^2 - 8z + 8)}{z^3}$$

and the output excited by $r[k] = a$ is $y[0] = 0$, $y[1] = a$, $y[2] = -3a$, and $y[k] = a$ for $k \geq 3$.

8.11   Two-dimensional estimator:

$$\dot{z} = \begin{bmatrix} -2 & 2 \\ -2 & -2 \end{bmatrix} z + \begin{bmatrix} 0.6282 \\ -0.3105 \end{bmatrix} u + \begin{bmatrix} 1 \\ 0 \end{bmatrix} y$$

$$\hat{x} = \begin{bmatrix} -12 & -27.5 \\ 19 & 32 \end{bmatrix} z$$

One-dimesnional estimator:

$$\dot{z} = -3z + (13/21)u + y$$

$$\hat{x} = \begin{bmatrix} -4 & 21 \\ 5 & -21 \end{bmatrix} \begin{bmatrix} y \\ z \end{bmatrix}$$

8.13 Select

$$
\mathbf{F} = \begin{bmatrix} -4 & 3 & 0 & 0 \\ -3 & 4 & 0 & 0 \\ 0 & 0 & -5 & 4 \\ 0 & 0 & -4 & -5 \end{bmatrix}
$$

If $\bar{\mathbf{K}} = \begin{bmatrix} 1 & 0 & 1 & 0 \\ 0 & 0 & 0 & 0 \end{bmatrix}$ then $\mathbf{K} = \begin{bmatrix} 62.5 & 147 & 20 & 515.5 \\ 0 & 0 & 0 & 0 \end{bmatrix}$

If $\bar{\mathbf{K}} = \begin{bmatrix} 1 & 0 & 0 & 0 \\ 0 & 0 & 1 & 0 \end{bmatrix}$ then $\mathbf{K} = \begin{bmatrix} -606.2 & -168 & -14.2 & -2 \\ 371.1 & 119.2 & 14.9 & 2.2 \end{bmatrix}$

CHAPTER 9

9.2 $C(s) = (10s + 20)/(s - 6)$, $p = -0.2$.

9.4 $C(s) = (22s - 4)/(s - 16)$, $p = 1$. There is no need for a feed forward gain because $\hat{g}(s)$ contains $1/s$.

9.6 $C(s) = (7s^3 + 14s^2 + 34s + 25)/(s(s^2 + 4))$.

9.8 Yes, no. The transfer function from r to u is $\hat{g}_{ur}(s) = s/(s + 1)(s - 2)$, which is not BIBO stable.

9.10 Yes, no, no, no, yes, no (row-wise).

9.12 $C_1(s) = -2(s + 3)/(s - 21)$, $C_2(s) = (28s - 6)/(s - 21)$. $A(s) = s - 21$ is not Hurwitz. Its implementation in Fig. 9.4(a) will not be totally stable. A minimal realization of $[C_1(s) \quad -C_2(s)]$ is

$$
\dot{x} = 21x + [-48 \quad -582]\begin{bmatrix} r \\ y \end{bmatrix} \qquad y = x + [-2 \quad -28]\begin{bmatrix} r \\ y \end{bmatrix}
$$

from which an op-amp circuit can be drawn.

9.15 (1) $a > 0$ and $b_0 = -2b_1$. (2) Yes, $b_0 = -2$, $b_1 = -4$.

9.16 The 1×2 compensator

$$
\mathbf{C}(s) = \frac{1}{s + 3.5}[3.5s + 12 \quad -2]
$$

will place the closed-loop poles at $-2, -1 \pm j, -3$. No.

9.18 If we select

$$
\mathbf{F}(s) = \mathrm{diag}((s + 2)(s^2 + 2s + 2)(s + 3), \ (s^2 + 2s + 2))
$$

then we cannot find a feedforward gain to achieve tracking. If

$$
\mathbf{F}(s) = \begin{bmatrix} (s + 2)(s^2 + 2s + 2)(s + 3) & 0 \\ 1 & s^2 + 2s + 2 \end{bmatrix}
$$

then the compensator

$$\mathbf{C}(s) = \mathbf{A}^{-1}(s)\mathbf{B}(s) = \begin{bmatrix} s - 4.7 & -53.7 \\ -3.3 & s - 4.3 \end{bmatrix}^{-1} \begin{bmatrix} -30.3s - 29.7 & 4.2s - 12 \\ -0.7s - 0.3 & 4s - 1 \end{bmatrix}$$

and the feedforward gain matrix

$$\mathbf{P} = \begin{bmatrix} 0.92 & 0 \\ -4.28 & 1 \end{bmatrix}$$

will achieve the design.

9.20 The diagonal transfer matrix

$$\hat{\mathbf{G}}_o(s) = \begin{bmatrix} \dfrac{-2(s - 1)}{s^2 + 2s + 2} & 0 \\ 0 & \dfrac{-2(s - 1)}{s^2 + 2s + 2} \end{bmatrix}$$

is implementable. The proper compensators $\mathbf{A}^{-1}(s)\mathbf{L}(s)$ and $\mathbf{A}^{-1}(s)\mathbf{M}(s)$ with

$$\mathbf{A}(s) = \begin{bmatrix} s + 5 & -14 \\ 0 & s + 4 \end{bmatrix} \qquad \mathbf{L}(s) = \begin{bmatrix} -2(s + 3) & 2(s + 3) \\ 2 & -2s \end{bmatrix}$$

$$\mathbf{M}(s) = \begin{bmatrix} -6 & 13s + 1 \\ 2 & -2s \end{bmatrix}$$

in the two-parameter configuration will achieve the design.

Index

331